METHODS IN MOLECULAR BIOLOGY

Series Editor
John M. Walker
School of Life and Medical Sciences
University of Hertfordshire
Hatfield, Hertfordshire, AL10 9AB, UK

For further volumes:
http://www.springer.com/series/7651

Computational Systems Biology

Methods and Protocols

Edited by

Tao Huang

Shanghai Institutes for Biological Sciences, Chinese Academy of Sciences, Shanghai, China

Humana Press

Editor
Tao Huang
Shanghai Institutes for Biological Sciences
Chinese Academy of Sciences
Shanghai, China

ISSN 1064-3745 ISSN 1940-6029 (electronic)
Methods in Molecular Biology
ISBN 978-1-4939-7716-1 ISBN 978-1-4939-7717-8 (eBook)
https://doi.org/10.1007/978-1-4939-7717-8

Library of Congress Control Number: 2018935135

Printed on acid-free paper

This Humana Press imprint is published by the registered company Springer Science+Business Media, LLC part of Springer Nature.
The registered company address is: 233 Spring Street, New York, NY 10013, U.S.A.

Preface

With the rapid development of high-throughput technologies, such as next-generation sequencing and single-cell sequencing, many tough biomedical questions can be answered since it is no longer impossible to get a whole picture of the biological system. Complex diseases, such as tuberculous meningitis and leukemia, involve dysfunctions on multiple levels, including DNA variants, mRNA differential expression, and protein fluctuation. Accurately measuring these molecules is the first step of understanding the biological system.

But even if we can get all these multi-omics data, the bioinformatics analysis of such big data is still very challenging. There are two types of analysis for deciphering the mechanism hidden behind the biomed big data. One method is machine learning. It can analyze various features and build a predictive model which can predict the response of a biological system to a perturbation or classify the subtypes of samples. In recent years, one of the machine learning methods, deep learning, is extremely popular and has become a powerful tool for big data analysis.

Another effective method is network analysis based on graph theories. Network is how we understand the complex world. It starts from a node. And a connection in real life is abstracted as an edge. It can grow fast and become more and more complex. Eventually, it will exhibit unique properties and reflect the complex system. It inspires the development of many algorithms, such as the neural network in deep learning. And in biomedicine, it is a wonderful way of integrating diverse big data and transforming the biological questions into mathematical questions, especially graph theory questions. The graph theory empowers the network analysis to see the hidden truth underneath the hairy ball we see. The visualization of a large-scale network can help us get a sense of the network, but it can't really give us the useful information that we are interested in, such as which genes are the key drivers and which genes are novel disease genes or possible drug targets.

In this book, we introduce the latest experimental and bioinformatics methods for DNA sequencing, RNA sequencing, cell-free tumor DNA sequencing, single-cell sequencing, and single-cell proteomics and metabolomics. Then, we review the advanced analysis methods, such as genome-wide association studies (GWAS), machine learning, reconstruction and analysis of gene regulatory networks, and differential coexpression network analysis, and give a practical guide for how to choose and use the right algorithm or software to handle specific high-throughput data or multi-omics data. A powerful novel RNA-seq data analysis and visualization tool, iSeq, is released in this book. The last parts of the book are the applications of these high-throughput technologies and advanced analysis methods in complex diseases, such as tuberculous meningitis and leukemia.

We hope that after reading this book, the readers can understand: how the biomed big data is generated, which tools can be used to process them, which advanced machine learning and network analysis are optional for data integration and knowledge discovery, and what achievements have been made nowadays.

Shanghai, China *Tao Huang*

Contents

Contributors

ADENIYI C. ADEOLA • *State Key Laboratory of Genetic Resources and Evolution, Kunming, Yunnan, China; China-Africa Centre for Research and Education & Yunnan Laboratory of Molecular Biology of Domestic Animals, Kunming, Yunnan, China; Animal Branch of the Germplasm Bank of Wild Species, Kunming Institute of Zoology, Chinese Academy of Sciences, Kunming, Yunnan, China*

LEI CAI • *Key Laboratory for the Genetics of Developmental and Neuropsychiatric Disorders (Ministry of Education), Collaborative Innovation Center for Genetics and Development, Bio-X Institutes, Shanghai Jiao Tong University, Shanghai, China*

MICHELLE CHANG • *Key Laboratory for the Genetics of Developmental and Neuropsychiatric Disorders (Ministry of Education), Collaborative Innovation Center of Genetics and Development, Bio-X Institutes, Shanghai Jiao Tong University, Shanghai, China*

JIAN CHEN • *State Key Laboratory of Transducer Technology, Institute of Electronics, Chinese Academy of Sciences, Beijing, China; University of Chinese Academy of Sciences, Beijing, China*

QINGFENG CHEN • *School of Computer, Electronics and Information, Guangxi University, Nanning, China; State Key Laboratory for Conservation and Utilization of Subtropical Agro-bioresources, Guangxi University, Nanning, China*

SHIFU CHEN • *HaploX Biotechnology, Shenzhen, Guangdong, China*

XING CHEN • *State Key Laboratory of Genetic Resources and Evolution, Kunming, Yunnan, China*

JING CHENG • *Department of Medical Instrument, Shanghai University of Medicine and Health Sciences, Shanghai, China*

HAO DAI • *School of Life Sciences and Biotechnology, Shanghai Jiao Tong University, Shanghai, China*

ZHAO-LI DING • *Kunming Biological Diversity Regional Centre of Large Apparatus and Equipments, Kunming, Yunnan, China; Public Technology Service Centre, Kunming, Yunnan, China*

BEIYUAN FAN • *State Key Laboratory of Transducer Technology, Institute of Electronics, Chinese Academy of Sciences, Beijing, China; University of Chinese Academy of Sciences, Beijing, China*

CAOQI FAN • *Peking-Tsinghua Center for Life Sciences, Academy for Advanced Interdisciplinary Studies; Center for Bioinformatics, School of Life Sciences, Peking University, Beijing, China*

ZUNYUN FU • *State Key Laboratory of Drug Research, Drug Discovery and Design Center, Shanghai Institute of Materia Medica, Chinese Academy of Sciences, Shanghai, China*

JINGBO GAN • *Peking-Tsinghua Center for Life Sciences, Academy for Advanced Interdisciplinary Studies; Center for Bioinformatics, School of Life Sciences, Peking University, Beijing, China*

SHAN GAO • *College of Life Sciences, Nankai University, Tianjin, People's Republic of China; Institute of Statistics, Nankai University, Tianjin, People's Republic of China*

XIANFU GAO • *Key Laboratory of Systems Biology, Institute of Biochemistry and Cell Biology, Shanghai Institutes for Biological Sciences, Chinese Academy of Sciences, Shanghai, China*

KAI HE • *State Key Laboratory of Genetic Resources and Evolution, Kunming, Yunnan, China*
LIN HE • *Key Laboratory for the Genetics of Developmental and Neuropsychiatric Disorders (Ministry of Education), Collaborative Innovation Center for Genetics and Development, Bio-X Institutes, Shanghai Jiao Tong University, Shanghai, China*
CUIHONG HUANG • *HaploX Biotechnology, Shenzhen, Guangdong, China*
LIYU HUANG • *Information and Network Center, Guangxi University, Nanning, China*
TAO HUANG • *Shanghai Institutes for Biological Sciences, Chinese Academy of Sciences, Shanghai, China*
CANGZHI JIA • *Department of Mathematics, Dalian Maritime University, Dalian, China*
LEI KONG • *Peking-Tsinghua Center for Life Sciences, Academy for Advanced Interdisciplinary Studies; Center for Bioinformatics, School of Life Sciences, Peking University, Beijing, China*
DEHUAN LAI • *School of Computer, Electronics and Information, Guangxi University, Nanning, China*
WEI LAN • *School of Computer, Electronics and Information, Guangxi University, Nanning, China*
WENXIAN LAN • *State Key Laboratory of Bio-Organic and Natural Product Chemistry, Shanghai Institute of Organic Chemistry, Chinese Academy of Sciences, Shanghai, China*
CHENG LI • *Peking-Tsinghua Center for Life Sciences, Academy for Advanced Interdisciplinary Studies; Center for Bioinformatics, School of Life Sciences, Peking University, Beijing, China; Center for Statistical Science, Peking University, Beijing, China*
GUI-MEI LI • *Kunming Biological Diversity Regional Centre of Large Apparatus and Equipments, Kunming, Yunnan, China; Public Technology Service Centre, Kunming, Yunnan, China*
JUN LI • *HaploX Biotechnology, Shenzhen, Guangdong, China*
BAO-HONG LIU • *State Key Laboratory of Veterinary Etiological Biology; Key Laboratory of Veterinary Parasitology of Gansu Province; Lanzhou Veterinary Research Institute, Chinese Academy of Agricultural Sciences, Lanzhou, Gansu, People's Republic of China; Jiangsu Co-Innovation Center for Prevention and Control of Animal Infectious Diseases and Zoonoses, Yangzhou, People's Republic of China*
MING LIU • *HaploX Biotechnology, Nanshan District, Shenzhen, Guangdong, China*
RENZHONG LIU • *HaploX Biotechnology, Shenzhen, Guangdong, China*
KEYI LONG • *Key Laboratory for the Genetics of Developmental and Neuropsychiatric Disorders (Ministry of Education), Collaborative Innovation Center for Genetics and Development, Bio-X Institutes, Shanghai Jiao Tong University, Shanghai, China*
DONG LU • *State Key Laboratory of Drug Research, Drug Discovery and Design Center, Shanghai Institute of Materia Medica, Chinese Academy of Sciences, Shanghai, China; University of Chinese Academy of Sciences, Beijing, China*
JING LU • *Key Laboratory of Molecular Pharmacology and Drug Evaluation (Yantai University), Ministry of Education, Collaborative Innovation Center of Advanced Drug Delivery System and Biotech Drugs in Universities of Shandong, School of Pharmacy, Yantai University, Yantai, China*
XIAOMIN LUO • *Drug Discovery and Design Center, State Key Laboratory of Drug Research, Shanghai Institute of Materia Medica, Chinese Academy of Sciences, Shanghai, China*

ROBERT W. MURPHY • *State Key Laboratory of Genetic Resources and Evolution, Kunming, Yunnan, China; Centre for Biodiversity and Conservation Biology, Royal Ontario Museum, Toronto, ON, Canada*

GANG NI • *State Key Laboratory of Genetic Resources and Evolution, Kunming, Yunnan, China; Yunnan Laboratory of Molecular Biology of Domestic Animals, Kunming, Yunnan, China*

YANG SHU • *Precision Medicine Center, State Key Laboratory of Biotherapy, Precision Medicine Key Laboratory of Sichuan Province, West China Hospital, Sichuan University, Chengdu, Sichuan, China*

JUNBO WANG • *State Key Laboratory of Transducer Technology, Institute of Electronics, Chinese Academy of Sciences, Beijing, China; University of Chinese Academy of Sciences, Beijing, China*

LU WANG • *Key Laboratory of Systems Biology, Institute of Biochemistry and Cell Biology, Chinese Academy Science, Shanghai, China*

WEN-ZHI WANG • *State Key Laboratory of Genetic Resources and Evolution, Kunming Institute of Zoology, Chinese Academy of Sciences, Kunming, Yunnan, China; Animal Branch of the Germplasm Bank of Wild Species, Kunming Institute of Zoology, Chinese Academy of Sciences, Kunming, Yunnan, China; Wildlife Forensics Science Services, Kunming, Yunnan, China; Guizhou Academy of Testing and Analysis, Guiyang, Guizhou, China*

YI-YI WANG • *Department of Neurology, Tianjin Haihe Hospital, Tianjin, P.R. China*

DONG-QING WEI • *School of Life Sciences and Biotechnology, Shanghai Jiao Tong University, Shanghai, China*

BING-DI XIE • *Department of Neurology, Tianjin Medical University General Hospital, Tianjin, P.R. China*

YI XIONG • *School of Life Sciences and Biotechnology, Shanghai Jiao Tong University, Shanghai, China*

HENG XU • *State Key Laboratory of Biotherapy, Precision Medicine Key Laboratory of Sichuan Province, Precision Medicine Center, West China Hospital, Sichuan University, Chengdu, Sichuan, China*

MINGYAN XU • *HaploX Biotechnology, Shenzhen, Guangdong, China*

YING XU • *Key Laboratory of Cell Differentiation and Apoptosis of Ministry of Education, Department of Pathophysiology, Shanghai Jiao-Tong University School of Medicine, Shanghai, China*

YUNGANG XU • *Center for Systems Medicine, School of Biomedical Informatics, UTHealth at Houston, Houston, TX, USA; Center for Bioinformatics and Systems Biology, Wake Forest School of Medicine, Winston-Salem, NC, USA*

XIANG-TIAN YU • *Key Laboratory of Systems Biology, Institute of Biochemistry and Cell Biology, Chinese Academy Science, Shanghai, China*

TAO ZENG • *Key Laboratory of Systems Biology, Institute of Biochemistry and Cell Biology, Chinese Academy Science, Shanghai, China*

CHAO ZHANG • *PKU-Tsinghua-NIBS Graduate Program, School of Life Sciences, Peking University, Beijing, China*

HONG ZHANG • *Key Laboratory for the Genetics of Developmental and Neuropsychiatric Disorders (Ministry of Education), Collaborative Innovation Center for Genetics and Development, Bio-X Institutes, Shanghai Jiaotong University, Shanghai, China*

YA-PING ZHANG • *State Key Laboratory of Genetic Resources and Evolution, Kunming, Yunnan, China; Yunnan Laboratory of Molecular Biology of Domestic Animals, Kunming, Yunnan, China; Animal Branch of the Germplasm Bank of Wild Species, Kunming Institute of Zoology, Chinese Academy of Sciences, Kunming, Yunnan, China; Laboratory for Conservation and Utilization of Bio-resource and Key Laboratory for Microbial Resources of the Ministry of Education, Yunnan University, Kunming, Yunnan, China*

GUANGYONG ZHENG • *Key Laboratory of Computational Biology, Bio-Med Big Data Center, CAS-MPG Partner Institute for Computational Biology, Shanghai Institutes for Biological Sciences, Chinese Academy of Sciences, Shanghai, China*

MINGYUE ZHENG • *State Key Laboratory of Drug Research, Drug Discovery and Design Center, Shanghai Institute of Materia Medica, Chinese Academy of Sciences, Shanghai, China*

XIAOBO ZHOU • *Center for Systems Medicine, School of Biomedical Informatics, UTHealth at Houston, Houston, TX, USA; Center for Bioinformatics and Systems Biology, Wake Forest School of Medicine, Winston-Salem, NC, USA*

YANQING ZHOU • *HaploX Biotechnology, Nanshan District, Shenzhen, Guangdong, China*

PING ZHU • *Peking-Tsinghua Center for Life Sciences, Academy for Advanced Interdisciplinary Studies; Center for Bioinformatics, School of Life Sciences, Peking University, Beijing, China*

XIAOLEI ZHU • *School of Life Sciences and Biotechnology, Shanghai Jiao Tong University, Shanghai, China*

YUN ZUO • *Department of Mathematics, Dalian Maritime University, Dalian, China*

Check for updates

Chapter 1

DNA Sequencing Data Analysis

Keyi Long, Lei Cai, and Lin He

Abstract

Among various biological data, DNA sequence is doubtlessly a fundamental datum. By obtaining particular DNA sequence data and analyzing, biologists get to understand life science more precisely. This chapter is an overview of DNA sequencing technology and its data analysis methods, providing information about DNA sequencing, several different methods, and tools applied in data analysis. Both advantages and disadvantages are discussed.

Key words DNA sequence, DNA sequencing, Data analysis, Sequence comparison, Methods and tools

1 DNA Sequencing

Three essential elements of life science are DNA, RNA, and protein; they lay the foundation of all living creatures. Millions of scientists make joint efforts to understand the mystery of life, and tons of work have been done to figure out relations between structures and their properties.

For molecular biologists, information encoded in the sequences of nucleic acid molecules is of vital importance since it not only passes the genetic information from generation to generation but also influences function by transcription and translation. Research at the frontiers of life science cannot be done without obtaining and analyzing certain DNA sequences, which means determining the particular order and number of the four bases—adenine, guanine, cytosine, and thymine—in a strand of DNA. Advances in recombinant DNA technology have allowed the isolation of large numbers of biologically interesting fragments of DNA [1].

1.1 Methods of DNA Sequencing

With the help of restriction endonucleases, large DNA molecules can be cut into small fragments in an orderly fashion. Also, recombinant DNA techniques aid in purifying and characterizing

Tao Huang (ed.), *Computational Systems Biology: Methods and Protocols*, Methods in Molecular Biology, vol. 1754,
https://doi.org/10.1007/978-1-4939-7717-8_1, © Springer Science+Business Media, LLC, part of Springer Nature 2018

individual restriction fragments from mixtures. And most importantly, to do DNA sequencing, at least three steps are needed: cloning, sequencing, and analyzing.

In the year 1970, Ray Wuat from Cornell University first tapped a location-specific primer extension strategy into determining DNA sequences. And ever since the year 1977 Sanger and his colleagues established the chain termination method and accomplish the first full DNA genome sequencing of bacteriophage ϕX174, DNA sequencing methods have been developed and improved.

1.1.1 Traditional Methods

There are two basic methods in DNA sequencing—the Maxam-Gilbert sequencing (also known as chemical sequencing) as well as the chain termination method (also known as Sanger sequencing). The former method attaches radioactive labels to the 5′ end of DNA, and by using chemical treatment, it generates subsequent breaks at particular bases. Autoradiography helps yield a series of dark bands, which represent the radiolabeled DNA fragments. On the other hand, Sanger's method requires modified di-deoxynucleoside triphosphates (ddNTPs). Due to the fact that DNA polymerase I cannot distinguish normal deoxynucleoside triphosphates (dNTPs) and ddNTPs, those new strands with ddNTPs lack a 3′-OH group required for the formation of a phosphodiester bond between two nucleotides, thus stopping the elongation of DNA. By labeling ddNTPs, we get to know the DNA sequence [2].

Although Sanger's way is effective in many aspects, it can only read 450 bp in a single reaction, and the process is time-consuming, limiting its use in large fragment sequencing. After prevailing for decades, other methods are invented and widely used on the basis of their work, like the shotgun strategy and bridge PCR. More importantly, with the rapid development of science and technology, high-throughput sequencing methods are established; they then play an essential role in modern DNA sequencing with the ability to process mass data in a short time.

1.1.2 High-Throughput (HTP) Sequencing Methods

Since the 1990s, a handful of new methods of DNA sequencing were invented—454 pyrosequencing, Illumina (Solexa) sequencing, and SOLiD sequencing are three most used technologies. Other methods include the massively parallel signature sequencing (MPSS), the polony sequencing, DNA nanoball sequencing, etc. These methods all share common characteristics of high throughput and low costs, and together they were known as the "next-generation" sequencing (NGS) methods. The core thought of HTP methods is to do DNA sequencing while synthesizing the new strand.

Nowadays, genomic questions are so complex that a depth of information is needed. In ultra-high-throughput sequencing, as many as 500,000 sequencing-by-synthesis operations may be run in parallel [3]. With its unprecedented throughput, speed, and scalability compared with traditional DNA sequencing, NGS enables researchers to study biological problems at a new level and has been widely implemented in commercial DNA sequencers. Table 1 makes comparisons between several high-throughput sequencing methods [4].

Among those NGS methods, 454 pyrosequencing is doubtlessly the most classic one. It does not require ddNTPs for chain termination. Instead, it mainly utilizes emulsion PCR to accomplish DNA elongation. By detecting the pyrophosphate released during nucleotide incorporation, the sequencer can analyze the sequence. Data will be stored in standard flowgram format (SFF) files for downstream analysis.

The process can be divided into the following steps:

1. Library construction. The library DNAs with 454-specific adaptors are denatured to be single strand.
2. Surface attachment and bridge amplification.
3. Denaturation and complete amplification. For example, by emulsion PCR.
4. Single base extension and sequencing.

The theory can be concluded as follows:

When one dNTP (dATP, dGTP, dCTP, dTTP) complements to the bases of the template strand with the help of DNA polymerase, one pyrophosphate (PPi) is released. Catalyzed by ATP sulfurylase, PPi can bind to adenosine-5′-phosphosulfate (APS) to generate ATP. With luciferase, the ATP drives the luciferin into oxyluciferin and generates visible light, which then be captured by CDD system. The signal will be then analyzed by computers and finally show the exact DNA sequence.

Although the next-generation sequencing methods are still the most prevailing technologies, the third-generation sequencing (TGS), also known as the single molecule sequencing (SMS), is developing rapidly. This kind of technology depends on detecting single molecule signal and no longer needs PCR, aiming to increase throughput and decrease the time to result and cost by eliminating the need for excessive reagents and harnessing the processivity of DNA polymerase [5].

2 Methods for DNA Sequencing Data Analysis

After obtaining the exact sequences of the nucleic acid, it is usually necessary to identify the quality of the outcome, to extract target

Table 1
Comparison of several high-throughput sequencing methods

Method	Read length	Accuracy (single read not consensus)	Reads per run	Time per run	Cost per 1 million bases (in US$)	Advantages	Disadvantages
Pyrosequencing (454)	700 bp	99.9%	1 million	24 h	$10	Long read size. Fast	Runs are expensive. Homopolymer errors
Sequencing by synthesis (Illumina)	MiniSeq, NextSeq, 75–300 bp; MiSeq, 50–600 bp; HiSeq 2500, 50–500 bp; HiSeq 3/4000, 50–300 bp; HiSeq X, 300 bp	99.9% (Phred30)	MiniSeq/MiSeq, 1–25 million; NextSeq, 130-400 million; HiSeq 2500, 300 million–2 billion; HiSeq 3/4000, 2.5 billion; HiSeq X, 3 billion	1–11 days, depending upon sequencer and specified read length	$0.05–0.15	Potential for high sequence yield, depending upon sequencer model and desired application	Equipment can be very expensive. Requires high concentrations of DNA
Sequencing by ligation (SOLiD sequencing)	50 + 35 or 50 + 50 bp	99.9%	1.2–1.4 billion	1–2 weeks	$0.13	Low cost per base	Slower than other methods. Has issues sequencing palindromic sequences
Nanopore sequencing	Dependent on library prep, not the device, so user chooses read length (up to 500 kb reported)	~92–97% single read (up to 99.96% consensus)	Dependent on read length selected by user	Data streamed in real time. Choose 1 min to 48 h	$500–999 per flow cell, base cost-dependent on expt	Very long reads, portable (palm sized)	Lower throughput than other machines, single read accuracy in 90 s

Table source: https://en.wikipedia.org/wiki/DNA_sequencing

fragments, and to compare the sequence with a reference genome. Also, biologists pay attention to other characteristics of the sequence that might determine its biological features. That is why the work of data analysis should be done for further study.

2.1 General Steps of DNA Sequencing Data Analysis

Generally, DNA sequencing data analysis includes these four steps:

- Trimming of overlapping sequences.
- Multiple alignments of template sequences.
- Consistency check between reading text and chromatogram peak data.
- Review and correction of software misreads.

To be more precise, by using DNA sequencing technology, especially the Sanger sequencing, we obtain data in the form of chromatogram—a series of four differently colored peaks. Usually, after opening the result file in a software such as Chromas Lite, there shows red, black, green, and blue peaks, each color corresponding to a different DNA base. On both ends of the chromatogram, there exist about 50 bases that are difficult to recognize. This is because of impurities and is a normal phenomenon.

When screening the chromatogram, we are likely to find two overlapping peaks. It seems that this spot represents a heterozygosity locus. However, things get more complicated when the two overlapping peaks have different axes or when the two peaks share one axis but are of the same height. This spot is not a heterozygosity locus since one peak is the interference peak. Mostly, one or two spots before a big base peak exists an interference peak whose height is approximately half of the big peak. The closer they are, the more interference they have. And under these circumstances, the computer often makes mistakes; that is where humans step in and correct those misreads.

When checking the outcome of the software, we conclude some rules to help us determine whether the results are accurate after tons of work:

1. The main peak mostly sits on the right side of the interference peak.
2. The interference peak can be higher or lower or of the same height than the main peak.

As a result, in order to reduce misreads, we often do several procedures:

1. Consistency check among reading text and results in gene pool and chromatogram peak data must be done.

2. When finding a possible spot, compare it with multiple samples.
3. Calculate the mutation rate of your finding, and compare it with data in authoritative publications or databases.

2.2 Procedure for NGS Data Analysis

2.2.1 Quality Control

When it comes to analyzing the results of next-generation DNA sequencing (NGS) data, the situation is more complicated. This is because the results are determined by varied DNA library constructing process and adaptors-adding process. Since the modern high-throughput sequencers can generate hundreds of millions of sequences in a single run, before analyzing this sequence to draw biological conclusions, we are prone to perform some simple quality control checks to ensure that the raw data looks good and there are no problems or biases in the data.

Although many sequencers will generate a QC report, this is usually not enough since it only focused on identifying problems which were generated by the sequencer itself. FastQC is a widely used software that aims to provide a more detailed QC report, which can spot problems which originate either in the sequencer or in the starting library material. When using FastQC, we should know the following steps:

1. Use the Linux system and install FastQC:

   (http://www.bioinformatics.babraham.ac.uk/projects/fastqc/).
2. Type in command "fastqc [-o output dir] [--(no)extract] [-f fastq|bam|sam] [-c contaminant file]." "output dir" means the output path, the parameter "extract" determines the output unpacking, and the parameter "-f" represents the format of input.
3. Run FastQC and read the result files:
   - The HTML report shows a summary of the modules which were run and a quick evaluation of whether the results of the module seem entirely normal (green tick), slightly abnormal (orange triangle), or very unusual (red cross).
   - View the per base sequence quality. Quality can be seen as the value of Fred. In "$-10 \times \log10(p)$," "p" stands for the possibility of a mistake. Values of the lower quartile and the median should be considered. If the value of the lower quartile exceeds 30, the quality can be regarded as very good.
   - View the per sequence quality scores. Normally, if 90% of the reads have the quality value of more than 35 scores, the quality can be regarded as very good.
   - View the distribution of A,T,G,C. In most cases, the amount of A/T (28%) outweighs that of G/C (22%).

2.2.2 Data Analysis

For data analysis, we choose Illumina system as an example. Illumina offers a variety of next-generation sequencing (NGS) data analysis software tools. Push-button tools for DNA sequence alignment, variant calling, and data visualization are all included. Data generated on Illumina sequencing instruments are automatically transferred and stored securely in BaseSpace Sequence Hub. And the analyzing procedure should be done as follows:

Primary Analysis

1. Judge the results' quality. If the outcome is not in good quality, the analyzing process will be meaningless.
2. Searching for your aim fragments.
3. Real-time analysis and base calling by the Illumina system.

Secondary Analysis

1. After real-time analysis (RTA) in the primary analysis, use MiSeq Reporter, an online software, to analyze data.
2. After opening MiSeq Reporter, click "analysis" to see different modules including A (assembly), E (enrichment), G (generate FASTQ), M (metagenomics), R (Resequencing), etc.
3. Choose the analyzing module you need and run the procedure.
4. Read the MiSeq Reporter report. For example, if you choose module R, after running the resequencing procedure, the detailed report will show a list of samples, a table of targets, a list of SNPs and their corresponding scores, Q score, as well as the depth of sequencing.
5. The output is in demultiplex (*.demux) and FASTQ (*.fastq) formats. You can use third-party software programs to further analyze the data.
6. Compare the results with the reference genome.

2.3 Several Tools to Facilitate Data Analysis

2.3.1 Artemis R5

It is a DNA sequence viewer and annotation tool written in Java. User can download it for free and run it under systems including UNIX, GNU/Linux, Macintosh, and Windows.

First, import information from EMBL and GenBank, as well as files in FASTA format. Then it gives visualization of sequence features, next-generation data and the results of analyses within the context of the sequence, and also its six-frame translation.

2.3.2 Arlequin

It is an integrated software package for population genetics data analysis. Arlequin provides methods to analyze patterns of genetic diversity within and between population samples [6].

The software is freely available on http://cmpg.unibe.ch/software/arlequin3. It can recognize data including DNA sequences, standard multilocus genotypes, RFLP data, microsatellite data, etc. It is a powerful software that is capable of many functions including molecular diversity, mismatch distribution, computation of standard genetic diversity indices, as well as the estimation of allele and haplotype frequencies. Also it can run tests of departure from linkage equilibrium and do thorough analyses of population subdivision under the AMOVA framework.

When imported data is in the RFLP data:

"1" means there exist restriction sites, while "0" means none, and "-" means a lack of restriction sites.

When imported data is DNA sequences:

"-" stands for a lack of nucleotide, while "?" stands for an unknown nucleotide.

"R" means A/G (purine), while "Y" means C/T (pyrimidine).

"M" means A/C, "W" means A/T; "S" means C/G, "K" means G/T, "B" means C/G/T, "D" means A/C/T; "H" means A/C/T; "V" means A/C/G; "N" means A/C/G/T.

2.3.3 DnaSP

It is a software for comprehensive analysis of DNA polymorphism data. As a powerful tool, it helps us to understand the evolutionary process and to establish the functional significance of particular genomic regions [7].

Remarkably, the DnaSP v5 can handle and analyze multiple data files in batch. It can identify conserved DNA regions, which can contribute to phylogenetic footprint-based analysis. Also, it allows exhaustive DNA polymorphism analysis, and the results can be illustrated graphically and in a text format.

2.3.4 SSAHA2 (Sequence Search and Alignment by Hashing Algorithm)

It is a pairwise sequence alignment program designed for the efficient mapping of sequencing reads onto genomic reference sequences.

It can recognize a range of output formats concluding SAM, CIGAR, PSL, etc. And this tool reads data from most sequencing platforms like ABI-Sanger, Roche 454, and Illumina-Solexa.

There are many other tools for researchers to use, facilitating them to better analyze data generated. Table 2 illustrates tools of different kinds.

3 Extension: Methods and Tools for DNA Sequence Analysis

3.1 Background

In the past decades, many manual methods have been applied to analyzing DNA sequence data. However, the drawbacks of these methods are apparent—when the data is in extraordinary amount, it takes lots of time and energy. Fortunately, computers are well-used in solving the problem. By establishing DNA sequence databases storing data information of magnanimity, researchers are able to adopt statistical approaches for analysis.

Table 2
Several tools for data analysis

Function	Name	Site
Plot	ggplot2 circos	http://docs.ggplot2.org/current/ http://circos.ca/
Mapping	BWA Bowtie2	http://bio-bwa.sourceforge.net/ http://bowtie-bio.sourceforge.net/bowtie2/index.shtml
SNP/indel calling	samtools gatk pindel	http://samtools.sourceforge.net/samtools.shtml http://www.broadinstitute.org/gatk/ http://gmt.genome.wustl.edu/pindel/0.2.4/index.html
Analysis tools	plink ngsTools	http://pngu.mgh.harvard.edu/~purcell/plink/index.shtml https://github.com/mfumagalli/ngsTools
Structure analysis	frappe structure ngsAdmix	http://med.stanford.edu/tanglab/software/frappe.html http://pritchardlab.stanford.edu/structure.html http://www.popgen.dk/software/index.php/NgsAdmix
Databases	DDBJ ENA KEGG ensembl	http://www.ddbj.nig.ac.jp/index-e.html http://www.ebi.ac.uk/ena/home http://www.genome.jp/kegg/ http://asia.ensembl.org/index.html

The key to data analysis is data mining, of which the basis is sequence similarities. The most common approach to similarity research is DNA sequence alignment which can find the optimal match between sequences according to similar matrix given, as well as probable insertion, deletion, and mutation.

3.1.1 Two Stages of DNA Sequence Analysis

Analyzing nucleic acid sequences with computer programs can be divided into two stages:

1. The first stage is the straightforward search for sequences with known properties, which involves position determination.
2. The second stage aims to detect subtle, less straightforward sequence patterns including controlling elements like promoters. The results can be presented by catalogs of sequence patterns.

3.1.2 Two Categories of Computational Approaches

Computational approaches to sequence alignment generally fall into two categories: global alignments and local alignments.

1. Calculating a global alignment is a form of global optimization that "forces" the alignment to span the entire length of all query sequences.
2. Local alignments identify regions of similarity within long sequences that are often widely divergent overall. Local

alignments are often preferable but can be more difficult to calculate because of the additional challenge of identifying the regions of similarity.

3.2 Methods and Tools

There are various widely used DNA sequencing data analysis tools; some are more familiar to us while some may not.

3.2.1 Two Types of DNA Sequence Alignment

DNA sequence alignment can be divided into different types:

1. Pairwise alignment: it can only compare two sequences.
2. Multiple sequence alignment: it is an extension of pairwise alignment to incorporate more than two sequences at a time.

Several software are chosen to be discussed as follows.

3.2.2 BLAST

BLAST, also known as Basic Local Alignment Search Tool (site: blast.ncbi.nlm.nih.gov/Blast.cgi), is an algorithm to compare primary biological sequence information. Usually, you don't have to download and install it. All you have to do is to visit the website stated above.

BLAST is actually a family of programs that is widely used in bioinformatics; it enables us to make comparison between the query sequence and a database of sequences. Those sequences can belong to DNA, RNA, or protein. By selecting particular BLAST tool and determining a certain threshold, we can identify sequences that resemble the input sequence. For nucleic acid, there is nucleotide-nucleotide BLAST (blastn). After putting in a DNA query and setting certain parameters, we get results showing the most similar DNA sequences.

Blastn does its job by locating short matches. Usually, there is a threshold score T. If the score is higher than a predetermined T, the alignment will be included in the results given by BLAST and vice versa. Therefore, choosing a proper value of T means getting a proper amount of results.

This tool is highly sensitive and can be utilized for several purposes: species identification, domains location, phylogeny establishment, etc.

1. Visit the site blast.ncbi.nlm.nih.gov/Blast.cgi and choose blastn.
2. Upload your DNA sequence in proper format like FASTA.
3. Set proper parameters including T.
4. Click BLAST.
5. Reviewing your alignment results; mismatches can be a frame-shift in the query sequence.
6. If any error exists, go back, check the sequence file, change values of parameters, and BLAST again.

3.2.3 CLUSTAL

Clustal is an effective tool for multiple alignment of nucleic acid and protein sequences. After downloading the software, we input data containing DNA sequences, then set certain parameters and wait for the results. When multiple sequence alignment is needed, we use Clustal X.

The proper input formats conclude NBRF/PIR, FASTA, EMBL/Swiss-Prot, Clustal, GCC/MSF, GCG9 RSF, and GDE, while the output format can be Clustal, NBRF/PIR, GCG/MSF, PHYLIP, GDE, or NEXUS.

When using Clustal for data analysis, the bigger the input file is, the longer it takes for alignment. The results obtained from Clustal can be further utilized by loading the output file into other software like MEGA, which will be soon discussed.

1. Download the desktop application and open it.
2. Upload file containing DNA sequences in proper format; at this stage, you can have a look at the colored bases.
3. Select different tools for different purposes:

   Select "do complete alignment" for a pairwise alignment.

   Select "do alignment from guide tree and phylogeny" to create a guide tree (or use a user-defined tree).

   Select "produce guide tree only" to use the guide tree to carry out a multiple alignment.
4. Review the results, save it in a favorable format.
5. The results can be used for further studies.

3.2.4 MEGA

MEGA is short for Molecular Evolutionary Genetics Analysis. As a desktop application released in 1993, it has continuously helped users conduct statistical analysis of biological macromolecules to study molecular evolution and construct phylogenetic trees [8].

MEGA is multifunctional. In addition to sequence alignment construction, it performs outstandingly in distance estimation and tree-making. To be more precise, the fact that MEGA has included likelihood methods for estimating evolutionary distances between sequence pairs as well as distance-based and maximum parsimony methods for inferring phylogenetic trees is historical.

1. Download the desktop application MEGA and open it.
2. Upload file containing DNA sequences in proper format, or open a file conducted by Clustal or any other tools that have results MEGA can recognize and further analyze.
3. Click "do complete alignment" if you upload DNA sequences; otherwise skip this step.
4. Select tools like "compare pairwise distances" or "construct neighbor-joining tree" for certain purposes.
5. Save the results in required format.

4 Applications of DNA Sequencing Data Analysis

DNA sequencing data analysis is of vital importance for multiple reasons [9–12]; one strong point is it can be applied in multiple conditions:

1. Obtaining information encoded in gene.

   We can compare sequences, predict the sequence of promoters and enhancers, and identify the order of amino acids in certain proteins.

2. Discovering new genes.

   We can discover new genes by analyzing EST (expressed sequence tag) sequences and using DNA chip technology.

3. Analyzing gene polymorphism.

   We can analyze gene polymorphism, especially SNP (single-nucleotide polymorphism) to identify and locate functional genes, which can be targets of human evolution or diseases.

4. Predicting advanced structures.

   We can use the information of the primary structure to predict advanced structures of nucleic acids and proteins, thus predicting their functions.

5. Achieving personalized medicine.

   With the soaring need for personalized medicine, health-care providers are capable of using DNA sequencing data to give medical suggestions to patients.

Besides all these applications above, next-generation sequencing data analysis distinguishes itself in the following aspects:

1. Sequence the whole genomes rapidly and can zoom in to deeply sequence target regions.
2. Analyze genome-wide methylation or DNA-protein interactions.
3. Help researchers to dig into microbial diversity in humans or in the environment.

Although using computers for data analysis has obvious advantages, there still exist weaknesses:

1. Processing DNA sequencing data requires time and experience.
2. Although results are illustrated by those tools, researchers are supposed to analyze DNA sequencing results, to see whether the outcome is reasonable, and to plan future experiments.
3. We cannot depend on computer analysis totally; the software also make mistakes.

References

1. Gingeras TR, Roberts RJ (1980) Steps toward computer analysis of nucleotide sequences. Science 209(4463):1322–1328
2. Sanger F, Air GM, Barrell BG, Brown NL, Coulson AR, Fiddes CA, Hutchison CA, Slocombe PM, Smith M (1977) Nucleotide sequence of bacteriophage phi X174 DNA. Nature 265(5596):687–695. https://doi.org/10.1016/0022-2836(78)90346-7
3. ten Bosch JR, Grody WW (2008) Keeping up with the next generation: massively parallel sequencing in clinical diagnostics. J Mol Diagn 10(6):484–492. https://doi.org/10.2353/jmoldx.2008.080027
4. Liu L, Li Y, Li S, Hu N, He Y, Pong R, Lin D, Lu L, Law M (2012) Comparison of next-generation sequencing systems. J Biomed Biotechnol 2012:251364. https://doi.org/10.1155/2012/251364
5. Schadt EE, Turner S, Kasarskis A (2010) A window into third-generation sequencing. Hum Mol Genet 19(R2):R227–R240. https://doi.org/10.1093/hmg/ddq416
6. Excoffier L, Laval G, Schneider S (2005) Arlequin (version 3.0): an integrated software package for population genetics data analysis. Evol Bioinformatics Online 1:47–50
7. Librado P, Rozas J (2009) DnaSP v5: a software for comprehensive analysis of DNA polymorphism data. Bioinformatics 25 (11):1451–1452
8. Kumar S, Nei M, Dudley J, Tamura K (2008) MEGA: a biologist-centric software for evolutionary analysis of DNA and protein sequences. Brief Bioinform 9(4):299–306. https://doi.org/10.1093/bib/bbn017
9. Cai L, Yuan W, Zhang Z, He L, Chou GC (2016) In-depth comparison of somatic point mutation callers based on different tumor next-generation sequencing depth data. Sci Rep 6:36540. https://doi.org/10.1038/srep36540
10. Huang T, Liu CL, Li LL, Cai MH, Chen WZ, Xu YF, O'Reilly PF, Cai L, He L (2016) A new method for identifying causal genes of schizophrenia and anti-tuberculosis drug-induced hepatotoxicity. Sci Rep 6:32571. https://doi.org/10.1038/srep32571
11. Fang S, Zhang Y, Xu M, Xue C, He L, Cai L, Xing X (2016) Identification of damaging nsSNVs in human ERCC2 gene. Chem Biol Drug Des 88(3):441–450. https://doi.org/10.1111/cbdd.12772
12. Cai L, Deng SL, Liang L, Pan H, Zhou J, Wang MY, Yue J, Wan CL, He G, He L (2013) Identification of genetic associations of SP110/MYBBP1A/RELA with pulmonary tuberculosis in the Chinese Han population. Hum Genet 132:265–273. https://doi.org/10.1007/s00439-012-1244-5

Chapter 2

Transcriptome Sequencing: RNA-Seq

Hong Zhang, Lin He, and Lei Cai

Abstract

RNA sequencing (RNA-seq) can not only be used to identify the expression of common or rare transcripts but also in the identification of other abnormal events, such as alternative splicing, novel transcripts, and fusion genes. In principle, RNA-seq can be carried out by almost all of the next-generation sequencing (NGS) platforms, but the libraries of different platforms are not exactly the same; each platform has its own kit to meet the special requirements of the instrument design.

Key words Next-generation sequencing, RNA sequencing, Messenger RNA, Library construction, Data analysis

1 Introduction

In a broad sense, transcriptome refers to the collection of all transcripts under certain physiological condition, including messenger RNA (mRNA), ribosomal RNA (rRNA), transfer RNA (tRNA), and noncoding RNA (ncRNA), while in a narrow sense, it refers to collection of all mRNA transcripts [1]. Transcriptome sequencing, also called RNA-seq or whole-transcriptome shotgun sequencing (WTSS), uses high-throughput sequencing technology to rapidly and comprehensively obtain the transcriptional status of biological samples at a specific time [2]. At present, RNA-seq is mainly used in the study of mRNA, small RNA, noncoding RNA, or microRNAs. Different types of RNA can be obtained by adding additional separation and enrichment steps before cDNA synthesis. Illumina TruSeq is a method using conjugated magnetic beads (oligo-dT) to capture ploy A+ from total RNA and then contract mRNA library. During the ploy A+ enrichment process, non-ploy A+ RNA, including miRNA, rRNA, and other noncoding RNA, were removed [3, 4]. The mRNA library preparation steps contain five steps: (1) RNA fragmentation, (2) reverse transcription, (3) adapter ligation, (4) library cleanup and amplification, and (5) library quantification, quality control [5] (Fig. 1). Here, we show the method of

Tao Huang (ed.), *Computational Systems Biology: Methods and Protocols*, Methods in Molecular Biology, vol. 1754,
https://doi.org/10.1007/978-1-4939-7717-8_2,

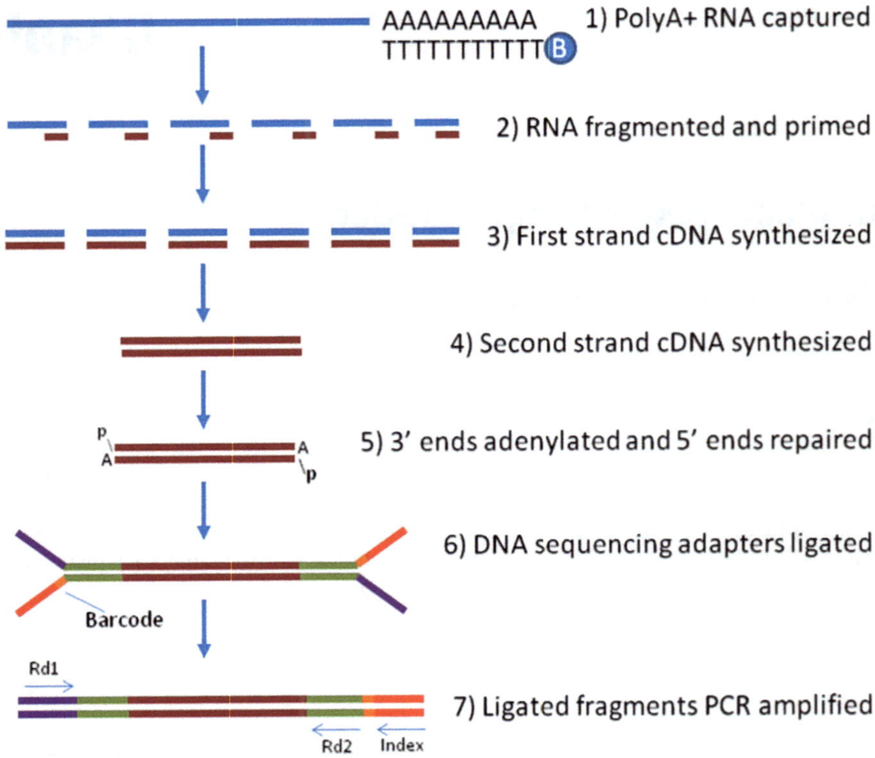

Fig. 1 mRNA library construction workflow for Illumina (from David Corney 2013)

the RNA-seq from total RNA extraction, library construction, and data analysis.

2 Materials

Prepare all solutions using ultrapure water and analytical grade reagents. Prepare and store all reagents at room temperature (unless indicated otherwise).

2.1 Total RNA Extraction

1. Liquid nitrogen.
2. 70% ethanol.
3. Tissue: keep the tissue in the liquid nitrogen until the procedure is completed.
4. TRIzol Reagent (Invitrogen).
5. DEPC-treated water (Ambion).
6. Chloroform: trichloromethane.
7. Isopropanol.
8. Thermo Scientific NanoDrop 2000 spectrophotometer: RNA quantification
9. Agilent 2100 Bioanalyzer system: RNA quality control.

2.2 mRNA Library Construction

1. RNA Purification Beads: purifying the poly-A containing mRNA molecules using oligo-dT attached magnetic bead, stored at 4 °C (Illumina, San Diego, CA).
2. Bead Washing Buffer (BWB), Elution Buffer (ELB), Bead-Binding Buffer (BBB): 1 tube per 48 reactions, stored at −20 °C (Illumina, San Diego, CA).
3. Elute, Prime, Fragment Mix (EPF): 1 tube per 48 reactions, stored at −20 °C (Illumina, San Diego, CA).
4. First-Strand Master Mix (FSM): 1 tube, stored at −20 °C (Illumina, San Diego, CA).
5. SuperScript II Reverse Transcriptase: 1 tube, stored at −20 °C.
6. Second-Strand Master Mix (SSM): 1 tube per 48 reactions, stored at −25 °C to −15 °C (Illumina, San Diego, CA).
7. AMPure XP beads: stored at 4 °C.
8. 80% ethanol.
9. Resuspension Buffer (RSB): 1 tube, stored at −20 °C.
10. End-Repair Mix: add 5′-phosphate groups needed for downstream ligation, 1 tube per 48 reactions, stored at −20 °C (Illumina, San Diego, CA).
11. A-Tailing Mix: make fragments compatible with adapters and prevent self-ligation by adding a 3′-A overhang, 1 tube per 48 reactions, stored at −20 °C (Illumina, San Diego, CA).
12. Ligation Mix: join 3′-T overhang adapters to 3′-A overhang inserts, 1 tube per 48 reactions, stored at −20 °C (Illumina, San Diego, CA).
13. Stop Ligation Buffer: inactivate the ligation. 1 tube per 48 reactions, stored at −20 °C (Illumina, San Diego, CA).
14. Resuspension Buffer (RSB): 1 tube, stored at −20 °C (Illumina, San Diego, CA).
15. PCR Master Mix (PMM): 1 tube per 48 reactions, stored at −20 °C (Illumina, San Diego, CA).
16. PCR Primer Cocktail (PPC): 1 tube per 48 reactions, stored at −20 °C (Illumina, San Diego, CA).
17. Sequencing chip: flow cell.
18. Illumina HiSeq system.

2.3 Data Analysis

1. Raw data processing: Trimmomatic.
2. Mapping: TopHat (Bowtie).
3. Quality control: RSeQC.
4. Differentially expressed gene analysis: htseq-count, DEseq, DAVID, KEGG.

5. Differential alternative splicing analysis: MISO (a mixture of isoforms).
6. Fusion gene analysis: TopHat-Fusion.

3 Methods

3.1 Total RNA Extraction

1. Remove the tissue sample from −80 °C refrigerator, and immediately put it in the thermos cup with liquid nitrogen (*see* **Note 1**).
2. Remove the sample from the liquid nitrogen and put into a 1.5 mL EP tube; add 300 μL TRIzol reagent, fully grinding with an electric tissue grinder; then add 700 L TRIzol; and place the tube on the ice for 30 min to ensure that sufficient crushing of the cells.
3. Add 200 μL chloroform, vortex, and then centrifuge at 13,000 × *g* for 10 min.
4. Remove supernatant to a new EP tube (*see* **Note 2**).
5. Add 500 μL isopropanol, vortex, place at −20 °C for 20 min, and then centrifuge at 13,000 × *g* for 10 min.
6. Discard supernatant; add 1 mL 70% ethanol solution, mild concussion for 10s; and then centrifuge at 8000 × *g* for 2 min.
7. Discard supernatant, and repeat **step 6** one time.
8. Discard supernatant, centrifuge at 8000 × *g* for 15 s, remove excess liquid, and place the EP tube on ice for 2 min to make ethanol fully volatile.
9. According to the precipitation size, add 30–200 μL ultrapure water.
10. Determine the concentration of RNA solution by using Nano-Drop 2000 spectrophotometer.
11. Use the Agilent 2100 Bioanalyzer system to detect the RNA integrity (*see* **Note 3**).
12. RNA solution should be stored in the −80 °C refrigerator.

3.2 Library Construction

1. Add 2 μg total RNA samples (less than 50 μL) to a 200 μL EP tube, dilute to 50 μL, then add 50 μL RNA Purification Beads (*see* **Note 4**), and gently pipette the entire volume up and down eight times to mix thoroughly.
2. Place the EP tube on PCR thermal cycler (65 °C for 5 min, 4 °C hold) to denature the RNA.
3. Place the EP tube at room temperature for 5 min to facilitate binding of the polyA RNA to the beads.

4. Place the EP tube on the magnetic stand for 5 min to separate the polyA RNA beads from the solution.
5. Discard the liquid, wash the beads by adding 200 μL Bead Washing Buffer, gently pipette the entire volume up and down eight times to mix thoroughly, and place the EP tube on the magnetic stand for 5 min.
6. Discard the liquid, add 50 μL of Elution Buffer, gently pipette, and place the EP tube on PCR thermal cycler (80 °C for 2 min, 25 °C hold).
7. Add 50 μL Bead-Binding Buffer, gently pipette, place the EP tube at room temperature for 5 min, then place the EP tube on the magnetic stand for 5 min, and discard the liquid (*see* **Note 5**).
8. Add 200 μL Bead Washing Buffer, gently pipette for eight times, and place the tube on the magnetic stand for 5 min.
9. Discard the liquid; add 19.5 μL Elute, Prime, Fragment Mix; gently pipette for eight times; and place the EP tube on PCR thermal cycler (94 °C for 8 min, 4 °C hold) (*see* **Note 6**).
10. Place the tube on the magnetic stand for 5 min, and remove 17 μL solution into a new EP tube.
11. Add 1 μL SuperScript II to 79.6 μL First-Strand Master Mix, and mix thoroughly (*see* **Note 7**).
12. Add 8 μL solution configured in **step 11** to the EP tube in **step 10**, and mix thoroughly.
13. Place the EP tube on PCR thermal cycler (25 °C for 10 min, 42 °C for 50 min, 70 °C for 15 min, 4 °C hold).
14. Add 25 μL Second-Strand Master Mix to the EP tube in **step 13**, mix thoroughly, and place the EP tube on PCR thermal cycler (16 °C for 1 h, 4 °C hold).
15. Add 90 μL AMPure XP purification beads, gently pipette for eight times, place the EP tube at room temperature for 15 min, and place the tube on the magnetic stand for 5 min.
16. Discard the liquid, add 200 μL 80% ethanol solution with the EP tube on the magnetic stand, and incubate the EP tube at room temperature for 30s.
17. Repeat **step 16** one time.
18. Discard the liquid, let the EP tube at room temperature for about 15 min till the full evaporation of the ethanol, and then remove the EP tube from the magnetic stand.
19. Add 62.5 μL Resuspension Buffer, place the EP tube at room temperature for 2 min, and then place it on the magnetic stand.
20. Remove 60 μL supernatant to a new EP tube.
21. Add 40 μL End-Repair Mix, mix thoroughly, and incubate the EP tube at 30 °C for 30 min.

22. Add 160 L AMPure XP purification beads, gently pipette for eight times, place the EP tube at room temperature for 15 min, and then place the tube on the magnetic stand for 5 min.
23. Discard the liquid, add 200 μL 80% ethanol solution with the EP tube on the magnetic stand, and incubate the EP tube at room temperature for 30s.
24. Repeat **step 23** one time.
25. Discard the liquid, let the EP tube at room temperature for about 15 min till the full evaporation of the ethanol, and then remove the EP tube from the magnetic stand.
26. Add 20 μL Resuspension Buffer, place the EP tube at room temperature for 2 min, and then place it on the magnetic stand.
27. Transfer 17.5 μL of the supernatant to a new EP tube, add 12.5 μL A-Tailing Mix, gently pipette for eight times, and incubate the EP tube at 37 °C for 30 min.
28. Add 2.5 μL Ligation Mix, 2.5 μL Resuspension Buffer, and 2.5 L RNA Adapter Index, gently pipette for eight times, and incubate the EP tube at 30 °C for 10 min.
29. Add 5 μL Stop Ligation Buffer, and gently pipette for eight times.
30. Add 42.5 μL AMPure XP purification beads, gently pipette for eight times, place the EP tube at room temperature for 15 min, and then place it on the magnetic stand.
31. Discard the liquid, add 200 μL 80% ethanol solution with the EP tube on the magnetic stand, and incubate the EP tube at room temperature for 30s.
32. Repeat **step 31** one time.
33. Discard the liquid, let the EP tube at room temperature for about 15 min till the full evaporation of the ethanol, and then remove the EP tube from the magnetic stand.
34. Add 22.5 L Resuspension Buffer, placed the EP tube at room temperature for 2 min, and then place it on the magnetic stand.
35. Transfer 20 μL of the supernatant to a new EP tube.
36. Add 25 μL PCR Master Mix and 5 μL PCR Primer Cocktail.
37. Place the EP tube on PCR thermal cycler (98 °C for 30s n; 15 cycles of98 °C for 10 s, 60 °C for 30 s, 72 °C for 30 s; 72 °C for 5 min; 10 °C hold).
38. Remove the EP tube from the PCR thermal cycler, adding 50 μL AMPure XP purification beads, gently pipette for eight times, place the EP tube at room temperature for 15 min, and then place it on the magnetic stand.
39. Discard the liquid, add 200 μL 80% ethanol solution with the EP tube on the magnetic stand, and incubate the EP tube at room temperature for 30s.

40. Repeat **step 39** one time.
41. Discard the liquid, let the EP tube at room temperature for about 15 min till the full evaporation of the ethanol, and then remove the EP tube from the magnetic stand.
42. Add 27.5 L Resuspension Buffer, place the EP tube at room temperature for 2 min, and then place it on the magnetic stand.
43. Transfer 25 μL of the supernatant to a new EP tube.
44. Dilute 2 μL library to 20 μL, and then use the 2% E-gel gel to detect the quality of the library; the final library size should be a band 200–400 bp.
45. Take appropriate library samples according to the requirements of the Illumina sequencing platform.

3.3 Data Analysis

3.3.1 Raw Reads Processing and Mapping

1. According to a base-calling pipeline, the images generated by the sequencing system (Illumina) are translated into nucleotide sequences. The raw reads are saved in fastq format, and Trimmomatic [6] can be used to filter the raw data prior to analyzing the data. There are three criteria: (1) discard reads shorter than 36 bases; (2) remove reads containing sequencing adaptors; and (3) remove bases with a quality score less than 15. The commands are as follows.

   For paired-end reads, the command is:

```
java -classpath trimmomatic.jar org.usadellab.trimmomatic.TrimmomaticPE
(see Note 8) -t threads $p -phred33 $input1 $input2 $output1_paired.fq.gz
$output1_unpaired.fq.gz $output2_paired.fq.gz $output2_unpaired.fq.gz IL-
LUMINACLIP:$WORKPATH/adapter.fa:2:40:15 SLIDINGWINDOW:4:15 MINLEN:36
```

   For single-end reads, the command is:

```
java -classpath trimmomatic.jar org.usadellab.trimmomatic.TrimmomaticSE
(see Note 8) -t threads $p -phred33 $input1 $output1_paired.fq.gz
$output1_unpaired.fq.gz ILLUMINACLIP:$WORKPATH/adapter.fa:2:40:15
SLIDINGWINDOW:4:15 MINLEN:36
```

2. Then, the clean sequencing reads should be aligned with UCSC hg19 reference genome using TopHat [7], which incorporates the Bowtie to perform the alignment. The command is:

```
tophat -p $p -G genes.gtf -o $tophat_out
$Reference/Sequence/BowtieIndex/genome
$output1_paired.fq.gz $output2_paired.fq.gz
```

3. RSeQC is a RNA-seq quality control package; it provides a number of useful modules that can comprehensively evaluate

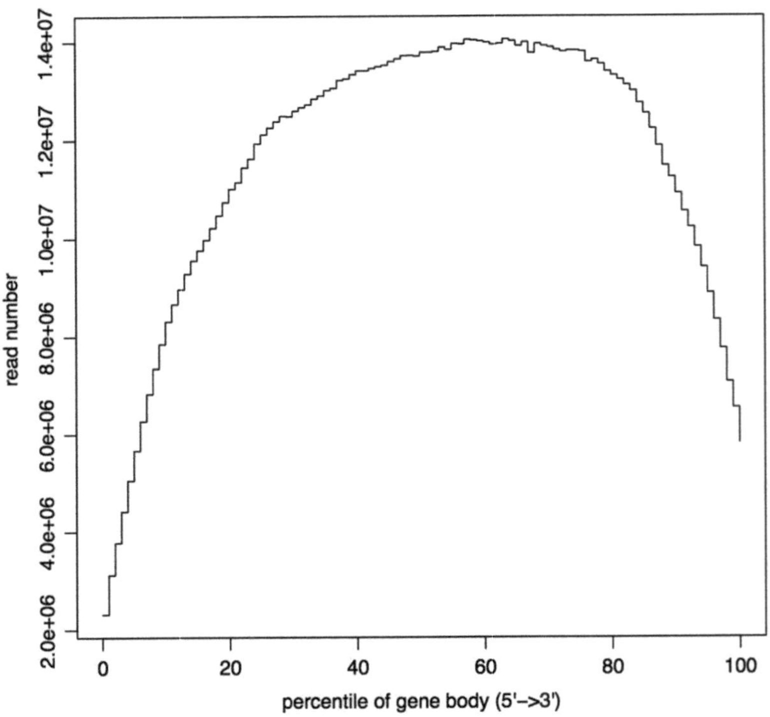

Fig. 2 Coverage uniformity over gene body without bias. Abscissa means the relative position of gene

RNA-seq data. geneBody_coverage.py can be used to calculate the RNA-seq reads coverage over gene body. The command is:

```
geneBody_coverage.py -r $Reference_BED -i $input -o $output
```

This module scales all transcripts to 100 nucleotides and calculates the number of reads covering each nucleotide position. Finally, it generates a plot illustrating the coverage profile along the gene body [8]. An example for the output is shown in Fig. 2, the reads coverage is uniform, and there is no 5′/3′ bias.

The commands for other packages of RSeQC are:

```
clipping_profile.py -i $input -o $output
infer_experiment.py -r $Reference_BED -i $input -o $output
inner_distance.py -r $Reference_BED -i $input -o $output
junction_annotation.py -r $Reference_BED -i $input -o $output
junction_saturation.py -r $Reference_BED -i $input -o $output
read_distribution.py -r $Reference_BED -i $input >$output_read_distribution.txt
read_duplication.py -i $input -o $output
read_GC.py -i $input -o $output
read_NVC.py -i $input -o $output
read_quality.py -i $input -o $output
RPKM_count.py -d '1++,1--,2+-,2-+' -i $input -o $output
RPKM_saturation.py -d '1++,1--,2+-,2-+' -r $Reference_BED -i $input -o $output
```

```
split_bam.py -r $Reference_BED -i $input -o $output
```

3.3.2 Differentially Expressed Genes Detection

1. The pipeline for the analysis of differentially expressed genes (DEGs) is shown in Fig. 3.
2. Use TopHat to align the reads on the genome, and then count reads in features with htseq-count; the commands are:

```
samtools view -h -o $tophat_out/accepted_hits.sam
$tophat_thout/accepted_hits.bam
htseq-count -s no $tophat_thout/accepted_hits.sam $genes.gtf > $htseq-count.out
```

3. Use DEseq or edgeR to detect the DEGs. An example for R commands of DEseq is:

```
    #!/user/bin/R/bin/Rscript
datafile = system.file("htseq-coun.txt",package="pasilla")
pasillaCountTable = read.table("htseq-coun.txt",header=TRUE, row.names=1)
pasillaDesign = data.frame(
row.names = colnames(pasillaCountTable),
condition = c("condition1"," condition2"),
libtype = c("paired-end","paired-end"))
condition = factor ( c( "control","case"))
library( "DESeq" )
cds = newCountDataSet( pasillaCountTable,condition )
```

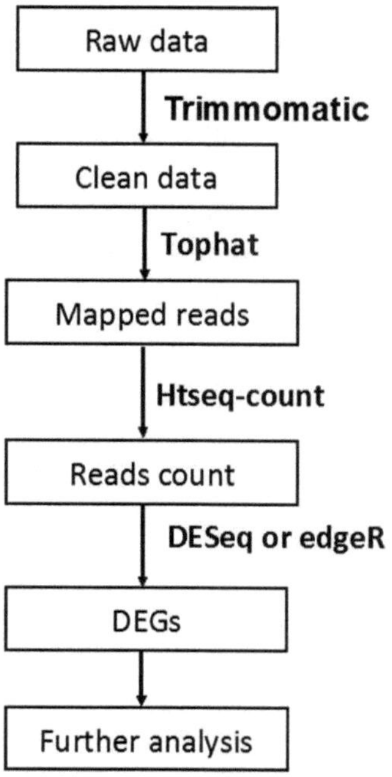

Fig. 3 The pipeline for the analysis of differentially expressed genes

```
cds = estimateSizeFactors( cds )
sizeFactors( cds )
head(counts( cds, normalized=TRUE )
cds = estimateDispersions( cds )
str( fitInfo(cds) )
plotDispEsts( cds )
res = nbinomTest( cds, "control", "case")
plotMA(res)
addmargins( table(res_sig = res$padj < .1, res_sig = res$padj < .1 ) )
write.csv (res, file=" diff_exp.csv")
```

4. Then we can use DAVID [9] to detect the functional enrichments and KEGG [10] to identify the significantly changed pathways for the differentially expressed genes.

3.3.3 Differential Exon-Skipping Event Detection

A mixture-of-isoform (MISO) [11] analysis adopted Bayesian inference algorithm to calculate the probability that a read came from a specific isoform. The software computes the percentage of transcripts that are spliced to include exons and is called the PSI (Ψ):

$$\text{Percentage splicing in (PSI)} = \frac{\#\ \text{of reads supporting inclusion}}{\#\ \text{of reads supporting inclusion} + \text{skipping}}$$

1. Compute the insert length:

```
misopy/pe_utils.py --compute-insert-len tophat_out/accepted_ hits.bam
ensGene.min_1000.const_exons.gff --output-dir insert-dist
```

2. Run MISO:

```
misopy/run_events_analysis.py --compute-genes-psi ref/hg19_miso_v2.0/
indexed_SE_events $tophat_thout/accepted_ hits.bam
--output-dir $OUTDIR/${SAMPLE}_output --read-len 101 --paired-end 157 33
```

3. Summarize MISO inferences:

```
misopy/run_miso.py --summarize-samples $OUTDIR/${SAMPLE}_output
$OUTDIR/${SAMPLE}_summaries/
```

4. Make pairwise comparisons between samples (*see* **Note 9**):

```
misopy/run_miso.py --compare-samples $OUTDIR/$control_output $OUTDIR/
$case_output $OUTDIR/
```

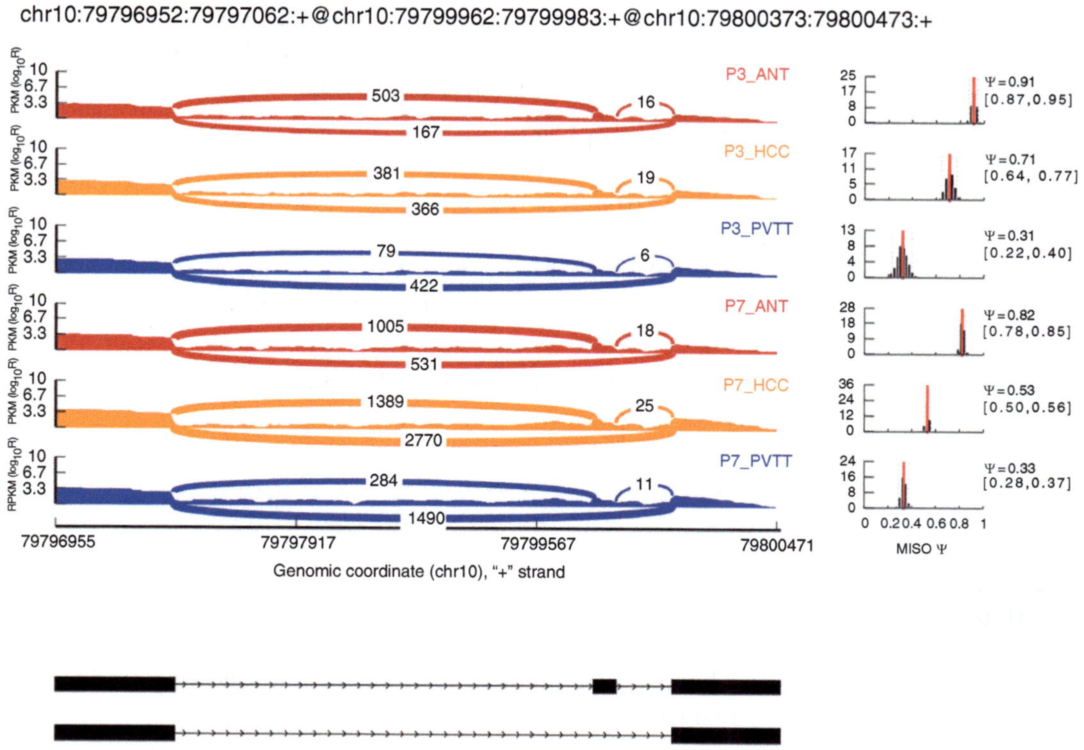

Fig. 4 RNA-seq reads coverage of the gene *RPS24* in P3 and P7. The RNA-seq reads were mapping to the UCSC reference genome (hg19) of *RPS24*. The PVTT tissue tracks were shown in blue, the HCC tissue in orange, and the ANT tissue in red. The counts of reads spanning the junction of exons were shown. Ψ = PSI (percentage splicing in)

5. Filter significant events (*see* **Note 10**):

```
misopy/filter_events.py --filter
$OUTDIR/$control_output_vs_$case_output/bayes-factors/$con-
trol_output_vs_$cas
e_output.miso_bf   --num-inc 1 --num-exc 1 --num-sum-inc-exc
10 --delta-psi 0.2
--bayes-factor 1000 --output-dir $OUTDIR/${SAMPLE}_filtered/
```

6. Visualizing and plotting MISO output by sashimi_plot, an example for *RSP24* gene [4] is shown in Fig. 4; the command is:

```
plott.py --plot-event
"chr10:79796952:79797062:+@chr10: 79799962:79799983:+@chr10:7
9800373:79800473:+"  ref/hg19_miso_v2.0/indexed_SE_events/
sample_sashimi_plot_settings.txt  --output-dir  test-plot-
RSP24/
```

3.3.4 Detecting Gene Fusions

Use TopHat to map the clean reads to the downloaded reference transcript sequences. Then, the mapped BAM files are passed to TopHat-Fusion [12] to detect the candidate fusions.

1. Running TopHat

```
tophat -o ${OUTDIR}/tophat_${SAMPLE} -p $p --fusion-search --keep-fasta-order --bowtie1 --no-coverage-search -r 0 --mate-std-dev 80 --max-intron-length 100000 --fusion-min-dist 100000 --fusion-anchor-length 13 --fusion-ignore-chromosomes chrM $BowtieIndex/genome $R1.fastq $R2.fastq
```

2. Run TopHat-fusion-post to filter out fusion candidates

```
tophat-fusion-post -p $p -o $OUTDIR/fusion_post_out --num-fusion-reads 1 --num-fusion-pairs 2 --num-fusion-both 5 $BowtieIndex/genome
```

4 Notes

1. The tissue should be stored immediately following sacrifice and extraction.
2. Be careful not to absorb liquids in the middle or lower part.
3. Illumina recommends that the total RNA integrity following isolation using Agilent 2100 with an RNA integrity number (RIN) value ≥ 8.
4. Vortex RNA Purification Beads tube vigorously to resuspend the oligo-dT beads.
5. This step can make mRNA combine with the beads and remove most of the rRNA.
6. The first strand of cDNA can be synthesized in this step by using reverse transcriptase and random primers.
7. The mixed solution should be prepared when used and can not be kept for too long after prepared.
8. PE, paired-end mode; SE, single-end mode.
9. To detect differentially expressed events or isoforms.
10. The default Bayes factor value is 10, and $|\Delta PSI| > 20\%$.

References

1. Villegas VE, Zaphiropoulos PG (2015) Neighboring gene regulation by antisense long non-coding RNAs. Int J Mol Sci 16 (2):3251–3266. https://doi.org/10.3390/ijms16023251
2. Conesa A, Madrigal P, Tarazona S, Gomez-Cabrero D, Cervera A, McPherson A, Szczesniak MW, Gaffney DJ, Elo LL, Zhang X, Mortazavi A (2016) A survey of best practices for RNA-seq data analysis. Genome Biol 17:13. https://doi.org/10.1186/s13059-016-0881-8

3. Zhang H, Weng X, Ye J, He L, Zhou D, Liu Y (2015) Promoter hypermethylation of TERT is associated with hepatocellular carcinoma in the Han Chinese population. Clin Res Hepatol Gastroenterol 39(5):600–609. https://doi.org/10.1016/j.clinre.2015.01.002
4. Zhang H, Ye J, Weng X, Liu F, He L, Zhou D, Liu Y (2015) Comparative transcriptome analysis reveals that the extracellular matrix receptor interaction contributes to the venous metastases of hepatocellular carcinoma. Cancer Genet 208(10):482–491. https://doi.org/10.1016/j.cancergen.2015.06.002
5. Head SR, Komori HK, LaMere SA, Whisenant T, Van Nieuwerburgh F, Salomon DR, Ordoukhanian P (2014) Library construction for next-generation sequencing: overviews and challenges. Biotechniques 56(2):61.-64, 66, 68, passim. https://doi.org/10.2144/000114133
6. Bolger AM, Lohse M, Usadel B (2014) Trimmomatic: a flexible trimmer for Illumina sequence data. Bioinformatics 30 (15):2114–2120. https://doi.org/10.1093/bioinformatics/btu170
7. Trapnell C, Roberts A, Goff L, Pertea G, Kim D, Kelley DR, Pimentel H, Salzberg SL, Rinn JL, Pachter L (2012) Differential gene and transcript expression analysis of RNA-seq experiments with TopHat and cufflinks. Nat Protoc 7(3):562–578. https://doi.org/10.1038/nprot.2012.016
8. Wang L, Wang S, Li W (2012) RSeQC: quality control of RNA-seq experiments. Bioinformatics 28(16):2184–2185. https://doi.org/10.1093/bioinformatics/bts356
9. Huang da W, Sherman BT, Lempicki RA (2009) Systematic and integrative analysis of large gene lists using DAVID bioinformatics resources. Nat Protoc 4(1):44–57. https://doi.org/10.1038/nprot.2008.211
10. Ogata H, Goto S, Fujibuchi W, Kanehisa M (1998) Computation with the KEGG pathway database. Biosystems 47(1–2):119–128
11. Katz Y, Wang ET, Airoldi EM, Burge CB (2010) Analysis and design of RNA sequencing experiments for identifying isoform regulation. Nat Methods 7(12):1009–1015. https://doi.org/10.1038/nmeth.1528
12. Kim D, Salzberg SL (2011) TopHat-fusion: an algorithm for discovery of novel fusion transcripts. Genome Biol 12(8):R72. https://doi.org/10.1186/gb-2011-12-8-r72

Chapter 3

Capture Hybridization of Long-Range DNA Fragments for High-Throughput Sequencing

Xing Chen, Gang Ni, Kai He, Zhao-Li Ding, Gui-Mei Li, Adeniyi C. Adeola, Robert W. Murphy, Wen-Zhi Wang, and Ya-Ping Zhang

Abstract

Capture hybridization coupled with high-throughput sequencing (HTS) has become one of the most popular approaches to address some scientific problems not only for fundamental evolution but also for ecology and human disease in recent years. However, the technical problem of limited probe capture ability affects its widespread application. Here, we propose to capture hybridize long-range DNA fragments for HTS (termed LR-LCH). We provide a case of three amphibian samples to examine LR-LCH with 2 kb libraries and comparison of standard capture hybridization with 480 bp libraries. Capture sensitivity increased from an average 13.57% of standard capture hybridization to an average 19.80% of LR-LCH; capture efficiency also increased from an average 72.56% of standard capture hybridization to an average 97.71% of LR-LCH. These indicate that longer fragments in the library generally contain both relatively variable regions and relatively conservative regions. The divergent parts of target DNA are enriched along with conservative parts of DNA sequence that effectively captured during hybridization. We present a protocol that allows users to overcome the low capture sensitivity problem for high divergent regions.

Key words Mitochondrial genome, Capture hybridization, Long-range PCR, NGS

List of Abbreviations

12s rRNA	12s ribosomal RNA
16s rRNA	16s ribosomal RNA
apt6	ATP synthase subunit 6
apt8	ATP synthase subunit 8
bp	Base pair
CO1	Barcoding gene from COX1
COX1	Cytochrome c oxidase subunit I
COX2	Cytochrome c oxidase subunit II
COX3	Cytochrome c oxidase subunit III
cytb	Cytochrome b

Authors Xing Chen, Gang Ni, Kai He, and Zhao-Li Ding contributed equally to this chapter.

Tao Huang (ed.), *Computational Systems Biology: Methods and Protocols*, Methods in Molecular Biology, vol. 1754,
https://doi.org/10.1007/978-1-4939-7717-8_3, © Springer Science+Business Media, LLC, part of Springer Nature 2018

HTS	High-throughput sequencing
K2P	Kimura 2-parameter
kb	Kilobase
LR-LCH	Long-range library capture hybridization
LR-PCR	Long-range polymerase chain reaction
mtDNA	Mitochondrial DNA
Mitogenome	Mitochondrial genome
ND1	NADH dehydrogenase subunit 1
ND2	NADH dehydrogenase subunit 2
ND3	NADH dehydrogenase subunit 3
ND4	NADH dehydrogenase subunit 4
ND5	NADH dehydrogenase subunit 5
ND6	NADH dehydrogenase subunit 6
PGM	Ion Torrent Personal Genome Machine

1 Introduction

Capture hybridization coupled with high-throughput sequencing (HTS) has become one of the most popular approaches to address some scientific problems not only for fundamental evolution but also for ecology and human disease in recent years [1–5]. Except restriction-site-associated DNA sequencing (RAD-seq, [5–7]) and whole-transcriptome shotgun sequencing (RNA-seq, [8]), capture hybridization coupled with HTS is a powerful approach that enables user cost- and time-effective for enriching and sequencing a massive amount of target loci (10s–1000s) distributed separately in a whole simultaneously [1, 5, 9]. According to target loci of interest, there exist versatile approaches, including exon capture, ultraconservative elements, hyRAD, and homemade probe capture for any specific gene [1, 6, 10–12]. As compared to RAD-seq and RNA-seq, the combined approaches have some merits: (1) it generated high-quality data, such as longer assembled contigs, lower variance in target coverage, and more accurate SNP calling [5, 12, 13]; (2) it allows user to align DNA/RNA sequence from widely taxonomic samples and infer complex and higher-level phylogenies [5, 14]. However, the combined method is not perfect. One shortage of the method is limitation of probe ability for cross-species capture, especially in divergent gene [10, 15].

Two decades ago, Jacobsen used capture hybridization method to enrich a single locus *lux* gene in microorganism genome from soil samples (*Pseudomonas fluorescens*, [16]). The number of locus for the capture hybridization greatly increased a decade ago because it incorporated HTS [17]. To date, the combined method has simplified and consists of four general steps: (1) prepare probe;

(2) prepare RNA/DNA library; (3) mix the probe and library to hybridize; (4) sequence hybridized library fragments. As compared to the old protocol, adding the second step has two obvious merits. The first is the copy number of target DNA from low level to moderate or high level. This results in great improvement of probe capture ability. The final process of library construction in step 2 is amplify the adapter-ligated DNA within 18 cycles. The quantity of library rapidly increased about 100–1000 times (from less than 1 ng to about 100–1000 ng depended on the sample quality). Tsangaras et al. demonstrated that the reduction of quantity of DNA library from 1500 to 15 ng results in the rapid decrease of rate of on-target reads in raw data (termed capture specificity) and the coverage rate of target genome region (termed capture sensitivity, [18]). Second, shorter than 500 bp of library length has been considered the best choice for both capture hybridization and sequencing experiments in order to get the best capture specificity and sensitivity. Gnirke et al. suggested that in-solution hybridization may be more efficient for libraries with fragment shorter than 500 bp [12, 19]. This length of libraries facilitated the most popular HTS platforms, such as less than 650 bp for MiSeq, less than 350 bp for HiSeq, and less than 480 bp for Ion Torrent PGM (the length included adaptor and index for these sequence machines). But when using single molecular sequence technology, the length of 500 bp was no longer a best parameter setting anymore since it enables sequencing the DNA fragment larger than 5 kb. Karamitros and Magiorkinis examined the efficiency when using 5 and 10 kb libraries for two long loci of interest from *phage lambda* and *Escherichia coli* and followed sequencing by using Oxford Nanopore MinION. The efficiency of their method is very well with 92.5% capture specificity and 99.73% capture sensitivity [20].

Researchers always require to balance the probe accessibility and sequence distance between probes and target DNA problem in order to increase capture sensitivity and capture specificity as high as possible [20–22]. There are some studies that examined the capture ability of various capture hybridization methods for divergent DNA sequence. Hedtke et al. designed an exon probe referring the reference genome of western clawed frog (*Xenopus tropicalis*) to capture other 16 frog samples (diversification dates to about 250 Mya). There is a negative correlation between divergence time and the number of on-target reads [10]. Bi et al. designed probe by referring the de novo assembled transcriptome contigs from the alpine chipmunk (*Tamias alpinus*). Within 1.5% sequence divergence in coding regions, there is no decline of capture specificity (24.4–29.1%) and capture sensitivity (about 90%). But both parameters decreased in the more divergent loci (about 9% sequence diversity, 30 Mya) from another genome of squirrel (*Ictidomys tridecemlineatus*, [15]).

Some studies provided solutions for increasing probe capture ability or reducing sequence distance between probes and target DNA. Mason et al. utilized PCR product to make probe from a single extant species Sunda colugo (*Galeopterus variegatus*). The probes successfully capture up to 13% divergence of target mitochondrial DNA (mtDNA), on average 76.92% capture specificity, and on average 56.29% capture sensitivity for 13 museum specimens of Sunda colugo [1]. Penalba et al. used long-range PCR (LR-PCR) product to make probe, and its capture efficiency increased rapidly for 27% divergence of target mtDNA from lizards [21]. Li et al. tested the various hybridization temperatures, touchdown strategy, and second-bait strategy for five pairs of animal groups (up to 298.6 Mya for western clawed toad). The study shows improvement of capture specificity to some degree [22]. Paijmans et al. examined the various temperatures, such as standard hybridization temperature 65 °C and low temperatures 60, 50, 48, and 45 °C, and touchdown strategy for both fresh and degraded DNA from carnivoran family Felidae. They concluded that capture specificity improved when applying 65 °C for degraded samples and touchdown strategy for fresh samples. There is no effect on improving capture sensitivity by adjusting hybridization, suggesting the temperature is not a crucial parameter to get divergent DNA [23].

Based on the observation of previous studies for divergent distance analysis across entire mitochondrial genome (mitogenome) and chromosome from animals, we noticed that a gene usually consists of relatively conservative and divergent parts. In other words, not all parts in a long locus have constant divergence. The distributional pattern of divergent parts intercepted by the conservative parts, despite the length of the former, was variable from one to another. For example, Mason et al. plotted a DNA sequence identity picture across the complete mitogenomes for Sunda colugo. Divergent part in conservative gene *16s* RNA is in the middle part and the beginning of *CO1* is much more divergent. Conservative region of divergent genes ND1, *ND2*, and *ND5* located in the middle part and for control region is at its beginning (Fig. 4c in ref. 1). Li et al. plotted a phylogenetic signal density picture across entire chromosome A1 and chromosome X for leopard (*Panthera* genus). These divergent and conservative parts were alternatively distributed in chromosome A1. Almost all the regions in chromosome X matched the pattern with some exceptions near the regions about 10, 25, 90, and 180 Mb (Fig. 4d in ref. 24).

In this study, to address the problem of capturing divergent DNA, we propose to modify a library by extending it to a longer length during the step 2 described above. We hypothesize that divergent DNA can be enriched along with its adjacent conservative DNA which can be captured easily (hereafter termed this strategy as LR-LCH). The general pipeline is shown in Fig. 1. First, we

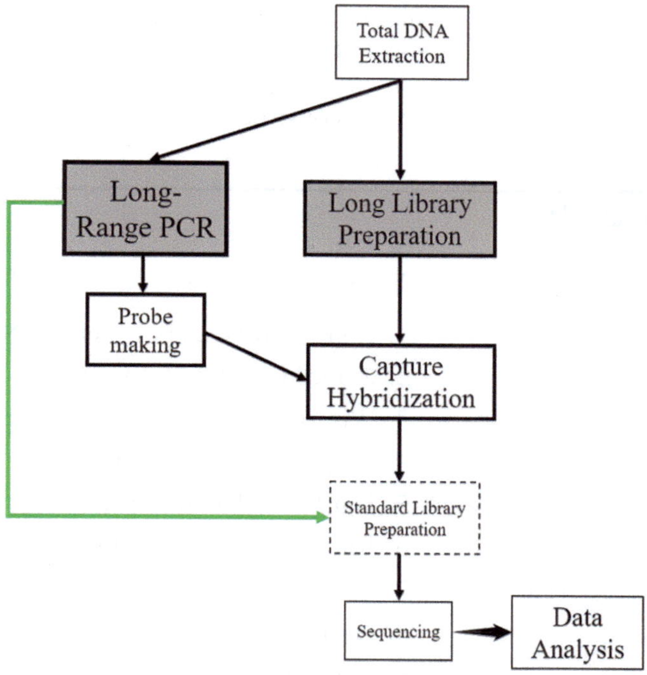

Fig. 1 Schematic pipeline for enriching mitochondrial DNA to high-throughput sequencing. The green line represents using the pair of long-range PCR (LR-PCR) amplicons to directly construct a library. Compared to standard library, the long-range library capture hybridization strategy (LR-LCR) has modification in library construction 1 and 2. LR-LCR requires a long fragment during shearing in library construction 1 for capturing high variable loci and additional library construction 2 for PGM sequencing. In standard hybridization, there is no construction library 2 and the enriched fragments directly sequenced

designed a new set of universal primers for vertebrate mitogenomes by referring its most conservative regions and then amplify mitogenomes, and its PCR product was used to make probe. Second, we prepared 2 kb length of libraries. Third, the libraries and probe were mixed to enrich target mtDNA. Then we sheared the enriched long-range library to approximately 300–700 bp and followed downstream library construction steps for sequencing. In the downstream sequence experiment, an Ion Torrent Personal Genome Machine (PGM) was used to sequence because it is fast and relatively inexpensive in terms of each run (not price per base). Each run using 316 chip generated over 800 Mb for 60 samples, and the data size for each sample is more than 10 Mb in general. These generated data was sufficient for de novo assembly of a complete mitochondrial genome. Although we sequenced with Ion Torrent platform, the protocol could also be applied to

Illumina platform in accordance with its library construction protocol. The detailed LR-LCH protocol is as follows. This protocol can be carried out in any molecular biology lab with standard library construction equipment.

2 Materials

All reagents and plasticware should be sterile.

1. LongAmp DNA polymerase (New England Biolabs).
2. 2.5 mM dNTP (Takara).
3. Wizard gel extraction kit (Promega).
4. Ampure beads (Beckman).
5. Ion Xpress barcode adapter kits from 1 to 96 (Thermo Fisher).
6. Ion Plus Fragment Library Kit (Thermo Fisher).
7. IonShear kit (Thermo Fisher).
8. Agarose gel.
9. Human Cot-1 DNA (Agilent).
10. Hybridization buffer and blocking agent (from an Agilent aCGH kit).
11. Streptavidin beads (M-270, Invitrogen).
12. Tween-20.
13. 3 M sodium acetate.
14. TE buffer (10 mM Tris, 1 mM EDTA, pH 8.0).
15. EBT and TET: 1× TE buffer, 0.05% Tween-20.
16. 1× bind and wash (BWT) buffer: 1 M NaCl, 10 mM Tris-Cl, 1 mM EDTA, 0.05% Tween-20, pH 8.0.
17. Hot wash (HW) buffer: 200 mL 10× PCR buffer, 200 mL $MgCl_2$ (25 mM), 1.6 mL H_2O.
18. Library Amplification Kit (KAPA).
19. 2100 Bioanalyzer (Agilent).
20. Qubit 2.0 (Invitrogen).
21. 2% E-gel (Invitrogen).
22. Focused-ultrasonicator M220 (Covaris).
23. PCR reaction tubes.
24. Covaris microTUBE.
25. NanoDrop (Thermo Fisher).
26. Magnetic rack.
27. Hybridization oven.
28. A thermal cycler.

3 Methods

3.1 Prepare Probe

1. Primes for amplifying mitogenome

   To achieve universality, we designed degenerate primers on the conservative regions as shown in Table 1. Primer pairs of F1/F2 and R1 were used to amplify a fragment from *12s rRNA* passed *COX1* to *COX3* (termed TF1, expected length: 5–9 kb). Primer pairs of F3 and R2/R3 were used to amplify remaining part of a mitogenome from *COX3* passed *cytb* to *12s/16s rRNA* (termed TR1; expected length, 5–9 kb).

2. Long-range PCR

   Long-range PCR is conducted in 25 μL reactions and mix the following reagents:

   (a) 0.8 μL forward primers (10 μM).

   (b) 0.8 μL reverse primers (10 μM).

   (c) 3 μL dNTP (2.5 mM).

   (d) 1 μL LongAmp DNA polymerase.

   (e) 5 μL 5× PCR buffer.

   (f) 50–200 ng template.

LR-PCR condition is as follows: initially incubate at 95 °C for 1 min, 30–32 cycles at 94 °C for 10 s, 58 °C for 40 s, and 65 °C extensive for variable times, and a final extension at 65 °C for 10 min, and hold 10 °C forever. Extension times are 10 min for TF1 and TR1. Check PCR product by using 0.8% agarose gel.

3. Purify LR-PCR product by using Wizard gel extraction kit.
4. Measure the concentration with a NanoDrop. The amounts of products should be up to 0.1–1.2 μg.
5. Mix PCR products according to amplicon length (and empirically adjusted according to sequence depth). The ratio of the

Table 1
Primer information

Primer name	Sequences	Location	ID
MtG_12s_480_F	GCTAGGAAACAAACTGGGAT**TAGATACC**	12s rRNA	F1
MtG_12s_270_F	TCGTGCCAGCCACCGCGG**TTAnAC**	12s rRNA	F2
MtG_cox3_R	AGCTGCGGCTTCAAAkCCr**AArTGrTG**	COX3	R1
MtG_cox3_F	ATGGCACACCAAGCACAyGChTwy**CAyATAGT**	COX3	F3
MtG_16s_1075_R	AGAGGACArGTGATTryGC**TACCTT**	16s rRNA	R2
MtG_12s_600_R	GGACACCGCCAAGTCCTTTGGG**TTTTAA**	12s rRNA	R3

TF1 to TR1 amplicon is 5:8. Probe making is conducted in 50 μL reactions and mix the following reagents:

(a) PCR product mixture 1.3 μg.

(b) 5 μL 10× dNTP mix.

(c) 5 μL 10× enzyme mix.

6. Mix and centrifuge briefly (15,000 × *g* for 5 s).
7. Incubate at 16 °C for 90 min.
8. Add 5 μL stop buffer.
9. Add 1/10 volume 3 M sodium acetate and 2 volumes cold (−20 °C) ethanol to the reaction tube. Freeze at −70 °C for 30 min.
10. Centrifuge at 15,000 × *g* for 10 min. Carefully remove the supernatant with a pipettor and dry the pellet.
11. Resuspend the pellet in 50 μL H_2O and precipitate the probe with sodium acetate and ethanol as described above.
12. Resuspend the probe in TE buffer and store at −20 °C.

3.2 Long DNA Library Preparation

1. LR-PCR products are mixed at a certain ratio the same with **step 5** in Subheading 3.1.
2. Shear the mixture in a Focused-ultrasonicator M220 (Covaris) by selecting the method DNA_2000bp_200_μL_Clear_microTUBE for 12 min. The shear volume is 200 μL.
3. End-repair reaction is carried out in 100 μL reactions and mix the following reagents:

   (a) 130 ng sheared DNA.

   (b) 20 μL 5× end-repair buffer.

   (c) 1 μL end-repair enzyme.

4. Adaptor ligation is carried out in 100 μL reactions and mix the following reagents in Ion Plus Fragment Library Kit:

   (a) 130 ng of sheared DNA.

   (b) 1.6 μL (Ion Xpress barcode adapter kits from 1 to 96).

   (c) 10 μL 10 × ligase buffer.

   (d) 2 μL dNTP mix.

   (e) 2 μL DNA ligase.

   (f) 8 μL nick repair polymerase.

   Incubate for 20 min at 25 °C in a thermal cycler followed by 72 °C incubation for 5 min.

5. Select long DNA fragment by using 0.4 volume of Ampure beads (i.e., 100 μL sample of DNA gets 40 μL of Ampure beads).

6. Library amplification is carried out in 100 μL reactions and mix the following reagents:
   (a) Size-selected library.
   (b) 10 μL 5× PCR buffer.
   (c) 5 μL 2.5 mM dNTP.
   (d) 2 μL of 10 μM forward and reverse primers.
   (e) 2 μL LongAmp DNA polymerase.

   Incubate 95 °C for 1 min and then 15 cycles of 94 °C for 10 s, 58 °C for 40 s, 65 °C for 3 min, and finally 65 °C for 10 min followed by holding at 4 °C.

7. Purify with Ampure bead and add 15 μL 1× TE buffer.

3.3 In-Solution Capture Hybridization

1. In-solution capture hybridization is carried out in 100 μL reactions:
   (a) 25 μL 2× hybridization buffer.
   (b) 5 μL 10× blocking agent.
   (c) 2 μL human Cot-1 DNA.
   (d) 2 μL of blocking adaptors (from Ion Plus Fragment Library Kit, Thermo Fisher).
   (e) 10–100 ng of bait and 100–1000 ng library (certain ratio of library and probe is 1:10).

   Incubate for 5 min at 95 °C and then incubate for 72 h at 65 °C.

2. After hybridization, incubate the mixture with 5 μL magnetic streptavidin beads (M-270, Invitrogen) for 20 min at room temperature.
3. Place the mixture into a magnetic rack to separate the magnetic beads from the supernatant.
4. Discard the supernatant.
5. Wash the beads using 200 μL of 1× BWT buffer, and vortex the mixture for 30 s each time.
6. Discard the supernatant.
7. Repeat **steps 5** and **6** for four times.
8. Wash the beads once with warmed HW buffer at 50 °C for 2 min.
9. Wash the beads once with 200 μL of 1× BWT buffer, and vortex the mixture for 30 s.
10. Wash the beads once with 100 μL of TET, and vortex the mixture for 30 s.

11. Separate hybridized target molecules from the bait in 30 μL TE by incubation at 95 °C for 5 min in a thermal cycler.
12. The PCR condition is the same with **step 6** in Subheading 3.2 for capture hybridization.

3.4 Standard Library Preparation

1. Shear the enriched libraries for 120 s using an IonShear kit (Thermo Fisher) in an open thermocycler. The 2 kb DNA fragments will be sheared to 300–500 bp.
2. Adaptor ligation is the same with **step 4** in Subheading 3.2 for capture hybridization.
3. Select 450–500 bp reads by using 2% E-gel.
4. Library amplification is carried out in a PCR volume of 50 μL by using a Library Amplification Kit:
   (a) 25 μL HiFi mix.
   (b) 21 μL selected fragment solution.
   (c) 4 μL primer mix (from Ion Plus Fragment Library Kit, Thermo Fisher).
5. Concentration is measured by using Qubit. Length of library is measured by using 2100 Bioanalyzer (Agilent).

4 A Case for Capturing Mitochondrial Genome from Amphibia

Three samples were selected to illustrate the performance of LR-LCR and standard capture hybridization method, including *Rana* sp.1, *Rana* sp.2, and *Onychodactylus* sp. To get all the three mitogenomes as reference, we amplified them by using the primers in Table 1 and sequenced LR-PCR products directly. To get probe, we use LR-PCR products from *Rana* sp.1 and total DNAs from other two species to prepare libraries. DNA from *Rana* sp.2 and *Onychodactylus* sp. are separated into two parts, respectively. One was used to construct libraries with length of 500 bp and another with length of 2 kb. The length of 500 bp library was prepared by using standard capture hybridization method; the length of 2 kb library according to LR-LCR strategy was processed and sequenced. All the libraries were listed in Table 2.

We used the probe of *Rana* sp.1 to capture a closely related mtDNA from *Rana* sp.2 (*CO1* K2P = 8.2%) and a relatively distant mtDNA from *Onychodactylus* sp. (*CO1* K2P = 25.5%). Standard capture hybridization and LR-LCH yielded an average depth per base of 48.65 and 156.15 coverage for *Rana* sp.2 and *Onychodactylus* sp., respectively (Table 2). On the one hand, to evaluate the capture sensitivity and specificity for standard capture hybridization method, we observed that two gaps existed in the MtG of *Rana* sp.2 at the end of *ND5* and in the noncoding regions. In *Onychodactylus* sp.1, five gaps occurred in relatively divergent regions at

Table 2
Results from Ion Torrent PGM of seven samples mixed together

Library name[a]	Total number of reads	Unique mapped reads (%)	Error (%)	Sequence depth	Coverage percentage (>2)[b]	Coverage percentage (> 10)[c]	Chimeric reads (%)
R.sp.1_P_480	30,453	90.93	0.54	426.8	>95[d]	>95[d]	0.74
R.sp.2_P_480	33,259	97.22	0.53	502.5	100	100	2.91
O.sp._P_480	37,450	98.34	0.59	685.4	100	100	1.33
R.sp.2_D_480	21,369	11.49	0.57	36.5	86.97	78.62	0.11
R.sp.2_D_2kb	45,605	21.32	0.63	80.9	99.49	95.42	0.73
O.sp._D_480	24,727	15.65	0.49	60.8	83.12	66.5	0.1
O.sp._D_2kb	84,599	18.25	0.62	231.4	100	100	0.12

[a]There are three parameters combined with underline in the library name: species name, source DNA, and library length. R.sp.1 represents *Rana* sp.1; R.sp.2 represents *Rana* sp.2; O.sp. represents *Onychodactylus* sp.; P represents using PCR products to construct library; D represents using total DNA. The number 480 represents the length of 480 base pairs of the libraries; 2 kb represents the length of 2000 base pairs of the libraries

[b]Sites with >2 of depth sequence are included to statistic genome coverage

[c]Sites with depth sequence >10 are included to statistic genome coverage

[d]The mitochondrial genome of *Rana* sp.1 was incompletely recovered for its control region which had long repetitive DNA

the end of *ND2*, at the beginning of *ND4* and *ND5*, at the end of *ND5*, and in control region. The capture sensitivity are 11.49% and 15.65% for *Rana* sp.2 and *Onychodactylus* sp., respectively; the capture efficiency are 21.32% and 18.25% for *Rana* sp.2 and *Onychodactylus* sp., respectively. On the other hand, to evaluate the capture sensitivity and specificity for LR-LCH, we observed that there is no gap for *Onychodactylus* sp. (Fig. 2b: green) and only one 400 bp gap in the repetitive region for *Rana* sp.2 (Fig. 2a: green). Capture sensitivity increased from an average of 13.57% to an average of 19.80%; capture efficiency also increased from an average of 72.56% to 97.71% (by using sites with sequence depth larger than 10).

5 Discussion

Above, we were able to demonstrate that LR-LCH is effective for capturing divergent loci in the two amphibian metagenomes. Below, we discuss the following: (1) only using sequence diversity rate to weigh capture ability of probe is not very accurate; (2) the advantage and disadvantage of the LR-LCH; and (3) its prospect.

Conserved sequences are similar or identical DNA sequences across species. In mitogenomes, *16s rRNA*, *12s rRNA*, *COX1*, *COX2*, and *COX3* are relatively conservative with comparison of *apt6*, *apt8*, NADH dehydrogenase genes, and control region. We noticed that the conservative parts of mitogenomes were captured easier than unconservative parts even though the diversity rate of former is much higher than the latter one. For example, *CO1* between *Rana* sp.1 and *Rana* sp.2 differed by approximately 25.5% (K2P distance); this variation was greater than the K2P of 15% for the regions between *ND5* and *ND6* of the two species, which is a gap. This indicated that capture ability is not fixed in a specific threshold for any DNA fragment. We may first consider the conservativeness of DNA sequence when evaluating and/or designing probe.

The first advantage is it reduces difficulty of de novo assembly with on-target reads. We use bait and iterative mapping strategy to recover these mitogenomes or other small-sized genome regions (for detailed steps, refer to [25]) and evaluate the performance according to computational time and iterative number. Although the number of on-target reads does not increase greatly, the reads distributed to the genome become even as compared to reads generated by using the standard capture hybridization. This reduced iterative number to two to three times and saved computational time of de novo assembly during each iteration, especially using MIRA assembler [26].

Another advantage is the length of libraries can be adjusted according to the sample set of interest. Here in these amphibian

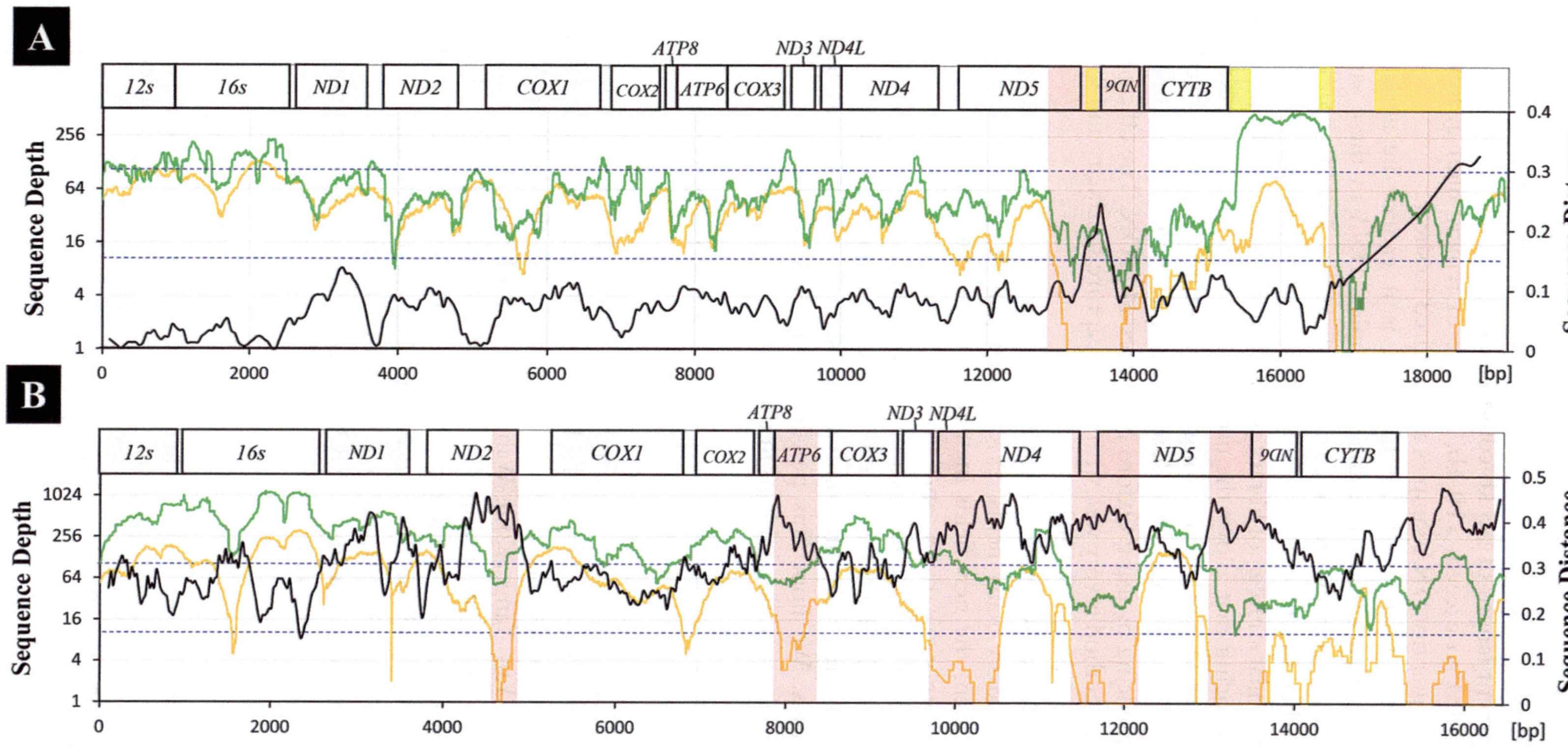

Fig. 2 Coverage distributions for 480 bp and 2 kb library. (**a**) represents *Rana* sp.2 results by using standard capture hybridization (orange line) and LR-HY (green line). Black line represents DNA sequence distance between *Rana* sp.1 and *Rana* sp.2. The sliding window length is 50 bp and the step length is 5 bp (below is the same). Dashed lines in **a** and **b** are constant at 0.15 and 0.3 sequence distance. The repetitive regions in *Rana* sp.2 which is labeled with yellow ranged from 13,424 to 13,572 bp, 15,402 to 15,660 bp, 16,593 to 16,770 bp, and 17,382 to 18,498 bp. (**b**) represents *Onychodactylus* sp. results by using standard capture hybridization (orange line) and LR-LCR (green line). Black line represents DNA sequence distance between *Rana* sp.1 and *Onychodactylus* sp. Dashed lines in **a** and **b** are constant at 0.15 and 0.3 of K2P. The regions with greatest sequence depth improvement are highlighted with red box

samples, DNA fragment with a length of 2 kb to cross-species capture is suitable for enrichment of high variable loci since the control regions are less than the length of 2 kb. It is possible that a fragment length of >3 kb could be captured, but it is not recommended to exceed >10 kb, because extremely high-quality and high-quantity DNA samples are required to shear.

One shortage of the LR-LCH is it cannot be applied to museum specimens and ancient samples. DNA in these samples already naturally degraded to small pieces (<500 bp, [27]). Another shortage is the complicated experiment as compared to the similar strategy of library preparation for single-molecule DNA sequencing that allows the user to sequence long library. Karamitros and Magiorkinis use long library capture method for two long loci of interest from *phage lambda* and *Escherichia coli* and followed sequencing by using Oxford Nanopore MinION. The efficiency of their method is very well with 92.5% capture specificity and 99.73% capture sensitivity [20].

We see capture hybridization becomes a more important method in many fields, except phylogenetics and ancient DNA research; the new fields include metagenomics, genetic mapping of phenotypic traits, and structural variation of some function region in pathogenic bacteria and human cancer cell lines [2, 4, 20, 28–31]. LR-LCH as a new kind of capture hybridization may provide the user obtaining better sequence quality and longer contigs. We hope this pipeline could help biologist insight of various scientific problems in these fields.

Acknowledgments

We thank Jing Che's research group for specimen collection and identification, and especially Hong-man Chen who examined species information and its CO1 sequence. We thank Dong Wang, Chun-Yan Yang, Kong-Wah Sing, and Elizabeth Georgian for reviewing the manuscript. This work was supported by the Ministry of Science and Technology of China (MOST no. 2012FY110800 to W.W.) and the National Natural Science Foundation of China (NSFC no. 31090251 to Y.Z.). Raw data from next-generation sequencing is available at SRA (http://www.ncbi.nlm.nih.gov/sra) under accession number SRP090718.

References

1. Mason VC et al (2011) Efficient cross-species capture hybridization and next-generation sequencing of mitochondrial genomes from noninvasively sampled museum specimens. Genome Res 21:1695–1704
2. Jones MR, Good JM (2016) Targeted capture in evolutionary and ecological genomics. Mol Ecol 25:185–202
3. Tsangaras K et al (2014) Hybridization capture reveals evolution and conservation across the

entire koala retrovirus genome. PLoS One 9: e95633

4. Gasc C et al (2016) Sequence capture by hybridization to explore modern and ancient genomic diversity in model and nonmodel organisms. Nucleic Acids Res 44:4504–4518
5. Leaché AD et al (2015) Phylogenomics of Phrynosomatid lizards: conflicting signals from sequence capture versus restriction site associated DNA sequencing. Genome Biol Evol 7(3):706–719. https://doi.org/10.1093/gbe/evv026
6. Suchan T et al (2016) Hybridization capture using RAD probes (hyRAD), a new tool for performing genomic analyses on collection specimens. PLoS One 11(3):e0151651. https://doi.org/10.1371/journal.pone.0151651
7. Miller MR et al (2007) Rapid and cost-effective polymorphism identification and genotyping using restriction site associated DNA (RAD) markers. Genome Res 17:240–248
8. Wang Z et al (2009) RNA-seq: a revolutionary tool for transcriptomics. Nat Rev Genet 10:57–63
9. Bekaert B et al (2016) In-solution hybridization for the targeted enrichment of the whole mitochondrial genome. Methods Mol Biol 1420:173–183
10. Hedtke SM et al (2013) Targeted enrichment: maximizing orthologous gene comparisons across deep evolutionary time. PLoS One 8: e67908
11. Faircloth BC et al (2012) Ultraconserved elements anchor thousands of genetic markers spanning multiple evolutionary timescales. Syst Biol 61:717–726
12. Gnirke A et al (2009) Solution hybrid selection with ultra-long oligonucleotides for massively parallel targeted sequencing. Nat Biotechnol 27:182–189
13. Ku CS et al (2012) Exome versus transcriptome sequencing in identifying coding region variants. Expert Rev Mol Diagn 12:241–251
14. Manthey JD et al (2016) Comparison of target-capture and restriction-site associated DNA sequencing for phylogenomics: a test in Cardinalid tanagers (Aves, genus: Piranga). Syst Biol 65(4):640–650. https://doi.org/10.1093/sysbio/syw005
15. Bi K et al (2012) Transcriptome based exon capture enables highly cost-effective comparative genomic data collection at moderate evolutionary scales. BMC Genomics 13:403
16. Jacobsen CS (1995) Microscale detection of specific bacterial DNA in soil with a magnetic capture-hybridization and PCR amplification assay. Appl Environ Microbiol 61 (9):3347–3352
17. Noonan JP et al (2006) Sequencing and analysis of Neanderthal genomic DNA. Science 314:1113–1118
18. Tsangaras K et al (2014) Hybridization capture using short PCR products enriches small genomes by capturing flanking sequences (CapFlank). PLoS One 9(10):e109101. https://doi.org/10.1371/journal.pone.0109101
19. Horn S (2012) Target enrichment via DNA hybridization capture. Methods Mol Biol 840:177–188
20. Karamitros T, Magiorkinis G (2015) A novel method for the multiplexed target enrichment of MinION next generation sequencing libraries using PCR-generated baits. Nucleic Acids Res 43(22):e152. https://doi.org/10.1093/nar/gkv773
21. Penalba JV et al (2014) Sequence capture using PCR-generated probes: a cost-effective method of targeted high-throughput sequencing for nonmodel organisms. Mol Ecol Resour 14:1000–1010
22. Li C et al (2013) Capturing protein-coding genes across highly divergent species. BioTechniques 54(6):321–326. https://doi.org/10.2144/000114039
23. Paijmans JL et al (2016) Impact of enrichment conditions on cross-species capture of fresh and degraded DNA. Mol Ecol Resour 16:42–55
24. Li G et al (2016) Phylogenomic evidence for ancient hybridization in the genomes of living cats (Felidae). Genome Res 26(1):1–11. https://doi.org/10.1101/gr.186668.114
25. Hahn C et al (2013) Reconstructing mitochondrial genomes directly from genomic next-generation sequencing reads - a baiting and iterative mapping approach. Nucleic Acids Res 41(13):e129
26. Burlibasa C, et al (1999) Genome sequence assembly using trace signals and additional sequence information. German conference on bioinformatics, vol. 99, pp. 45–56
27. Sawyer S et al (2012) Temporal patterns of nucleotide misincorporations and DNA fragmentation in ancient DNA. PLoS One 7: e34131
28. Eckert SE et al (2016) Enrichment by hybridisation of long DNA fragments for Nanopore sequencing. Microb Genom. https://doi.org/10.1099/mgen.0.000087

29. Chevalier FD et al (2014) Efficient linkage mapping using exome capture and extreme QTL in schistosome parasites. BMC Genomics 15:617
30. Denonfoux J et al (2013) Gene capture coupled to high throughput sequencing as a strategy for targeted metagenome exploration. DNA Res 20:185–196
31. Fan PF et al (2017) Description of a new species of Hoolock gibbon (primates: Hylobatidae) based on integrative taxonomy. Am J Primatol 79:e22631. https://doi.org/10.1002/ajp.22631

Chapter 4

The Introduction and Clinical Application of Cell-Free Tumor DNA

Jun Li, Renzhong Liu, Cuihong Huang, Shifu Chen, and Mingyan Xu

Abstract

Cell-free tumor DNA (ctDNA) is a kind of potential tumor biomarkers originated from cancer lesion in the circulating liquids. Liquid biopsy, as a minimally invasive or noninvasive manner, is a cutting-edge technology to detect ctDNA and other circulating biomarkers in the blood or other body fluids. ctDNA is mostly used for cancer patients to select targeted drugs in clinical application. In addition, ctDNA could also be applied to monitor tumor progression and recurrence. In conclusion, ctDNA is a very promising tumor biomarker for diagnosis and monitoring, which would increasingly become a routine clinical application in recent years.

Key words ctDNA, Tumor biomarker, Liquid biopsy, Targeted drug, EGFR, TKI

1 Introduction

Liquid biopsy is a kind of cutting-edge technology to analyze a range of tumor material in the blood or other body fluids in a minimally invasive or noninvasive manner. The tumor material includes circulating tumor cells (CTCs), cell-free tumor DNA (ctDNA), messenger RNA (mRNA), microRNA (miRNA), and exosomes. Of these tumor biomarkers, ctDNA are most widely recognized in clinical application.

In this chapter, we will focus on the introduction and clinical application of ctDNA. Comparing to ctDNA, cell-free DNA (cfDNA) is a broader term which describes DNA that is freely circulating, but is not necessarily of tumor origin. However, ctDNA is a term much better known in the cancer field and it has been adopted by clinical experts. So in this chapter, we will use both cfDNA and ctDNA. When we mention ctDNA, be noted that it refers to the cfDNA from tumor patients. In next section, we will also briefly introduce circulating tumor cells (CTCs), circulating exosomes and circulating mRNA as they are the other major

Tao Huang (ed.), *Computational Systems Biology: Methods and Protocols*, Methods in Molecular Biology, vol. 1754, https://doi.org/10.1007/978-1-4939-7717-8_4,

circulating tumor biomarkers. And in section 3 and 4, we will introduce the ctDNA detection methods and its clinical application.

2 Circulating Tumor Biomarkers

2.1 CTCs

CTCs originate from the primary and metastatic tumor tissue. Most CTC assays work by selecting tumor cells based on their biological (protein expression) and physical properties (size, density, deformability, electric charge). Similar to ctDNA, CTCs have been used to monitor tumor burden. CellSearch (Veridex, New Jersey, USA) is an FDA-approved assay for the enumeration of CTCs. Although commercially available, the clinical application is limited due to low detection rates. Based on the characteristics of CTC, it is applicable to assess the risk of progression after treatment, rather than diagnosis before treatment [1]. Moreover, the clinical value of CTC is limited to certain cancer types. For example, CTCs can be detected in over 60% of patients with metastatic breast and prostate cancer but only in 30–40% of patients with metastatic colorectal cancer (mCRC) [2–4]. Compared with the CTCs detected in metastatic breast and prostate cancer patients (median 6–7 CTC/7.5 mL), the CTCs detected in mCRC patients are typically at a lower number (median 1–2 CTC/7.5 mL) [5–7]. There are several reasons for this discrepancy. First, CRC CTCs often travel in adherent clusters due to increased surface adhesion molecules. Second, the circulation anatomy may affect cell collection. Specifically, hepatic drainage of portal blood may act as a filter, resulting in fewer CTCs in peripheral blood. Tumor location also plays a role. For example, lower rectal tumors have more CTCs in the central venous blood than tumors in the upper rectum and colon. Finally, as CTCs undergo epithelial-mesenchymal transition (EMT), cells lose their epithelial markers and become more difficult to detect. CellSearch relies on the expression of EpCAM, a marker of epithelial phenotype. Even when the cells do express EpCAM, it has been shown that the specific EpCAM antibody can greatly change the detection rate.

2.2 ctDNA

ctDNA is comprised of fragments of tumor-derived cfDNA from necrotic or apoptotic tumor cells. ctDNA can originate from CTCs, metastases, or the primary tumor. There are two challenges in ctDNA detection. First, a tumor-encoded somatic aberration (e.g., mutation, translocation, or methylation event) must be present in the tumor genome if it is ever to be detected in the blood. While genes such as APC, TP53, and KRAS are frequently mutated in GI malignancies, none is mutated in 100% of them. Second, the amount of ctDNA is very likely related with tumor burden in a nonlinear manner. Small amount of ctDNA is present in early-stage tumor, sometimes occupying only 0.01% of total cell-free DNA (cfDNA) [8]. For example, for localized GI tumors [colorectal cancer (CRC), gastric or gastroesophageal (GE) cancer, and pancreatic cancer], ctDNA can be detected in 48–73% of patients [9]. However, as tumor burden increases, deep surveillance of sufficient genomic space enhances ctDNA detection to nearly 100% [10]. Thus, ctDNA is currently most applicable to patients with advanced tumor. The traditional DNA sequencing approaches such as Sanger sequencing or pyrosequencing are only able to detect ctDNA in patients with significant tumor burden. To improve the ability to detect and analyze ctDNA, a variety of technologies have been developed, including digital PCR, NGS, BEAMing (beads, emulsion, amplification, and magnetics), PAP (pyrophosphorolysis-activated polymerization), CAPP-Seq (cancer personalized profiling by deep sequencing), and TAm-Seq (tagged-amplicon deep sequencing). The optimal liquid biopsy assay remains an area of active investigation.

Nonetheless, compared with traditional CTC assays, ctDNA is arguably more sensitive [11, 12]. For example, in a recent study of advanced solid tumors, ctDNA was always detected when CTCs were present [10]. Moreover, ctDNA was often detected when CTCs were absent. However, it should be noted that this study did not utilize enrichment methods prior to CTC detection, which is now the standard step among CTC. In addition to the ongoing research regarding early detection of malignancy, ctDNA is also being applied to identify clinically actionable somatic point mutations or deletions, monitor treatment response, and determine therapeutic resistance to targeted therapeutics and disease progression. It may also be useful in measuring minimal residual disease.

2.3 Circulating Exosomes

Exosomes are extracellular vesicles containing proteins and nucleic acids. They are secreted by all cells and circulate in the blood. Cancer cell-derived exosomes may have unique cell surface proteins, distinguishable from normal exosomes. For example, glypican-1 (GPC1), a membrane-anchored protein overexpressed in breast [13] and pancreatic cancer [14], was recently revealed to be detected exclusively in cancer exosomes. In a validation cohort of 56 patients with pancreatic ductal adenocarcinoma (PDA) (with patients from all four stages of cancer including carcinoma in situ), 6 patients with benign pancreatic disease such as chronic pancreatitis, and 20 healthy people, GPC1+ circulating exosomes (crExos) were 100% sensitive and specific for distinguishing patients with PDAC [15]. This study reveals that exosomes may serve as a potential noninvasive diagnostic and screening biomarker to detect early stages of cancer to facilitate possible curative surgical therapy. More validation studies are necessary to develop exosomes as a novel cancer biomarker.

2.4 Circulating mRNA

Several circulating mRNA candidates have been studied for CRC screening and prognosis after surgical resection [16–18]. mRNA biomarkers in the blood, such as CEA, CK20, CK19, human telomerase reverse transcriptase (hTERT), and guanylyl cyclase C (GC-C), have all been examined in the perioperative setting [17]. In most cases, the persistence of tumor-associated mRNA within 24 h after tumor resection has been a strong predictor of relapse [19]. In a meta-analysis of nine studies with CRC patients undergoing curative surgery, CTC detection based on CEA, CK19, and CK20 mRNA correlated with the development of hepatic metastases and decreased disease-free survival [20]. Recently, Rodia et al. utilized a novel Transcriptome Mapper (TRAM) to identify some specific RNAs in the blood as the candidate mRNA markers to detect CRC. After validation by qRT-PCR, the expression of genes including TSPAN8, LGALS4, COL1A2, and CEACAM6 was found to be statistically different in the blood of 67 patients and 67 healthy controls [19]. However, circulating mRNA also has its limitations. Extracellular mRNA is susceptible to degradation by RNase in the plasma. Attempts to use RNase inhibitors have been unsuccessful at protecting mRNA from degradation [21]. Thus, the application of mRNA in cancer screening and prognosis requires more investigation.

3 Methods of Detecting Genomic Changes from cfDNA

Nowadays, there are various techniques to detect genetic alterations in cfDNA, including real-time polymerase chain reaction (RT-PCR), digital PCR, amplification protocols with magnetic beads in oil emulsions (beads, emulsion, amplification, and magnetics (BEAMing)), next-generation sequencing (NGS), and mass spectrometry (MS) genotyping. Currently there are few standardized methods for processing cfDNA samples and analyzing the results. Thus, it is necessary to establish evidence-based guidelines in order to select the most reliable and cost-effective method for clinical application. Here, these techniques will be described below.

3.1 RT-PCR

Three types of RT-PCR are routinely employed, based on the probes utilized: TaqMan probes, TaqMan Detection Mutation Assay, and Scorpion probes. TaqMan probes have a mutation detection limit of about 10%, whereas TaqMan Detection Mutation Assay may detect mutations as low as 0.1% [22, 23]. These variations of sensitivity are due to the probe design: TaqMan probes consist of a fluorophore covalently attached to the 5′-end of the oligonucleotide probe and a quencher at the 3′-end, whereas TaqMan Detection Mutation Assay contains an allele-specific primer and an MGB blocker oligonucleotide that suppresses the wild-type background. Scorpion probes are covalently bound to a primer, a fluorophore, and a quencher. In the absence of the mutation, during the Scorpion PCR reaction, the quencher close to the fluorophore absorbs its fluorescence. However, the presence of a mutation leads to separation of the quencher from the fluorophore, resulting in an increase of the emitted fluorescence. In lung cancer, the majority of cfDNA studies have utilized RT-PCR. Particularly, studies aiming at the detection of epidermal growth factor receptor (EGFR) mutations in cfDNA have reported wide variability in their results. An Asian study including 86 advanced NSCLC patients compared EGFR mutation status in pretreatment cfDNA and tissue samples, which showed sensitivity of 43.1% and specificity of 100% [24]. Another study including 76 Caucasian advanced NSCLC patients also evaluated the detection of EGFR mutation in pretreatment blood samples and cfDNA, revealing greater sensitivity (78%) and equal specificity (100%) than the previous study [25]. However, the RT-PCR technology is not suitable for detecting many mutations since it's designed for the detection of only a few targets.

3.2 Digital PCR

Digital PCR and RT-PCR share the same principle. However, the key difference between both techniques lies in the procedure to quantify nucleic acids. RT-PCR works with a unique solution and performs one reaction per sample, whereas digital PCR produces thousands of replicates from a single sample and performs one reaction per replicate. Therefore, the sensitivity of digital PCR is much higher than RT-PCR, being capable to detect cfDNA within a germline DNA background in a ratio up to 1:100,000 [26]. Nevertheless, the main disadvantage of digital PCR is that the standard thresholds of determining the presence and abundance of mutant cfDNA have not been established.

3.3 BEAMing

The principle of this technology lies in the association of digital PCR and flow cytometry, using beads, emulsification, amplification, and magnetics to achieve the necessary level of sensitivity. The BEAMing procedure begins with isolating and purifying the cfDNA, followed by preamplification by conventional PCR. These DNA templates are then amplified again by emulsion PCR using primers, which are covalently connected to magnetic microbeads via streptavidin-biotin interactions. At the end of the reaction, the amplicons generated in each emulsion droplet will remain physically attached to the microbeads, making it easier to separate and purify them using a magnet. Subsequently, the DNA attached to the microbeads is analyzed to evaluate the presence and the amount of mutations using flow cytometry. This technology is able to detect a low percentage of mutant DNA in a higher amount of fragments comprising wild-type DNA, approximately one single mutant allele in a background of 10,000 wild-type alleles. Thus, it is also able to provide a digital readout of copy number quantification [9]. One study, enrolling 44 advanced NSCLC patients with activating EGFR mutations in tumor tissue, revealed the status of EGFR in the plasma of 32 cases (72.7%; 95% CI, 58.0–83.6%) [27]. Nevertheless, the procedure of BEAMing is a complex system, which limits its feasibility and reproducibility.

3.4 NGS

To date, NGS has demonstrated the highest sensitivity and specificity of detection in clinical molecular oncology. Remarkably, published studies have demonstrated that NGS is a feasible, accurate, and sensitive technique for identifying tumor-derived mutations in cfDNA, with sensitivity and specificity of more than 85% for NSCLC (I–IV stages). Nowadays there are many specialized and modified NGS techniques available, but they share the same principle [28–30]. They are based on the production of short sequences from single molecule of DNA and their comparison to a reference sequence, which results in the sequencing of a significant portion of the genome. Deep sequencing also allows the targeted investigation of specific candidate loci, even if mutated alleles are highly diluted. Furthermore, the main advantage of this technology is that it may identify rearrangements and regions of copy number aberrations, genetic alterations that are not detectable with other techniques [28]. However, this methodology is expensive to implement, requiring expert personnel to comprehensively analyze and interpret the results and a well-developed infrastructure for storage of large amount of sequencing data. These limitations have hampered the application of NGS for routine clinical practice.

3.5 Mass Spectrometry

To date, the leading method for detection of mutations based on mass spectrometer (MS) is matrix-assisted laser desorption-ionization time-of-light (MALDI-TOF). This technology is able to detect different alleles based on different masses of the primer extension products. The analysis process of MALDI-TOF MS is based on the following steps: amplification, primer extension reaction and ionization, separation by size, and detection of nucleic acids. In the end, a software provides a mass spectrum of the extension products [31]. Thus, this methodology could be applied to analyze multiple mutations. In a study, MALDI-TOF MS was used to detect EGFR and KRAS (KRAS proto-oncogene, GTPase) mutations in 2387 NSCLC patients, demonstrating that the MALDI-TOF platform reduced the rate of missed mutations compared with Sanger sequencing [32]. However, it is more expensive and time-consuming than other techniques. Moreover, it only returns genotypic data.

4 Application

4.1 Screening

Current common cancer screening methods include direct visualization, imaging, histologic evaluation, and blood protein tests. While these methods could help patients to identify potentially curable early-stage malignancies, there are still several barriers for cancer screening [33, 34]. Many tests are uncomfortable for patients (colonoscopies, pap smears, mammograms) or expose patients to radiation (CT scans, mammography). Some are unpleasant to accomplish (stool studies). And almost all tests require a schedule of additional healthcare visits. Thus, the optimal cancer screening assay remains an area of active research.

Noninvasive technique is a hot spot of cancer screen research. Two novel noninvasive techniques were recently approved by FDA for CRC screening. In 2014, the FDA approved fecal DNA testing with Cologuard (Exact Sciences Corporation, Madison, WI, USA). Cologuard is the first noninvasive screening test for colorectal cancer that analyzes both stool DNA and blood biomarkers. It has been proven to find 92% of cancers and 69% of the most advanced precancerous polyps in average-risk patients. In 2016, the FDA approved Epi proColon (Epigenomics, Berlin, Germany). It is a blood-based test that detects methylated Septin9 (SEPT9) DNA to identify CRC patients who choose not to undergo colonoscopy and stool-based fecal immunochemical tests [35]. Other assays for detecting early-stage GI malignancies are currently under development.

Liquid biopsy, as a kind of noninvasive techniques, has been widely applied to detect malignancies in clinical setting. In a study of 125,426 women undergoing noninvasive prenatal testing (NIPT) for fetal aneuploidy via Verifi Prenatal Test (Illumina, San Diego, California, USA), 3757 women were positive for one or more aneuploidies involving chromosome 13, 18, 21, X, or Y. Of these women, 39 had multiple aneuploidies in which 7 (18%) were diagnosed with occult malignancy [36, 37]. Three other women with single aneuploidy were also diagnosed with malignancies. In three of the ten cases with malignancy, the discordant NIPT test prompted the evaluation of malignancy. One limitation of this approach for cancer screening is that it requires a comparison between two different types of tissue—in this case mother and fetus. One approach to overcome this barrier is "nucleosome footprint"—ctDNA nucleosome spacing is unique with respect to the nuclear architecture and gene expression. Thus, it may inform the cell type [38].

4.2 Surveillance for Recurrent Disease

ctDNA is an effective biomarker for detecting tumor recurrence. For example, ctDNA has been shown to be reliable in detecting tumor progression or metastasis for breast cancer, often with clinical lead time of up to almost 1 year [39]. Similar studies have found that liquid biopsies can also be used for surveillance of CRC. For CRC patients, the BEAMing and patient-specific somatic structural variant (SSV) techniques have been investigated as tools to identify cancer recurrence [40]. Based on both methods, ctDNA was found to be more reliable and sensitive at detecting tumor recurrence than CEA. Diehl et al. measured ctDNA in patients after complete and incomplete resections. Those patients with undetectable ctDNA had 100% recurrence-free survival (RFS) at 2 years, compared to less than 15% RFS for patients with detectable ctDNA. Frattini et al. reported similar findings in a cohort of 70 CRC patients undergoing surgery. Their plasma DNA and CEA levels were followed for 16 months. Prior to surgery, circulating plasma DNA was elevated in all patients, whereas only 30% of patients had an elevated CEA. However, plasma DNA levels increased in all patients who ultimately developed metastases or loco-regional relapse. On the contrary, CEA levels did not increase in all patients with recurrent disease. The comparison revealed the sensitivity advantage of plasma DNA than CEA.

For patients with stage II CRC, ctDNA is being evaluated in the postoperative setting to detect disease recurrence. Preliminary data of 190 patients demonstrate cancer recurrence in 5 of 6 patients with detectable ctDNA and 5 of 72 patients with undetectable ctDNA [41]. Patients with detectable ctDNA had a shorter recurrence-free survival (median 234 days vs. undefined, HR 23.09, $P < 0.0001$). For patients with stage II colon cancer whose overall prognosis is good and the benefit of adjuvant chemotherapy is limited, ctDNA might eventually be used to identify patients with the greatest risk of recurrence. In addition to detection of recurrent disease, the presence of circulating tumor products may eventually guide adjuvant chemotherapy decisions [42].

4.3 Guidance of Targeted Drugs

Lung cancer is the leading cause of cancer-related death among men and women in the world. Elucidation of molecular landscapes of lung cancer is pivotal to guide treatment choices in both clinical practice and trials [21]. The inherent molecular heterogeneity and dynamic evolution of cancer genomes are not properly revealed by tissue biopsy. Evaluation of tumor-derived cfDNA in plasma offers the potential to overcome some barriers such as difficult anatomical locations, dangerous tumors surrounding major vessels or in eloquent regions of the brain, and difficulty in continuous monitoring. Noninvasive liquid biopsy technology is useful for rapid identification of tumor-specific mutations in NSCLC. Detection of targetable tumor-specific mutation in the blood may facilitate personalized treatment selection for individual patients with lung cancer [28].

4.3.1 EGFR TKIs

NSCLC is the most common type of lung cancer. NSCLC tumors may shed tumor DNA into a patient's blood, making it possible to detect specific mutations in blood samples. The US FDA and the European Medicines Agency approved blood-based tests to detect EGFR gene mutations in the NSCLC patients if no tissue sample is available [43]. NSCLC patients harboring the somatic mutations of

EGFR have been shown to be hyperresponsive to the EGFR tyrosine kinase inhibitor (TKI) gefitinib (IRESSA).

The clinical utility of liquid biopsy-based techniques for treatment decision-making is well exemplified by the determination of EGFR mutations in blood ctDNA to guide the use of EGFR TKIs for advanced-stage NSCLC patients. Several studies have compared the status of EGFR mutations in blood ctDNA and matched tissue samples of lung cancer patients [27, 44, 45]. The first attempt to detect EGFR mutations in blood sample was reported in 2006 [46]. Since then, numerous studies, including NCT01788163 and NCT01785888 (two large, multinational, diagnostic, noncomparative intervention trials), as well as two meta-analyses [47, 48], have confirmed that EGFR mutations are detectable in ctDNA with high specificity (>93%), but improvable sensitivity (<70%), indicating the diagnostic value of ctDNA.

EURTAC and FASTACT-2 are two randomized clinical trials of first-generation EGFR TKIs, which have addressed the clinical utility of liquid biopsy and ctDNA analysis for EGFR genotyping [25, 49]. EURTAC was the first trial to assess the feasibility of cfDNA from blood samples as a surrogate of tumor biopsy for determining EGFR mutation status and to correlate EGFR mutations in cfDNA with outcome. The trial assigned advanced NSCLC patients to receive erlotinib or a platinum-based chemotherapy doublet. These patients could offer baseline serum or plasma samples harboring oncogenic *EGFR* mutations in the tumor. EGFR exon 19 deletions and L858R mutation were examined in cfDNA by a multiplex RT-PCR assay. The result revealed that blood and tissue biomarkers had comparable predictive power. It was found that EGFR L858R mutation, no matter detected in tissue DNA or ctDNA, negatively affected median PFS and overall survival in erlotinib-treated patients compared with those whose tumors harbor exon 19 deletions [25].

In the FASTACT-2 trial, cobas EGFR blood test was used to evaluate predictive biomarkers of induction chemotherapy plus either erlotinib or placebo, followed by erlotinib or placebo maintenance treatment [49]. This retrospective analysis also revealed that baseline EGFR mutation-positive patients who became EGFR-negative in plasma ctDNA at the end of the induction period would have better PFS and overall survival than those who remained EGFR mutation-positive. This trial enabled the detection of G719A/C/S, L858R, L861Q, S768I, and T790M mutations, 19 deletions, and 5 insertions in exon 20 afterward in another clinical trial (NCT00815971). The trial result also came to the conclusion that the EGFR gene in plasma samples is feasible and represents a noninvasive supplement to tissue biopsy analysis [50].

EGFR T790M mutation occurs in half of NSCLC patients with acquired EGFR TKI (TKI) resistance, which could be dynamically

Table 1
First-, second-, and third-generation EGFR TKIs

	Drug		Indications	Targets	Approval
First-generation EGFR TKI	Gefitinib	IRESSA	First line for EGFR exon 19 deletions or exon 21 (L858R) substitution mutations, second line or third line for patients with regionally advanced or metastatic NSCLC who had failed one or two prior treatment regimens	19 del L858R	2003
	Erlotinib	TARCEVA	First line for EGFR exon 19 deletions or exon 21 (L858R) substitution mutations, second line or third line for patients with regionally advanced or metastatic NSCLC who had failed one or two prior treatment regimens	19 del L858R	2004
	Icotinib	CONMANA	First line for EGFR-sensitive mutations in locally advanced or metastatic NSCLC patients with first-line treatment		2011
Second-generation EGFR TKI	Afatinib	GILOTRIF	First-line metastatic NSCLC whose tumors have EGFR exon 19 deletions or exon 21 (L858R) substitution mutations Metastatic, squamous NSCLC progressing after platinum-based chemotherapy	19 del L858R	2013
Third-generation EGFR TKI	Osimertinib	TAGRISSO	Metastatic EGFR T790 M-positive NSCLC, as detected by an FDA-approved test, whose disease has progressed on or after EGFR TKI therapy	T790M	2015

monitored by circulating tumor DNA (ctDNA) using serial plasma samples from NSCLC patients. Monitoring T790M gatekeeper mutation will further impact on treatment decision and patient survival. Importantly, the standard care in the clinical practice is rapidly changing owing to the availability of third-generation EGFR TKI, which is designed to have low activity against wild-type EGFR while retaining activity against EGFR isoforms with resistant mutations (including T790 M mutation) (Table 1) [51, 52]. A seminal study retrospectively established that patients with T790M-positive tumors treated with osimertinib, a third-generation EGFR TKI that binds irreversibly to EGFRs harboring common activating mutations and the T790M resistance mutation, had equivalent outcomes whether the mutation was detected in tumor tissue or blood [53]. They further prospectively confirmed and expanded these findings in patients with newly diagnosed NSCLC and those with acquired resistance to EGFR TKI therapy.

All patients underwent concomitant tissue and blood-based DNA genotyping using the cobas blood test, followed by longitudinal blood-based monitoring using the same test for EGFR exon 19 deletions, L858R and T790M mutations, and/or all KRAS codon 12 (G12X) alterations. The positive predictive value of plasma ctDNA analysis was 100% for EGFR19 deletions, EGFR L858R, and KRAS G12X and 79% for T790M. Interestingly, sensitivity for both EGFR and KRAS was higher for patients with multiple metastatic sites and those with hepatic or bone metastases than those with a single metastatic site on these organs.

Indeed, liquid biopsy technologies are considered by regulatory agencies, such as the FDA and the European Medicines Agency (EMA), as in vitro diagnostic devices. Therefore, it is subject to approval based on the criteria of safety and effectiveness. To accelerate the development, adoption, and reimbursement of blood (and other body fluids) liquid biopsy, these generic rules must be translated into standards for clinical concordance, validation, and utility studies. The FDA has published a draft guidance document to facilitate this process [54]. Liquid biopsy is crucial in view of the fact that Illumina, a prominent manufacturer of NGS apparatus, and other large companies, such as Johnson & Johnson, Qiagen, and Roche, as well as at least 30 other new companies, are steadily developing diagnostic blood tests for multiple cancer types. To date, however, liquid biopsy has merely entered the clinical practice of NSCLC management. In January 2015, the EMA granted the approval of In Vitro Diagnostic Medical Device (IVD) marketing in Europe to the therascreen EGFR RGQ PCR Kit. And in 2016, the FDA approved the cobas EGFR Mutation Test v2 (an updated version of the prior cobas EGFR Mutation Test) for application in the USA. Both of these IVDs can detect EGFR mutations in plasma ctDNA with comparable accuracy to those in DNA from tumor tissue specimens by Sanger sequencing, as revealed by large clinical trials of small-molecule EGFR TKIs in patients with EGFR-mutated NSCLC. The therascreen assay enables the detection of 19 different exon 19 deletions and 3 distinct exon 20 insertions (although it cannot be used to distinguish which specific deletion or insertion is present), as well as L858R, L861Q, G719X, S768I, and T790M mutations in EGFR.

In the Lung-LUX3 trial, the overall percentage agreement (OPA) between the therascreen test and the tissue-based test results was 92.2% (95% CI, 89.0–94.8%) [55]. For 27 of 28 discordant sample pairs (96%), the presence of EGFR mutation was detected in plasma ctDNA, but not in DNA from tumor tissue samples [56]. On the one hand, the clinical utility of the therascreen test was confirmed in a phase IV trial of gefitinib [57], in which 12 of 201 patients with no tumor tissue available for genotyping were found to harbor EGFR mutations by this liquid biopsy assay.

On the other hand, the test had a 34% false-negative rate. 36 out of 105 patients tested positive for EGFR mutations in tissue DNA, but had no EGFR mutations detected in plasma cfDNA [57]. Therefore, the EMA has amended the drug label for gefitinib to state that the detection of EGFR mutations in ctDNA should only be attempted for patients without an evaluable tumor sample [58].

The cobas EGFR Mutation Test v2 enables identification of 42 EGFR mutations, including exon 19 deletions, exon 20 insertions, and S768I, L861Q, L585R, and T790M mutations, in both tissue and plasma. The plasma test, however, is approved by FDA as companion diagnostics for determining the eligibility of NSCLC patients for erlotinib treatment based only on the presence of EGFR exon 19 deletions and the L858R substitution mutation; patients who test negative for these alterations must undergo routine biopsy sampling and testing for EGFR mutations in formalin-fixed paraffin-embedded (FFPE) tissue samples. The approval of this test was based on the results of the phase III ENSURE trial (NCT01342965) conducted in China and Southern Asia to evaluate the efficacy and safety of erlotinib versus that of chemotherapy as first-line treatment in 217 patients with stage IIIB/IV NSCLC. Plasma samples were available for 517 of 610 patients (86.0%) screened for inclusion, in which 214 of 217 patients (98.6%) ultimately enrolled in this study [59]. Plasma samples were negative for EGFR exon 19 deletions and L585R mutation in 98.2% (95% CI, 95.4–99.3%) of tissue-negative patients. However, only 76.7% (95% CI, 70.5–81.9%) of the patients who tested positive for EGFR exon 19 deletions and L858R mutation in tissue also had these alterations detected in plasma samples. This relatively low positive percentage agreement (76.7%) between tissue and plasma test results prompted the FDA to specify that patients without detectable mutations in plasma should undergo retesting using tumor tissue if available. The efficacy of erlotinib, based on the detection of EGFR mutations in plasma using the cobas EGFR Mutation Test v2, was also evaluated by "bridging" to the drug efficacy results of the same trial based on tissue assessment using the cobas EGFR Mutation Test v1 among the patients who tested positive for exon 19 deletion and/or an L858R mutations in plasma, and those treated with erlotinib had improved progression-free survival (PFS) compared with those who received chemotherapy. The cobas EGFR Mutation Test v2 has also been approved by the FDA for the detection of T790M mutation in tissue samples, but not yet in plasma samples. Interestingly, the performance of the cobas EGFR Mutation Test v2 in detecting the T790M resistance mutation in ctDNA has been evaluated in a pooled retrospective analysis of two single-arm phase II registration studies of osimertinib (AURA extension, NCT01802632;

AURA2, NCT02094261). The positive percentage agreement and negative percentage agreement between tissue and plasma testings were 61.4% and 78.6%, respectively [60]. It is not dissimilar to the results obtained for detection of exon 19 deletions and L585R mutations in ctDNA using the same test. Notably, liquid biopsy analysis of EGFR mutations (as well as KRAS mutation) at diagnosis or relapse was prospectively validated in 180 patients with advanced-stage NSCLC in a study conducted at the Dana-Farber Cancer Institute (Boston, Massachusetts, USA). Mutations were detected using a digital PCR platform with high specificity (>79%) and with a median turnaround time of only 3 days, compared with 12 days and 27 days for tissue genotyping of newly diagnosed and relapsed tumors, respectively. In addition to NSCLC, the eligibility of targeted therapy has extended to many cancer types, including melanoma and colorectal and breast cancers. Currently, most decisions are made based on biopsy results from tumor tissue specimens. But, such assessments could potentially be replaced with liquid biopsy approaches in the future. This replacement might be particularly important when surgery is not indicated and tissue biopsy samples are difficult to obtain or when genetic heterogeneity could complicate decision-making based on limited tissue specimens. However, extensive researches are necessary to prove the utility and reliability of ctDNA analysis in such diverse settings. Of note, the preliminary results from a large validation trial of Guardant360 were presented at the 2016 ASCO Annual Meeting [61]. The trial was a study of a digital DNA sequencing technology encompassing a broad panel of 70 genes, including all current clinically actionable genes with approved targeted drugs. The study included 15,191 patients with advanced-stage lung, breast, colorectal, or other cancers [62]. Accuracy was assessed indirectly by comparing the frequencies of specific DNA changes in ctDNA with those of tissue-based DNA changes from patients included in The Cancer Genome Atlas (TCGA) and directly by matched plasma and tissue samples from a subset of almost 400 patients with NSCLC or colorectal cancer. Correlations between the TCGA and ctDNA data ranged from 92 to 99% across multiple cancer genes and different classes of gene alterations. However, the EGFR T790M resistance mutation was only detected in plasma DNA from patients treated with EGFR TKIs, which was not detectable in pretreatment samples. The matched plasma-tissue comparison showed a PPV of 87%. Importantly, the PPV increased to 98% when the plasma and tissue samples were collected <6 months apart. The Guardant360 test is quite sensitive, since mutated ctDNA accounting for 0.4% of the total cfDNA in the blood could be detected. The Guardant360 test enabled the detection of resistance mutations in EGFR, ALK, and KRAS that were not detectable in the matched tumor biopsy samples in almost one-third of the patients. Overall, the test identified a molecular alteration in 64% of patients, including 362 NSCLC patients with no available tumor tissue [62]. In several instances, mutations were identified after an initial response to a targeted agent that can be successfully inhibited with additional lines of therapy, emphasizing the potential clinical significance of repeated assessment of ctDNA for targetable mutations throughout the disease course. For example, EGFR mutations that emerge during the treatment of NSCLC patients with erlotinib or gefitinib can be successfully targeted with rociletinib. Similarly, second-generation inhibitors are available for ALK mutations that emerge after initial ALK blockade. The same principles apply to other mutations associated with a range of solid and hematological cancers.

4.3.2 ALK TKI

ALK is a validated molecular target in several ALK-rearranged malignancies, including NSCLC cancer, anaplastic large-cell lymphoma, and pediatric neuroblastoma. Anaplastic lymphoma kinase (ALK) gene rearrangements represent the fusion between ALK and various partner genes, in which the most widely recognized rearrangement is the echinoderm microtubule-associated protein-like 4 (EML4)-ALK fusion detected in NSCLC. ALK rearrangements represent a unique subset of NSCLC patients. ALK inhibitors may represent a very effective therapeutic strategy, which could be applied to approximately 5% of NSCLC [63] (Table 2).

Patients with ALK rearrangements are resistant to EGFR TKIs but share many of the clinical features of NSCLC patients likely to harbor EGFR mutations (e.g., adenocarcinoma histology, never smokers, light smokers) except they are more likely to be men and may be younger [64, 65]. However, ALK rearrangements and EGFR mutations are mutually exclusive [66]. Collaborative efforts between basic and clinical scientists in academia and pharmaceutical industry have led to the development of numerous ALK tyrosine kinase inhibitors (TKIs). Preclinical studies established the potential of this ALK fusion. Transgenic mice expressing EML4-ALK in lung alveolar epithelial cells developed innumerable adenocarcinoma nodules in their lungs soon after birth [67]. So far, three oral ALK TKIs (crizotinib, ceritinib, and alectinib) have received the approval by the US FDA for treatment of advanced or metastatic NSCLC with ALK gene rearrangement. Clinical trials have demonstrated remarkable responses in this patient population [68–73].

Although originally developed as a potent MET inhibitor, crizotinib was the first ALK TKI to be applied in the clinical practice [74]. In 2011, the results of phase I/II trial demonstrated the robust clinical activity in advanced ALK-rearranged NSCLC. Thus, crizotinib was granted accelerated approval by the FDA [63]. Thereafter,

Table 2
First-, second-, and third-generation ALK TKIs

	Drug		Indications	Targets	Approval
First generation ALK-TKI	Crizotinib	XALKORI	Metastatic NSCLC whose tumors are ALK-positive as detected by an FDA-approved tests. Metastatic NSCLC whose tumors are ROS1-positive.	MET amplification, 14 exon skipping ROS 1 fusion and ALK fusion	2011
Second generation ALK-TKI	Ceritinib	ZYKADIA	ALK-positive metastatic NSCLC who have progressed on or are intolerant to crizotinib.	ALK fusion	2014
	Alectinib	ALECENSA	ALK-positive, metastatic NSCLC who have progressed on or are intolerant or crizotinib.	ALK fusion	2015
	Brigatinib	ALUNBRIG	ALK-positive metastatic NSCLC who have progressed on or are intolerant to crizonitib.	ALK fusion	2017
Third generation ALK-TKI	Lorlatinib		FDA grants Lorlatinib breakthrough therapy designation for the treatment of patients with ALK-positive metastatic NSCLC who were previously treated with one or more ALK inhibitors.	ALK fusion	2017

two phase III trials showed that crizotinib was superior to first- and second-line cytotoxic chemotherapy in advanced ALK-rearranged NSCLC [72, 73]. The promising result was that the objective response rate (ORR) in ALK-positive NSCLC patients ranged from 60 to 74%, with median progression-free survival (PFS) of 8–-11 months across all crizotinib trials. In addition, two second-generation ALK inhibitors, ceritinib and alectinib, thereafter, received accelerated approval for patients with ALK-positive NSCLC previously treated with crizotinib. In the phase I ASCEND-1 trial, ceritinib demonstrated an ORR of 56% and median PFS of 6.9 months in crizotinib-pretreated patients with ALK-rearranged NSCLC [68, 71]. In two phase II studies, alectinib induced objective responses in ~50% of crizotinib-resistant NSCLCs with ALK rearrangement, in which median PFS in both studies was 8–9 months [69, 70]. Brigatinib, another second-generation inhibitor, has been associated with confirmed ORR of 62% and median

PFS of 13.2 months in crizotinib-pretreated patients in one phase I/II trial [75]. On April 28, 2017, brigatinib was accelerated approval by the FDA for the treatment of patients with ALK-positive metastatic NSCLC who have progressed on or are intolerant to crizotinib. These second-generation inhibitors are more potent than crizotinib. Thus, they can overcome the most common crizotinib-resistant mutations including the L1196M gatekeeper mutation, which are active in crizotinib-resistant tumors without ALK resistance mutations. These second-generation inhibitors are more effective than crizotinib against central nervous system (CNS) metastases [68–71].

5 Perspective

The finding of circulating biomarkers originated from cancer lesion in the peripheral blood is a cutting-edge scientific discovery. The clinical application of cfDNA for diagnosis and treatment offers a minimal invasive method to sample. However, there are still several challenges to overcome before routine application by clinicians. Due to the low level of cfDNA in the peripheral blood, the application of liquid biopsy to monitor tumor progression and recurrence is not so well. Advances in genomics approaches are essential to enhance the specificity and sensitivity of liquid biopsy. In addition, it is necessary to establish a standard analysis procedure of liquid biopsy to ensure the accuracy of analysis. With the solution of these challenges, liquid biopsy is possible to be widely used in routine clinical application. In conclusion, cfDNA is a very promising biomarker for liquid biopsy. The clinical value will become much more obvious with the exploration of cfDNA application.

References

1. Allard WJ, Matera J, Miller MC, Repollet M, Connelly MC, Rao C, Tibbe AG, Uhr JW, Terstappen LW (2004) Tumour cells circulate in the peripheral blood of all major carcinomas but not in healthy subjects or patients with nonmalignant diseases. Clin Cancer Res 10 (20):6897–6904. https://doi.org/10.1158/1078-0432.CCR-04-0378
2. Hiraiwa K, Takeuchi H, Hasegawa H, Saikawa Y, Suda K, Ando T, Kumagai K, Irino T, Yoshikawa T, Matsuda S, Kitajima M, Kitagawa Y (2008) Clinical significance of circulating tumour cells in blood from patients with gastrointestinal cancers. Ann Surg Oncol 15(11):3092. https://doi.org/10.1245/s10434-008-0122-9
3. Rink M, Chun FK, Minner S, Friedrich M, Mauermann O, Heinzer H, Huland H, Fisch M, Pantel K, Riethdorf S (2007) Detection of circulating tumour cells in peripheral blood of patients with metastatic breast cancer: a validation study of the CellSearch system. Clin Cancer Res 13(3):920–928. https://doi.org/10.1111/j.1464-410X.2010.09562.x.
4. Stott SL, Lee RJ, Nagrath S, Yu M, Miyamoto DT, Ulkus L, Inserra EJ, Ulman M, Springer S, Nakamura Z, Moore AL, Tsukrov DI, Kempner ME, Dahl DM, Wu CL, Iafrate AJ, Smith MR, Tompkins RG, Sequist LV, Toner M, Haber DA, Maheswaran S (2010) Isolation and characterization of circulating tumour cells from patients with localized and metastatic prostate cancer. Sci Transl Med 2 (25):25ra23-25ra23. https://doi.org/10.1126/scitranslmed.3000403

5. Cohen SJ, Alpaugh RK, Gross S, O'Hara SM, Smirnov DA, Terstappen LW, Allard WJ, Bilbee M, Cheng JD, Hoffman JP, Lewis NL, Pellegrino A, Rogatko A, Sigurdson E, Wang H, Watson JC, Weiner LM, Meropol NJ (2006) Isolation and characterization of circulating tumour cells in patients with metastatic colorectal cancer. Clin Colorectal Cancer 6(2):125–132. https://doi.org/10.3816/CCC.2006.n.029
6. Cristofanilli M, Budd GT, Ellis MJ, Stopeck A, Matera J, Miller MC, Reuben JM, Doyle GV, Allard WJ, Terstappen LW, Hayes DF (2004) Circulating tumour cells, disease progression, and survival in metastatic breast cancer. N Engl J Med 351(8):781–791. https://doi.org/10.1056/NEJMoa040766
7. Moreno JG, O'Hara SM, Gross S, Doyle G, Fritsche H, Gomella LG, Terstappen LW (2001) Changes in circulating carcinoma cells in patients with metastatic prostate cancer correlate with disease status. Urology 58 (3):386–392. https://doi.org/10.1016/S0090-4295(01)01191-8
8. Diehl F, Schmidt K, Choti MA, Romans K, Goodman S, Li M, Thornton K, Agrawal N, Sokoll L, Szabo SA, Kinzler KW, Vogelstein B, Diaz LA Jr (2008) Circulating mutant DNA to assess tumour dynamics. Nat Med 14 (9):985–990. https://doi.org/10.1038/nm.1789
9. Diehl F, Li M, Dressman D, He Y, Shen D, Szabo S, Diaz LA Jr, Goodman SN, David KA, Juhl H, Kinzler KW, Vogelstein B (2005) Detection and quantification of mutations in the plasma of patients with colorectal tumours. Proc Natl Acad Sci U S A 102 (45):16368–16373. https://doi.org/10.1073/pnas.0507904102.
10. Bettegowda C, Sausen M, Leary RJ, Kinde I, Wang Y, Agrawal N, Bartlett BR, Wang H, Luber B, Alani RM, Antonarakis ES, Azad NS, Bardelli A, Brem H, Cameron JL, Lee CC, Fecher LA, Gallia GL, Gibbs P, Le D, Giuntoli RL, Goggins M, Hogarty MD, Holdhoff M, Hong SM, Jiao Y, Juhl HH, Kim JJ, Siravegna G, Laheru DA, Lauricella C, Lim M, Lipson EJ, Marie SK, Netto GJ, Oliner KS, Olivi A, Olsson L, Riggins GJ, Sartore-Bianchi A, Schmidt K, Shih lM, Oba-Shinjo SM, Siena S, Theodorescu D, Tie J, Harkins TT, Veronese S, Wang TL, Weingart JD, Wolfgang CL, Wood LD, Xing D, Hruban RH, Wu J, Allen PJ, Schmidt CM, Choti MA, Velculescu VE, Kinzler KW, Vogelstein B, Papadopoulos N, Diaz LA Jr (2014) Detection of circulating tumour DNA in early-and late-stage human malignancies. Sci Transl Med 6(224):224ra224-224ra224. https://doi.org/10.1126/scitranslmed.3007094
11. Bidard FC, Madic J, Mariani P, Piperno-Neumann S, Rampanou A, Servois V, Cassoux N, Desjardins L, Milder M, Vaucher I, Pierga JY, Lebofsky R, Stern MH, Lantz O (2014) Detection rate and prognostic value of circulating tumour cells and circulating tumour DNA in metastatic uveal melanoma. Int J Cancer 134(5):1207–1213. https://doi.org/10.1002/ijc.28436
12. Freidin MB, Freydina DV, Leung M, Fernandez AM, Nicholson AG, Lim E (2015) Circulating tumour DNA outperforms circulating tumour cells for KRAS mutation detection in thoracic malignancies. Clin Chem 61 (10):1299–1304. https://doi.org/10.1373/clinchem.2015.242453
13. Matsuda K, Maruyama H, Guo F, Kleeff J, Itakura J, Matsumoto Y, Lander AD, Korc M (2001) Glypican-1 is overexpressed in human breast cancer and modulates the mitogenic effects of multiple heparin-binding growth factors in breast cancer cells. Cancer Res 61 (14):5562–5569
14. Kleeff J, Ishiwata T, Kumbasar A, Friess H, Büchler MW, Lander AD, Korc M (1998) The cell-surface heparan sulfate proteoglycan glypican-1 regulates growth factor action in pancreatic carcinoma cells and is overexpressed in human pancreatic cancer. J Clin Investig 102 (9):1662. https://doi.org/10.1172/JCI4105
15. Melo SA, Luecke LB, Kahlert C, Fernandez AF, Gammon ST, Kaye J, LeBleu VS, Mittendorf EA, Weitz J, Rahbari N, Reissfelder C, Pilarsky C, Fraga MF, Piwnica-Worms D, Kalluri R (2015) Glypican-1 identifies cancer exosomes and detects early pancreatic cancer. Nature 523(7559):177–182. https://doi.org/10.1038/nature14581
16. Bujanda L, Sarasqueta C, Cosme A, Hijona E, Enríquez-Navascués JM, Placer C, Villarreal E, Herreros-Villanueva M, Giraldez MD, Gironella M, Balaguer F, Castells A (2013) Evaluation of alpha 1-antitrypsin and the levels of mRNA expression of matrix metalloproteinase 7, urokinase type plasminogen activator receptor and COX-2 for the diagnosis of colorectal cancer. PLoS One 8(1):e51810. https://doi.org/10.1371/journal.pone.0051810
17. Lim SH, Spring KJ, de Souza P, MacKenzie S, Bokey L (2015) Circulating tumour cells and circulating nucleic acids as a measure of tumour dissemination in non-metastatic colorectal cancer surgery. Eur J Surg Oncol 41(3):309–314. https://doi.org/10.1016/j.ejso.2014.12.005
18. Rodia MT, Ugolini G, Mattei G, Montroni I, Zattoni D, Ghignone F, Veronese G, Marisi G, Lauriola M, Strippoli P, Solmi R (2016)

Systematic large-scale meta-analysis identifies a panel of two mRNAs as blood biomarkers for colorectal cancer detection. Oncotarget 7 (21):30295–30306. https://doi.org/10.18632/oncotarget.8108

19. Allen-Mersh TG, McCullough TK, Patel H, Wharton RQ, Glover C, Jonas SK (2007) Role of circulating tumour cells in predicting recurrence after excision of primary colorectal carcinoma. Br J Surg 94(1):96–105. https://doi.org/10.1002/bjs.5526
20. Katsuno H, Zacharakis E, Aziz O, Rao C, Deeba S, Paraskeva P, Ziprin P, Athanasiou T, Darzi A (2008) Does the presence of circulating tumour cells in the venous drainage of curative colorectal cancer resections determine prognosis? A meta-analysis. Ann Surg Oncol 15(11):3083–3091. https://doi.org/10.1245/s10434-008-0131-8
21. El-Hefnawy T, Raja S, Kelly L, Bigbee WL, Kirkwood JM, Luketich JD, Godfrey TE (2004) Characterization of amplifiable, circulating RNA in plasma and its potential as a tool for cancer diagnostics. Clin Chem 50 (3):564–573. https://doi.org/10.1373/clinchem.2003.028506
22. Bao Y, Ching B, Mouanoutoua M, Yu W, Keys D, Desai S, Stevens J (2012) Cancer biomarker research using castPCR technology. AACR
23. Wilkening S, Hemminki K, Thirumaran RK, Bermejo JL, Bonn S, Försti A, Kumar R (2005) Determination of allele frequency in pooled DNA: comparison of three PCR-based methods. BioTechniques 39(6):853–858
24. Goto K, Ichinose Y, Ohe Y, Yamamoto N, Negoro S, Nishio K, Itoh Y, Jiang H, Duffield E, McCormack R, Saijo N, Mok T, Fukuoka M (2012) Epidermal growth factor receptor mutation status in circulating free DNA in serum: from IPASS, a phase III study of gefitinib or carboplatin/paclitaxel in non-small cell lung cancer. J Thorac Oncol 7 (1):115–121. https://doi.org/10.1097/JTO.0b013e3182307f98
25. Karachaliou N, Mayo-de las Casas C, Queralt C, de Aguirre I, Melloni B, Cardenal F, Garcia-Gomez R, Massuti B, Sánchez JM, Porta R, Ponce-Aix S, Moran T, Carcereny E, Felip E, Bover I, Insa A, Reguart N, Isla D, Vergnenegre A, de Marinis F, Gervais R, Corre R, Paz-Ares L, Morales-Espinosa D, Viteri S, Drozdowskyj A, Jordana-Ariza N, Ramirez-Serrano JL, Molina-Vila MA, Rosell R, Spanish Lung Cancer Group (2015) Association of EGFR L858R mutation in circulating free DNA with survival in the EURTAC trial. JAMA Oncol 1(2):149–157. https://doi.org/10.1001/jamaoncol.2014.257
26. Hindson BJ, Ness KD, Masquelier DA, Belgrader P, Heredia NJ, Makarewicz AJ, Bright IJ, Lucero MY, Hiddessen AL, Legler TC, Kitano TK, Hodel MR, Petersen JF, Wyatt PW, Steenblock ER, Shah PH, Bousse LJ, Troup CB, Mellen JC, Wittmann DK, Erndt NG, Cauley TH, Koehler RT, So AP, Dube S, Rose KA, Montesclaros L, Wang S, Stumbo DP, Hodges SP, Romine S, Milanovich FP, White HE, Regan JF, Karlin-Neumann GA, Hindson CM, Saxonov S, Colston BW (2011) High-throughput droplet digital PCR system for absolute quantitation of DNA copy number. Anal Chem 83(22):8604–8610. https://doi.org/10.1021/ac202028g
27. Taniguchi K, Uchida J, Nishino K, Kumagai T, Okuyama T, Okami J, Kato K (2011) Quantitative detection of EGFR mutations in circulating tumour DNA derived from lung adenocarcinomas. Clin Cancer Res 17 (24):7808–7815. https://doi.org/10.1158/1078-0432.CCR-11-1712
28. Taniguchi K, Uchida J, Nishino K, Kumagai T, Okuyama T, Okami J, Higashiyama M, Kodama K, Imamura F, Kato K (2012) Noninvasive identification and monitoring of cancer mutations by targeted deep sequencing of plasma DNA. Sci Transl Med 4 (136):136ra168-136ra168. https://doi.org/10.1158/1078-0432.CCR-11-1712
29. Kinde I, Wu J, Papadopoulos N, Kinzler KW, Vogelstein B (2011) Detection and quantification of rare mutations with massively parallel sequencing. Proc Natl Acad Sci 108 (23):9530–9535. https://doi.org/10.1073/pnas.1105422108
30. Newman AM, Bratman SV, To J, Wynne JF, Eclov NC, Modlin LA, Liu CL, Neal JW, Wakelee HA, Merritt RE, Shrager JB, Loo BW Jr, Alizadeh AA, Diehn M (2014) An ultrasensitive method for quantitating circulating tumour DNA with broad patient coverage. Nat Med 20(5):548–554. https://doi.org/10.1038/nm.3519
31. Griffin TJ, Smith LM (2000) Single-nucleotide polymorphism analysis by MALDI–TOF mass spectrometry. Trends Biotechnol 18(2):77–84
32. Giannini R, Lupi C, Sensi E, Alì G, Proietti A, Boldrini L, Servadio A, Giordano M, Macerola E, Bruno R, Borrelli N, Chella A, Melfi F, Lucchi M, Ribechini A, Vasile E, Cappuzzo F, Mussi A, Fontanini G (2016) EGFR and KRAS mutational analysis in a large series of Italian non-small cell lung cancer patients: 2,387 cases from a single center.

Oncol Rep 36(2):1166–1172. https://doi.org/10.3892/or.2016.4874

33. Finney Rutten LJ, Nelson DE (2004) Meissner HI. Examination of population-wide trends in barriers to cancer screening from a diffusion of innovation perspective (1987–2000). Prev Med 38:258–268. https://doi.org/10.1016/j.ypmed.2003.10.011
34. Lo SH, Waller J, Wardle J (2013) Comparing barriers to colorectal cancer screening with barriers to breast and cervical screening a population-based survey of screening age women in great Britain. J Med Screen 20 (2):73–79
35. Potter NT, Hurban P, White MN, Whitlock KD, Lofton-Day CE, Tetzner R, Koenig T, Quigley NB, Weiss G (2014) Validation of a real-time PCR-based qualitative assay for the detection of methylated SEPT9 DNA in human plasma. Clin Chem 60(9):1183–1191. https://doi.org/10.1373/clinchem.2013.221044
36. Romero R, Mahoney MJ (2015) Noninvasive prenatal testing and detection of maternal cancer. JAMA 314(2):131–133. https://doi.org/10.1001/jama.2015.7523
37. Bianchi DW, Chudova D, Sehnert AJ (2015) Noninvasive prenatal testing and incidental detection of occult maternal malignancies. JAMA 314(2):162–169. https://doi.org/10.1001/jama.2015.7120
38. Snyder MW, Kircher M, Hill AJ (2016) Cell-free DNA comprises an in vivo nucleosome footprint that informs its tissues-of-origin. Cell 164(1):57–68. https://doi.org/10.1016/j.cell.2015.11.05.
39. Olsson E, Winter C, George A, Chen Y, Howlin J, Tang MH, Dahlgren M, Schulz R, Grabau D, van Westen D, Fernö M, Ingvar C, Rose C, Bendahl PO, Rydén L, Borg Å, Gruvberger-Saal SK, Jernström H, Saal LH (2015) Serial monitoring of circulating tumour DNA in patients with primary breast cancer for detection of occult metastatic disease. EMBO Mol Med 7(8):1034–1047. https://doi.org/10.15252/emmm.201404913
40. Reinert T, Schøler LV, Thomsen R, Tobiasen H, Vang S, Nordentoft I, Lamy P, Kannerup AS, Mortensen FV, Stribolt K, Hamilton-Dutoit S, Nielsen HJ, Laurberg S, Pallisgaard N, Pedersen JS, Ørntoft TF, Andersen CL (2016) Analysis of circulating tumour DNA to monitor disease burden following colorectal cancer surgery. Gut 65:gutjnl-2014. https://doi.org/10.1136/gutjnl-2014-308859
41. Tie J, Kinde I, Wang Y (2014) Circulating tumour DNA (ctDNA) as a marker of recurrence risk in stage II colon cancer (CC). ASCO annual meeting proceedings, Chicago
42. Quasar Collaborative Group, Gray R, Barnwell J, McConkey C, Hills RK, Williams NS, Kerr DJ (2007) Adjuvant chemotherapy versus observation in patients with colorectal cancer: a randomised study. Lancet 370:2020–2029. https://doi.org/10.1016/S0140-6736(07)61866-2
43. www.fda.gov/NewsEvents/Newsroom/PressAnnouncements/ucm504488.htm.
44. Kuang Y, Rogers A, Yeap BY, Wang L, Makrigiorgos M, Vetrand K, Thiede S, Distel RJ, Jänne PA (2009) Noninvasive detection of EGFR T790M in gefitinib or erlotinib resistant non–small cell lung cancer. Clin Cancer Res 15 (8):2630–2636. https://doi.org/10.1158/1078-0432.CCR-08-2592
45. Nakamura T, Sueoka-Aragane N, Iwanaga K, Sato A, Komiya K, Kobayashi N, Hayashi S, Hosomi T, Hirai M, Sueoka E, Kimura S (2012) Application of a highly sensitive detection system for epidermal growth factor receptor mutations in plasma DNA. J Thorac Oncol 7(9):1369–1381. https://doi.org/10.1097/JTO.0b013e31825f2821
46. Kimura H, Kasahara K, Kawaishi M, Kunitoh H, Tamura T, Holloway B, Nishio K (2006) Detection of epidermal growth factor receptor mutations in serum as a predictor of the response to gefitinib in patients with non--small-cell lung cancer. Clin Cancer Res 12 (13):3915–3921. https://doi.org/10.1158/1078-0432.CCR-05-2324
47. Luo J, Shen L, Zheng D (2014) Diagnostic value of circulating free DNA for the detection of EGFR mutation status in NSCLC: a systematic review and meta-analysis. Sci Rep 4:6269. https://doi.org/10.1038/srep06269
48. Qiu M, Wang J, Xu Y, Ding X, Li M, Jiang F, Xu L, Yin R (2014) Circulating tumour DNA is effective for the detection of EGFR mutation in non–small cell lung cancer: a meta-analysis. Cancer Epidemiol Prev Biomarkers 24 (1):206–212. https://doi.org/10.1158/1055-9965.EPI-14-0895
49. Mok T, Wu YL, Lee JS, Yu CJ, Sriuranpong V, Sandoval-Tan J, Ladrera G, Thongprasert S, Srimuninnimit V, Liao M, Zhu Y, Zhou C, Fuerte F, Margono B, Wen W, Tsai J, Truman M, Klughammer B, Shames DS, Wu L (2015) Detection and dynamic changes of EGFR mutations from circulating tumour DNA as a predictor of survival outcomes in NSCLC patients treated with first-line intercalated erlotinib and chemotherapy. Clin Cancer Res 21(14):3196–3203. https://doi.org/10.1158/1078-0432.CCR-14-2594

50. Weber B, Meldgaard P, Hager H, Wu L, Wei W, Tsai J, Khalil A, Nexo E, Sorensen BS (2014) Detection of EGFR mutations in plasma and biopsies from non-small cell lung cancer patients by allele-specific PCR assays. BMC Cancer 14(1):294. https://doi.org/10.1186/1471-2407-14-294.
51. Cross DA, Ashton SE, Ghiorghiu S, Eberlein C, Nebhan CA, Spitzler PJ, Orme JP, Finlay MR, Ward RA, Mellor MJ, Hughes G, Rahi A, Jacobs VN, Red Brewer M, Ichihara E, Sun J, Jin H, Ballard P, Al-Kadhimi K, Rowlinson R, Klinowska T, Richmond GH, Cantarini M, Kim DW, Ranson MR, Pao W (2014) AZD9291, an irreversible EGFR TKI, overcomes T790M-mediated resistance to EGFR inhibitors in lung cancer. Cancer Discov 4(9):1046–1061. https://doi.org/10.1158/2159-8290.CD-14-0337
52. Zhou W, Ercan D, Chen L, Yun CH, Li D, Capelletti M, Cortot AB, Chirieac L, Iacob RE, Padera R, Engen JR, Wong KK, Eck MJ, Gray NS, Jänne PA (2009) Novel mutant-selective EGFR kinase inhibitors against EGFR T790M. Nature 462 (7276):1070–1074. https://doi.org/10.1038/nature08622
53. Oxnard GR, Thress KS, Alden RS, Lawrance R, Paweletz CP, Cantarini M, Yang JC, Barrett JC, Jänne PA (2016) Association between plasma genotyping and outcomes of treatment with osimertinib (AZD9291) in advanced non–small-cell lung cancer. J Clin Oncol 34 (28):3375–3382. https://doi.org/10.1200/JCO.2016.66.7162
54. http://www.fda.gov/ucm/groups/fdagov-public/@fdagov-meddev-gen/documents/document/ucm510824.pdf
55. Sequist LV, Yang JC, Yamamoto N, O'Byrne K, Hirsh V, Mok T, Geater SL, Orlov S, Tsai CM, Boyer M, Su WC, Bennouna J, Kato T, Gorbunova V, Lee KH, Shah R, Massey D, Zazulina V, Shahidi M, Schuler M (2013) Phase III study of afatinib or cisplatin plus pemetrexed in patients with metastatic lung adenocarcinoma with EGFR mutations. J Clin Oncol 31(27):3327–3334. https://doi.org/10.1200/JCO.2012.44.2806
56. FDA http://www.accessdata.fda.gov/cdrh_docs/pdf12/P120022c.pdf
57. Douillard J, Ostoros G, Cobo M, Ciuleanu T, McCormack R, Webster A, Milenkova T (2014) First-line gefitinib in Caucasian EGFR mutation-positive NSCLC patients: a phase-IV, open-label, single-arm study. Br J Cancer 110(1):55–62. https://doi.org/10.1038/bjc.2013.721
58. http://www.ema.europa.eu/docs/en_GB/document_library/EPAR_Product_Information/human/001016/WC500036358.pdf
59. http://www.fda.gov/Drugs/InformationOnDrugs/ApprovedDrugs/ucm504540.htm
60. Jenkins S, Yang J, Ramalingam S, Yu K, Patel S, Weston S, Lawrance R, Cantarini M, Jänne P, Mitsudomi T (2016) 134O_PR: plasma ctDNA analysis for detection of EGFR T790M mutation in patients (pts) with EGFR mutation-positive advanced non-small cell lung cancer (aNSCLC). J Thorac Oncol 11((4): S153–S154. https://doi.org/10.1016/S1556-0864(16)30327-6
61. Lanman RB, Mortimer SA, Zill OA, Sebisanovic D, Lopez R, Blau S, Collisson EA, Divers SG, Hoon DS, Kopetz ES, Lee J, Nikolinakos PG, Baca AM, Kermani BG, Eltoukhy H, Talasaz A (2015) Analytical and clinical validation of a digital sequencing panel for quantitative, highly accurate evaluation of cell-free circulating tumour DNA. PLoS One 10(10):e0140712. https://doi.org/10.1371/journal.pone.0140712
62. Zill OA, Mortimer S, Banks K (2016) Somatic genomic landscape of over 15,000 patients with advanced-stage cancer from clinical next-generation sequencing analysis of circulating tumour DNA. J Clin Oncol 34(suppl)
63. Kwak EL, Bang YJ, Camidge DR, Shaw AT, Solomon B, Maki RG, Ou SH, Dezube BJ, Jänne PA, Costa DB, Varella-Garcia M, Kim WH, Lynch TJ, Fidias P, Stubbs H, Engelman JA, Sequist LV, Tan W, Gandhi L, Mino-Kenudson M, Wei GC, Shreeve SM, Ratain MJ, Settleman J, Christensen JG, Haber DA, Wilner K, Salgia R, Shapiro GI, Clark JW, Iafrate AJ (2010) Anaplastic lymphoma kinase inhibition in non–small-cell lung cancer. N Engl J Med 363(18):1693–1703. https://doi.org/10.1200/JCO.2016.34.15_suppl.LBA11501
64. Koivunen JP, Kim J, Lee J, Rogers AM, Park JO, Zhao X, Naoki K, Okamoto I, Nakagawa K, Yeap BY, Meyerson M, Wong KK, Richards WG, Sugarbaker DJ, Johnson BE, Jänne PA (2008) Mutations in the LKB1 tumour suppressor are frequently detected in tumours from Caucasian but not Asian lung cancer patients. Br J Cancer 99(2):245–252. https://doi.org/10.1038/sj.bjc.6604469
65. Shaw AT, Yeap BY, Mino-Kenudson M, Digumarthy SR, Costa DB, Heist RS, Solomon B, Stubbs H, Admane S, McDermott U, Settleman J, Kobayashi S, Mark EJ, Rodig SJ, Chirieac LR, Kwak EL, Lynch TJ, Iafrate AJ (2009) Clinical features and outcome of patients with non–small-cell lung cancer who

harbor EML4-ALK. J Clin Oncol 27 (26):4247–4253. https://doi.org/10.1200/JCO.2009.22.6993

66. Inamura K, Takeuchi K, Togashi Y, Hatano S, Ninomiya H, Motoi N, Mun MY, Sakao Y, Okumura S, Nakagawa K, Soda M, Choi YL, Mano H, Ishikawa Y, Nakagawa K (2009) EML4-ALK lung cancers are characterized by rare other mutations, a TTF-1 cell lineage, an acinar histology, and young onset. Mod Pathol 22(4):508–515. https://doi.org/10.1038/modpathol.2009.2

67. Soda M, Takada S, Takeuchi K, Choi YL, Enomoto M, Ueno T, Haruta H, Hamada T, Yamashita Y, Ishikawa Y, Sugiyama Y, Mano H (2008) A mouse model for EML4-ALK-positive lung cancer. Proc Natl Acad Sci 105 (50):19893–19897. https://doi.org/10.1073/pnas.0805381105

68. Kim DW, Mehra R, Tan DS, Felip E, Chow LQ, Camidge DR, Vansteenkiste J, Sharma S, De Pas T, Riely GJ, Solomon BJ, Wolf J, Thomas M, Schuler M, Liu G, Santoro A, Sutradhar S, Li S, Szczudlo T, Yovine A, Shaw AT (2016) Activity and safety of ceritinib in patients with ALK-rearranged non-small-cell lung cancer (ASCEND-1): updated results from the multicentre, open-label, phase 1 trial. Lancet Oncol 17(4):452–463. https://doi.org/10.1016/S1470-2045(15)00614-2

69. Ou SH, Ahn JS, De Petris L, Govindan R, Yang JC, Hughes B, Lena H, Moro-Sibilot D, Bearz A, Ramirez SV, Mekhail T, Spira A, Bordogna W, Balas B, Morcos PN, Monnet A, Zeaiter A, Kim DW (2015) Alectinib in crizotinib-refractory ALK-rearranged non--small-cell lung cancer: a phase II global study. J Clin Oncol 34(7):661–668. https://doi.org/10.1200/JCO.2015.63.9443.

70. Shaw AT, Gandhi L, Gadgeel S, Riely GJ, Cetnar J, West H, Camidge DR, Socinski MA, Chiappori A, Mekhail T, Chao BH, Borghaei H, Gold KA, Zeaiter A, Bordogna W, Balas B, Puig O, Henschel V, Ou SH (2016) Alectinib in ALK-positive, crizotinib-resistant, non-small-cell lung cancer: a single-group, multicentre, phase 2 trial. Lancet Oncol 17(2):234–242. https://doi.org/10.1016/S1470-2045(15)00488-X

71. Shaw AT, Kim DW, Mehra R, Tan DS, Felip E, Chow LQ, Camidge DR, Vansteenkiste J, Sharma S, De Pas T, Riely GJ, Solomon BJ, Wolf J, Thomas M, Schuler M, Liu G, Santoro A, Lau YY, Goldwasser M, Boral AL, Engelman JA (2014) Ceritinib in ALK-rearranged non–small-cell lung cancer. N Engl J Med 370(13):1189–1197. https://doi.org/10.1056/NEJMoa1311107.

72. Shaw AT, Kim DW, Nakagawa K, Seto T, Crinó L, Ahn MJ, De Pas T, Besse B, Solomon BJ, Blackhall F, Wu YL, Thomas M, O'Byrne KJ, Moro-Sibilot D, Camidge DR, Mok T, Hirsh V, Riely GJ, Iyer S, Tassell V, Polli A, Wilner KD, Jänne PA (2013) Crizotinib versus chemotherapy in advanced ALK-positive lung cancer. N Engl J Med 368(25):2385–2394. https://doi.org/10.1056/NEJMoa1214886

73. Solomon BJ, Mok T, Kim DW, Wu YL, Nakagawa K, Mekhail T, Felip E, Cappuzzo F, Paolini J, Usari T, Iyer S, Reisman A, Wilner KD, Tursi J, Blackhall F, PROFILE 1014 Investigators (2014) First-line crizotinib versus chemotherapy in ALK-positive lung cancer. N Engl J Med 371 (23):2167–2177. https://doi.org/10.1056/NEJMoa1408440

74. Zou HY, Li Q, Lee JH, Arango ME, McDonnell SR, Yamazaki S, Koudriakova TB, Alton G, Cui JJ, Kung PP, Nambu MD, Los G, Bender SL, Mroczkowski B, Christensen JG (2007) An orally available small-molecule inhibitor of c-met, PF-2341066, exhibits cytoreductive antitumor efficacy through antiproliferative and antiangiogenic mechanisms. Cancer Res 67(9):4408–4417. https://doi.org/10.1158/0008-5472.CAN-06-4443

75. Gettinger SN, Bazhenova LA, Langer CJ, Salgia R, Gold KA, Rosell R, Shaw AT, Weiss GJ, Tugnait M, Narasimhan NI, Dorer DJ, Kerstein D, Rivera VM, Clackson T, Haluska FG, Camidge DR (2016) Activity and safety of brigatinib in ALK-rearranged non-small-cell lung cancer and other malignancies: a single-arm, open-label, phase 1/2 trial. Lancet Oncol 17(12):1683–1696. https://doi.org/10.1016/S1470-2045(16)30392-8

Chapter 5

Bioinformatics Analysis for Cell-Free Tumor DNA Sequencing Data

Shifu Chen, Ming Liu, and Yanqing Zhou

Abstract

As a major biomarker of liquid biopsy, cell-free tumor DNA (ctDNA), which can be extracted from blood, urine, or other circulating liquids, is able to provide comprehensive genetic information of tumor and better overcome the tumor heterogeneity problem comparing to tissue biopsy. Developed in recent years, next-generation sequencing (NGS) is a widely used technology for analyzing ctDNA. Although the technologies of processing ctDNA samples are mature, the task to detect low mutated allele frequency (MAF) variations from noisy sequencing data remains challenging. In this chapter, the authors will first explain the difficulties of analyzing ctDNA sequencing data, review related technologies, and then present some novel bioinformatics methods for analyzing ctDNA NGS data in better ways.

Key words Liquid biopsy, Circulating tumor DNA, ctDNA, Gene fusion, CNV, Mutation visualization, OpenGene

1 Introduction

Cell-free DNA (cfDNA) is the extracellular DNA fragments, which is mainly derived from apoptosis, as well as part from necrosis and active cell release. For cancer patients, both the normal and tumor cells are the source of cfDNA, and the fraction from tumor cells are called ctDNA (cell-free tumor DNA), which can be isolated from secretions, excretions, and body fluids [1–3]. The existence of cfDNA was first reported by Mandel and Metais in 1948, but it was not until in 1977 that it was found to have a correlation with cancer [4].

1.1 ctDNA and Its Applications

The possibility to use cfDNA as materials for liquid biopsy for cancer bases on the fact that all cancers are a disease of DNA alterations, including genomic variations like mutations, translocations, amplifications, gene loss, etc., and epigenetic variations. Collectively, these DNA changes determine a set of tumor-specific variations of each tumor against the background of normal DNA

Tao Huang (ed.), *Computational Systems Biology: Methods and Protocols*, Methods in Molecular Biology, vol. 1754, https://doi.org/10.1007/978-1-4939-7717-8_5,

from other tissues. Since both normal and tumor cells shed DNA into the circulation, in general there is a relatively low abundance of tumor-specific cfDNA in the total cfDNA. Before a DNA fragment is sequenced, we do not know whether it is derived from a tumor cell. So we can only do cfDNA sequencing in all at this stage and have no way to merely do ctDNA sequencing. However, for applications in cancer field, ctDNA is a term much better known and more widely used by researchers. So in this chapter, we will use both ctDNA sequencing and cfDNA sequencing. When we mention ctDNA sequencing for cancer patients, be noted that it is the same as cfDNA sequencing.

In general, cfDNA dynamically reflects the overall state of the body. The tumor-specific ctDNA are good biomarkers for early detection and prevention, assessing minimal residual disease and prognosis, monitoring tumor burden, and guiding therapies. It allows a relatively noninvasive repeated serial sampling for continuous monitoring of disease. Moreover, the ability to identify specific drug-sensitizing or resistance mutations in the blood of cancer patients makes liquid biopsy a good alternative method for tissue biopsies, which is often difficult or impossible to obtain for late-stage cancer patients using the conventional methods. Although lots of studies support these concepts, the use of cfDNA in clinical practice is still in its infancy and requires rigorous prospective validation studies to demonstrate the benefit of this promising analyte to facilitate the clinical decision making. Figure 1 illustrates how ctDNA can be applied for cancer diagnosis and therapy.

1.2 How Is ctDNA Sequenced

As there is a relatively low abundance of ctDNA in the total cfDNA, highly sensitive techniques are required for ctDNA detection. Advancements in next-generation sequencing (NGS) have made it possible to detect low occurrence mutations in a heterogeneous population [5].

The main steps for ctDNA sequencing include sample processing and cfDNA extraction, NGS library construction, hybrid capture, and sequencing.

CtDNA can be extracted from bodily fluids like plasma, cerebral spinal fluid, urine, pleural effusion, etc. The choice of material is related to the clinical or research purpose. The amount of cfDNA extracted from 1 ml plasma is relatively low, and an accurate quantification of cfDNA is necessary for the subsequent library construction procedure. A classical library construction process on Illumina platform includes end repair to produce blunt-ended and 5′-phosphorylated dsDNA fragments; A-tailing, during which dAMP is added to the 3′-ends of blunt-ended dsDNA library fragments; adapter ligation, during which dsDNA adapters with 3′-dTMP overhangs are ligated to 3′-A-tailed library fragments; and library amplification, which employs PCR to amplify library fragments carrying appropriate adapter sequences on both ends.

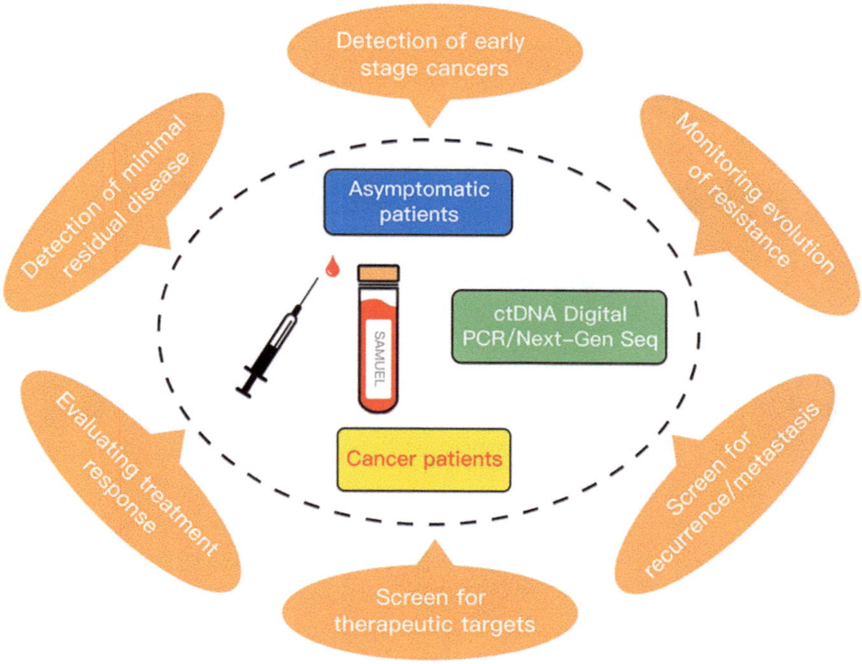

Fig. 1 Applications of ctDNA for cancer diagnosis and therapy. Cell-free DNA can be evaluated by different techniques like digital PCR (dPCR) or next-generation sequencing (NGS). Digital PCR can only cover a few target loci, while NGS can cover millions of loci, including the whole exome or even whole genome. CtDNA testing can be applied to search for targeted therapy drugs, evaluate treatment response, and detect early-stage cancers

The adapter ligation efficiency and the number of PCR cycles are two critical factors for this procedure. For hybridization capture, first, one should design a panel consistent with the particular application and then order the probes to a manufacturer; especially NimbleGen SeqCap EZ could be a good choice. The Illumina sequencing platform is usually applied in the last sequencing step. Figure 2 demonstrates how blood samples are processed and how the libraries are prepared and sequenced.

1.3 Difficulties of Analyzing ctDNA NGS Data

Cell-free tumor DNA is only a small fraction of cell-free DNA, especially for samples from early-stage cancer patients. This fact makes it hard to detect tumor-specific mutations. Furthermore, PCR and sequencing errors, DNA oxidative damages, and software-introduced artifacts can produce a high level of noise and introduce many false-positive mutations.

The amount of tumor-specific DNA can vary greatly from less than 0.01% to more than 90% [3]. The variability of ctDNA abundance is associated with tumor burden, stage, vascularity, cellular turnover, and response to therapy. Theoretically DNA alteration of any fraction is detectable via deep sequencing with sufficient number of molecules. However, amplification bias during PCR of

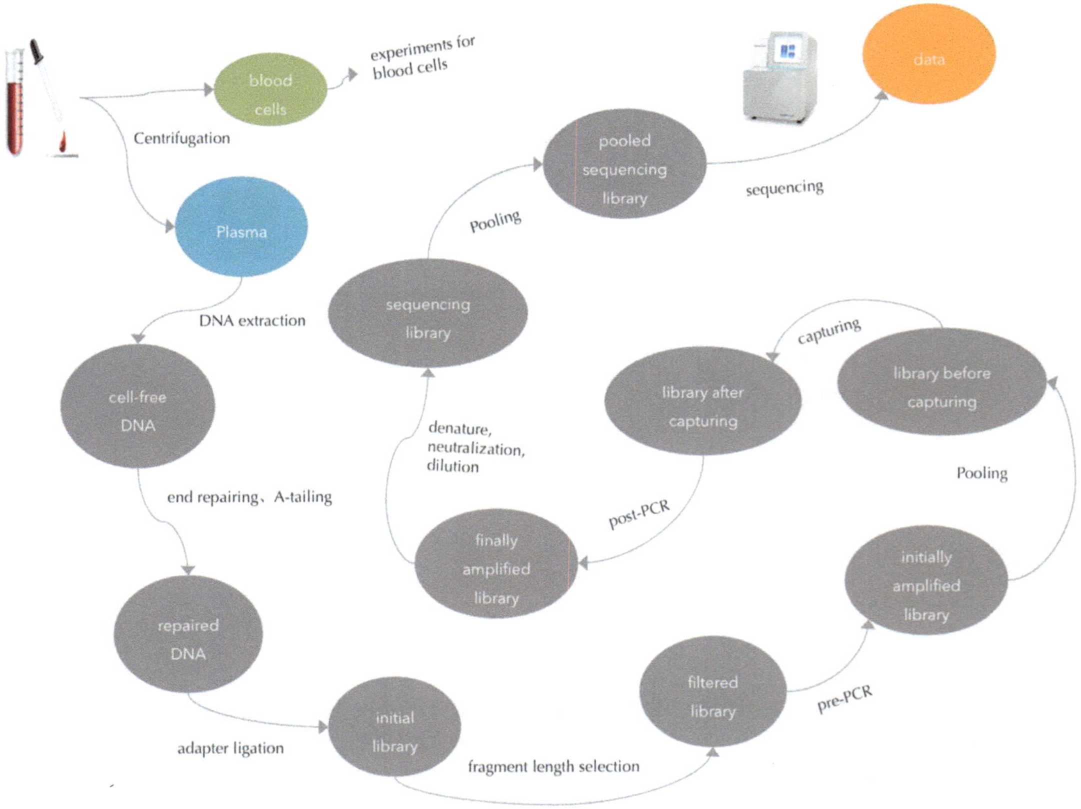

Fig. 2 The workflow of ctDNA sequencing. In this workflow, a blood sample is first processed, then DNA is extracted from the blood plasma, and an initial library is prepared. This initial library will then be amplified by PCR and be enriched with target capturing methods. The captured library can be amplified again and be processed to a sequencing library, which can be pooled and then be sequenced to generate the data

heterogeneous mixtures can result in skewed populations, and polymerase errors can result in wrong base incorporations and rearrangements. Furthermore, errors arising during sequencing process can result in approximately 0.1–1% incorrect bases calling [6], which are known as sequencing errors. Table 1 shows the error ratios of different major NGS platforms.

Library preparation can also introduce significant errors. For instance, guanine oxidation is an important source of artificial mutations because 8-oxoG tends to pair with adenine instead of cytosine [7]. Long-time heat incubations, which are common in many DNA extraction and hybrid capture protocols, can significantly increase the number of G → T substitutions. Recently a study showed that a DNA repairing process can eliminate 77% and 82% of G → T and C → A errors, respectively [8]. This study indicates DNA lesions can cause a large amount of errors.

Besides errors introduced during sample preparation and sequencing, software and analysis tools can also introduce errors. Particularly, false-positive variants can be called in the reference

Table 1
A comparison of sequencing error ratios of different sequencing platforms

Platform	Most frequent error types	Error ratio
Capillary sequencing	Single nucleotide substitutions	10^{-1}
454 GS Junior	Deletions	10^{-2}
PacBio RS	CG deletions	10^{-2}
Ion Torrent PGM	Short deletions	10^{-2}
Solid	A-T bias	2×10^{-2}
Illumina MiSeq	Single nucleotide substitutions	10^{-3}
Illumina HiSeq	Single nucleotide substitutions	10^{-3}
Illumina NextSeq	Single nucleotide substitutions	10^{-3}

genome regions with homologous sequences and repetitive sequences.

Cell-free DNA fragments are usually short and have a compact peak near 167 bp [9]. This fact increases the possibility that two different original cfDNA fragments share an identical sequence and consequently increases the difficulty to remove these duplications since the deduplication algorithms will not be able to differentiate such identical and duplicated reads caused by amplification.

In summary, detecting low-frequency mutations from the noisy ctDNA sequencing data is challenging. Conventional tools cannot handle well the ctDNA analysis tasks, and more specialized tools are therefore needed.

1.4 ctDNA Sequencing Data Analysis Pipeline

To analyze ctDNA sequencing data, a series of software tools needs to be involved. For example, the raw sequencing data from Illumina sequencers are obtained in a base calling (BCL) format. This BCL file needs to be de-multiplexed to separate FASTQ files according to sample barcodes. Then the FASTQ files would be measured with quality control tools to guarantee they fulfill the quality requirement and be filtered to remove low-quality and wrongly represented reads. Next, the filtered FASTQ files would be aligned to the reference genome with aligners, and the output should be SAM/BAM files. Then the BAM files need to be sorted and duplications removed. Then variant callers are required to process the BAM file and generate a VCF with raw variant records. Next, this VCF file should be annotated with databases like dbSNP and COSMIC. A baseline technology will be applied to mark some false-positive mutations, and then the unique reads supporting each mutation will be counted to make a complete VCF. This VCF file will then be filtered to generate a clean one and visualized with tools for interactive analysis. Finally the target mutations will

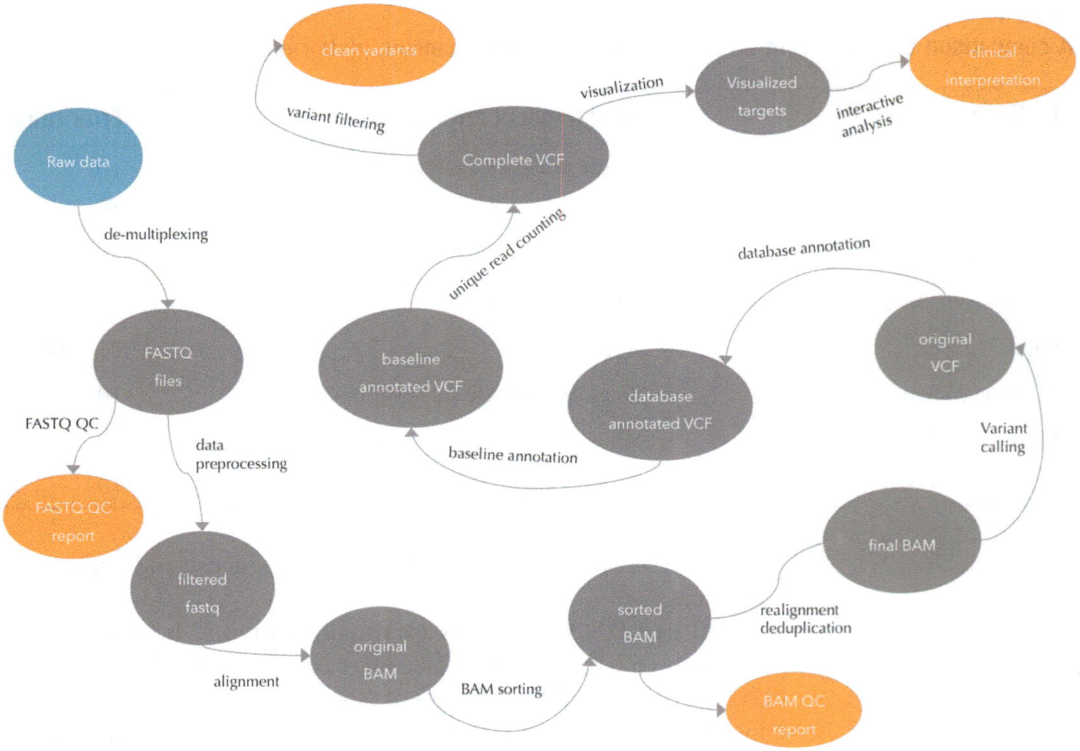

Fig. 3 A typical pipeline for analyzing ctDNA sequencing data. The raw data will be de-multiplexed to separate FASTQ files by sample indexes and then be filtered to remove bad reads. The FASTQ reads will be aligned to reference genome to generate BAM files. Variant callers will scan the sorted and processed BAM files to generate original VCF files, which can then be annotated with databases and baseline data. After a complete VCF is generated, it can be filtered to create clean variants or be visualized for variant validation

be interpreted and reported. These tools can be arranged into a pipeline. Figure 3 demonstrates a ctDNA sequencing data analysis pipeline regularly used by the authors.

For Illumina platforms, the tool bcl2fastq is used to convert BCL format files to FASTQ files. Illumina platforms support multiplexing by using different barcodes for different samples, so de-multiplexing is performed along with the conversion.

Some additional tools can perform quality control and data filtering over FASTQ files, e.g., FastQC and Trimmomatic [10]. The authors suggest using AfterQC [11], which is highly optimized for ctDNA sequencing data processing. AfterQC will be introduced in the next section.

A lot of aligners can be used to map DNA sequencing reads to reference genome, such as bowtie2 [12] and BWA [13]. According to our practice, BWA provides a better performance both in alignment quality and speed. BWA is a software package for mapping low-divergent sequences against a large reference genome. It consists of three algorithms: BWA-backtrack, BWA-SW, and

BWA-MEM. BWA-MEM is generally recommended for high-quality queries, as it is faster and more accurate. But be aware that BWA and any other aligners may still introduce misalignments, especially in reference genome regions with repetitive or homologous sequences.

The alignment process will generate a SAM file containing the alignment information that can be immediately converted to BAM, which is the binary identity of SAM. This BAM file is usually disordered and should be sorted and then indexed. The most commonly used tool to sort and index BAM files is Samtools [14], and there exist some other tools that can sort BAM faster. For example, Sambamba [15] is a high performance tool working with SAM/BAM data. Sambamba is written in D language, and its source is available at: https://github.com/lomereiter/sambamba.

After BAM file is sorted and indexed, an optional process is to apply realignment to improve the detection of insertions and deletions (INDELs). Some tools like ABRA [16] can perform assembly-based realignment to output cleaner INDELs, but these tools are usually slow. Quality control of BAM files can be applied now to evaluate the data's alignment quality and detect unwanted biases. This process can be done with tools like Qualimap [17].

The subsequent process is deduplication. Samtools rmdup and Picard MarkDuplicates (http://picard.sourceforge.net) are commonly used to identify and collapse read duplication based on reads' mapping coordinates and quality scores. Since cfDNA fragments are short and their length distribution is compactly close to 167 bp, lots of reads derived from different original DNA fragments may share identical mapping coordinates, and they should not be considered as duplication. So we do not suggest using Samtools rmdup or Picard MarkDuplicates for deduplication, and we will discuss new methods and strategies in the next section.

Variant calling is the key process following the BAM operations (sort, realign, dedup). Cancer genomes are known to harbor a wide range of mutations, including single nucleotide variants (SNVs), multiple nucleotide variants (MNVs), small insertions and deletions (INDELs), and complex variants, such as copy number variants (CNVs) and gene fusions. A number of variant callers, such as GATK HaplotypeCaller [18], FreeBayes (https://github.com/ekg/freebayes), MuTect2 [19], and VarScan2 [20], can be used to call SNV, MNV, and small INDELs. According to our experience, GATK HaplotypeCaller and FreeBayes are not good at calling ctDNA's low-frequency somatic mutations from ultra-deep sequencing data, since they are originally designed for genotyping and discovering genetic polymorphism. MuTect2 is much better in calling somatic mutations, especially with tumor-normal paired data. However, it just works well with tissue sequencing data but is not sensitive enough to detect low-frequency mutations in ctDNA sequencing data. VarScan2 is very sensitive in detecting low-frequency mutations but likely to report a large amount of false-positive mutations. So we could not find a perfect variant caller for detecting low-frequency mutations in such ultra-deep NGS data like ctDNA sequencing data. Currently we suggest VarScan2, combined with strict variant filtering. Be aware that some variant callers, like GATK HaplotypeCaller, cannot scale well with depth and typically downsample (randomly remove portions of data) to improve their computational performance. However downsampling can significantly reduce the sensitivity to detect low allele frequency mutations and is not suggested for ctDNA sequencing data analysis.

After the variant calling process is done, the original VCF file is obtained. This VCF file can be annotated with annotation tools like ANNOVAR [21] to obtain coding sequence and protein changes and compare with databases like dbSNP, ClinVar, and COSMIC.

A mutation baseline will be used to annotate each variant for how many times this variant was recorded in the past data. This information can be used to filter false-positive mutations caused by software artifacts and other regular systematic errors. Baseline technology will be introduced in next section.

To calculate the supporting read number for each mutation more accurately, we can consider the reads with the same mapping coordinate as a single unique read. A tool called MrBam (https://github.com/OpenGene/MrBam) is used to count each mutation's unique reference support and unique alternative support.

After the unique read counting is done, we obtain a complete VCF file. The records in this VCF file can be added into the mutation baseline. This VCF file can be filtered according to different conditions to remove as many false-positive mutations as possible. A white list, which consists of the important clinical targets (i.e., cancer druggable mutation targets), is usually used in this filtering process to avoid the important target mutations being filtered out unexpectedly.

On another track, the called variants can be visualized with tools like MutScan (https://github.com/OpenGene/MutScan) to produce mutation visualization for interactive analysis. Mutations that are important for cancer diagnosis and therapy will be manually interpreted.

Besides SNVs and INDELs, another two important kinds of variants for cancer diagnosis are gene fusions and copy number variants (CNV). Most of these tools can only work with sorted BAM files. For example, DELLY [22] and Factera [23] can be used to detect gene fusions, and CNVkit (https://github.com/etal/cnvkit) can be used to detect gene amplifications from targeted DNA sequencing. One exception is that FusionDirect, a tool developed by the authors, can work with FASTQ files directly to detect target fusions.

The authors have created an open source project to demonstrate this pipeline, which is available at GitHub (https://github.com/OpenGene/ctdna-pipeline). By studying it, the readers can learn how to install the tools, prepare required databases and reference data, and try the pipeline with FASTQ files for testing.

In the pipeline presented above, more than a half of the tools are commonly used software (i.e., BWA, Samtools, and VarScan2), while the rest ones are developed by the authors (i.e., MutScan, AfterQC, and MrBam). These newly developed tools are highly optimized for ctDNA sequencing data analysis. Most of these tools are open source projects under the GitHub organization OpenGene (https://github.com/OpenGene). We will introduce some of them in the next section.

2 New Methods

Since tumor-specific DNA is only a small part of cfDNA, the mutated allele frequency (MAF) of somatic mutations in ctDNA is usually very low [24]. To detect mutations with such low MAF, we should apply target capturing and ultra-deep sequencing (i.e., 10,000× or deeper). However, sequencing errors and experiment errors (i.e., PCR errors) in such ultra-deep sequencing can cause high-level background noise and make it difficult to detect mutations from ctDNA NGS data with both high sensitivity and specificity. Furthermore, the detection of gene fusions is also difficult since cfDNA fragments are usually short and tumor-specific DNA fragments are too few. Since the copy number change in tumor cells only results in a slight difference of total cfDNA's copy number, detecting copy number variation (CNV) is even more challenging than detecting fusions.

In this section, we will present some new methods to partially address the problems listed above. Some of them are developed by the authors and has been used in our regular pipelines.

2.1 Better Data Preprocessing

Data preprocessing is an important step to obtain cleaner data for downstream analysis. For NGS raw data (in FASTQ format), it is necessary to discard low-quality reads, cut adapters, and apply other filters. Furthermore, quality control (QC) methods are also needed to make sure the data fulfill the quality requirements.

Some good tools can perform quality control, such like FastQC with per-base and per-sequence quality profiling functions and PRINSEQ [25] with FASTA/FASTQ statistics capability, while some other tools can perform read trimming, such like Trimmomatic [10] and SolexaQA [26]. Since the way to do data filtering depends on the QC result and the filtered data also need a post-filtering QC, a tool with both rich QC and filtering functions is still wanted.

Since cfDNA fragments are usually short (~167 bp) [9], 2 × 150 paired-end sequencing will result in overlapped read pairs. Based on this fact, we can perform overlapping analysis for paired-end sequencing data. When the DNA template length is less than twice of the sequencing length, the pair of reads will be overlapped. Note that each base in the overlapped region is actually sequenced twice, so the inconsistency of these pairs may reflect the sequencing errors.

AfterQC [11] is a tool developed by authors to address lots of practical sequencing data quality control and filtering problems. In addition to regular quality control functions like per-cycle base content and quality statistics, AfterQC also provides lots of new functions like automatic trimming and overlapping analysis. For example, we found that some sequencers (like Illumina NextSeq series) may output lots of polyX reads with high-quality scores. AfterQC can remove them using its polyX filter, whereas normal quality filters cannot. We also found that if the amplification or sequencing process has a serious strand bias, the sequence reads will show K-MER count bias (i.e., the counts of ATCGATCG and its reverse complement CGATCGAT are significantly different). Based on this finding, AfterQC provides K-MER counting based strand bias profiling. Another major contribution of this tool is overlapping analysis for paired-end sequencing data, which can be used to profile the sequencing error rate and use it for error base correction or removing. For every input of a single or pair of FASTQ files, AfterQC outputs an HTML report, which contains the quality control and data filtering summary, and a list of interactive figures. Table 2 shows feature comparison of AfterQC and other NGS quality control or filtering tools.

AfterQC is designed to process FASTQ files in batches. It goes through a folder with all FASTQ files (can be single-end or paired-end output), which are typically data of a sequencing run for different samples, and passes each FASTQ file or pair into the QC and filtering pipeline. First, AfterQC will run a bubble detection to find the bubbles raised in the sequencing process; second, a pre-filtering QC will be conducted to profile the data with per-cycle base content and quality curves; third, AfterQC will perform automatic read trimming based on data quality profiling; fourth, each read will be filtered by bubble filter, polyX filter, quality filter, and overlapping analysis filters, and the ones failed to pass these filters will be discarded as bad reads; fifth, an error correction based on overlapping analysis will be applied for paired-end sequencing data; finally, AfterQC will store the good reads, perform post-filtering QC profiling, and generate HTML reports.

AfterQC can handle automatic trimming of FASTQ data. There are two strategies for trimming, local strategy and global strategy. Some tools, like Trimmomatic, apply local strategy, which perform trimming read by read. However, local trimming strategy

Table 2
Feature comparison of FastQC, Trimmomatic, Cutadapt, and AfterQC

	FastQC	Trimmomatic	Cutadapt	AfterQC
Quality control	Rich functionality	Few functionality	Few functionality	Rich functionality
Auto trimming	None	Read by read	Read by read	Global trimming
Cutting adapter	None	Single-end/pair-end	Single-end/pair-end	Paired-end only
PolyX filtering	None	None	None	Supported
Figure plotting	Static	Static	Static	Interactive
Overlap analysis	None	Cutting adapter only	None	Supported with error correction
Sequence error profiling	None	None	None	Supported
Bubble detection	None	None	None	Supported
Programming language	Java	Java	Python	Python, C
Speed	Fast	Fast	Fast	Fast only for single-end data

has some drawbacks. The first drawback is that local trimming only uses the quality information for trimming, but it cannot utilize the global statistical information to discover the abnormal cycles. The second drawback is local trimming results in unaligned trimming, which means duplicated reads may be trimmed differently and consequently causes some deduplication tools like Picard to fail. Most of these deduplication tools detect duplications only by clustering reads with identical mapping positions. In contrast, AfterQC implements a global trimming strategy, i.e., it trims all the reads in the same manner. An algorithm is used to determine how many cycles to trim in the front and tail, which is based on the segmentation of the per-cycle base content curves and base quality curves.

A major advantage of AfterQC is the overlapping analysis. Let T denote the length of a sequenced DNA template, and S denote the length of paired-end sequencing length, then the pair of reads will totally overlap if $T \leq S$, will overlap with a length of $2S - T$, if $S < T < 2S$, and will not overlap if $2S \leq T$. AfterQC checks how does each pair of reads overlap based on edit distance optimization. For a pair of reads $R1$ and $R2$, let O be the offset, and we place $R2$ under $R1$, then we will have vertically aligned subsequences $R1o$ and $R2o$, and we can calculate their edit distance $ed.(R1o, R2o)$. The method optimizes offset O to obtain the minimal edit distance,

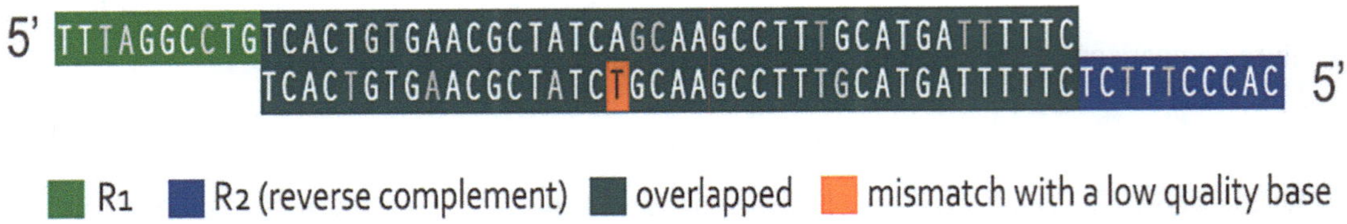

Fig. 4 How AfterQC's overlapping analysis works. The edit distance of the overlapped subsequences is 1. A mismatch pair is found with a high-quality base *A* and a very low-quality base *T*. This *T* will be recognized as wrongly represented and can be corrected

$ed.(R1o - 1, R2o - 1) > ed.(R1o, R2o) < ed.(R1o + 1, R2o + 1)$. Figure 4 shows an example of how AfterQC's overlapping analysis works.

Based on overlapping analysis, AfterQC can detect mismatches. If the mismatched pair has unbalanced quality scores, which means one base has high-quality score (i.e., >Q30) and the other has very low-quality score (i.e., <Q15), AfterQC can automatically correct the base with low quality. If the quality scores are not unbalanced, AfterQC can mask them by changing the bases to N or assigning zero quality scores to them. Based on the mismatches, AfterQC can evaluate the sequencing error rate and profile the sequencing error transform distribution (i.e., how many bases are T but sequenced as C).

Overlapping analysis can be used for automatic adapter cutting. In the overlapping analysis process, we get the optimal offset *O* for the best local alignment of each pair. The overlapping length of this pair can be directly calculated using the offset *O*. If *O* is found negative, the bases outside overlapping region will be considered as a part of adapter sequences and then be cut automatically.

AfterQC is an open source tool: https://github.com/OpenGene/AfterQC. It is implemented in Python and C++, with PyPy support enabled. AfterQC generates a standalone HTML report for each input, with figures plot by Plotly. A sample report can be found at: http://opengene.org/AfterQC/report.html.

2.2 Molecular Barcoding Sequencing and Its Data Analysis

The potential of NGS deep sequencing for ctDNA was hampered by systemic errors introduced by PCR and sequencing methods [27, 28]. Molecular indexing combined with deep sequencing holds great promise to break the limit imposed by PCR and sequencing errors and enables the detection of rare and ultra-rare mutations [29, 30].

Tagging individual templates with molecular barcodes has been proposed and reported since 2007 [31]. The molecular barcodes or molecular indexes have been given various names, such as unique identifiers (UID) [29], unique molecular identifiers (UMI) [32], primer ID [30], duplex barcodes [33], etc. They are usually designed as a string of totally random nucleotides (such as NNNNNNNN), partially degenerate nucleotides (such as

NNNRNYNN), or defined nucleotides (when template molecules are limited). UID or UMI could be introduced to targeted templates by ligation or through primers during PCR or reverse transcription.

Tagging DNA fragments with UIDs or duplex barcodes has been shown to reduce errors and improve sequencing accuracy, as true mutations could be distinguished from PCR errors or sequencing errors based on the consensus reads sharing the same UID. At present, classic tag-based methods are SafeSeq, CircleSeq, and duplex sequencing [34]. SafeSeq is a single-stranded tagging method based on "barcoding." An alternative to single-stranded tagging based on shear points is the circle sequencing methodology which utilizes the strand displacement activity of Phi29's DNA polymerase to generate multiple copies of circularized DNA molecules in tandem prior to amplification. However, both of these two methods cannot distinguish true variants from artificial variants introduced during the initial rounds of PCR amplification. In contrast duplex sequencing resolved these types of errors by tagging both strands of dsDNA, exploiting the fact that DNA naturally exists as a double-stranded entity, with one molecule reciprocally encoding the sequence information of its complement. Table 3 compares claimed error ratios of SafeSeq, CircleSeq, and duplex sequencing.

The analysis of molecular barcoding-enabled sequencing data can be divided into three steps.

The first step is extracting the UID. Be noted that the barcodes ligated to the original DNA template are usually made by DNA synthesis technology, which usually has high error ratio. For example, if 8-nt barcode is designed, we still have a chance to get 7-nt or 9-bt barcode due to synthesis error. To address this issue, a fixed sequence that consists of a few bases (usually three to five bases) is usually used to indicate the boundary of UID and original DNA sequence. Splitting algorithms should seek for this flag near the designed position, and typically the algorithm should allow one base mismatch to enable DNA synthesis or sequencing error tolerance. By using special adapters, some molecular barcoding methods place the UID on the multiplexing index positions (I7 or I5 index

Table 3
A comparison of different molecular barcoding methods.

Method	Claimed error ratio
SafeSeq	1.4×10^{-5}
CircleSeq	7.6×10^{-6}
Duplex sequencing	5×10^{-8}

for Illumina TrueSeq). UID extraction is much easier in this case since it can be taken directly from the sample index. This process is done with FASTQ data.

The second step is clustering the reads derived from the same original DNA. These reads should share very similar UID and mapping coordination. But due to the presence of PCR and sequencing errors, they are not required to be completely identical. Usually one base substitution mismatch is tolerated, and loose clustering methods can allow mismatches of INDELs or more than one substitution. This process is usually done with sorted BAM files, but it can also be done with FASTQ files based on sequence clustering algorithms.

The final step is generating consensus read for each read cluster. First, the reads in same cluster should be aligned together. This process can be done with a multiple sequence alignment tool like Clustal [35]. The complete multiple sequence alignment is usually time-consuming, and if we limit the number of mismatched substitutions and INDELs, some naive methods can run much faster. After the alignment is done, the consensus read can be generated by scanning it from front to tail. For each position, all bases in this position will be used to vote for the consensus base, according to their quality scores. For the positions with completely identical bases, the quality score of this consensus base can be adjusted a bit higher, and, vice versa, for a position that shows no consensus, the quality score of result base can be adjusted to be lower. In case when only two reads are clustered, if the two bases in the same positions are different but both have high-quality scores, this position can then be masked with N or zero quality score.

2.3 Baseline Methods

NGS data have different kinds of errors. Some errors, like sequencing error and PCR error, are random and can happen with any nucleotide at any genome position, although with some biases. Some errors are more regular, such like errors caused by misalignment usually happening in genome's high repetitive regions. These regular errors can be eliminated with baseline technologies.

Baseline technology is to combine and store all related detected mutations and other related information from as many samples as possible and then make statistics of these data and provide interfaces for querying and updating. Baseline data is usually stored in database, so it can utilize the standard SQL language for inserting, updating, deleting, and querying. Two different types of databases can be used: row-oriented database and column-oriented database. Row-oriented database is the mainstream form of relational database, like MySQL and PostgreSQL, whereas column-oriented database is less known, like Infobright and MonetDB. Row-oriented databases can support online transaction processing (OLTP) and are highly optimized for relational queries, whereas column-oriented databases can provide higher data compression ratio.

The baseline should store each mutation with its chromosome, position, reference, and alternative bases, combined with numbers of mutated reads and total depth. With this baseline, we then can count how many times a mutation of specific location with specific alteration has been detected, what its average MAF is, and what the mutated read number is.

Since some mutations can be detected in many different types of cancers, a better solution is to build a specific baseline with data sequenced from healthy people. Then this baseline can be used to filter false-positive mutations. When a mutation is called, its baseline-repeating number will be evaluated. If baseline-repeating number is too high, then this mutation can be considered as a false positive and need to be evaluated carefully.

Another usage of baseline is to detect hotspot mutations, both somatic and germline ones. By mining hot mutations from the baseline built with tumor individuals, we can find target mutations with potential to be biomarkers.

2.4 Target Variant Detection by Scanning FASTQ Data Directly

Regular mutation detection pipeline for NGS data usually involves many tools in different steps. These tools may cause information loss due to different filters applied and may finally cause miss detection of true mutations, especially the ones with low MAF. This kind of false negatives caused by data analysis is not acceptable in clinical applications, since it will make the patient miss an opportunity for better treatment.

On the contrary, false-positive detection of these key mutations should be also avoided since it can lead to an expensive but ineffective treatment and may even cause serious adverse reactions. Regular NGS pipeline can detect a lot of substitutions and INDELs and unavoidably raise false positives. Especially, caused by inaccurate reference genome mapping of aligners, a large percentage of the INDELs called in genome's high repetitive regions are false positives.

The authors have developed some tools that can detect target mutations by just scanning raw FASTQ data, without doing any alignment and variant calling. One tool is MutScan, which is built on error-tolerant string searching algorithms and is highly optimized for speed with rolling hash and bloom filters [36]. MutScan can run in reference free mode to detect target mutations, which are predefined in the program. With a VCF file and its corresponding reference FastA files provided, MutScan can scan all the variants in the VCF and visualize them by creating a HTML file for each variant.

MutScan is ultra-sensitive and ultra-fast. It can grab mutations with as few as one mutated read supported. It can run 50× faster than a regular pipeline (AfterQC + BWA + Samtools + VarScan2), if it only scans the predefined cancer druggable targets. Furthermore, the interactive HTML reports generated by MutScan can help to

Fig. 5 A demonstration of MutScan's interactive HTML. The demonstrated mutation is EGFR p.T790M (hg19 chr7:55,249,071 C > T), which is an important druggable target for lung cancer. The colors of the bases indicate the quality score (green and blue mean high quality, red means low quality). Due to page size limitation, this figure is an incomplete screenshot. The full report can be found at http:/opengene.org/MutScan/report.html

visualize and validate target mutations. Figure 5 shows a demonstration of MutScan's interactive mutation pileup.

MutScan is available at: https://github.com/OpenGene/MutScan. It is written in C++ with multi-threading support. It supports both single-end and paired-end data, and for latter one, it will try to merge each pair with quality adjustment and error correction.

Another tool developed by authors is FusionDirect, which can detect gene fusions directly from raw FASTQ data. This tool also works with FASTQ files directly and requires no alignment. It can output fusion sites (genes and positions), along with the reads supporting the fusions. Figure 6 gives an example of the output of FusionDirect.

```
#Fusion:ALK-EML4 (total: 3, unique: 2)
>2_merged_120_ALK:intron:19|+chr2:29446598_EML4:exon:21|-chr2:42553364/1
AATTGAACCTGTGTATTTATCCTCCTTAAGCTAGATTTCCATCATACTTAGAAATACTAATAAAATGATTAAAGAAGGTGTGTCTT
TAATTGAAGCATGATTTAAAGTAAATGCAAAGCTATGTCGTCCAATCAATGTCCTTACAATC
>2_merged_120_ALK:intron:19|+chr2:29446598_EML4:exon:21|-chr2:42553364/2
GCTGCAAACTAATCAGGAATCGATCGGATTGTAAGGACATTGATTGGACGACATAGCTTTGCATTTACTTAAAATCATGCTTCAA
TTAAAGACACACCTTCTTTAATCATTTTATTAGTATTTCTAAGTATGATGGAAATCTATCTTAA
>2_merged_120_ALK:intron:19|+chr2:29446598_EML4:exon:21|-chr2:42553364/merged
AATTGAACCTGTGTATTTATCCTCCTTAAGCTAGATTTCCATCATACTTAGAAATACTAATAAAATGATTAAAGAAGGTGTGTCTT
TAATTGAAGCATGATTTAAAGTAAATGCAAAGCTATGTCGTCCAATCAATGTCCTTACAATCCGATCGATTCCTGATTAGTTTGCA
GC
>1_merged_60_ALK:intron:19|+chr2:29446598_EML4:exon:21|-chr2:42553364/1
TAAAATGATTAAAGAAGGTGTGTCTTTAATTGAAGCATGATTTAAAGTAAATGCAAAGCTATGTCGTCCAATCAATGTCCTTACA
ATCCGATCGATTCCTGATTAGTTTGCAGCCATTTGGAATGTCCCCTTTAAATTTAGAAACAG
>1_merged_60_ALK:intron:19|+chr2:29446598_EML4:exon:21|-chr2:42553364/2
GTAAAAGTGGCTAGTTTGAATCAAGATGCACTTTCAAATACATTTGTACACAAGCACTATGATTATACTTCCTGTTTCTAAATTTA
AAGGGGACATTCCAAATGGCTGCAAACTAATCAGGAATCGATCGGATTGTAAGGACATTGATT
>1_merged_60_ALK:intron:19|+chr2:29446598_EML4:exon:21|-chr2:42553364/merged
TAAAATGATTAAAGAAGGTGTGTCTTTAATTGAAGCATGATTTAAAGTAAATGCAAAGCTATGTCGTCCAATCAATGTCCTTACA
ATCCGATCGATTCCTGATTAGTTTGCAGCCATTTGGAATGTCCCCTTTAAATTTAGAAACAGGAAGTATAATCATAGTGCTTGTGTA
CAAATGTATTTGAAAGTGCATCTTGATTCAAACTAGCCACTTTTAC
```

Fig. 6 FusionDirect result example. In the result, an EML4-ALK fusion is detected and reported with three supporting read pairs, while two of them are unique. The reads of each pair are overlapped so they are merged by pair before detection applied

FusionDirect needs a BED file containing four columns (chromosome, start position, end position, gene name). If this file is not provided, FusionDirect will use the built-in BED file, which contains most fusion genes of high clinical importance.

FusionDirect is available at: https://github.com/OpenGene/FusionDirect.jl. It is written in Julia, which is a fresh language allowing high performance technical computing. FusionDirect is built upon the OpenGene Julia library (https://github.com/OpenGene/OpenGene.jl), which provides basic sequence and variant representations and I/O functions of regular NGS-related file formats (i.e., FASTQ/FastA/VCF).

2.5 Deduplication and Unique Supporting Read Counting

When it comes to determine the confidence of a called variant, the most important evidence is the number and quality of its supporting reads. To calculate numbers of supporting reads, we need to identify and collapse duplicated reads.

There exist some tools to remove PCR duplication. Picard MarkDuplicates compares sequences in the five primary positions of both reads and read pairs in a SAM/BAM file. After duplicated reads are marked, this tool differentiates the primary and duplicated reads using an algorithm ranking reads by the summation of their base quality scores. However, this tool can result in unwanted removal of tumor-derived mutated reads, when it shares mapping coordination with some wild-type reads.

Another approach was introduced by CAPP-seq [37]. It collapses those reads with completely identical sequences except the reads with ultralow-quality scores. This method is less lossy since it removes fewer reads comparing with Picard MarkDuplicates. However, it is usually affected by sequencing errors, so the duplication level of processed data can still be very high.

Molecular barcoding sequencing, which has been introduced above, is a new approach that appears to be effective for removing PCR duplication. Since the UID ligation is performed before any amplification happens, the reads derived from the same original DNA will share the same UID. Based on the clustering of UID and read sequence, the PCR duplication can be detected and the consensus read generation process will remove the duplicated reads. Table 4 compares existing deduplication tools.

The methods described above detect duplication before calling variants. An alternative strategy is to detect duplication after variant calling is done, which collapses the reads with same mapping positions (start and end) as a unique read and gives the numbers of reads supporting reference and alternative base for each mutation. This unique read counting method can provide more accurate supporting read calculation. With this strategy applied, we can apply less lossy deduplication methods like CAPP-seq method to keep more information for variant calling. We can even skip deduplication before variant calling if the variant caller is able to handle the data with duplication.

MrBam is a tool designed for such unique read counting task. It differentiates the result reads generated by one single read or multiple reads sharing same mapping coordination. For paired-end sequencing data, it differentiates the cases where mutation is located in read pair's overlapped or non-overlapped region.

Table 4
Feature comparison of existing deduplication tools

	Information loss	Background noise	Error correction
Picard MarkDuplicates	High	Low	None
CAPP-seq	Low	High	None
Molecular barcodes	Low	Low	Yes

MrBam will give numbers of unique reads for a combination of following conditions: supporting reference or alternative, clustered by single or multiple reads, and locating in overlapped or non-overlapped region.

The result of MrBam can be used to filter variants called from ctDNA sequencing data. According to our experience, to report a mutation, we need at least two unique read pairs supporting it, and each pair should either have this mutation in its overlapped region or be a consensus pair generated by multiple pairs. Due to the high ratio sequencing error and extreme depth of ctDNA sequencing data, the mutations only supported by a few single reads at their non-overlapped regions are usually false positive.

MrBam is an open source project. It is developed in Python with its source available at: http://githubs.com/OpenGene/MrBam.

2.6 Methylation Analysis of Cell-Free DNA

Methylation changes are common for different cancer types and usually occur early in cancer development, typically repressing the expression of tumor suppressor genes [38]. Aberrant DNA methylation may offer a more consistent and hence broadly applicable marker of tumor DNA in blood than mutations [39].

There is a very large amount of published information describing DNA methylation patterns in tumor tissue and their impact on patient prognosis. When tumor DNA is shed into the blood stream, these patterns are also detectable in plasma and serum [40].

Tumor-specific ctDNA methylation can be used to quantitate tumor DNA, providing information about the level of tumor burden, as well as revealing the methylation patterns in the tumor. DNA methylation-based biomarkers could be incorporated into patient care and management with only very minor changes to clinical practice, such as recent applications of methylated ctDNA in determining cancer prognosis and in disease monitoring following surgery or during chemotherapy treatment. Methylated ctDNA assays are also developed to meet the stringent criteria required for cancer screening.

Next-generation sequencing platforms allow the construction of genomic maps of DNA methylation at a single-base resolution [41]. Treating genomic DNA with sodium bisulfite deaminates unmethylated cytosine (C) to uracil (U), while methylated C residues remain unaffected [42]. The U eventually converts to thymine (T) in a subsequent polymerase chain reaction (PCR). Whole-genome bisulfite sequencing (WGBS) and reduced representation bisulfite sequencing (RRBS) are two classic methods for genome-wide methylation study.

WGBS (BS-seq; MethylC-seq) theoretically covers all the C information [43]. In this method, genomic DNA is purified and sheared into fragments. The fragmented DNAs are end-repaired; adenine bases are added to the 3′ end (A-tailing) of the DNA fragments, and methylated adapters are ligated to the DNA fragments. The DNA fragments are size-selected before sodium bisulfite treatment and PCR amplification, and the resulting library is sequenced. The major advantage of WGBS is its ability to assess the methylation state of nearly every CpG site, including low CpG-density regions, such as intergenic "gene deserts," partially methylated domains, and distal regulatory elements. It can also determine absolute DNA methylation level and reveal methylation sequence context.

RRBS was developed cheaper than WGBS, which integrates Msp1 restriction enzyme digestion, bisulfite conversion, and next-generation sequencing for the analysis of methylation patterns of specific fragments. A size selection of MspI-digested fragments between 40 and 220 bps was found to cover 85% of CGIs, mostly in promoters, which compose only 1–3% of the mammalian genome, thereby significantly decreasing the amount of sequencing [44]. RRBS-based protocols are more cost-effective than WGBS because these methods focus on the enrichment of CpG-rich regions in close proximity to the restriction enzyme's recognition sequence. However, these protocols may exhibit a lack of coverage at intergenic and distal regulatory elements that are relatively less studied.

Recently, target capturing-based bisulfite sequencing methods have also been developed, and some kits like NimbleGen SeqCap Epi have been commercialized to provide targeted methylation analysis. Since ultra-deep sequencing is usually needed due to low fraction of tumor DNA in cfDNA, the ability of doing target capturing bisulfite sequencing is very important for analyzing methylation information of ctDNA samples.

One of the major applications of ctDNA methylation analysis is to detect early-stage cancers. Circulating methylated SEPT9 DNA in plasma was developed as a biomarker of colorectal cancer [45], and methylation at the SHP-1 promoter 2 (SHP1P2) was reported as a biomarker of non-small cell lung cancer (NSCLC). These biomarkers are usually more sensitive than protein biomarkers (i.e., carcinoembryonic antigen, CEA) and have the potential to be applied in cancer screening or early-stage cancer detection.

Another major application of ctDNA methylation analysis is identifying tissue of origin for carcinoma of unknown primary (CUP). This application is based on the fact that different human tissues and cells have different DNA methylation patterns. Recently, a method of identifying methylation haplotype blocks was developed to perform tumor tissue-of-origin mapping from plasma DNA [46].

The bioinformatics pipeline to analyze bisulfite sequencing (BS-seq) data is different from analyzing normal sequencing data. The key steps of analyzing BS-seq data are quality control, mapping, methylation scoring, differential methylation assessment, etc.

The QA process for BS-seq data is like the same process for normal sequencing data, including quality profiling, adapter trimming, and low-quality reads filtering. However, be aware that bisulfite treatment will result in overrepresentation of T and underrepresentation of C, which may be considered biased by conventional QC tools. Therefore conventional QC tools, like FastQC, are not a good choice to handle quality control for BS-seq data. BseQC [47] and MethyQA [48] are a better choice since they are specialized for BS-seq data.

Mapping BS-seq reads to reference genome is challenging since the sequences do not exactly match the reference, and the library complexity is reduced due to bisulfite treatment [49]. Furthermore, every given T could either be a genuine genomic T or a converted unmethylated C. Due to these reasons, conventional alignment tools such as BWA and Bowtie are unsuitable for mapping BS-seq reads to reference [50]. Some BS-seq specialized aligners have been developed, and typically they can be categorized into two wildcard aligners and three-letter aligners. Wild-card aligners like BSMAP [51] operate by replacing C with Y (IUPAC code for cytosine or thymine), while three-letter aligners like Bismark [52] convert C to T in both sequenced reads and reference.

Once alignment is done, methylation scores can be calculated for cytosines or genomic regions to find differentially methylated cytosines (DMCs) and differentially methylated regions (DMRs). Cytosine methylation scores are calculated by aggregating overlapping reads and calculating the proportion of C or T, which is called β-score. This process can be achieved by tools like Bismark and GBSA [53]. Software like Methylkit [54] provides a strategy of dividing the genome into small bins, and the mean β-score is taken as bin score. Then statistical tests like Fisher's exact test (FET) can be applied to assess the statistical relevance of DMCs/DMRs between samples. This part of work can also be done with Methylkit, which is a comprehensive R package for analyzing DNA methylation (https://code.google.com/p/methylkit).

Recently some novel methylation analysis methods for BS-seq data have been published. For instance, Gao et al. presented a method to search for genomic regions with highly coordinated methylation. This method is based on blocks of tightly coupled CpG sites, which is called methylation haplotype block (MHB). Then methylation analysis can be done in block level (MHL), and the results based on MHL analysis are much better than those based on analyzing single-CpG sites, which means this method can be applied for identifying tissue of origin [46].

Bisulfite sequencing, as the golden method for analyzing DNA methylation, has been studied for many years, and lots of methods and tools have been developed. Due to the urgent needs of establishing methylation analysis for cancer screening and tissue-of-origin identification, BS-seq data analysis will draw more attention of researchers. We cannot discuss all the aspects of BS-seq in this chapter. A collection of BS-seq data analysis tools and pipelines can be found in OMIC tools online (https://omictools.com/bs-seq-category).

2.7 Machine Learning Methods

Machine learning (ML) technologies are very popular for creating data models in lots of domains, and it can also be applied into ctDNA data analysis. Most applicable methods are supervised learning methods, which build classifiers based on training from labeled data. In this subsection, we will show how to use ML technology to build classifiers with ctDNA sequencing data.

One ML application is to classify cfDNA data and non-cfDNA data. CfDNA has certain fragmentation patterns, which can bring nonrandom base content curves of the sequencing data's beginning cycles. The cfDNA fragmentation patterns were first reported by Chandrananda et al. at one nucleotide resolution in 2014 [55]. They found some high frequency 10-nucleotide motifs on either side of cfDNA fragments, and the first two bases of the cfDNA at cleavage site could determine most of the other eight bases. His further study in 2015 indicated that these fragmentation patterns were related to the nonrandom biological cleavage over chromosomes. The ten positions on either side of the DNA cleavage site show consistent patterns with preference of specific nucleotides for nucleosomal cores and linker regions. Figure 7 shows the fragmentation pattern of plasma cfDNA sequencing data.

Since this fragmentation pattern of cfDNA is stable and unique, it can be used to differentiate data of cfDNA and data of other kinds of samples. The authors have developed an open source tool, called CfdnaPattern, to train classifiers like SVM, KNN, or random forest to predict whether a FASTQ is sequenced from cfDNA or not. Cross validation using 0.632+ bootstrapping [56] with more than 3000 FASTQ files gave a result of 99.8% average accuracy, obtained with random forest, linear SVM, or KNN classifiers. This tool is written in Python, with the widely used Python machine learning package scikit-learn. This tool is available at: https://github.com/OpenGene/CfdnaPattern.

Another ML application is to predict whether a mutation is somatic or germline. Typically, tumor and normal samples are both

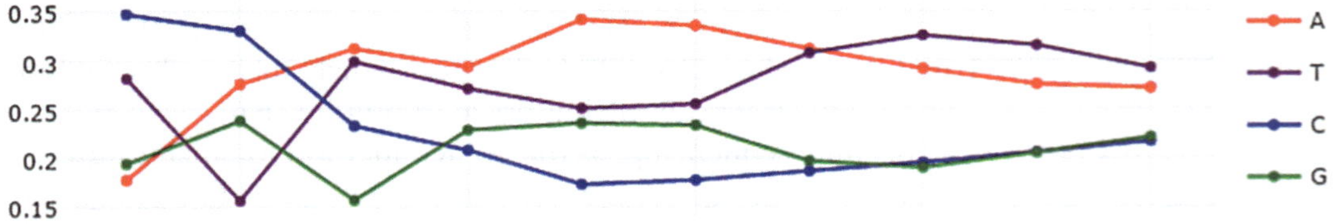

Fig. 7 The cfDNA fragmentation pattern. This figure shows content curves at the first ten cycles of plasma cfDNA sequencing data. This pattern is found stable and can be repeated by different plasma cfDNA samples

sequenced, and the normal sample can be used as a reference to determine the mutations called in tumor sample to be germline or somatic mutations. But for some cases, we may not have matched normal samples for tumor samples, and then we can apply an ML method to classify mutations based on the reads supporting references and the mutations.

DeepSomatic is a tool providing such functions. It can classify somatic and germline mutations with deep neural networks. All reads covering the mutation are extracted and sampled to 256 reads if the read number is greater than 256. Then these reads' bases around the mutation site are coded as a 2D image, with each pixel containing following channels: the read base and its quality score, the reference base, and the lengths of insertion or deletion. Then a deep convolutional neural network (CNN) is constructed with five conventional layers. The model was trained and validated with the tumor-normal paired data, and then cross validation evaluation suggested that this model has an average accuracy higher than 99.9%. DeepSomatic is also an open source tool available at: https://github.com/OpenGene/DeepSomatic.

2.8 Data Simulation

Tuning bioinformatics pipelines and training software parameters require sequencing data with known ground truth, which are actually difficult to get from real sequencing data. Particularly, for ctDNA sequencing applications, which aim to detect low-frequency variations from ultra-deep sequencing data, it is hard to tell whether a called variation is a true positive or a false positive caused by errors from sequencing or other processes. In these cases, simulated data with configured variations can be used to troubleshoot and validate bioinformatics programs.

Although many next-generation sequencing simulators have already been developed, most of them lack of capability to simulate some practical features, such as target capturing sequencing, copy number variations, gene fusions, amplification bias, and sequencing errors. The authors developed SeqMaker, a modern NGS simulator with capability to simulate different kinds of variations, with amplification bias and sequencing errors integrated. Target capturing sequencing is simply supported by using a capturing panel description file, other characteristics like sequencing error rate, average duplication level, DNA template length distribution, and quality distribution can be easily configured with a simple JSON format profile file. With the integration sequencing errors and amplification bias, SeqMaker is able to simulate more real next-generation sequencing data. The configurable variants and capturing regions make SeqMaker very useful to generate data for training bioinformatics pipelines for applications like somatic mutation calling. Table 5 compares the features of SeqMaker and other NGS simulators.

Table 5
A comparison of SeqMaker and other NGS simulators

	SNV	INDEL	INV	TRA	CNV	UMI
SeqMaker	Yes	Yes	Yes	Yes	Yes	Yes
BEAR	No	No	No	No	No	No
dwgsim	Yes	Yes	Yes	Yes	No	No
GemSIM	Yes	No	No	No	No	No
Grinder	Yes	Yes	No	No	No	No
Mason	Yes	Yes	No	No	No	No
pIRS	Yes	Yes	Yes	No	No	No
SInC	Yes	Yes	No	No	Yes	No
wgsim	Yes	Yes	No	No	No	No

SeqMaker is a tool which generates sequencing reads with SNV, INDEL, CNV, and gene fusion enabled, with sequencing error and PCR bias integrated. This tool uses a JSON format profile file to describe the sequencing simulation settings, and a BED format like TSV file to configure the target regions of capturing. First, the simulator samples DNA fragments from whole genome or the target regions configured by the panel file, and CNVs are simulated in this process. Second, the DNA fragments will be altered to simulate SNVs, INDELs, and gene fusions according to the variation list configured in the profile file. Third, a sequencing process will be simulated on each DNA fragment to generate NGS reads, and sequencing errors and amplification bias are also simulated in this process. Finally, generated reads are written into FASTQ files.

SeqMaker is written in Julia, and the source code is available at GitHub: https://github.com/OpenGene/SeqMaker.jl. Currently, it only supports Illumina platforms. More efforts are needed to build simulators for other platforms, especially the new generations of sequencers like PacBio and Nanopore platforms.

3 Discussion

As an innovative method in cancer field, liquid biopsy has current or potential applications in cancer diagnosis, monitoring, and screening. Cell-free tumor DNA, as a major component of liquid biopsy, has been widely used in personalized drug guidance for tumor patients. For those patients not suitable for taking tissue samples by surgery or needle puncture, ctDNA sequencing gives them new opportunities for diagnosis of tumors.

Since ctDNA should be sequenced very deeply, typically target capturing with small gene panels is applied with cost consideration. However, small panels have some disadvantages. Small panels do not allow to detect mutations out of the target regions, difficult to detect large-scale copy number variations, and hard to calculate total mutation burden (TMB) which usually require large panels or whole exome sequencing. As the sequencing cost goes down, it is not difficult to speculate that the whole exome or even whole genome deep sequencing will become affordable and more widely adopted for ctDNA sequencing. Then very big sequencing data will be acquired, and data processing and analysis for such data would be very challenging.

3.1 Conclusion

In this chapter, we introduced the concept and applications of ctDNA, explained the difficulties of analyzing ctDNA NGS data, reviewed some related tools and presented some new methods or tools. One should realize that somatic mutations in cfDNA usually have very low MAF since tumor-specific DNA fragments are usually a small fraction of whole cfDNA. One should be also aware that errors may happen during the experiments and sequencing steps, and software can also introduce artifacts like misalignment or false-positive variant calling.

3.2 Future Work

Although we have discussed so many aspects of bioinformatics for ctDNA NGS data analysis, there still exist topics that have not been discussed above.

Data compression is a key topic we have not discussed in this chapter. Since ctDNA usually requires ultra-deep sequencing, it usually produces very big data. Imagine that if 10,000× WES is applied, we would obtain more than 500 Gb data for a single sample, giving an uncompressed raw file bigger than 1 TB. Storing or transferring such big files will be very challenging, and methods offering high compress ratio will be urgently needed. From signal processing's perspective, the ctDNA sequencing data is highly redundant since it is very deep and has the potential to be compressed with high ratio. However, it is still not easy to compress such kind of data due to three reasons: inconsistent reads due to sequencing errors, varying quality scores, and the requirement of lossless compression. Current methods like DSRC have shown better performance comparing to universal compressors like gzip and bzip2, but the compression ratio improvement is still not satisfactory. Some new compressors like gtz (https://github.com/Genetalks/gtz) have been developed, but they are still not optimized for deep sequencing data. In our opinion, the perfect deep sequencing data compressor should implement local de novo assembly or apply reference-based strategies to achieve much higher compression ratio.

Another topic that remains to be discussed is CNV detection. Since tumor-specific DNA is only a small part of cfDNA, copy number change in tumor cells only leads to slight copy number difference in the ctDNA sequencing data. For instance, if tumor-specific DNA is 1% of the whole cfDNA, and copy number fold in the tumor cells is five, the copy number in whole cfDNA data will be 104%, which is just slightly higher than average level. Current CNV detectors, like CNVkit, are not designed to deal with ctDNA sequencing data and are not sensitive enough to detect such subtle changes in CNV. Better CNV detectors remain to be developed, which should provide better normalization for deep and target-captured ctDNA sequencing data.

Some new methods targeting for cancer immunology are drawing attraction recently. One topic is to predict the outcome of cancer immunotherapies, especially PD-1/PD-L1 checkpoint inhibitors. Tumor mutation burden (TMB) has been shown to be associated with the response of cancer immunotherapies. However, TMB is usually calculated with tissue whole exome sequencing data, and calculating TMB with ctDNA is still challenging due to the low MAF and high level of noises. Methods optimized for ctDNA-based TMB calculation are needed, and this topic can be discussed in future. Another topic related to cancer immunotherapy is neoantigen discovery. In December 2016, Parker Institute for Cancer Immunotherapy and others announced the formation of the Tumor Neoantigen Selection Alliance. This alliance involves researchers from 30 nonprofit institutions and aims to identify software that can best predict neoantigens from patient tumor DNA. For now, computational prediction of neoantigens capable of eliciting efficacious antitumor responses in patients remains a hit-or-miss affair. It is even much more challenging to do the same prediction from patient's ctDNA. The neoantigen prediction study will be a hot topic in both academic and industrial communities, and the progress and outcome can be discussed in the future.

References

1. Kohler CBZ, Radpour R et al (2011) Cell-free DNA in the circulation as a potential cancer biomarker. Anticancer Res 31:2623–2628
2. Diehl F, Schmidt K, Choti MA, Romans K, Goodman S, Li M, Thornton K, Agrawal N, Sokoll L, Szabo SA, Kinzler KW, Vogelstein B, Diaz LA Jr (2008) Circulating mutant DNA to assess tumor dynamics. Nat Med 14 (9):985–990. https://doi.org/10.1038/nm.1789
3. Heitzer E, Ulz P, Geigl JB (2015) Circulating tumor DNA as a liquid biopsy for cancer. Clin Chem 61(1):112–123. https://doi.org/10.1373/clinchem.2014.222679
4. Leon SASB, Sklaroff DM et al (1977) Free DNA in the serum of cancer patients and the effect of therapy. Cancer Res 37:646–650
5. Beaver JA, Jelovac D, Balukrishna S, Cochran RL, Croessmann S, Zabransky DJ, Wong HY, Valda Toro P, Cidado J, Blair BG, Chu D, Burns T, Higgins MJ, Stearns V, Jacobs L, Habibi M, Lange J, Hurley PJ, Lauring J, VanDenBerg DA, Kessler J, Jeter S, Samuels ML, Maar D, Cope L, Cimino-Mathews A, Argani P, Wolff AC, Park BH (2014) Detection of cancer DNA in plasma of patients with early-stage breast cancer. Clin Cancer Res 20 (10):2643–2650. https://doi.org/10.1158/1078-0432.CCR-13-2933

6. Fox EJ, Reid-Bayliss KS, Emond MJ, Loeb LA (2014) Accuracy of next generation sequencing platforms. Next Gener Seq Appl 1. https://doi.org/10.4172/jngsa.1000106
7. Arbeithuber B, Makova KD, Tiemann-Boege I (2016) Artifactual mutations resulting from DNA lesions limit detection levels in ultrasensitive sequencing applications. DNA Res 23 (6):547–559. https://doi.org/10.1093/dnares/dsw038
8. Lixin Chen PL (2017) DNA damage is a pervasive cause of sequencing errors, directly confounding variant identification. Science 355 (6326):752–756
9. Underhill HR, Kitzman JO, Hellwig S, Welker NC, Daza R, Baker DN, Gligorich KM, Rostomily RC, Bronner MP, Shendure J (2016) Fragment length of circulating tumor DNA. PLoS Genet 12(7):e1006162. https://doi.org/10.1371/journal.pgen.1006162
10. Bolger AM, Lohse M, Usadel B (2014) Trimmomatic: a flexible trimmer for Illumina sequence data. Bioinformatics 30 (15):2114–2120. https://doi.org/10.1093/bioinformatics/btu170
11. Chen S, Huang T, Zhou Y, Han Y, Xu M, Gu J (2017) AfterQC: automatic filtering, trimming, error removing and quality control for fastq data. BMC Bioinformatics 18(Suppl 3; 80):91–100. https://doi.org/10.1186/s12859-017-1469-3
12. Langmead B, Salzberg SL (2012) Fast gapped-read alignment with Bowtie 2. Nat Methods 9 (4):357–359. https://doi.org/10.1038/nmeth.1923
13. Li H, Durbin R (2010) Fast and accurate long-read alignment with Burrows-Wheeler transform. Bioinformatics 26(5):589–595. https://doi.org/10.1093/bioinformatics/btp698
14. Li H, Handsaker B, Wysoker A, Fennell T, Ruan J, Homer N, Marth G, Abecasis G, Durbin R, Genome Project Data Processing S (2009) The sequence alignment/map format and SAMtools. Bioinformatics 25 (16):2078–2079. https://doi.org/10.1093/bioinformatics/btp352
15. Tarasov A, Viella AJ, Cuppen E, Nijman IJ, Prins P (2015) Sambamba: fast processing of NGS alignment formats. Bioinformatics. https://doi.org/10.5281/zenodo.13200
16. Mose LE, Wilkerson MD, Hayes DN, Perou CM, Parker JS (2014) ABRA: improved coding indel detection via assembly-based realignment. Bioinformatics 30(19):2813–2815. https://doi.org/10.1093/bioinformatics/btu376
17. Garcia-Alcalde F, Okonechnikov K, Carbonell J, Cruz LM, Gotz S, Tarazona S, Dopazo J, Meyer TF, Conesa A (2012) Qualimap: evaluating next-generation sequencing alignment data. Bioinformatics 28 (20):2678–2679. https://doi.org/10.1093/bioinformatics/bts503
18. McKenna A, Hanna M, Banks E, Sivachenko A, Cibulskis K, Kernytsky A, Garimella K, Altshuler D, Gabriel S, Daly M, DePristo MA (2010) The genome analysis toolkit: a MapReduce framework for analyzing next-generation DNA sequencing data. Genome Res 20 (9):1297–1303. https://doi.org/10.1101/gr.107524.110
19. Cibulskis K, Lawrence MS, Carter SL, Sivachenko A, Jaffe D, Sougnez C, Gabriel S, Meyerson M, Lander ES, Getz G (2013) Sensitive detection of somatic point mutations in impure and heterogeneous cancer samples. Nat Biotechnol 31(3):213–219. https://doi.org/10.1038/nbt.2514
20. Koboldt DC, Zhang Q, Larson DE, Shen D, McLellan MD, Lin L, Miller CA, Mardis ER, Ding L, Wilson RK (2012) VarScan 2: somatic mutation and copy number alteration discovery in cancer by exome sequencing. Genome Res 22(3):568–576. https://doi.org/10.1101/gr.129684.111
21. Wang K, Li M, Hakonarson H (2010) ANNOVAR: functional annotation of genetic variants from high-throughput sequencing data. Nucleic Acids Res 38(16):e164. https://doi.org/10.1093/nar/gkq603
22. Rausch T, Zichner T, Schlattl A, Stutz AM, Benes V, Korbel JO (2012) DELLY: structural variant discovery by integrated paired-end and split-read analysis. Bioinformatics 28(18):i333–i339. https://doi.org/10.1093/bioinformatics/bts378
23. Newman AM, Bratman SV, Stehr H, Lee LJ, Liu CL, Diehn M, Alizadeh AA (2014) FACTERA: a practical method for the discovery of genomic rearrangements at breakpoint resolution. Bioinformatics 30(23):3390–3393. https://doi.org/10.1093/bioinformatics/btu549
24. Wang K, Ma Q, Jiang L, Lai S, Lu X, Hou Y, Wu CI, Ruan J (2016) Ultra-precise detection of mutations by droplet-based amplification of circularized DNA. BMC Genomics 17:214. https://doi.org/10.1186/s12864-016-2480-1
25. Schmieder R, Edwards R (2011) Quality control and preprocessing of metagenomic datasets. Bioinformatics 27(6):863–864. https://doi.org/10.1093/bioinformatics/btr026

26. Cox MP, Peterson DA, Biggs PJ (2010) SolexaQA: at-a-glance quality assessment of Illumina second-generation sequencing data. BMC Bioinformatics 11(1):485. https://doi.org/10.1186/1471-2105-11-485
27. Meldrum C, Doyle MA, Tothill RW (2011) Next-generation sequencing for cancer diagnostics a practical perspective. Clin Biochem Rev 32(4):177–195
28. Tindall KRKT (1988) Fidelity of DNA synthesis by the Thermus aquaticus DNA polymerase. Biochemistry 27:6008–6013
29. Kinde IWJ, Papadopoulos N, Kinzler KW, Vogelstein B (2011) Detection and quantification of rare mutations with. Proc Natl Acad Sci U S A 108(23):9530–9535
30. Liang RH, Mo T, Dong W, Lee GQ, Swenson LC, McCloskey RM, Woods CK, Brumme CJ, Ho CK, Schinkel J, Joy JB, Harrigan PR, Poon AF (2014) Theoretical and experimental assessment of degenerate primer tagging in ultra-deep applications of next-generation sequencing. Nucleic Acids Res 42(12):e98. https://doi.org/10.1093/nar/gku355
31. Hoffmann C, Minkah N, Leipzig J, Wang G, Arens MQ, Tebas P, Bushman FD (2007) DNA bar coding and pyrosequencing to identify rare HIV drug resistance mutations. Nucleic Acids Res 35(13):e91. https://doi.org/10.1093/nar/gkm435
32. Kivioja T, Vaharautio A, Karlsson K, Bonke M, Enge M, Linnarsson S, Taipale J (2011) Counting absolute numbers of molecules using unique molecular identifiers. Nat Methods 9(1):72–74. https://doi.org/10.1038/nmeth.1778
33. Michael W, Schmitta SRK, Salka JJ, Foxa EJ, Hiattb JB, Loeba LA (2012) Detection of ultra-rare mutations by next-generation sequencing. Proc Natl Acad Sci U S A 109:14508–14513
34. Kennedy SR, Schmitt MW, Fox EJ, Kohrn BF, Salk JJ, Ahn EH, Prindle MJ, Kuong KJ, Shen JC, Risques RA, Loeb LA (2014) Detecting ultralow-frequency mutations by Duplex Sequencing. Nat Protoc 9(11):2586–2606. https://doi.org/10.1038/nprot.2014.170
35. Larkin MA, Blackshields G, Brown NP, Chenna R, McGettigan PA, McWilliam H, Valentin F, Wallace IM, Wilm A, Lopez R, Thompson JD, Gibson TJ, Higgins DG (2007) Clustal W and Clustal X version 2.0. Bioinformatics 23(21):2947–2948. https://doi.org/10.1093/bioinformatics/btm404
36. Kirsch A, Mitzenmacher M (2008) Less hashing, same performance: building a better bloom filter. Random Struct Algor 33 (2):187–218. https://doi.org/10.1002/rsa.20208
37. Newman AM, Bratman SV, To J, Wynne JF, Eclov NC, Modlin LA, Liu CL, Neal JW, Wakelee HA, Merritt RE, Shrager JB, Loo BW Jr, Alizadeh AA, Diehn M (2014) An ultrasensitive method for quantitating circulating tumor DNA with broad patient coverage. Nat Med 20(5):548–554. https://doi.org/10.1038/nm.3519
38. Jones SBBPA (2011) A decade of exploring the cancer epigenome – biological and translational implications. Nat Rev Cancer 11 (10):726–734. https://doi.org/10.1038/nrc3130
39. Warton K, Samimi G (2015) Methylation of cell-free circulating DNA in the diagnosis of cancer. Front Mol Biosci 2:13. https://doi.org/10.3389/fmolb.2015.00013
40. Heyn H, Esteller M (2012) DNA methylation profiling in the clinic: applications and challenges. Nat Rev Genet 13(10):679–692. https://doi.org/10.1038/nrg3270
41. Laird PW (2010) Principles and challenges of genomewide DNA methylation analysis. Nat Rev Genet 11(3):191–203. https://doi.org/10.1038/nrg2732
42. Frommer MML, Millar DS, Collis CM, Watt F, Grigg GW et al (1992) A genomic sequencing protocol that yields a positive display of 5-methylcytosine residues in individual DNA strands. Proc Natl Acad Sci U S A 89 (18):27–31
43. Urich MA, Nery JR, Lister R, Schmitz RJ, Ecker JR (2015) MethylC-seq library preparation for base-resolution whole-genome bisulfite sequencing. Nat Protoc 10(3):475–483. https://doi.org/10.1038/nprot.2014.114
44. Gu H, Smith ZD, Bock C, Boyle P, Gnirke A, Meissner A (2011) Preparation of reduced representation bisulfite sequencing libraries for genome-scale DNA methylation profiling. Nat Protoc 6(4):468–481. https://doi.org/10.1038/nprot.2010.190
45. deVos T, Tetzner R, Model F, Weiss G, Schuster M, Distler J, Steiger KV, Grutzmann R, Pilarsky C, Habermann JK, Fleshner PR, Oubre BM, Day R, Sledziewski AZ, Lofton-Day C (2009) Circulating methylated SEPT9 DNA in plasma is a biomarker for colorectal cancer. Clin Chem 55 (7):1337–1346. https://doi.org/10.1373/clinchem.2008.115808
46. Guo S, Diep D, Plongthongkum N, Fung HL, Zhang K, Zhang K (2017) Identification of methylation haplotype blocks aids in deconvolution of heterogeneous tissue samples and

tumor tissue-of-origin mapping from plasma DNA. Nat Genet 49(4):635–642. https://doi.org/10.1038/ng.3805

47. Lin X, Sun D, Rodriguez B, Zhao Q, Sun H, Zhang Y, Li W (2013) BSeQC: quality control of bisulfite sequencing experiments. Bioinformatics 29(24):3227–3229. https://doi.org/10.1093/bioinformatics/btt548
48. Sun S, Noviski A, Yu X (2013) MethyQA: a pipeline for bisulfite-treated methylation sequencing quality assessment. BMC Bioinformatics 14:259
49. Krueger F, Kreck B, Franke A, Andrews SR (2012) DNA methylome analysis using short bisulfite sequencing data. Nat Methods 9 (2):145–151
50. Adusumalli S, Mohd Omar MF, Soong R, Benoukraf T (2014) Methodological aspects of whole-genome bisulfite sequencing analysis. Brief Bioinform 16(3):369–379. https://doi.org/10.1093/bib/bbu016
51. Xi Y, Li W (2009) BSMAP: whole genome bisulfite sequence MAPping program. BMC Bioinformatics 10:232. https://doi.org/10.1186/1471-2105-10-232
52. Krueger F, Andrews SR (2011) Bismark: a flexible aligner and methylation caller for Bisulfite-seq applications. Bioinformatics 27 (11):1571–1572. https://doi.org/10.1093/bioinformatics/btr167
53. Benoukraf T, Wongphayak S, Hadi LH, Wu M, Soong R (2013) GBSA: a comprehensive software for analysing whole genome bisulfite sequencing data. Nucleic Acids Res 41(4): e55. https://doi.org/10.1093/nar/gks1281
54. Akalin A, Kormaksson M, Li S, Garrett-Bakelman FE, Figueroa ME, Melnick A, Mason CE (2012) methylKit: a comprehensive R package for the analysis of genome-wide DNA methylation profiles. Genome Biol 13: R87
55. Chandrananda D, Thorne NP, Bahlo M (2015) High-resolution characterization of sequence signatures due to non-random cleavage of cell-free DNA. BMC Med Genet 8:29. https://doi.org/10.1186/s12920-015-0107-z
56. Efron B, Tibshirani R (1997) Improvements on cross-validation: the .632+ bootstrap method. J Am Stat Assoc 92(438):548–560

Check for updates

Chapter 6

An Overview of Genome-Wide Association Studies

Michelle Chang, Lin He, and Lei Cai

Abstract

Genome-wide association study (GWAS) is a powerful study design to identify genetic variants of a trait and, in particular, detect the association between common single-nucleotide polymorphisms (SNPs) and common human diseases such as heart disease, inflammatory bowel disease, type 2 diabetes, and psychiatric disorders. The standard strategy of population-based case-control studies for GWAS is illustrated in this chapter. We provide an overview of the concepts underlying GWAS, as well as provide guidelines for statistical methods performed in GWAS.

Key words Genome-wide association studies, SNPs, Linkage disequilibrium, Case-control, Two-stage analysis, Genotyping, Common disease common variant hypothesis

1 Introduction

In the early age, linkage analysis and gene mapping had been successfully used for detecting gene variants affecting Mendelian traits (i.e., single-gene disorders, such as cystic fibrosis). However, for complex diseases such as heart disease and inflammatory bowel disease, genetic association studies are more efficient to identify common and penetrant variants implicated in common disease than the classical approaches, such as linkage analysis or candidate gene (CG), due to the idea of common diseases related to multiple susceptibility variants. This is also termed "common disease-common variant hypothesis." This hypothesis states that common diseases are likely influenced by genetic variation that is also common in the population. Over the last 5 years, it has been tested for a variety of common disorders, and while some of the heritability for these conditions are not yet explained, common alleles certainly play a role in susceptibility, though it should not be assumed that the entire genetic component of any common disease is due to common alleles only. The National Human Genome Research Institute GWAS Catalog lists over 3600 SNPs identified for common diseases or traits, and in general, common diseases have

Tao Huang (ed.), *Computational Systems Biology: Methods and Protocols*, Methods in Molecular Biology, vol. 1754, https://doi.org/10.1007/978-1-4939-7717-8_6,

multiple susceptibility alleles, each with small effect sizes (typically increasing disease risk between 1.2 and 2 times the population risk) [1]. GWAS started several years ago because the technologies developed have reached the requirements to perform GWAS, including the catalog of human genetic variants, low-cost and accurate methods of genotyping to identify gene variants, large numbers of informative samples, and efficient statistical design for analysis.

Since the Human Genome Project (HGP) was completed in 2003, many DNA sequence variants have been gradually identified and thus capable to be used for GWAS. With the development of the International HapMap Project that provided the location of ~4 million common SNPs in population of different geographical origins and the allelic association between SNPs, which also termed linkage disequilibrium (LD), we now can find disease-predisposing genetic variants for complex traits [2]. By high-throughput genotyping technology, GWAS reveals the association between hundreds and thousands of SNPs (usually called tag SNPs) and clinical conditions and measurable traits.

1.1 Concepts Underlying the Study Design

The ultimate goal of GWAS is to identify a large portion of the common single-nucleotide genetic variation for association with a complex disease or variation in a quantitative trait. To develop new prevention and treatment strategies to who is at risk, it is important to understand the biological basis of genetic effects in developing new medical therapies.

Single-nucleotide polymorphisms (SNPs) are single base-pair changes in the DNA sequence that occur with high frequency in the human genome, also known as modern units of genetic variation. These genetic polymorphisms have proven to be very useful as genetic markers and can be used to detect the disease variants via LD. This relationship among SNPs, genotyping merely a set of informative SNPs to serve as proxy markers (usually called tagging SNPs, with r2 > 0.8), is sufficient to capture most of the genetic information of SNPs, which are not genotyped with only slight loss of statistical power. r2 is a measurement of "correlation" or LD between two SNPs whose value ranges from 0 to 1 (r2 of one indicates complete LD). r2 depends on both allele frequencies and recombination between the two SNPs.

Linkage analysis was subsequently applied successfully to identify genetic variants that contribute to rare disorders like Huntington disease. When applied to more common disorders, like heart disease or various forms of cancer, linkage analysis has not fared as well. This implies the genetic mechanisms that influence common disorders are different from those that cause rare disorders.

1.2 Study Designs and Strategies for GWAS

There are two major types of GWAS: population-based studies and family-based studies. Here we will focus on population-based case-control studies, which have been the method of choice due to the ease of data collection [3]. Case-control study is one of the strategies when selecting the individuals to perform analysis, meaning ascertaining cases that are affected with or without disease. The most obvious advantage of the case-control study design is that large numbers of case and control participants can be obtained in a short time. However, this design is prone to biases that are mainly caused by population stratification. It usually follows a two-stage study design that is cost-effective since markers unlikely to be associated with the disease can be eliminated in the first stage [4]. The two-stage approach for GWAS is a common strategy that maximizes statistical power while still maintaining reasonable costs [5].

GWAS have been performed and many loci have been identified where genetic factors are associated with the trait or disease. The interactive GWAS diagram [6] (Fig. 1) is a visualization cataloged by the NHGRI. This picture shows the various chromosomes and with some colored dots representing positions of some of these loci, and most recently there is at least 100,000 SNPs and published SNP-trait associations with P-values $< 1.0 \times 10^{-5}$ [7].

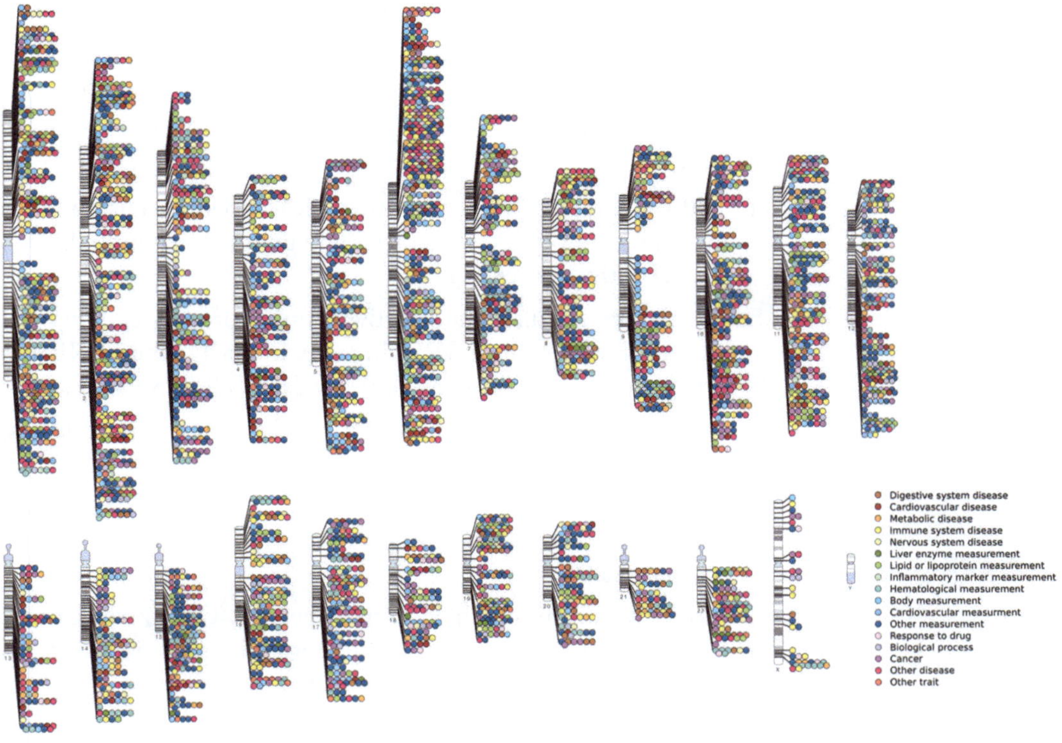

Fig. 1 The NHGRI GWAS Catalog, a curated resource of SNP-trait associations [7]

This chapter provides an overview of statistical methods for genome-wide association studies. Here we describe how to perform basic statistical analysis in a population-based genetic association case-control study. The steps described involve (1) determining the disease of interest for GWAS, (2) collecting study samples and genotyping, (3) quality control for individuals and SNPs, (4) adjustment for multiple testing, and (5) designing replication studies.

2 Materials

A standard GWAS analysis requires genotypic information, phenotypic information, and a model that describes the specification of prior distributions.

1. Computer workstations with Unix/Linux and Windows operating systems:

   PLINK for association analysis: [8]
   http://pngu.mgh.harvard.edu/~purcell/plink/download.shtml.
2. Product for chip-based microarray from (*see* **Note 1**):

   Illumina platform:
   http://www.ocimumbio.com/illumina-platform/.
   Affymetrix platform: http://www.affymetrix.com/catalog/131406/AFFY/Complete+GeneChip+Instrument+System#1_1.
3. Statistical software for data analysis and graphing such as:

   R: http://cran.r-project.org/.
   Haploview: https://www.broadinstitute.org/haploview/haploview.
4. 1000 Genomes Project resources for access to current data, including genotypes, sequences, and genome mapping: http://www.1000genomes.org/.
5. Program to calculate the effective number of independent SNPs among a collection of SNPs in LD with each other, SNPSpD: http://fraser.qimr.edu.au/general/daleN/SNPSpD/.
6. Program to convert HapMap format files to pedigree and map format files:
   http://bioinformatics.well.ox.ac.uk/resources.shtml.
7. Genetic Power Calculator for two-stage GWA case-control studies, CaTS:

   http://www.sph.umich.edu/csg/abecasis/CaTS/index.html.
8. Files: HapMap genotypic and phenotypic information, to be downloaded from http://www.hapmap.org.

9. Genotyped cases and controls data from the Wellcome Trust Case Control Consortium (WTCCC):

   https://www.wtccc.org.uk/.

10. Genotyped cases and controls data from UK Biobank:

    www.ukbiobank.ac.uk/.

3 Methods

The goal of GWAS is to understand the variation in complex traits and diseases by relating genotypes of SNPs to observed phenotypes. To do this, it is necessary to detect which markers are associated with variation in the traits. In this section we aim to provide an overview of the methods available for case-control designs in GWAS.

3.1 Determine the Disease of Interest for GWAS

Defining the disease or phenotype of interest as accurately and specifically as possible is the first step in a case-control design. Without specific case definitions, both the genetic and the environmental heterogeneity will increase in underlying causal factors, while the power of detection of an effect drastically decreases. In this part, we introduce the fundamental thoughts before you perform experimental design of GWAS with sufficient power to robustly detect genomic effects.

1. Define a consensus case definition of the disease of interest from relevant organizations, such as the World Health Organization or recognized disease-specified associations to standardize phenotype [9] (*see* **Note 2**).
2. Determine the heritability of the disease with available evidence from familial studies (*see* **Note 3**).
3. Consider whether the etiology of the disease includes one or more common underlying polymorphisms (allele frequency > 0.01) to ensure the population-based study is an appropriate design for the study.

3.2 Collect Study Samples and Genotyping

Two main companies have provided a number of fixed content panels available for genotyping arrays or chips with set SNPs that are being evaluated on them. The approaches used to select the SNPs for these panels, including random SNPs, selected haplotype tag SNPs, and the nucleotides chosen to be on these panels but are actually copy number variations which cause the signal differs.

Ideally, further phenotypic information should be available for such a panel so as to exclude known cases and to enable matching of controls to cases on the adjustment in analysis.

If no such panel is available for the population from which cases were derived, check if there are other epidemiological studies that included population-based controls with phenotypic information for whom DNA may already have been collected.

1. Recruit healthy control samples and disease-affected case samples with available information of their ethnicity, age, sex, physical condition, and geographical area (*see* **Note 4**). A large sample size is required in genome-wide association studies (*see* **Note 5**).
2. Extract the genomic DNA of all case and control individuals.
3. Conduct genome-wide genotyping analysis by using chip-based microarray technology to assay 1 million or more SNPs. In some of the arrays, if particular variants of interest are missing from the panel, such as less common or rare variants, user is allowed to add an additional 10,000 or 50,000 single-nucleotide variants. Two main platforms are used for most GWAS: Affymetrix and Illumina platforms (*see* **Note 1**). You can skip **steps 2** and **3** if you have accessed to samples that are already genotyped (*see* **Note 6**).
4. Compare case group and control group and collect the SNPs that are significant (i.e., P-value < 1e-7) (*see* **Note 7**).

3.3 Quality Control for Individuals and SNPs

Identifying and excluding poor quality samples are valuable to avoid having errors in the data that might lead to false-positive or false-negative associations. The samples that had a success rate less than some level, maybe at the 95% of the SNPs, are successful. The more SNPs that fail, the more SNPs that succeed are called into questions as to perhaps be generating inaccurate genotypes. It could be that heterozygotes are being miscalled as homozygotes for particular alleles. An excess of heterozygous genotypes suggest that a DNA sample might be a mixture of two DNA samples. Here we overview the QC issues for the samples and SNP-based genotyping methods used in GWAS.

1. Screen the samples with (1) per-individual call rate >80~90% and (2) per-individual heterozygosity between 23 and 30%, and remove those which are not qualified.
2. Screen the SNPs that were generated with (1) per-SNP call rate > 90% (*see* **Note 8**), (2) minor allele frequency (MAF) >3% (*see* **Note 9**), and (3) Hardy-Weinberg equilibrium (HWE) which existed in both controls and cases (*see* **Note 10**).
   (a) One can detect SNPs that are of poor quality by looking for a genotyping success rate less than 95%, which is a threshold commonly used; often the analyses are done using a small percentage of samples that are duplicated and present twice within the set of samples being

genotyped. The genotyping results from the duplicate samples are compared, and mismatches between the identical samples are bad for a SNP.

(b) Do a test for HW equilibrium and look for the expected proportions of genotype frequencies which are not consistent with the observed allele frequencies. These statistical tests can be used to identify the errors. If there are related individuals within samples such as parents and a child, trios, then one can look for Mendelian inheritance of alleles from the parents to the child.

(c) Some groups will add additional quality control samples to their genotyping to allow specific type of SNP error to be detected.

(d) Avoid sample switches that can happen during the process of moving DNA samples to be genotyped. Use genotype data to evaluate whether the sex of the sample matches the expected sex of the individual.

(e) A fully automated pipeline for analysis and reporting of QC results for Illumina SNP data is available at http://www-personal.une.edu. au/~cgondro2/CGhomepage.

3.4 Adjustment for Multiple Testing

A commonly used threshold of significance assumes the number of common variants being tested across the population. A *P*-value threshold of 0.5 declares that a particular result is significant. Achieving a threshold like this requires either a large effect of that particular variant or a large sample size to detect a more modest effect. Here we will discuss the multiple testing methods for an adjustment in GWAS.

1. Perform Bonferroni correction for multiple testing (*see* **Note 11**). If the SNP markers with *P*-values are less than 0.05 after Bonferroni correction, they are then considered significant of the association between the SNPs and the traits of the disease (*see* **Note 12**).
2. Determine the false discovery rate (FDR) to estimate the proportion of significant results (usually at alpha = 0.05) that are false positives and maintain the true results.
3. Perform permutation testing by software packages such as PLINK software or PRESTO to generate the empirical distribution of test statistics for a given dataset.
4. You can also obtain per-SNP significance thresholds for a given family-wise error rate (FWER) from Hoggart et al. [10].

3.5 Design Replication Studies

1. After the SNPs that might be susceptible sites of the disease are screened in the above steps, replication studies are needed to distinguish between “statistical artifacts” and “true

associations" [11]. The replication samples should ideally be larger to be well powered to identify spuriously associated SNPs (*see* **Note 13**).

2. The guidelines to conduct replication studies were suggested by NCI-NHGRI Working Group on Replication in Association Studies [12].

3.6 Statistical Analysis

Larger sample sizes will have a greater possibility of identifying genetic factors that have a more modest effect. The common way is that, for instance, each group does their own GWA analysis, and then the data from several studies is combined together by performing a meta-analysis of the results for each genetic variant. To obtain a statistical significant result, we will show a typical procedure for GWAS.

1. Create the Manhattan plot by HaploView v4.1 [13] or the quantile-quantile (Q-Q) plot generated by R 2.8.1 for data visualization, which can provide a visual summary of association test results that draws immediate attention to any significant regions.

   Q-Q plot: One way to evaluate whether there is an excess of significant results at a given threshold is to plot the *P*-values that result from the test of association against the *P*-values from a uniform distribution. The correction for population stratification can reduce the excess result and excess associations that are false positive and that are not due to true genetic signals.
2. Perform fine mapping around the newly identified susceptibility gene locus by genotyping tag SNPs and performing imputation [14] (*see* **Note 14**).
3. Meta-analysis is useful for the replication of initial association results, and can increase power and opportunity to identify novel signals associated with a disease. When performing meta-analysis, one has to concern about heterogeneity between the studies. For example, when the WTCCC performed a GWAS of T2D, they showed strong evidence of association of variants at the FTO locus. However, a couple of other studies that were doing association analysis of T2D at the same time did show the same result. It is because the WTCCC were more obese than the controls in that study, whereas in the other diabetes studies, their case-control selection had been more balanced with respect to body size. Identifying this source of heterogeneity between the studies led to the identification of this BMI gene.
   (a) Meta-analysis can be conducted by PLINK to combine multiple data from GWA studies and provide a quantitative evaluation of the consistency/inconsistency or heterogeneity of the results across multiple datasets (*see* **Note 15**).

(b) Potential issues for performing a meta-analysis across studies: one is that different genotyping platforms may be used, and different analysis strategies might have been used in the beginning, and the definition of cases and controls may differ as well.

4 Notes

1. Two platforms offer different approaches to assay SNP variation. Illumina uses bead-based technology to recognize SNP alleles, while Affymetrix uses DNA sequence as a spot on the chip to detect a specific SNP allele [15]. Take into consideration that different SNPs should be selected by the platform for different study aim of human population. For example, more SNPs are required with higher genomic coverage for African population study, because African genomes have been existed a longer time and therefore have less LD between alleles at different SNPs. The genomic coverage in Africans was lower due to greater genetic diversity and weaker LD.
2. Vague or nonexistent consensus definition may increase etiological heterogeneity of a disease. An evidence-based and well-defined phenotype is the best used for a genome-wide association study.
3. If the population-based heritability of common disease is lower than 20%, a large sample size is needed to identify the propensity of genetic variants and hence increase the difficulty of the study.
4. Well-characterized samples have long been used in the golden rule of epidemiological study design for case-control analysis. Population stratification refers to the most important type of bias that may produce false-positive association results in case-control studies. It can be minimized if the controls are matched to cases when cases and controls are restricted to a particular ethnic group.
5. Large sample size is one of the determinant factor for a successful GWAS that ensure the genuine associations rank at the top of the SNPs according to the *P*-values [16].
6. You can also obtain genotyped controls and cases from database, such as the Wellcome Trust Case Control Consortium or UK Biobank, which is a more efficient way but may lack data that you need in your study.
7. For a standard two-stage study design, genotyping a large scale of markers on available samples and using *P*-value threshold to identify potential signals of association is the end of the first stage. In stage two, we usually genotype the top signals on the

remaining samples while a combined analysis of data from both stages is conducted [1]. Significant signals, as termed SNPs in GWAS, are subsequently tested for replication. Power calculation software such as CaTS can also be used to easily determine the required sample size and thresholds and receive power calculations for two-stage genome-wide association studies [17].

8. A recommended threshold for removing SNPs with low call rate is approximately 98–99%; however this threshold may vary from study to study [18].

9. The frequency of a SNP is given in terms of the minor allele frequency or the frequency of the less common allele. For instance, a SNP with a minor allele (*A*) frequency of 0.60 implies that 60% of a population has the *A* allele versus the more common allele (the major allele), which is found in 60% of the population.

   Also, consider the low-frequency and rare variants that are not common variants (minor allele frequency $< 5\%$) showing in your study. Poor coverage for rare variants and low LD with SNP markers cause lack of success in identifying rare variants. A larger sample and the discovery of the relative proportion of common variants and rare variants in the total genetic contribution can increase statistical power. In recent years, rare variant association studies (RVASs) have become a growing field of genome-associated study [19].

10. Hardy-Weinberg equilibrium allows allele and genotype frequencies to be estimated from one generation to the next. Departure from this equilibrium will be indicative of potential genotyping errors, population stratification, or actual association to the trait under study [18].

11. In GWA approach, it's very important for multiple testing to avoid false-positive results in your studies. There are few factors that need to be aware of to minimize the false-positive results: (1) statistical adjustment such as Bonferroni correction, false discovery rate (FDR), or permutation testing; (2) stringent *P*-value that indicates that the allele frequency is significantly altered between two sample groups; and (3) large sample size for both genome-wide scan and replication studies.

12. The Bonferroni correction adjusts the alpha value from $\alpha = 0.05$ to $\alpha = (0.05/n)$ where n is the number of statistical tests conducted, which is also the number of GWAS markers to be investigated. Be aware that Bonferroni correction will be too conservative when some of the SNPs are in LD while it assumes that each association test of the SNPs is independent of all other tests. You can use LD information from SNPSpD or

HapMap to estimate the effective number n of independent SNPs [20].

13. Well-designed replication studies are crucial to validate or refute the initial positive association.
14. Fine mapping is a process of refining the associated variants to a credible set most likely to include the causal variant, which requires (1) that all common SNPs in the associated region are genotyped or imputed with high confidence, (2) very stringent quality control, and (3) large sample sizes sufficient in separating out SNPs in high LD.
15. Meta-analysis methodology of GWA studies has successfully improved the power for detecting and validating disease-gene associations in some conditions, such as type 2 diabetes. There is a wide array of approaches, including fixed effects, random effects, Bayesian meta-analysis, and trans-ethnic meta-analysis, you can apply in your study for particular needs [21].

References

1. Skol AD, Scott LJ, Abecasis GR, Boehnke M (2007) Optimal designs for two-stage genome-wide association studies. Genet Epidemiol 31(7):776–788. https://doi.org/10.1002/gepi.20240
2. Visscher PM, Brown MA, McCarthy MI, Yang J (2012) Five years of GWAS discovery. Am J Hum Genet 90(1):7–24. https://doi.org/10.1016/j.ajhg.2011.11.029
3. Haldar T, Ghosh S (2011) Power comparison between population-based case-control studies and family-based transmission-disequilibrium tests: an empirical study. Indian J Hum Genet 17(Suppl 1):S27–S31. https://doi.org/10.4103/0971-6866.80355
4. Satagopan JM (2004) Two-stage designs for gene-disease association studies with sample size constraints. Biometrics 60(3):589–597
5. Kronenberg F (2008) Genome-wide association studies in aging-related processes such as diabetes mellitus, atherosclerosis and cancer. Exp Gerontol 43(1):39–43. https://doi.org/10.1016/j.exger.2007.09.005
6. MacArthur J, Bowler E et al (2017) The new NHGRI-EBI catalog of published genome-wide association studies (GWAS catalog). Nucleic Acids Res 45(Database issue): D896–D901
7. Welter D, MacArthur J, Morales J, Burdett T, Hall P, Junkins H, Klemm A, Flicek P, Manolio T, Hindorff L, Parkinson H (2014) The NHGRI GWAS catalog, a curated resource of SNP-trait associations. Nucleic Acids Res 42 (Database issue):D1001–D1006. https://doi.org/10.1093/nar/gkt1229
8. Purcell S, Neale B, Todd-Brown K, Thomas L, Ferreira MA, Bender D, Maller J, Sklar P, de Bakker PI, Daly MJ, Sham PC (2007) PLINK: a tool set for whole-genome association and population-based linkage analyses. Am J Hum Genet 81(3):559–575. https://doi.org/10.1086/519795
9. Zondervan KT, Cardon LR (2007) Designing candidate gene and genome-wide case-control association studies. Nat Protoc 2 (10):2492–2501. https://doi.org/10.1038/nprot.2007.366
10. Hoggart CJ, Clark TG, De Iorio M, Whittaker JC, Balding DJ (2008) Genome-wide significance for dense SNP and resequencing data. Genet Epidemiol 32(2):179–185. https://doi.org/10.1002/gepi.20292
11. Seng KC, Seng CK (2008) The success of the genome-wide association approach: a brief story of a long struggle. Eur J Hum Genet 16 (5):554–564. Bush WS, Moore JH (2012) Chapter 11: Genome-wide association studies. PLoS Comput Biol 8 (12):e1002822. doi:10.1371/journal.pcbi.1002822
12. Chanock S, NCI-NHGRI Working Group on Replication in Association Studies et al (2007) Replicating genotype-phenotype associations. Nature 447(7145):655–660. https://doi.org/10.1038/447655a
13. Barrett JC, Fry B, Maller J, Daly MJ (2005) Haploview: analysis and visualization of LD

and haplotype maps. Bioinformatics 21 (2):263–265. https://doi.org/10.1093/bioinformatics/bth457

14. Spain SL, Barrett JC (2015) Strategies for fine-mapping complex traits. Hum Mol Genet 24 (R1):R111–R119. https://doi.org/10.1093/hmg/ddv260
15. Distefano JK, Taverna DM (2011) Review technological issues and experimental design of gene association studies. Methods Mol Biol 700:3–16
16. Clarke GM, Anderson CA, Pettersson FH, Cardon LR, Morris AP, Zondervan KT (2011) Basic statistical analysis in genetic case-control studies. Nat Protoc 6(2):121–133. https://doi.org/10.1038/nprot.2010.182
17. Skol AD, Scott LJ et al (2006) Joint analysis is more efficient than replication-based analysis for two-stage genome-wide association studies. Nat Genet 5:554–564. https://doi.org/10.1038/ejhg.2008.12
18. Turner S, Armstrong LL, Bradford Y, Carlson CS, Crawford DC, Crenshaw AT, de Andrade M, Doheny KF, Haines JL, Hayes G, Jarvik G, Jiang L, Kullo IJ, Li R, Ling H, Manolio TA, Matsumoto M, McCarty CA, McDavid AN, Mirel DB, Paschall JE, Pugh EW, Rasmussen LV, Wilke RA, Zuvich RL, Ritchie MD (2011) Quality control procedures for genome-wide association studies. Curr Protoc Hum Genet Chapter 1:Unit1 19. doi:https://doi.org/10.1002/0471142905.hg0119s68
19. Auer PL, Lettre G (2015) Rare variant association studies: considerations, challenges and opportunities. Genome Med 7(1):16. https://doi.org/10.1186/s13073-015-0138-2
20. Nyholt DR (2004) A simple correction for multiple testing for single-nucleotide polymorphisms in linkage disequilibrium with each other. Am J Hum Genet 74:765–769
21. Zeggini E, Ioannidis JP (2009) Meta-analysis in genome-wide association studies. Pharmacogenomics 10(2):191–201. https://doi.org/10.2217/14622416.10.2.191

Chapter 7

Integrative Analysis of Omics Big Data

Xiang-Tian Yu and Tao Zeng

Abstract

The diversity and huge omics data take biology and biomedicine research and application into a big data era, just like that popular in human society a decade ago. They are opening a new challenge from horizontal data ensemble (e.g., the similar types of data collected from different labs or companies) to vertical data ensemble (e.g., the different types of data collected for a group of person with match information), which requires the integrative analysis in biology and biomedicine and also asks for emergent development of data integration to address the great changes from previous population-guided to newly individual-guided investigations.

Data integration is an effective concept to solve the complex problem or understand the complicate system. Several benchmark studies have revealed the heterogeneity and trade-off that existed in the analysis of omics data. Integrative analysis can combine and investigate many datasets in a cost-effective reproducible way. Current integration approaches on biological data have two modes: one is "bottom-up integration" mode with follow-up manual integration, and the other one is "top-down integration" mode with follow-up in silico integration.

This paper will firstly summarize the combinatory analysis approaches to give candidate protocol on biological experiment design for effectively integrative study on genomics and then survey the data fusion approaches to give helpful instruction on computational model development for biological significance detection, which have also provided newly data resources and analysis tools to support the precision medicine dependent on the big biomedical data. Finally, the problems and future directions are highlighted for integrative analysis of omics big data.

Key words Integration, Omics, High throughput, Big data, Complex diseases, Bayesian, Matrix decomposition, Machine learning, Subtype, Precision medicine

1 Introduction

High-throughput screening is one of the primary technologies for exploring complex intracellular dynamics in modern biology, and the data produced by such approaches are usually called as omics data [1]. The intuitive omics on genome appeared from the Human Genome Project for obtaining the blueprint of complete human genetic information; after which, the transcriptome and proteome are also becoming available to measure the expression abundance of mRNA and protein, respectively [2]. Lately, the epigenomics was

Tao Huang (ed.), *Computational Systems Biology: Methods and Protocols*, Methods in Molecular Biology, vol. 1754, https://doi.org/10.1007/978-1-4939-7717-8_7, © Springer Science+Business Media, LLC, part of Springer Nature 2018

developed to investigate the previously thought "dark matter" on genome (e.g., the potential regulatory elements located at noncoding sequences) [3, 4]. Along with the deep understanding of genotype-phenotype association, the metabolites have been widely applied to bridge the genome and phenome due to their outcome role of regulation [5], so that the metabonomics is increased to available for more accurate phenotype indication [6]. Meanwhile, the interactions or associations among different molecules are also confirmed and gathered in databases, which provide the metadata on molecule networks, so called as interactome [7, 8]. These diversity and huge omics data take biology and biomedicine research and application into a big data era (*see* **Note 1**), just like that popular in human society a decade ago [9]. They are opening a new challenge from horizontal data ensemble (e.g., the similar types of data collected from different labs or companies) to vertical data ensemble (e.g., the different types of data collected for a group of person with match information), which provide distinct but often complementary information [10] and are also helpful to address the great changes from previous population-guided to newly individual-guided investigations [11].

Integration is an effective concept to solve the complex problem or understand the complicate system [12]. In computational viewpoints, the data integration can make full use of complementary information [13], carry on necessary noise deduction [14], supply abstract of hidden factor [15], realize bias correction in analysis [16], and introduce common and diversity of data pattern [17]. Meanwhile in biological fields, the data integration is a multi-view investigation on the completeness and complexity of the biological system. Especially in the high-throughput cancer genomic studies, results from the analysis of single datasets often suffer from a lack of reproducibility because of small sample sizes, and the benchmark studies have revealed the heterogeneity and trade-off existed in the analysis of omics data [18, 19]. To address these problems, integrative analysis can effectively combine and investigate many datasets in a cost-effective way to improve reproducibility.

Briefly, current integrative analysis methods on biological data (e.g., omics data discussed in this paper) have two modes: one is "bottom-up integration" (i.e., data combination with follow-up manual integration), and the other one is "top-down integration" (i.e., data fusion with follow-up in silico integration). In the "bottom-up integration," the combination of large amounts of public data may allow us to examine general dynamical relationships during gene regulations [20] [21], e.g., combining different types of data provides a more comprehensive model of the cancer cell than that offered by any single type [22]. These combinatory analyses are expected to integrate the diverse data to reconstruct biologically meaningful networks and potentially provide a more reliable insight into the underlying biological mechanisms [23]. By contrast, in the "top-down integration," the general integration idea is based on information fusion, where different data types can offer complementary perspectives on the same biological phenomenon. The integrative approaches would be more powerful when they can incorporate all data types simultaneously and generate a single integrated sample-cluster assignment, such as the statistic-based methods [10, 11, 24–26], the machine-learning-based methods [22, 27, 28], and the matrix-based methods [21, 29, 30]. Especially, the tensor structure is a basic feature of the multi-view data [31] to uncover shared signals across different high-dimensional data, and it is valuable to develop a model that applies a matrix decomposition to the gene expression matrix for each data type but with a linked individual (e.g., a set of latent components) [32].

Some review on integration study has shown the application potential of integrative analysis on high-dimensional genomic data [13, 33–37]. By contrast, this paper will firstly summarize the combinatory approaches to give candidate protocol on biological experiment design for effectively integrative study on genomics and next survey the data fusion approaches to give helpful instruction on computational model development for meaningful biological significance detection, which also provide new data resources and analysis tools to support the precision medicine dependent on big biomedical data. Below, we will introduce the data resources for integrative analysis, the batch effect removal in integration, the two integration modes, and the tool and visualization of integration analysis, respectively. Finally, we supply a few highlight notes on the problems and future directions for integrative analysis of omics big data.

2 Materials

Being the solid foundation of integrative biological analysis, the data sources, especially the online public data depositions, have provided enormous wealth of data and resources. According to the biological background of these data in databases, the widely accessible data can be summarized as several categories as shown in Table 1.

The genome sequencing technologies open the door to the high-throughput data in biology; thus, the human genome and other species' genomes have been sequenced and published with each passing year. The 1000 Genomes Project [38] has contributed great data on human genomics, and it is designed to supply the largest public human variation and genotype data. The ENCODE (Encyclopedia of DNA Elements) Consortium [39] is built to offer a comprehensive understanding on the functional elements in the human genome, which act/regulate at the DNA, RNA, or protein

Table 1
The category of data sources

Category	Database	URL
Genomics-focused	1000 Genomes [38] Encode [39] 3CDB [42] 4DGenome [43]	http://www.1000genomes.org/ https://www.encodeproject.org/ http://3cdb.big.ac.cn/ https://4dgenome.research.chop.edu/
Transcriptome-focused	NCBI GEO [44] TCGA [45] ICGC [46]	https://www.ncbi.nlm.nih.gov/geo/ https://cancergenome.nih.gov/ http://icgc.org/
Epigenomics-focused	miRBase [47] lncRNAdb [48] NGSmethDB [49] MethylomeDB [50]	http://www.mirbase.org/ http://www.lncrnadb.org/ http://bioinfo2.ugr.es:8080/NGSmethDB/ http://www.neuroepigenomics.org/methylomedb/
Metagenomics-focused	HMDB [51] EBI metagenomics [52]	http://www.hmdb.ca/ https://www.ebi.ac.uk/metagenomics/
Interactome-focused	BioGRID [58] STRING [59] KEGG [60] Reactome [61, 62]	https://thebiogrid.org/ http://www.string-db.org/ http://www.kegg.jp/ http://www.reactome.org/

levels when and where a gene is active. Beyond such coding information of biological sequences, the high-order structure of those sequences have also been resolved recently based on the development of Hi-C or similar technologies [40, 41]. A database of manually curated 3C data (3CDB) [42] is implemented to extract and store the contact frequencies between selected genomic sites in a cell population by literature review and manually extraction. Similarly, the 4DGenome [43] database stores chromatin interaction data compiled by literature curation or computational prediction, which would be efficient on investigating the spatial structure-and-function relationship of genomes.

After the sequence clarity, the detection and estimation of transcriptome have been widely studied based on the microarray or deep-sequencing technologies. NCBI GEO [44] is a well-known database to access the transcriptome data from many different biological experiments, focusing on different spices, different tissues, different cells, or different stresses. Particularly on the study of human cancer, the TCGA [45] and ICGC [46] have generated comprehensive, multidimensional maps of the key genomic changes in more than 30 types of cancer, which are public for assisting the cancer research community to improve the prevention, diagnosis, and treatment of cancer.

Recently, the conventional noncoding information or "dark matter" on genome has also been attractive and inspiring to recover many unknown regulatory factors. One is the miRNA, and the miRBase [47] database publishes predicted hairpin portion of a miRNA transcript, with information on the location and sequence of the mature miRNA sequence. Second is the lncRNA, and the lncRNAdb [48] is a manually curated reference database dependent on capturing a great proportion of the literature describing functions for individual eukaryotic lncRNAs. Third is the methylation, and the NGSmethDB [49] is a repository with single-base whole-genome methylome maps on the best-assembled eukaryotic genomes and the reliable and high-quality methylomes; meanwhile, the MethylomeDB [50] is an expert database containing genome-wide brain DNA methylation profiles of human and mouse brain specimens generated from in-house and collected from third-party publication.

Lately, along with the development of central dogma, the metabolism as the outcomes of regulation can reflect more phenotype-associated genetic information. For example, the Human Metabolome Database (HMDB) [51] is a free database gathering human-source small molecule metabolites, which contains or links chemical data, clinical data, and molecular biology/ biochemistry data, and can be applied in biomarker discovery. Similarly, EBI metagenomics [52] is a freely available center for the storage and analysis of WGS sequenced meta-genomic/meta-transcriptomic data and also provides a standardized analysis workflow to produce rich taxonomic diversity and functional annotations with great consistence on different types of data.

In addition, from the systematical viewpoint on all biological elements, their associations or interactions can be summarized and abstracted as a network form, which inspire the network biology [53–57], and the integrative resources of such biological network knowledge can be obtained from several public databases, such as:

The Biological General Repository for Interaction Datasets (BioGRID) [58] which is an open access database dedicated to the annotation and archival of protein, genetic, and chemical interactions for all major model organism species and humans, by reviewing the biomedical literature for major model organism species.

The STRING database [59] which tries to provide a critical assessment and integration of protein-protein interactions, including direct (physical) as well as indirect (functional) associations, especially the inferred protein-protein associations from co-expression data.

The KEGG [60] which is an encyclopedia of genes and genomes, designed to assign functional meanings to genes and genomes both at the molecular and network level in the form of molecular interactions, reactions, and relations.

Table 2
The category of data structure

Data structure	Experimental protocol	Cases with visualization
Vector	Nucleic acid or amino acid Modification site	The UCSC Genome Browser database [63] MEXPRESS visualizing TCGA [64]
Matrix	Gene-sample Gene-time	Co-expression of gene profiles [65] AIE for cell cycle pattern [66]
Tensor	Gene-sample-source Gene-sample-time	Pan-cancer analysis on TCGA [68] Edge network modeling virus infection [113]
High-order cube	Gene-sample-source-time	Cross-tissue and cross-species transcriptome analysis [70]

The Reactome [61, 62] which plays both as an archive of biological processes by modeling signal transduction, transport, DNA replication, metabolism, and other cellular processes in an ordered network of molecular transformations and as a bioinformatics tool to discover unexpected functional relationships in biological data.

On the other hand, the data from above resources can have different data structures as listed in Table 2, which will determine the direction of follow-up integrative analysis. In mathematical terms, the data structures of such high-throughput data can usually take as a vector, a matrix, a tensor, and their combinations (Fig. 1). Simply, any sequence data (e.g., DNA sequencing) can transform to a (sequenced) vector; each element in a vector represents a nucleic acid or an amino acid or a modification site on the particular location of one sequence, e.g., the string consisted of (A,C,G,T) from 5′ to 3′ on DNA sequence [63], or the barcode-like signal of methylation level on CpG islands along the DNA sequence [64]. Meantime, the expression data of genes from a large cohort study can be organized as a matrix, where a row indicates a gene and a column indicates a sample, so that each element in a matrix represents one gene's expression level in one sample, e.g., the expression of genes in a group of individuals with the same disease [65] or the gene expression of cell cycle at consecutive time points [66]. Next, the triple-way biological experiment can produce data viewed in a cubic form and always be formalized as tensor, and there are two general types of such data [67]: one is "gene-sample-source", which collects the expression data from multiple samples under several biological conditions, e.g., an element in such tensor can point the expression level of one gene from one tissue of the same sample [68]; and the other one is "gene-sample-time", which gathers the expression data from a sample at a particular time point,

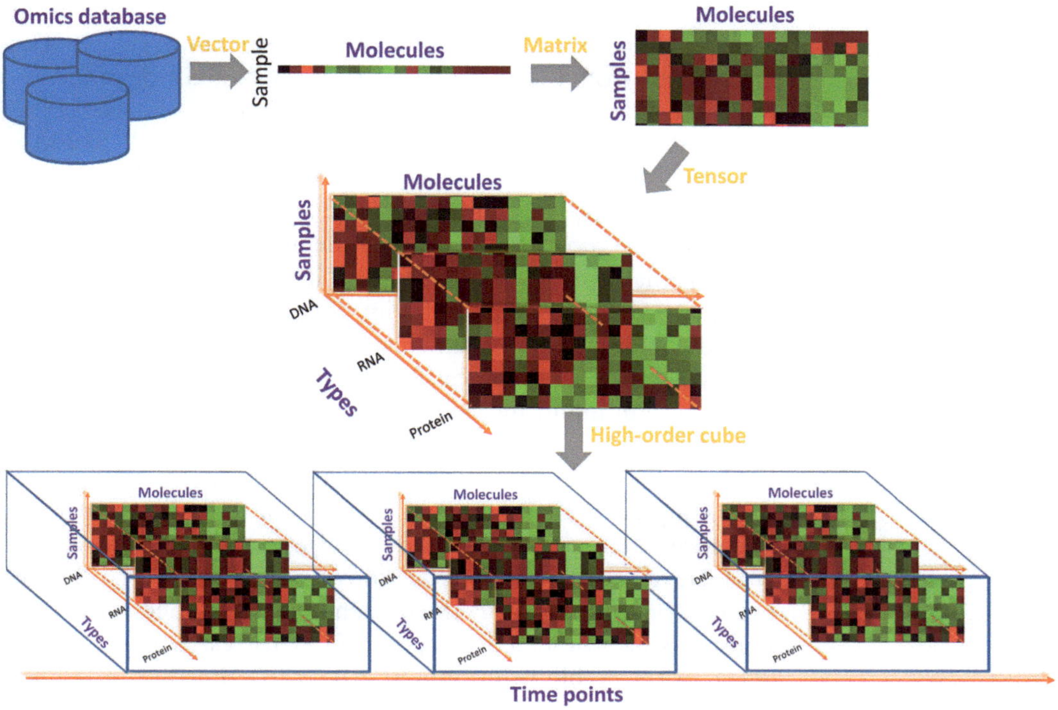

Fig. 1 The organization of data structures in omics big data

e.g., an element in such tensor can reflect the expression value of one gene of the same individual at an early or latter time point during virus infection [69]. Furthermore, nowadays, known as the era of big data, more delicate biological experiments can be carried on, and more complex data structure would be faced, e.g., the combination of tensor as "gene-sample-source-time", whose representative case is the cross-tissue gene expression analysis on evolutionary [70].

3 Methods

3.1 Batch Effect Removal Before Integration

The removal of batch effect should be an important step ahead of many integrative analyses on biological big data. Many variables will play in any given research, such as the influence of age or sex on the diseases. Especially there are many sources of variation when the expressions of thousands of genes are measured at once, so that the batch effects become more critical due to the complexity of genomes inside and environments outside [71]. In practice, the sequencing and microarray samples are usually collected or processed in multiple batches (e.g., at different times), which are easy to produce technical biases and incorrect results in the downstream analysis [72]. For example, to estimate expression correlation over thousands of samples is possible nowadays because large amounts of expression data can be publicly available; however, extracting information from the correlation data is not straightforward due to the expression data generated by different laboratories from different cell types under different biological conditions [73]. To address those issues from batch effect, many computational approaches have been proposed. The "surrogate variable analysis" (SVA) is introduced to recover the effects of the important missed variables and essentially produce an analysis as if all relevant variables were included, which has shown the improved biological accuracy and reproducibility [71]. Meanwhile, the ComBat removes batch effects based on an empirical Bayes framework, which centers data to the overall grand mean of all samples and obtains an adjusted data without coinciding with the location of any original batches [74]. And a modified version of ComBat (M-ComBat) adopts to shift samples to the mean and variance of a "gold standard" as reference batch rather than the grand mean and pooled variance [75]. Next, an extension of PCA known as guided PCA (gPCA) has been proposed to quantify the existence of batch effects, and a new statistic is also designed to apply gPCA to test whether a batch effect exists in high-throughput data [76]. Further, a software pipeline, BatchQC, is implemented to use interactive visualizations and statistics to evaluate the impact of batch effects in a genomic dataset, which can also apply existing adjustment tools and allow to evaluate researchers' benefits interactively [72].

As an initiative integrative application related to batch effect removal, the conventional horizontal data ensemble needs to integrate the same type of data from different studies. For example, an integrative pre-screening is provided to reduce the dimensionality in cancer genomic studies for the analysis of multiple cancer genomic datasets, which can be coupled with existing analysis methods to identify cancer markers [77]. And by analyzing the accrued gene expression data in TCGA pan-cancer (PANCAN) data, the paired normal samples seem to be in general more informative on patient survival than tumors, whose analysis supports the importance of collecting and profiling matched normal tissues to gain more insights on disease etiology and patient progression [78].

3.2 Bottom-Up Integration

According to the combination of different types of high-throughput data, the "bottom-up integration" approaches have many particular analysis frameworks as summarized in Table 3. Generally, the mutation and transcriptome information are both considered, especially the mRNA expression is used in almost any analysis (*see* **Note 2**). Below, considering the usage of mutation or not, the integrative methods are introduced and discussed, respectively.

On one hand, the mutation-centered integration mainly tries to identify the genetic determinants of phenotype and its change,

Table 3
The representative approaches of bottom-up integration

Method	Level of omics								Biological purpose
	Mutation	mRNA	miRNA	Modification	Protein	Metabolite	Network	Annotation	
"eQTL-based" [80]	✓	✓							eQTL-based analysis
TieDIE [92]	✓	✓			✓				Robust synthetize of signaling network
"Network-based" [79]	✓						✓	✓	Identification of disrupted pathways
"Integrative network analysis" [114]	✓	✓		✓					Identifying important genetic and epigenetic features
"TCGA-based" [82–85]	✓	✓	✓	✓					Characterizing somatic alterations
"TCGA-based" [86–88]	✓	✓		✓					Characterizing the genomic/epigenomic landscape
"TCGA-based" [90, 91]	✓	✓	✓	✓	✓				Cancer subtypes caused by different subsets of genetic and epigenetic abnormalities
"Generalizable framework" [81]	✓	✓							Identifying pathogenetically relevant mutated genes
"Integrative framework" [89]	✓	✓			✓				Determining the prognostic, predictive, and therapeutic relevance of the functional proteome
"Pan-cancer initiative" [93]	✓	✓	✓	✓	✓				The Cancer Genome Atlas pan-cancer analysis project
dChip-GemiNI [98]		✓	✓				✓		Detecting feed-forward loops (FFLs) on TF-miRNA-mRNA network

(continued)

Table 3 (continued)

Method	Level of omics								Biological purpose
	Mutation	mRNA	miRNA	Modification	Protein	Metabolite	Network	Annotation	
"Semi-supervised normalization pipelines" [100]		✓			✓	✓	✓	✓	Training predictive cell models based on integrated data sources
"Integrative computational pipeline" [97]		✓	✓	✓					Dissecting the transcription factors (TFs) responsible for altered miRNA expression
"Data-driven discovery" [95]		✓						✓	Data-driven discovery of pain gene candidates
"Layers of regulation" [99]		✓			✓				Adaptive mechanisms include posttranscriptional and posttranslational events
pRSEM [96]		✓		✓					Estimating relative isoform abundances
SPIA [94]		✓						✓	Pan-cancer analysis on pathways

and other omics data can be assisted to recognize the sensitive mutation by removing the passenger mutations.

1. *The direct mapping of mutation information on pathway/network knowledge.* A network-based method has been used to integrate copy number alteration data with human protein-protein interaction networks and pathway databases to identify pathways that are commonly disrupted in many different types of cancer, which are likely essential for tumor formation in the majority of the cancers [79].
2. *The combination of mutation and transcriptome.* As a typical quantitative association approach, the eQTL (expression quantitative trait locus)-based analyses have been proposed to investigate the germ line determinants of gene expression in tumors by using the multilevel information from The Cancer Genome Atlas (TCGA) [80]. And in an investigation of the aggressive lung tumor subtype with poor prognosis, an integrated analyses have been conducted to identify pathogenetically relevant mutated genes by a generalizable framework for the identification of biologically relevant genes in the context of high mutational background [81].
3. *The further consideration of epigenetic influence.* A genome-scale analysis of 276 samples has been analyzed to characterize the somatic alterations in colorectal carcinoma, including exome sequence, DNA copy number, promoter methylation, and messenger RNA and microRNA expression [82–85]. Similarly, 178 lung squamous cell carcinomas have been deeply profiled to provide a comprehensive landscape of genomic and epigenomic alterations in squamous cell lung cancers and develop molecularly targeted agents for target treatment [86–88].
4. *The additional integration with protein expression.* The direct study of the functional proteome has the potential to provide a wealth of information that complements and extends genomic, epigenomic, and transcriptomic analysis. The resultant proteomic data in TCGA can be integrated with genomic and transcriptomic analyses of the same samples to identify commonalities, differences, emergent pathways, and network biology within and across tumor lineages [89]. By integrating information across platforms including reverse phase protein arrays, a hypothesis is held that much of the clinically observable plasticity and heterogeneity occurs within, and not across, the major biological subtypes of breast cancer [90, 91]. Besides, the Tied Diffusion through Interacting Events (TieDIE) is developed to integrate differentially expressed master transcriptional regulators, functionally mutated genes, and differentially activated kinases to synthesize a robust signaling

network consisting of druggable kinase pathways, which will be helpful for drug prioritization in individual patients [92].

5. *To provide basic data support.* The Cancer Genome Atlas (TCGA) Research Network has profiled and analyzed large numbers of human tumors to discover molecular aberrations at the DNA, RNA, protein, and epigenetic levels, whose resulting rich data provide a major opportunity to develop an integrated picture of commonalities, differences, and emergent themes across tumor lineages [45]. Dependent on TCGA, the pan-cancer initiative compares multiple tumor types, and the molecular aberrations and their functional roles across tumor types will enlighten how to extend therapies effective in one cancer type to others with a similar genomic profile [93].

   On the other hand, the transcriptome-centered integration mainly tries to identify the phenotype-associated genes by the complementary information from other omics data.

6. *The functional enrichment based on the expression abundance and its differential changes.* By a software package signaling pathway impact analysis (SPIA), all signaling pathways in the KEGG PATHWAY database have been widely investigated and obtained several notable findings concerning many pathways to be new discoveries, which imply many opportunities for laboratory and clinical follow-up studies [94]. Specially, a novel integrative paradigm has been applied for data-driven discovery of pain gene candidates, taking advantage of the vast amount of existing disease-related clinical literature and gene expression microarray data, which enables efficient biological studies validating additional candidates [95].

7. *The functional complementation between transcriptome and epigenome.* To improve the quantification accuracy of isoforms, a computational method as prior-enhanced RSEM (pRSEM) is proposed to use a complementary data type in addition to RNA-seq data, which shown to be superior than competing methods in estimating relative isoform abundances within or across conditions in qRT-PCR validations [96]. Another case is that an integrative computational pipeline has identified TFs with binding sites significantly overrepresented among miRNA genes overexpressed in ovarian carcinoma, and it can be applied to discover transcriptional regulatory mechanisms in other biological settings where analogous genomic data are available [97]. Besides, the dChip-GemiNI (Gene and miRNA Network-based Integration) method can statistically rank computationally predicted FFLs by accounting for differential gene and miRNA expression between two biological conditions such as normal and cancer and also derive potential TF-target genes and miRNA-mRNA interactions [98].

8. *The functional influence of protein on gene regulatory.* Stress responses were believed to be predominantly regulated at the transcriptional level; however, the adaptive mechanisms should include post-transcriptional and post-translational events. To address this issue, three layers of regulation have been integrated as transcriptome, translatome, and proteome, which is useful to gain a deeper understanding of how sophisticated regulation networks operate [99]. And semi-supervised normalization pipelines have been designed and performed experimental characterization to create a quality-controlled multi-omics compendium for *E. coli*, and a multi-scale model has further been trained by integrating four omics layers to predict genome-wide concentrations and growth dynamics [100].

3.3 Top-Down Integration

The standard "bottom-up integration" approach as above integrative clustering is usually to separate clustering followed by manual integration. By contrast, a more computational powerful approach would incorporate all data types simultaneously and generate a single integrated cluster assignment (*see* **Note 3**), which are thought as "top-down integration" as shown in Table 4.

1. *Statistic-based integration model.* One key integrative idea is unifying hidden factor from different types of data. A joint latent variable model as iCluster is developed for integrative clustering by incorporating flexible modeling of the associations between different data types and the variance-covariance structure within data types while simultaneously reducing the dimensionality of the datasets [24]. To extend the scope of integrative analysis for the inclusion of somatic mutation data, an expanded framework iCluster+ is further proposed to ensemble discrete and continuous variables that arise from integrated genomic, epigenomic, and transcriptomic profiling [11]. Similarly, a novel algorithm termed moCluster employs a multiblock multivariate analysis to define a set of latent variables representing joint patterns across input datasets, which is passed to an ordinary clustering algorithm in order to discover joint clusters [101]. The other important integrative idea is unifying data distribution under the theoretical framework around Bayesian principles. An integrative Bayesian analysis of genomics data (iBAG) framework is proposed to identify important genes/biomarkers by using hierarchical modeling to combine the data obtained from multiple platforms into one model [25]. And a Bayesian method referred as MDI (Multiple Dataset Integration) has been presented for the unsupervised integrative modeling, where each dataset is modeled using a Dirichlet-multinomial allocation (DMA) mixture model, with dependencies between these models captured

Table 4
The representative approaches of top-down integration

Categories	Methods	Computational instructions
Statistic (factor-centered)	"Residuals" [115]	A two-stage approach based on regularized singular value decomposition, and regularized estimation of prediction model
	iCluster [24]	A joint latent variable model incorporating the variance-covariance structures
	iCluster+ [11]	Joint modeling is proposed to ensemble discrete and continuous variables
	moCluster [101]	Multiblock multivariate analysis and an ordinary clustering algorithm
	iBAG [25]	Hierarchical modeling within Bayesian analysis
	MDI [10]	Dirichlet-multinomial allocation (DMA) mixture model within Bayesian analysis
	"Nonparametric Bayesian model" [26]	A hierarchy of Dirichlet processes within a nonparametric Bayesian model
	"Factor analysis" [116]	Factor analysis
Optimization (matrix-centered)	"Joint matrix factorization" [21]	Joint nonnegative matrix factorization
	"Multi-view bi-clustering" [30]	Rank matrix factorization
	GSVD [104] [105]	Higher-order generalized singular value decomposition
	"Ping-pong" [29]	Ping-pong algorithm
Machine learning (pattern-centered)	"Linear discriminant analysis" [22]	Factor analysis, combined with linear discriminant analysis
	"Kernel-based" [27]	Multiple kernel learning
	JointCluster [28]	Simultaneous clustering of multiple networks
	SNF [102]	Similarity network fusion based on theoretical multi-view learning framework
	PFA [103]	Pattern fusion analysis based on local tangent space alignment (LTSA) theory

through parameters that describe the agreement among the datasets [10]. Meanwhile, a nonparametric Bayesian model has been introduced to discover prognostic cancer subtypes by constructing a hierarchy of Dirichlet processes and has shown a good ability to distinguish concordant and discordant signals within each patient sample [26].

2. *Machine-learning-based integration model.* The main idea under such methods is to extract significant data pattern along with integrative analysis. An extended multiple kernel learning has been applied for dimensionality reduction approaches, and several kernels per data type are applicable to avoid the unnecessary choice of the best kernel functions and

kernel parameters for each data type beforehand [27]. And in a biological application, the high-throughput screens for mRNA, miRNA, and proteins have been jointly analyzed using factor analysis, combined with linear discriminant analysis (LDA), to identify the molecular characteristics of cancer [22]. Especially when focused on characterizing biological network, an algorithm JointCluster is implemented to find sets of genes that cluster well in multiple networks of interest, such as co-expression networks summarizing correlations among the expression profiles of genes and physical networks describing protein-protein and protein-DNA interactions among genes or gene products [28]. To produce a comprehensive view of a given disease by diverse types of genome-wide data, similarity network fusion (SNF) has been inspired from the theoretical multi-view learning framework to construct the networks of samples (e.g., patients) for each data type and fuse them into one network, which can represent the sample patterns underlying data [102]. Recently, a new framework called "pattern fusion analysis" (PFA) has been proposed to perform automated information alignment and bias correction and to fuse local sample patterns (e.g., from each data type) into a global sample pattern corresponding to phenotypes (e.g., across most data types). Particular, PFA can identify common and complementary sample patterns from different omics profiles by optimally adjusting the effects of each data type based on the local tangent space alignment (LTSA) theory [103].

3. *Matrix-based integration model.* Previously, the integrative scheme of ping-pong algorithm was proposed to integrate more than one type of data from the same biological samples, which is dependent on the usage of co-modules describing coherent patterns across paired datasets [29]. Actually, these methods can be included into several classes according to the type of applied matrix decomposition: one is a joint (nonnegative) matrix factorization technique that projects multiple types of genomic data onto a common coordinate system, in which heterogeneous variables weighted highly in the same projected direction form a multidimensional module (md-module) [21]; two is higher-order generalized singular value decomposition (GSVD), which is designed for efficient, parameter-free and reproducible identification of network modules simultaneously across multiple conditions [104, 105]; and three is rank matrix factorization as multi-view bi-clustering to model subtyping and recognize subtype-specific features simultaneously, e.g., integrate mutational and expression data while taking into account the clonal properties of carcinogenesis [30].

3.4 Tool and Visualization of Integration

Currently, the academic studies not only develop the biological or computational techniques for integrative analysis but also provide many software tools and visualization resources for iteratively review by biologist or clinician as listed in Table 5, to easily understand the complicate structure and information in multi-view data and their meta-outcome (*see* **Note 4**).

As the general applications of integrative analysis and visualization tool public accessible, Ensembl Genomes is an integrative resource for genome-scale data from non-vertebrate species [106], which exploits and extends technology developed in the context of the Ensembl project and provides a complementary set of resources for non-vertebrate species through a consistent set of programmatic and interactive interfaces. Similarly, the cBioPortal for Cancer Genomics provides a Web resource for exploring, visualizing, and analyzing multidimensional cancer genomics data [107], whose portal reduces molecular profiling data from cancer tissues and cell lines into readily understandable genetic, epigenetic, gene expression, and proteomic events.

Meanwhile, as expert approaches of integrative analysis and visualization tool online, a Web tool, named Integrated Clustering of Multidimensional biomedical data (ICM), can provide an interface from which to fuse, cluster, and visualize multidimensional biomedical data and knowledge or can explore the heterogeneity of a disease or a biological process by identifying subgroups of patients [108]. Next, an integrative meta-analysis of expression data (INMEX) is designed to support meta-analysis of multiple gene expression datasets, as well as datasets from gene expression and metabolomics experiments, whose statistical analysis module allows researchers to combine multiple datasets based on *P* values, effect sizes, rank orders, and other features [109]. Then, a Web server, SteinerNet, establishes a framework for integrating transcriptional, proteomic, and interactome data by searching for the solution to the prize-collecting Steiner tree problem [110]. Besides, a new data integration framework, Anduril, is introduced for translating fragmented large-scale data into testable predictions, and it allows rapid integration of heterogeneous data with state-of-the-art computational methods and existing knowledge in bio-databases [111].

Similarly, when taking particular focus on integrative analysis and visualization on TCGA data, Web-TCGA, a Web-based, freely accessible online tool, can also be run in a private instance, for integrated analysis of molecular cancer datasets provided by TCGA [68]. And MEXPRESS is developed as a straightforward and easy-to-use Web tool for the integration and visualization of the expression, DNA methylation, and clinical TCGA data on a single-gene level, which offers clinical researchers a simple way to evaluate the TCGA data for their genes or candidate biomarkers of interest [64]. And CrossHub software is developed to enable

Table 5
The representative approaches of integrative visualization

Methods	Description	URL
ICM [108]	Integrated clustering of multiple types of omics data is essential for developing individual-based treatments and precision medicine	http://biotech.bmi.ac.cn/icm/
MEXPRESS [64]	Offers clinical researchers a simple way to evaluate the TCGA data for their genes or candidate biomarkers of interest	http://mexpress.be
SteinerNet [110]	For researchers who would like to integrate their high-throughput data for a specific condition or cellular response and to find biologically meaningful pathway	http://fraenkel-nsf.csbi.mit.edu/steinernet/
CrossHub [112]	The contribution of different mechanisms to the regulation of gene expression varies for different tissues and tumors	https://sourceforge.net/projects/crosshub/
Anduril [111]	To translate the fragmented and heterogeneous datasets into knowledge	http://csbi.ltdk.helsinki.fi/anduril/
Web-TCGA [68]	Integrated analysis of molecular cancer datasets provided by TCGA	https://sourceforge.net/projects/webtcga/
Ensembl Genomes [106]	Participants in a growing range of collaborations involved in the annotation and analysis of genomes	http://www.ensemblgenomes.org
INMEX [109]	Properly combining or integrating the datasets with similar basic hypotheses can help reduce study bias, increase statistical power, and improve overall biological understanding	http://www.inmex.ca
cBioPortal [107]	To provide a practical guide to the analysis and visualization features of the cBioPortal for cancer genomics	http://cbioportal.org

two-way identification of most possible TF-gene interactions: on the basis of ENCODE ChIP-Seq binding evidence or Jaspar prediction and co-expression according to the data of the largest cancer omics resource [112].

4 Notes

This paper has given a comprehensive summary of data resources, data analysis, and data visualization supporting the integration of big biological data. Finally, we would like to list several notes on this review:

1. Conventional big data from society would have a large number of samples, and each sample has a few features/attributes. By contrast, the big biological data would supply not large but enough samples and test tens of thousands of features for each sample simultaneously. *This small-sample high-dimensional data requires new analytic approaches, including the data integration.*
2. "Bottom-up integration" mode with follow-up manual integration is always the hypothesis-driven approaches to extract the significant enriched or observed biological knowledge in data. The key of these methods is there should be clear and suitable biological hints on the experiments and outcome data, and then the data combination can extract the biological signals in each type of data and explain the same preset biological hypothesis in a single analysis framework. Although for different combinations on data types, there is already corresponding integrative analysis framework, it is still short of more general and flexible scheme to deal with the existing data types and potential new data types. *It is urgently required to design quantitative evaluation on the confidence of driver hypothesis ahead of data analysis and also on the contribution of different data types to the biological hypothesis.*
3. Meanwhile, "top-down integration" mode with follow-up in silico integration is usually the data-driven approaches to extract the most probable feature signals or sample patterns in data. The key of these methods is there must be efficient correction to reduce the noise and bias in different types of data, and then the data fusion can identify the coordinate data distribution or data correlation in multiple types of data in a unified mathematical model. Many techniques are available; however, they are used solid constraint on the union of data coordination, which limit their application on the diverse biological systems. *Thus, the more relaxations, e.g., soft-constraint-based approaches, will expand the power of data fusion in biological study and detect unseen biological patterns.*

4. Besides, a few tools are available for general study or special application, but the platform of benchmarking the integration methods requires further development, both on the "gold-standard" data and criteria. And the databases on storage and reanalysis of metadata of integration outcome also ask for attention, design, and advance.

References

1. Field D, Sansone SA, Collis A, Booth T, Dukes P, Gregurick SK, Kennedy K, Kolar P, Kolker E, Maxon M, Millard S, Mugabushaka AM, Perrin N, Remacle JE, Remington K, Rocca-Serra P, Taylor CF, Thorley M, Tiwari B, Wilbanks J (2009) Megascience. 'Omics data sharing'. Science 326 (5950):234–236. https://doi.org/10.1126/science.1180598
2. Vo TV, Das J, Meyer MJ, Cordero NA, Akturk N, Wei X, Fair BJ, Degatano AG, Fragoza R, Liu LG, Matsuyama A, Trickey M, Horibata S, Grimson A, Yamano H, Yoshida M, Roth FP, Pleiss JA, Xia Y, Yu H (2016) A proteome-wide fission yeast interactome reveals network evolution principles from yeasts to human. Cell 164 (1–2):310–323. https://doi.org/10.1016/j.cell.2015.11.037
3. Madhani HD, Francis NJ, Kingston RE, Kornberg RD, Moazed D, Narlikar GJ, Panning B, Struhl K (2008) Epigenomics: a roadmap, but to where? Science 322 (5898):43–44. https://doi.org/10.1126/science.322.5898.43b
4. Romanoski CE, Glass CK, Stunnenberg HG, Wilson L, Almouzni G (2015) Epigenomics: roadmap for regulation. Nature 518 (7539):314–316. https://doi.org/10.1038/518314a
5. Lage K, Karlberg EO, Storling ZM, Olason PI, Pedersen AG, Rigina O, Hinsby AM, Tumer Z, Pociot F, Tommerup N, Moreau Y, Brunak S (2007) A human phenome-interactome network of protein complexes implicated in genetic disorders. Nat Biotechnol 25(3):309–316. https://doi.org/10.1038/nbt1295
6. Nicholson JK, Lindon JC (2008) Systems biology: metabonomics. Nature 455 (7216):1054–1056. https://doi.org/10.1038/4551054a
7. Rolland T, Tasan M, Charloteaux B, Pevzner SJ, Zhong Q, Sahni N, Yi S, Lemmens I, Fontanillo C, Mosca R, Kamburov A, Ghiassian SD, Yang X, Ghamsari L, Balcha D, Begg BE, Braun P, Brehme M, Broly MP, Carvunis AR, Convery-Zupan D, Corominas R, Coulombe-Huntington J, Dann E, Dreze M, Dricot A, Fan C, Franzosa E, Gebreab F, Gutierrez BJ, Hardy MF, Jin M, Kang S, Kiros R, Lin GN, Luck K, MacWilliams A, Menche J, Murray RR, Palagi A, Poulin MM, Rambout X, Rasla J, Reichert P, Romero V, Ruyssinck E, Sahalie JM, Scholz A, Shah AA, Sharma A, Shen Y, Spirohn K, Tam S, Tejeda AO, Trigg SA, Twizere JC, Vega K, Walsh J, Cusick ME, Xia Y, Barabasi AL, Iakoucheva LM, Aloy P, De Las Rivas J, Tavernier J, Calderwood MA, Hill DE, Hao T, Roth FP, Vidal M (2014) A proteome-scale map of the human interactome network. Cell 159(5):1212–1226. https://doi.org/10.1016/j.cell.2014.10.050
8. Friedel CC, Zimmer R (2006) Toward the complete interactome. Nat Biotechnol 24 (6):614–615.; Author reply 615. https://doi.org/10.1038/nbt0606-614
9. Buxton B, Hayward V, Pearson I, Karkkainen L, Greiner H, Dyson E, Ito J, Chung A, Kelly K, Schillace S (2008) Big data: the next Google. Interview by Duncan Graham-Rowe. Nature 455(7209):8–9. https://doi.org/10.1038/455008a
10. Kirk P, Griffin JE, Savage RS, Ghahramani Z, Wild DL (2012) Bayesian correlated clustering to integrate multiple datasets. Bioinformatics 28(24):3290–3297. https://doi.org/10.1093/bioinformatics/bts595
11. Mo Q, Wang S, Seshan VE, Olshen AB, Schultz N, Sander C, Powers RS, Ladanyi M, Shen R (2013) Pattern discovery and cancer gene identification in integrated cancer genomic data. Proc Natl Acad Sci U S A 110(11):4245–4250. https://doi.org/10.1073/pnas.1208949110
12. Rapport DJ, Maffi L (2013) A call for integrative thinking. Science 339(6123):1032. https://doi.org/10.1126/science.339.6123.1032-a
13. Wen Y, Wei Y, Zhang S, Li S, Liu H, Wang F, Zhao Y, Zhang D, Zhang Y (2016) Cell subpopulation deconvolution reveals breast

cancer heterogeneity based on DNA methylation signature. Brief Bioinform. https://doi.org/10.1093/bib/bbw028

14. Voillet V, Besse P, Liaubet L, San Cristobal M, Gonzalez I (2016) Handling missing rows in multi-omics data integration: multiple imputation in multiple factor analysis framework. BMC Bioinformatics 17(1):402. https://doi.org/10.1186/s12859-016-1273-5
15. Weischenfeldt J, Simon R, Feuerbach L, Schlangen K, Weichenhan D, Minner S, Wuttig D, Warnatz HJ, Stehr H, Rausch T, Jager N, Gu L, Bogatyrova O, Stutz AM, Claus R, Eils J, Eils R, Gerhauser C, Huang PH, Hutter B, Kabbe R, Lawerenz C, Radomski S, Bartholomae CC, Falth M, Gade S, Schmidt M, Amschler N, Hass T, Galal R, Gjoni J, Kuner R, Baer C, Masser S, von Kalle C, Zichner T, Benes V, Raeder B, Mader M, Amstislavskiy V, Avci M, Lehrach H, Parkhomchuk D, Sultan M, Burkhardt L, Graefen M, Huland H, Kluth M, Krohn A, Sirma H, Stumm L, Steurer S, Grupp K, Sultmann H, Sauter G, Plass C, Brors B, Yaspo ML, Korbel JO, Schlomm T (2013) Integrative genomic analyses reveal an androgen-driven somatic alteration landscape in early-onset prostate cancer. Cancer Cell 23(2):159–170. https://doi.org/10.1016/j.ccr.2013.01.002
16. Shen R, Mo Q, Schultz N, Seshan VE, Olshen AB, Huse J, Ladanyi M, Sander C (2012) Integrative subtype discovery in glioblastoma using iCluster. PLoS One 7(4):e35236. https://doi.org/10.1371/journal.pone.0035236
17. Zeng T, Wang DC, Wang X, Xu F, Chen L (2014) Prediction of dynamical drug sensitivity and resistance by module network rewiring-analysis based on transcriptional profiling. Drug Resist Updates 17(3):64–76. https://doi.org/10.1016/j.drup.2014.08.002
18. Shi X, Shen S, Liu J, Huang J, Zhou Y, Ma S (2014) Similarity of markers identified from cancer gene expression studies: observations from GEO. Brief Bioinform 15(5):671–684. https://doi.org/10.1093/bib/bbt044
19. Shi X, Yi H, Ma S (2015) Measures for the degree of overlap of gene signatures and applications to TCGA. Brief Bioinform 16 (5):735–744. https://doi.org/10.1093/bib/bbu049
20. Bebek G, Koyuturk M, Price ND, Chance MR (2012) Network biology methods integrating biological data for translational science. Brief Bioinform 13(4):446–459. https://doi.org/10.1093/bib/bbr075
21. Zhang S, Liu CC, Li W, Shen H, Laird PW, Zhou XJ (2012) Discovery of multi-dimensional modules by integrative analysis of cancer genomic data. Nucleic Acids Res 40(19):9379–9391. https://doi.org/10.1093/nar/gks725
22. Liu Y, Devescovi V, Chen S, Nardini C (2013) Multilevel omic data integration in cancer cell lines: advanced annotation and emergent properties. BMC Syst Biol 7:14. https://doi.org/10.1186/1752-0509-7-14
23. Hieke S, Benner A, Schlenl RF, Schumacher M, Bullinger L, Binder H (2016) Integrating multiple molecular sources into a clinical risk prediction signature by extracting complementary information. BMC Bioinformatics 17(1):327. https://doi.org/10.1186/s12859-016-1183-6
24. Shen R, Olshen AB, Ladanyi M (2009) Integrative clustering of multiple genomic data types using a joint latent variable model with application to breast and lung cancer subtype analysis. Bioinformatics 25(22):2906–2912. https://doi.org/10.1093/bioinformatics/btp543
25. Wang W, Baladandayuthapani V, Morris JS, Broom BM, Manyam G, Do KA (2013) iBAG: integrative Bayesian analysis of high-dimensional multiplatform genomics data. Bioinformatics 29(2):149–159. https://doi.org/10.1093/bioinformatics/bts655
26. Yuan Y, Savage RS, Markowetz F (2011) Patient-specific data fusion defines prognostic cancer subtypes. PLoS Comput Biol 7(10): e1002227. https://doi.org/10.1371/journal.pcbi.1002227
27. Speicher NK, Pfeifer N (2015) Integrating different data types by regularized unsupervised multiple kernel learning with application to cancer subtype discovery. Bioinformatics 31(12):i268–i275. https://doi.org/10.1093/bioinformatics/btv244
28. Narayanan M, Vetta A, Schadt EE, Zhu J (2010) Simultaneous clustering of multiple gene expression and physical interaction datasets. PLoS Comput Biol 6(4):e1000742. https://doi.org/10.1371/journal.pcbi.1000742
29. Kutalik Z, Beckmann JS, Bergmann S (2008) A modular approach for integrative analysis of large-scale gene-expression and drug-response data. Nat Biotechnol 26 (5):531–539. https://doi.org/10.1038/nbt1397
30. Le Van T, van Leeuwen M, Carolina Fierro A, De Maeyer D, Van den Eynden J, Verbeke L, De Raedt L, Marchal K, Nijssen S (2016) Simultaneous discovery of cancer subtypes

and subtype features by molecular data integration. Bioinformatics 32(17):i445–i454. https://doi.org/10.1093/bioinformatics/btw434

31. Seely JS, Kaufman MT, Ryu SI, Shenoy KV, Cunningham JP, Churchland MM (2016) Tensor analysis reveals distinct population structure that parallels the different computational roles of areas M1 and V1. PLoS Comput Biol 12(11):e1005164. https://doi.org/10.1371/journal.pcbi.1005164
32. Hore V, Vinuela A, Buil A, Knight J, McCarthy MI, Small K, Marchini J (2016) Tensor decomposition for multiple-tissue gene expression experiments. Nat Genet 48(9):1094–1100. https://doi.org/10.1038/ng.3624
33. Bersanelli M, Mosca E, Remondini D, Giampieri E, Sala C, Castellani G, Milanesi L (2016) Methods for the integration of multi-omics data: mathematical aspects. BMC Bioinformatics 17(Suppl 2):15. https://doi.org/10.1186/s12859-015-0857-9
34. Meng C, Zeleznik OA, Thallinger GG, Kuster B, Gholami AM, Culhane AC (2016) Dimension reduction techniques for the integrative analysis of multi-omics data. Brief Bioinform 17(4):628–641. https://doi.org/10.1093/bib/bbv108
35. Luo Y, Wang F, Szolovits P (2016) Tensor factorization toward precision medicine. Brief Bioinform. https://doi.org/10.1093/bib/bbw026
36. Vargas AJ, Harris CC (2016) Biomarker development in the precision medicine era: lung cancer as a case study. Nat Rev Cancer 16(8):525–537. https://doi.org/10.1038/nrc.2016.56
37. Lahti L, Schafer M, Klein HU, Bicciato S, Dugas M (2013) Cancer gene prioritization by integrative analysis of mRNA expression and DNA copy number data: a comparative review. Brief Bioinform 14(1):27–35. https://doi.org/10.1093/bib/bbs005
38. Genomes Project C, Abecasis GR, Auton A, Brooks LD, DePristo MA, Durbin RM, Handsaker RE, Kang HM, Marth GT, McVean GA (2012) An integrated map of genetic variation from 1,092 human genomes. Nature 491(7422):56–65. https://doi.org/10.1038/nature11632
39. Gerstein M (2012) Genomics: ENCODE leads the way on big data. Nature 489(7415):208. https://doi.org/10.1038/489208b
40. Nagano T, Lubling Y, Stevens TJ, Schoenfelder S, Yaffe E, Dean W, Laue ED, Tanay A, Fraser P (2013) Single-cell Hi-C reveals cell-to-cell variability in chromosome structure. Nature 502(7469):59–64. https://doi.org/10.1038/nature12593
41. Dekker J, Marti-Renom MA, Mirny LA (2013) Exploring the three-dimensional organization of genomes: interpreting chromatin interaction data. Nat Rev Genet 14(6):390–403. https://doi.org/10.1038/nrg3454
42. Yun X, Xia L, Tang B, Zhang H, Li F, Zhang Z (2016) 3CDB: a manually curated database of chromosome conformation capture data. Database (Oxford). https://doi.org/10.1093/database/baw044
43. Teng L, He B, Wang J, Tan K (2016) 4DGenome: a comprehensive database of chromatin interactions. Bioinformatics 32(17):2727. https://doi.org/10.1093/bioinformatics/btw375
44. Barrett T, Wilhite SE, Ledoux P, Evangelista C, Kim IF, Tomashevsky M, Marshall KA, Phillippy KH, Sherman PM, Holko M, Yefanov A, Lee H, Zhang N, Robertson CL, Serova N, Davis S, Soboleva A (2013) NCBI GEO: archive for functional genomics data sets--update. Nucleic Acids Res 41(Database issue):D991–D995. https://doi.org/10.1093/nar/gks1193
45. Kim HS, Minna JD, White MA (2013) GWAS meets TCGA to illuminate mechanisms of cancer predisposition. Cell 152(3):387–389. https://doi.org/10.1016/j.cell.2013.01.027
46. International Cancer Genome C, Hudson TJ, Anderson W, Artez A, Barker AD, Bell C, Bernabe RR, Bhan MK, Calvo F, Eerola I, Gerhard DS, Guttmacher A, Guyer M, Hemsley FM, Jennings JL, Kerr D, Klatt P, Kolar P, Kusada J, Lane DP, Laplace F, Youyong L, Nettekoven G, Ozenberger B, Peterson J, Rao TS, Remacle J, Schafer AJ, Shibata T, Stratton MR, Vockley JG, Watanabe K, Yang H, Yuen MM, Knoppers BM, Bobrow M, Cambon-Thomsen A, Dressler LG, Dyke SO, Joly Y, Kato K, Kennedy KL, Nicolas P, Parker MJ, Rial-Sebbag E, Romeo-Casabona CM, Shaw KM, Wallace S, Wiesner GL, Zeps N, Lichter P, Biankin AV, Chabannon C, Chin L, Clement B, de Alava E, Degos F, Ferguson ML, Geary P, Hayes DN, Hudson TJ, Johns AL, Kasprzyk A, Nakagawa H, Penny R, Piris MA, Sarin R, Scarpa A, Shibata T, van de Vijver M, Futreal PA, Aburatani H, Bayes M, Botwell DD, Campbell PJ, Estivill X, Gerhard DS, Grimmond SM, Gut I, Hirst M, Lopez-Otin C, Majumder P, Marra M, McPherson

JD, Nakagawa H, Ning Z, Puente XS, Ruan Y, Shibata T, Stratton MR, Stunnenberg HG, Swerdlow H, Velculescu VE, Wilson RK, Xue HH, Yang L, Spellman PT, Bader GD, Boutros PC, Campbell PJ, Flicek P, Getz G, Guigo R, Guo G, Haussler D, Heath S, Hubbard TJ, Jiang T, Jones SM, Li Q, Lopez-Bigas N, Luo R, Muthuswamy L, Ouellette BF, Pearson JV, Puente XS, Quesada V, Raphael BJ, Sander C, Shibata T, Speed TP, Stein LD, Stuart JM, Teague JW, Totoki Y, Tsunoda T, Valencia A, Wheeler DA, Wu H, Zhao S, Zhou G, Stein LD, Guigo R, Hubbard TJ, Joly Y, Jones SM, Kasprzyk A, Lathrop M, Lopez-Bigas N, Ouellette BF, Spellman PT, Teague JW, Thomas G, Valencia A, Yoshida T, Kennedy KL, Axton M, Dyke SO, Futreal PA, Gerhard DS, Gunter C, Guyer M, Hudson TJ, McPherson JD, Miller LJ, Ozenberger B, Shaw KM, Kasprzyk A, Stein LD, Zhang J, Haider SA, Wang J, Yung CK, Cros A, Liang Y, Gnaneshan S, Guberman J, Hsu J, Bobrow M, Chalmers DR, Hasel KW, Joly Y, Kaan TS, Kennedy KL, Knoppers BM, Lowrance WW, Masui T, Nicolas P, Rial-Sebbag E, Rodriguez LL, Vergely C, Yoshida T, Grimmond SM, Biankin AV, Bowtell DD, Cloonan N, deFazio A, Eshleman JR, Etemadmoghadam D, Gardiner BB, Kench JG, Scarpa A, Sutherland RL, Tempero MA, Waddell NJ, Wilson PJ, McPherson JD, Gallinger S, Tsao MS, Shaw PA, Petersen GM, Mukhopadhyay D, Chin L, DePinho RA, Thayer S, Muthuswamy L, Shazand K, Beck T, Sam M, Timms L, Ballin V, Lu Y, Ji J, Zhang X, Chen F, Hu X, Zhou G, Yang Q, Tian G, Zhang L, Xing X, Li X, Zhu Z, Yu Y, Yu J, Yang H, Lathrop M, Tost J, Brennan P, Holcatova I, Zaridze D, Brazma A, Egevard L, Prokhortchouk E, Banks RE, Uhlen M, Cambon-Thomsen A, Viksna J, Ponten F, Skryabin K, Stratton MR, Futreal PA, Birney E, Borg A, Borresen-Dale AL, Caldas C, Foekens JA, Martin S, Reis-Filho JS, Richardson AL, Sotiriou C, Stunnenberg HG, Thoms G, van de Vijver M, van't Veer L, Calvo F, Birnbaum D, Blanche H, Boucher P, Boyault S, Chabannon C, Gut I, Masson-Jacquemier JD, Lathrop M, Pauporte I, Pivot X, Vincent-Salomon A, Tabone E, Theillet C, Thomas G, Tost J, Treilleux I, Calvo F, Bioulac-Sage P, Clement B, Decaens T, Degos F, Franco D, Gut I, Gut M, Heath S, Lathrop M, Samuel D, Thomas G, Zucman-Rossi J, Lichter P, Eils R, Brors B, Korbel JO, Korshunov A, Landgraf P, Lehrach H, Pfister S, Radlwimmer B, Reifenberger G, Taylor MD, von Kalle C, Majumder PP, Sarin R, Rao TS, Bhan MK, Scarpa A, Pederzoli P, Lawlor RA, Delledonne M, Bardelli A, Biankin AV, Grimmond SM, Gress T, Klimstra D, Zamboni G, Shibata T, Nakamura Y, Nakagawa H, Kusada J, Tsunoda T, Miyano S, Aburatani H, Kato K, Fujimoto A, Yoshida T, Campo E, Lopez-Otin C, Estivill X, Guigo R, de Sanjose S, Piris MA, Montserrat E, Gonzalez-Diaz M, Puente XS, Jares P, Valencia A, Himmelbauer H, Quesada V, Bea S, Stratton MR, Futreal PA, Campbell PJ, Vincent-Salomon A, Richardson AL, Reis-Filho JS, van de Vijver M, Thomas G, Masson-Jacquemier JD, Aparicio S, Borg A, Borresen-Dale AL, Caldas C, Foekens JA, Stunnenberg HG, van't Veer L, Easton DF, Spellman PT, Martin S, Barker AD, Chin L, Collins FS, Compton CC, Ferguson ML, Gerhard DS, Getz G, Gunter C, Guttmacher A, Guyer M, Hayes DN, Lander ES, Ozenberger B, Penny R, Peterson J, Sander C, Shaw KM, Speed TP, Spellman PT, Vockley JG, Wheeler DA, Wilson RK, Hudson TJ, Chin L, Knoppers BM, Lander ES, Lichter P, Stein LD, Stratton MR, Anderson W, Barker AD, Bell C, Bobrow M, Burke W, Collins FS, Compton CC, DePinho RA, Easton DF, Futreal PA, Gerhard DS, Green AR, Guyer M, Hamilton SR, Hubbard TJ, Kallioniemi OP, Kennedy KL, Ley TJ, Liu ET, Lu Y, Majumder P, Marra M, Ozenberger B, Peterson J, Schafer AJ, Spellman PT, Stunnenberg HG, Wainwright BJ, Wilson RK, Yang H (2010) International network of cancer genome projects. Nature 464 (7291):993–998. https://doi.org/10.1038/nature08987

47. Kozomara A, Griffiths-Jones S (2014) miRBase: annotating high confidence microRNAs using deep sequencing data. Nucleic Acids Res 42(Database issue):D68–D73. https://doi.org/10.1093/nar/gkt1181

48. Quek XC, Thomson DW, Maag JL, Bartonicek N, Signal B, Clark MB, Gloss BS, Dinger ME (2015) lncRNAdb v2.0: expanding the reference database for functional long noncoding RNAs. Nucleic Acids Res 43 (Database issue):D168–D173. https://doi.org/10.1093/nar/gku988

49. Lebron R, Gomez-Martin C, Carpena P, Bernaola-Galvan P, Barturen G, Hackenberg M, Oliver JL (2017) NGSmethDB 2017: enhanced methylomes and differential methylation. Nucleic Acids Res 45(D1):D97–D103. https://doi.org/10.1093/nar/gkw996

50. Xin Y, Chanrion B, O'Donnell AH, Milekic M, Costa R, Ge Y, Haghighi FG (2012) MethylomeDB: a database of DNA methylation profiles of the brain. Nucleic Acids Res 40(Database issue): D1245–D1249. https://doi.org/10.1093/nar/gkr1193
51. Wishart DS, Jewison T, Guo AC, Wilson M, Knox C, Liu Y, Djoumbou Y, Mandal R, Aziat F, Dong E, Bouatra S, Sinelnikov I, Arndt D, Xia J, Liu P, Yallou F, Bjorndahl T, Perez-Pineiro R, Eisner R, Allen F, Neveu V, Greiner R, Scalbert A (2013) HMDB 3.0—the human metabolome database in 2013. Nucleic Acids Res 41(Database issue): D801–D807. https://doi.org/10.1093/nar/gks1065
52. Mitchell A, Bucchini F, Cochrane G, Denise H, ten Hoopen P, Fraser M, Pesseat S, Potter S, Scheremetjew M, Sterk P, Finn RD (2016) EBI metagenomics in 2016—an expanding and evolving resource for the analysis and archiving of metagenomic data. Nucleic Acids Res 44(D1):D595–D603. https://doi.org/10.1093/nar/gkv1195
53. Friedman A, Perrimon N (2007) Genetic screening for signal transduction in the era of network biology. Cell 128(2):225–231. https://doi.org/10.1016/j.cell.2007.01.007
54. Barabasi AL, Oltvai ZN (2004) Network biology: understanding the cell's functional organization. Nat Rev Genet 5(2):101–113. https://doi.org/10.1038/nrg1272
55. Goymer P (2008) Network biology: why do we need hubs? Nat Rev Genet 9(9):650
56. Hu JX, Thomas CE, Brunak S (2016) Network biology concepts in complex disease comorbidities. Nat Rev Genet 17 (10):615–629. https://doi.org/10.1038/nrg.2016.87
57. New AM, Lehner B (2015) Systems biology: network evolution hinges on history. Nature 523(7560):297–298. https://doi.org/10.1038/nature14537
58. Chatr-Aryamontri A, Oughtred R, Boucher L, Rust J, Chang C, Kolas NK, O'Donnell L, Oster S, Theesfeld C, Sellam A, Stark C, Breitkreutz BJ, Dolinski K, Tyers M (2017) The BioGRID interaction database: 2017 update. Nucleic Acids Res 45(D1):D369–D379. https://doi.org/10.1093/nar/gkw1102
59. Szklarczyk D, Franceschini A, Wyder S, Forslund K, Heller D, Huerta-Cepas J, Simonovic M, Roth A, Santos A, Tsafou KP, Kuhn M, Bork P, Jensen LJ, von Mering C (2015) STRING v10: protein-protein interaction networks, integrated over the tree of life. Nucleic Acids Res 43(Database issue):D447–D452. https://doi.org/10.1093/nar/gku1003
60. Kanehisa M, Furumichi M, Tanabe M, Sato Y, Morishima K (2017) KEGG: new perspectives on genomes, pathways, diseases and drugs. Nucleic Acids Res 45(D1):D353–D361. https://doi.org/10.1093/nar/gkw1092
61. Fabregat A, Sidiropoulos K, Garapati P, Gillespie M, Hausmann K, Haw R, Jassal B, Jupe S, Korninger F, McKay S, Matthews L, May B, Milacic M, Rothfels K, Shamovsky V, Webber M, Weiser J, Williams M, Wu G, Stein L, Hermjakob H, D'Eustachio P (2016) The reactome pathway knowledgebase. Nucleic Acids Res 44(D1): D481–D487. https://doi.org/10.1093/nar/gkv1351
62. Bohler A, Wu G, Kutmon M, Pradhana LA, Coort SL, Hanspers K, Haw R, Pico AR, Evelo CT (2016) Reactome from a WikiPathways perspective. PLoS Comput Biol 12(5): e1004941. https://doi.org/10.1371/journal.pcbi.1004941
63. Tyner C, Barber GP, Casper J, Clawson H, Diekhans M, Eisenhart C, Fischer CM, Gibson D, Gonzalez JN, Guruvadoo L, Haeussler M, Heitner S, Hinrichs AS, Karolchik D, Lee BT, Lee CM, Nejad P, Raney BJ, Rosenbloom KR, Speir ML, Villarreal C, Vivian J, Zweig AS, Haussler D, Kuhn RM, Kent WJ (2017) The UCSC Genome Browser database: 2017 update. Nucleic Acids Res 45(D1):D626–D634. https://doi.org/10.1093/nar/gkw1134
64. Koch A, De Meyer T, Jeschke J, Van Criekinge W (2015) MEXPRESS: visualizing expression, DNA methylation and clinical TCGA data. BMC Genomics 16:636. https://doi.org/10.1186/s12864-015-1847-z
65. van 't Veer LJ, Dai H, van de Vijver MJ, He YD, Hart AA, Mao M, Peterse HL, van der Kooy K, Marton MJ, Witteveen AT, Schreiber GJ, Kerkhoven RM, Roberts C, Linsley PS, Bernards R, Friend SH (2002) Gene expression profiling predicts clinical outcome of breast cancer. Nature 415(6871):530–536. https://doi.org/10.1038/415530a
66. Zeng T, Li J (2010) Maximization of negative correlations in time-course gene expression data for enhancing understanding of molecular pathways. Nucleic Acids Res 38(1):e1. https://doi.org/10.1093/nar/gkp822
67. Zeng T, Guo X, Liu J (2014) Negative correlation based gene markers identification in

integrative gene expression data. Int J Data Min Bioinform 10(1):1–17

68. Deng M, Bragelmann J, Schultze JL, Perner S (2016) Web-TCGA: an online platform for integrated analysis of molecular cancer data sets. BMC Bioinformatics 17:72. https://doi.org/10.1186/s12859-016-0917-9
69. Huang Y, Zaas AK, Rao A, Dobigeon N, Woolf PJ, Veldman T, Oien NC, McClain MT, Varkey JB, Nicholson B, Carin L, Kingsmore S, Woods CW, Ginsburg GS, Hero AO III (2011) Temporal dynamics of host molecular responses differentiate symptomatic and asymptomatic influenza a infection. PLoS Genet 7(8):e1002234. https://doi.org/10.1371/journal.pgen.1002234
70. Brawand D, Soumillon M, Necsulea A, Julien P, Csardi G, Harrigan P, Weier M, Liechti A, Aximu-Petri A, Kircher M, Albert FW, Zeller U, Khaitovich P, Grutzner F, Bergmann S, Nielsen R, Paabo S, Kaessmann H (2011) The evolution of gene expression levels in mammalian organs. Nature 478 (7369):343–348. https://doi.org/10.1038/nature10532
71. Leek JT, Storey JD (2007) Capturing heterogeneity in gene expression studies by surrogate variable analysis. PLoS Genet 3 (9):1724–1735. https://doi.org/10.1371/journal.pgen.0030161
72. Manimaran S, Selby HM, Okrah K, Ruberman C, Leek JT, Quackenbush J, Haibe-Kains B, Bravo HC, Johnson WE (2016) BatchQC: interactive software for evaluating sample and batch effects in genomic data. Bioinformatics. https://doi.org/10.1093/bioinformatics/btw538
73. Vandenbon A, Dinh VH, Mikami N, Kitagawa Y, Teraguchi S, Ohkura N, Sakaguchi S (2016) Immuno-Navigator, a batch-corrected coexpression database, reveals cell type-specific gene networks in the immune system. Proc Natl Acad Sci U S A 113(17): E2393–E2402. https://doi.org/10.1073/pnas.1604351113
74. Johnson WE, Li C, Rabinovic A (2007) Adjusting batch effects in microarray expression data using empirical Bayes methods. Biostatistics 8(1):118–127. https://doi.org/10.1093/biostatistics/kxj037
75. Stein CK, Qu P, Epstein J, Buros A, Rosenthal A, Crowley J, Morgan G, Barlogie B (2015) Removing batch effects from purified plasma cell gene expression microarrays with modified ComBat. BMC Bioinformatics 16:63. https://doi.org/10.1186/s12859-015-0478-3
76. Reese SE, Archer KJ, Therneau TM, Atkinson EJ, Vachon CM, de Andrade M, Kocher JP, Eckel-Passow JE (2013) A new statistic for identifying batch effects in high-throughput genomic data that uses guided principal component analysis. Bioinformatics 29 (22):2877–2883. https://doi.org/10.1093/bioinformatics/btt480
77. Song R, Huang J, Ma S (2012) Integrative prescreening in analysis of multiple cancer genomic studies. BMC Bioinformatics 13:168. https://doi.org/10.1186/1471-2105-13-168
78. Huang X, Stern DF, Zhao H (2016) Transcriptional profiles from paired normal samples offer complementary information on cancer patient survival—evidence from TCGA pan-cancer data. Sci Rep 6:20567. https://doi.org/10.1038/srep20567
79. Hwang TH, Atluri G, Kuang R, Kumar V, Starr T, Silverstein KA, Haverty PM, Zhang Z, Liu J (2013) Large-scale integrative network-based analysis identifies common pathways disrupted by copy number alterations across cancers. BMC Genomics 14:440. https://doi.org/10.1186/1471-2164-14-440
80. Li Q, Seo JH, Stranger B, McKenna A, Pe'er I, Laframboise T, Brown M, Tyekucheva S, Freedman ML (2013) Integrative eQTL-based analyses reveal the biology of breast cancer risk loci. Cell 152(3):633–641. https://doi.org/10.1016/j.cell.2012.12.034
81. Peifer M, Fernandez-Cuesta L, Sos ML, George J, Seidel D, Kasper LH, Plenker D, Leenders F, Sun R, Zander T, Menon R, Koker M, Dahmen I, Muller C, Di Cerbo V, Schildhaus HU, Altmuller J, Baessmann I, Becker C, de Wilde B, Vandesompele J, Bohm D, Ansen S, Gabler F, Wilkening I, Heynck S, Heuckmann JM, Lu X, Carter SL, Cibulskis K, Banerji S, Getz G, Park KS, Rauh D, Grutter C, Fischer M, Pasqualucci L, Wright G, Wainer Z, Russell P, Petersen I, Chen Y, Stoelben E, Ludwig C, Schnabel P, Hoffmann H, Muley T, Brockmann M, Engel-Riedel W, Muscarella LA, Fazio VM, Groen H, Timens W, Sietsma H, Thunnissen E, Smit E, Heideman DA, Snijders PJ, Cappuzzo F, Ligorio C, Damiani S, Field J, Solberg S, Brustugun OT, Lund-Iversen M, Sanger J, Clement JH, Soltermann A, Moch H, Weder W, Solomon B, Soria JC, Validire P, Besse B, Brambilla E, Brambilla C, Lantuejoul S, Lorimier P, Schneider PM, Hallek M, Pao W,

Meyerson M, Sage J, Shendure J, Schneider R, Buttner R, Wolf J, Nurnberg P, Perner S, Heukamp LC, Brindle PK, Haas S, Thomas RK (2012) Integrative genome analyses identify key somatic driver mutations of small-cell lung cancer. Nat Genet 44 (10):1104–1110. https://doi.org/10.1038/ng.2396

82. Cancer Genome Atlas N (2012) Comprehensive molecular characterization of human colon and rectal cancer. Nature 487 (7407):330–337. https://doi.org/10.1038/nature11252
83. Cancer Genome Atlas Research N (2013) Comprehensive molecular characterization of clear cell renal cell carcinoma. Nature 499 (7456):43–49. https://doi.org/10.1038/nature12222
84. Cancer Genome Atlas Research N (2014) Comprehensive molecular characterization of gastric adenocarcinoma. Nature 513 (7517):202–209. https://doi.org/10.1038/nature13480
85. Cancer Genome Atlas Research N (2014) Comprehensive molecular characterization of urothelial bladder carcinoma. Nature 507 (7492):315–322. https://doi.org/10.1038/nature12965
86. Cancer Genome Atlas N (2015) Comprehensive genomic characterization of head and neck squamous cell carcinomas. Nature 517 (7536):576–582. https://doi.org/10.1038/nature14129
87. Cancer Genome Atlas Research N (2008) Comprehensive genomic characterization defines human glioblastoma genes and core pathways. Nature 455(7216):1061–1068. https://doi.org/10.1038/nature07385
88. Cancer Genome Atlas Research N (2012) Comprehensive genomic characterization of squamous cell lung cancers. Nature 489 (7417):519–525. https://doi.org/10.1038/nature11404
89. Akbani R, Ng PK, Werner HM, Shahmoradgoli M, Zhang F, Ju Z, Liu W, Yang JY, Yoshihara K, Li J, Ling S, Seviour EG, Ram PT, Minna JD, Diao L, Tong P, Heymach JV, Hill SM, Dondelinger F, Stadler N, Byers LA, Meric-Bernstam F, Weinstein JN, Broom BM, Verhaak RG, Liang H, Mukherjee S, Lu Y, Mills GB (2014) A pan-cancer proteomic perspective on The Cancer Genome Atlas. Nat Commun 5:3887. https://doi.org/10.1038/ncomms4887
90. Ciriello G, Gatza ML, Beck AH, Wilkerson MD, Rhie SK, Pastore A, Zhang H, McLellan M, Yau C, Kandoth C, Bowlby R, Shen H, Hayat S, Fieldhouse R, Lester SC, Tse GM, Factor RE, Collins LC, Allison KH, Chen YY, Jensen K, Johnson NB, Oesterreich S, Mills GB, Cherniack AD, Robertson G, Benz C, Sander C, Laird PW, Hoadley KA, King TA, Network TR, Perou CM (2015) Comprehensive molecular portraits of invasive lobular breast cancer. Cell 163(2):506–519. https://doi.org/10.1016/j.cell.2015.09.033
91. Cancer Genome Atlas N (2012) Comprehensive molecular portraits of human breast tumours. Nature 490(7418):61–70. https://doi.org/10.1038/nature11412
92. Drake JM, Paull EO, Graham NA, Lee JK, Smith BA, Titz B, Stoyanova T, Faltermeier CM, Uzunangelov V, Carlin DE, Fleming DT, Wong CK, Newton Y, Sudha S, Vashisht AA, Huang J, Wohlschlegel JA, Graeber TG, Witte ON, Stuart JM (2016) Phosphoproteome integration reveals patient-specific networks in prostate cancer. Cell 166 (4):1041–1054. https://doi.org/10.1016/j.cell.2016.07.007
93. Cancer Genome Atlas Research N, Weinstein JN, Collisson EA, Mills GB, Shaw KR, Ozenberger BA, Ellrott K, Shmulevich I, Sander C, Stuart JM (2013) The Cancer Genome Atlas Pan-Cancer analysis project. Nat Genet 45 (10):1113–1120. https://doi.org/10.1038/ng.2764
94. Neapolitan R, Horvath CM, Jiang X (2015) Pan-cancer analysis of TCGA data reveals notable signaling pathways. BMC Cancer 15:516. https://doi.org/10.1186/s12885-015-1484-6
95. Ruau D, Dudley JT, Chen R, Phillips NG, Swan GE, Lazzeroni LC, Clark JD, Butte AJ, Angst MS (2012) Integrative approach to pain genetics identifies pain sensitivity loci across diseases. PLoS Comput Biol 8(6): e1002538. https://doi.org/10.1371/journal.pcbi.1002538
96. Liu P, Sanalkumar R, Bresnick EH, Keles S, Dewey CN (2016) Integrative analysis with ChIP-seq advances the limits of transcript quantification from RNA-seq. Genome Res 26(8):1124–1133. https://doi.org/10.1101/gr.199174.115
97. Knouf EC, Garg K, Arroyo JD, Correa Y, Sarkar D, Parkin RK, Wurz K, O'Briant KC, Godwin AK, Urban ND, Ruzzo WL, Gentleman R, Drescher CW, Swisher EM, Tewari M (2012) An integrative genomic approach identifies p73 and p63 as activators of miR-200 microRNA family transcription. Nucleic Acids Res 40(2):499–510. https://doi.org/10.1093/nar/gkr731

98. Yan Z, Shah PK, Amin SB, Samur MK, Huang N, Wang X, Misra V, Ji H, Gabuzda D, Li C (2012) Integrative analysis of gene and miRNA expression profiles with transcription factor-miRNA feed-forward loops identifies regulators in human cancers. Nucleic Acids Res 40(17):e135. https://doi.org/10.1093/nar/gks395
99. Berghoff BA, Konzer A, Mank NN, Looso M, Rische T, Forstner KU, Kruger M, Klug G (2013) Integrative "omics"-approach discovers dynamic and regulatory features of bacterial stress responses. PLoS Genet 9(6): e1003576. https://doi.org/10.1371/journal.pgen.1003576
100. Kim M, Rai N, Zorraquino V, Tagkopoulos I (2016) Multi-omics integration accurately predicts cellular state in unexplored conditions for Escherichia coli. Nat Commun 7:13090. https://doi.org/10.1038/ncomms13090
101. Meng C, Helm D, Frejno M, Kuster B (2016) moCluster: identifying joint patterns across multiple omics data sets. J Proteome Res 15 (3):755–765. https://doi.org/10.1021/acs.jproteome.5b00824
102. Wang B, Mezlini AM, Demir F, Fiume M, Tu Z, Brudno M, Haibe-Kains B, Goldenberg A (2014) Similarity network fusion for aggregating data types on a genomic scale. Nat Methods 11(3):333–337. https://doi.org/10.1038/nmeth.2810
103. Shi Q, Zhang C, Peng M, Yu X, Zeng T, Liu J, Chen L (2017) Pattern fusion analysis by adaptive alignment of multiple heterogeneous omics data. Bioinformatics. https://doi.org/10.1093/bioinformatics/btx176
104. Lee CH, Alpert BO, Sankaranarayanan P, Alter O (2012) GSVD comparison of patient-matched normal and tumor aCGH profiles reveals global copy-number alterations predicting glioblastoma multiforme survival. PLoS One 7(1):e30098. https://doi.org/10.1371/journal.pone.0030098
105. Xiao X, Moreno-Moral A, Rotival M, Bottolo L, Petretto E (2014) Multi-tissue analysis of co-expression networks by higher-order generalized singular value decomposition identifies functionally coherent transcriptional modules. PLoS Genet 10(1): e1004006. https://doi.org/10.1371/journal.pgen.1004006
106. Kersey PJ, Staines DM, Lawson D, Kulesha E, Derwent P, Humphrey JC, Hughes DS, Keenan S, Kerhornou A, Koscielny G, Langridge N, McDowall MD, Megy K, Maheswari U, Nuhn M, Paulini M, Pedro H, Toneva I, Wilson D, Yates A, Birney E (2012) Ensembl genomes: an integrative resource for genome-scale data from non-vertebrate species. Nucleic Acids Res 40 (Database issue):D91–D97. https://doi.org/10.1093/nar/gkr895
107. Gao J, Aksoy BA, Dogrusoz U, Dresdner G, Gross B, Sumer SO, Sun Y, Jacobsen A, Sinha R, Larsson E, Cerami E, Sander C, Schultz N (2013) Integrative analysis of complex cancer genomics and clinical profiles using the cBioPortal. Sci Signal 6(269):pl1. https://doi.org/10.1126/scisignal.2004088
108. He S, He H, Xu W, Huang X, Jiang S, Li F, He F, Bo X (2016) ICM: a web server for integrated clustering of multi-dimensional biomedical data. Nucleic Acids Res 44(W1): W154–W159. https://doi.org/10.1093/nar/gkw378
109. Xia J, Fjell CD, Mayer ML, Pena OM, Wishart DS, Hancock RE (2013) INMEX—a web-based tool for integrative meta-analysis of expression data. Nucleic Acids Res 41(Web Server issue):W63–W70. https://doi.org/10.1093/nar/gkt338
110. Tuncbag N, McCallum S, Huang SS, Fraenkel E (2012) SteinerNet: a web server for integrating 'omic' data to discover hidden components of response pathways. Nucleic Acids Res 40(Web Server issue): W505–W509. https://doi.org/10.1093/nar/gks445
111. Ovaska K, Laakso M, Haapa-Paananen S, Louhimo R, Chen P, Aittomaki V, Valo E, Nunez-Fontarnau J, Rantanen V, Karinen S, Nousiainen K, Lahesmaa-Korpinen AM, Miettinen M, Saarinen L, Kohonen P, Wu J, Westermarck J, Hautaniemi S (2010) Large-scale data integration framework provides a comprehensive view on glioblastoma multiforme. Genome Med 2(9):65. https://doi.org/10.1186/gm186
112. Krasnov GS, Dmitriev AA, Melnikova NV, Zaretsky AR, Nasedkina TV, Zasedatelev AS, Senchenko VN, Kudryavtseva AV (2016) CrossHub: a tool for multi-way analysis of The Cancer Genome Atlas (TCGA) in the context of gene expression regulation mechanisms. Nucleic Acids Res 44(7):e62. https://doi.org/10.1093/nar/gkv1478
113. Yu X, Li G, Chen L (2014) Prediction and early diagnosis of complex diseases by edge-network. Bioinformatics 30(6):852–859. https://doi.org/10.1093/bioinformatics/btt620
114. Zhang Q, Burdette JE, Wang JP (2014) Integrative network analysis of TCGA data for ovarian cancer. BMC Syst Biol 8:1338. https://doi.org/10.1186/s12918-014-0136-9

115. Zhu R, Zhao Q, Zhao H, Ma S (2016) Integrating multidimensional omics data for cancer outcome. Biostatistics 17(4):605–618. https://doi.org/10.1093/biostatistics/kxw010

116. Wang XV, Verhaak RG, Purdom E, Spellman PT, Speed TP (2011) Unifying gene expression measures from multiple platforms using factor analysis. PLoS One 6(3):e17691. https://doi.org/10.1371/journal.pone.0017691

Check for updates

Chapter 8

The Reconstruction and Analysis of Gene Regulatory Networks

Guangyong Zheng and Tao Huang

Abstract

In post-genomic era, an important task is to explore the function of individual biological molecules (i.e., gene, noncoding RNA, protein, metabolite) and their organization in living cells. For this end, gene regulatory networks (GRNs) are constructed to show relationship between biological molecules, in which the vertices of network denote biological molecules and the edges of network present connection between nodes (Strogatz, Nature 410:268–276, 2001; Bray, Science 301:1864–1865, 2003). Biologists can understand not only the function of biological molecules but also the organization of components of living cells through interpreting the GRNs, since a gene regulatory network is a comprehensively physiological map of living cells and reflects influence of genetic and epigenetic factors (Strogatz, Nature 410:268–276, 2001; Bray, Science 301:1864–1865, 2003). In this paper, we will review the inference methods of GRN reconstruction and analysis approaches of network structure. As a powerful tool for studying complex diseases and biological processes, the applications of the network method in pathway analysis and disease gene identification will be introduced.

Key words Gene regulatory network, Network reconstruction, Module detection, Pathway analysis, Disease gene identification

1 Introduction

In general, a gene regulatory network is established with a reverse engineering strategy, in which gene expression data is utilized as input and topology structure of network is generated as output [2, 3]. With the development of high-throughput technology, gene expression data is accumulated with an unprecedented speed and thus provides sufficient source data for GRN reconstruction. Nowadays, gene expression data can be collected from public database, for example, the Gene Expression Omnibus (GEO) database (www.ncbi.nlm.nih.gov/geo), a comprehensive microarray data repository [4]; the Sequence Read Archive (SRA) database (www.ncbi.nlm.nih.gov/sra), a data warehouse storing next-generation sequencing data [5]; and the Expression Atlas (www.ebi.ac.uk/

Tao Huang (ed.), *Computational Systems Biology: Methods and Protocols*, Methods in Molecular Biology, vol. 1754, https://doi.org/10.1007/978-1-4939-7717-8_8,

gxa), an integrative expression database providing differential and baseline expression information [6]. On the other hand, the gene expression data can be produced de novo through high-throughput omics methods, such as microarray technology, RNA-seq assay, and RT-PCR experiment. Please keep in mind that sample size of expression data is an important consideration for GRN reconstruction since various inference methods have different size requirement. In the following sections, we will show how these network methods work and how they can be applied in exploring disease mechanism.

2 Inference Methods of GRN Reconstruction

Recently, enormous network inference methods have been developed in computational biology field. These methods can be grouped into two categories, one for static network and another for dynamic network. In here, the static network means a GRN is constructed without limitation of spatial and temporal conditions, while a dynamic network describes a GRN under a spatiotemporal condition.

2.1 Methods for Static Networks

2.1.1 Information-Theoretic Models

A particular advantage of information-theoretic model is its minimal mathematical assumptions of network reconstruction [7, 8]. However, an information-theoretic model can only provide an undirected network commonly. For this model, a correlated measurement is calculated to capture correlation between genes for network inference. The most popular correlated measurements are the Pearson's correlation coefficient (PCC) and the Spearman's rank correlation coefficient (SCC) [9, 10]. While the former can detect linear correlation, the latter is suitable to nonlinear correlation inspection. The PCC and SCC measurements between genes X and Y can be expressed as follows.

$$\begin{aligned} \mathrm{PCC(X,Y)} &= \frac{\mathrm{cov}(X,\Upsilon)}{\sigma_X \sigma_\Upsilon} \\ &= \frac{\sum_{i=1}^{n} (X_i - \bar{X})(\Upsilon_i - \bar{\Upsilon})}{\sqrt{\sum_{i=1}^{n} (X_i - \bar{X})^2} \sqrt{\sum_{i=1}^{n} (\Upsilon_i - \bar{\Upsilon})^2}} \\ \mathrm{SCC(X,Y)} &= \frac{\mathrm{cov}(R_x, R_y)}{\sigma_{Rx}\sigma_{Ry}} \end{aligned} \tag{1}$$

where n is the number of experiment and i denotes the ith expression value of a gene, $\mathrm{cov}(X,\Upsilon)$ is covariance between genes X and Y, $\mathrm{cov}(R_x,R_y)$ is covariance between rank variables X and Υ, σ_X and σ_Υ

are standard deviations of genes X and Y, and σ_{Rx} and σ_{Ry} are standard deviations of rank variables *X* and *Y*.

In the recent years, the Gini correlation coefficient (GCC) measurement is used to detect correlation of genes for network reconstruction, which can compensate for the shortcomings of the PCC and SCC measurements and thus provide more reliable correlation of genes [11]. The GCC measurement between genes X and Y is described by the following equations.

$$\begin{aligned} \mathrm{GCC}(\mathrm{X},\mathrm{Y}) &= \frac{\sum_{i=1}^{n}(2i-n-1)\cdot v_x(i,y)}{\sum_{i=1}^{n}(2i-n-1)\cdot v_x(i,x)} \\ \mathrm{GCC}(\mathrm{Y},\mathrm{X}) &= \frac{\sum_{i=1}^{n}(2i-n-1)\cdot v_y(i,x)}{\sum_{i=1}^{n}(2i-n-1)\cdot v_y(i,y)} \end{aligned} \tag{2}$$

where n is the number of experiment, $v_x(i, x)$ is the ith expression value of gene X stored in an increasing order, $v_x(i, y)$ is the corresponding expression value of gene X in the gene pair (X, Y) for the ith expression value of gene Y stored in an increasing order, while $v_y(i, x)$ and $v_y(i, y)$ are defined similarly to $v_x(i, y)$ and $v_x(i, x)$, respectively. According to Eq. 2, the GCC measurement can be considered as differences between two curves weighted by the information deriving from the rank order of gene expression data. In practice, the GCC measurement between genes X and Y is designed as the higher value among GCC(X,Y) and GCC(Y, X).

In addition, the measurement of mutual information (MI) is also utilized to calculate correlation between genes widely. For the MI measurement, it can detect nonlinear and non-monotonic relationship [8, 10, 12]. The mutual information between two genes X and Y is defined as the following equation based on the information theory.

$$\mathrm{MI}(\mathrm{X},\mathrm{Y}) = \sum_{y\in Y}\sum_{x\in X} p(x,y)\log\left(\frac{p(x,y)}{p(x)p(y)}\right) \tag{3}$$

where $p(x,y)$ is the joint probability distribution of X and Y $p(x)$ and $p(y)$ are the marginal probability distribution of X and Y, respectively. According to information theory, assuming that the gene X is similar to gene Z, which is similar to gene Y simultaneously, then a similar conclusion between genes X and Y is drawn easily. Actually, whether the genes X and Y are similar or not should consider the gene Z. Therefore, in the recent years, the conditional mutual information (CMI) measurement is proposed to reduce false-positive rate of the MI measure in correlation computing [13, 14]. The

CMI measurement between genes X and Y given the gene Z as a condition is defined as follows.

$$\mathrm{CMI}(\mathrm{X},\mathrm{Y}|\mathrm{Z}) = \sum_{x\in X,\, y\in Y,\, z\in Z} p(x,y,z)\log\frac{p(x,y|z)}{p(x|z)p(y|z)} \tag{4}$$

where $p(x, y, z)$ are joint probability distribution of gene triple (X, Y, Z) while $p(x|z)$, $p(y|z)$, and $p(x,y|z)$ are conditional probabilities of genes X and Y and gene pair (X,Y) given gene Z as a condition. According to Eq. 4, the CMI measurement can inspect whether there is a direct correlation between genes X and Y and thus enhance the accuracy of relationship detection for gene pairs. However, when the expression pattern of gene X or Y is strongly similar to gene Z, performance of the CMI measurement is decreased dramatically. So a new measurement of partial mutual information (PMI) is proposed to refine the CMI measurement [15], which is presented as follows.

$$\begin{aligned}\mathrm{PMI}(\mathrm{X},\mathrm{Y}|\mathrm{Z}) &= \sum_{x\in \mathrm{X},\, y\in \mathrm{Y},\, z\in \mathrm{Z}} p(x,y,z)\log\frac{p(x,y|z)}{p^*(x|z)p^*(y|z)}\\ p^*(x|z) = \sum_{y\in \mathrm{Y}} p(x|z,y)p(y),&\quad p^*(y|z) = \sum_{x\in \mathrm{X}} p(y|z,x)p(x)\end{aligned} \tag{5}$$

where the $p(x,y,z)$, $p(x|z)$, $p(y|z)$, and $p(x,y|z)$ have the same definition with the CMI measurement. Numerical studies of simulated and realistic data demonstrate that the PMI does have higher performance compared to the CMI measurement in relationship detection.

2.1.2 Probabilistic Graphical Models

A gene regulatory network is presented as a graph model $G = \langle V, E\rangle$, where V stands for genes and E denotes links between genes [2, 16]. Assuming n is observation times of experiment and m is the total number of genes, then the expression data (D) can be presented as an n × m matrix ($D = (d_1, d_2 \ldots d_m)$). As for the problem of GRN reconstruction, it is equivalent to infer an optimize model (G) using the matrix data (D). In the following subsections, we will introduce the Bayesian network model and Gaussian graphical model for the network inference problem.

Bayesian network model is a directed acyclic graph (DAG), where an edge from gene X to Y indicates a regulation from X to Y [17]. In other words, the gene X is a parent node, and the gene Y is a target node of gene X. For this model, the probability distribution of a network is generally factored in terms of the conditional distributions of each node variable given its parents.

$$P(\mathrm{D}) = \Pi_{j=1}^{m} p(\mathrm{d_j}|\mathrm{Pa}(\mathrm{d_j})) \tag{6}$$

where d_j presents expression profile of gene j and $Pa(d_j)$ are parent nodes of gene j. For the GRN inference problem, this is done by maximizing the probability of P(G|D), which means Bayesian model identifies the optimal network topology that best explains the expression data. Because the number of possible network topologies increases with the number of genes in an exponential manner, it is not feasible to search for all possible networks. Therefore, some heuristic algorithms, like genetic algorithm and evolutionary algorithm, have been proposed for Bayesian network inference [18, 19]. One limitation of Bayesian models is that they can't present loop motif for networks since they are directed acyclic graphs. However, feedback loop motifs are prevalent in biological systems.

In Gaussian graphical model, the gene expression data (D) is assumed having a Gaussian (normal) distribution, and relationship between genes is expressed as conditional dependencies through calculating the partial correlation measurement [20]. Given the genes X and Y and their k correlated variables Z ($Z_1, Z_2 \ldots Z_k$) with covariance matrix W, then the relationship between genes X and Y, termed *Pxy.z*, can be computed with the following equations.

$$r_x = \mathrm{X} - \mathrm{W}\cdot\mathrm{Z}, \quad r_y = \mathrm{Y} - \mathrm{W}\cdot\mathrm{Z}$$

$$Pxy.z = \frac{\operatorname{cov}(r_x, r_y)}{\sigma r_x \sigma r_y} = \frac{n\sum_{i=1}^{n} r_{x,i} r_{y,i} - \sum_{i=1}^{n} r_{x,i} \sum_{i=1}^{n} r_{y,i}}{\sqrt{n\sum_{i=1}^{n} r_{x,i}^2 - \left(\sum_{i=1}^{n} r_{x,i}\right)^2}\sqrt{n\sum_{i=1}^{n} r_{y,i}^2 - \left(\sum_{i=1}^{n} r_{y,i}\right)^2}} \tag{7}$$

where r_x and r_y are residual variables X and Y given the k-dimension Z as controlling variables, n is the number of experiment, and $r_{x,i}$ and $r_{y,i}$ are the ith expression value of genes X and Y, respectively. Please keep in mind that when the number of genes greatly exceeds the number of experiments, the covariance matrix W can't be estimated certainly. Therefore, some regularized regression methods, such as the LASSO, two-stage adaptive LASSO, and ridge regression approaches, have been developed to help estimate the covariance matrix correctly [21, 22], which promotes the application of the Gaussian graphical model in network inference problem.

2.1.3 Integrative Inferring Models

Each inferring method has its strengths and weaknesses because of various mathematic assumptions, which lead to different bias of network reconstruction. For example, the information-theoretic models can detect feedback loop, while the probabilistic graphical models can't. Whereas the Bayesian model gives directionality of each links, information-theoretic models do not. Therefore, combining different inferring models can provide more reliable gene regulatory networks. For the integrative inferring models, different inferring methods are applied to reconstruct networks firstly. Then consensus gene connections from different networks are identified based on p-value calculated by the Fisher's inverse $\times^2$ test. Performance investigation by the Dialogue on Reverse Engineering Assessment and Methods (DREAM) project shows that the combining approach outperforms each of single methods [23].

2.2 Methods for Dynamic Networks

In previous section, we have discussed inferring methods for static networks, which are not limited to a certain physiological condition. Actually, many gene regulatory networks are conducted in a specific biological system with a certain physiological condition, in which gene expression data is generated at some discrete time points dynamically. In this scenario, GRNs inferred with these expression data are dynamic and present stochastic characteristics for a specific biological system. For dynamic GRNs, two types of equations are adopted to depict characteristics of these networks commonly. One type of formula is expression equation, which presents expression function for genes under a certain physiological condition (regarded as a state in mathematics) [24]. Another type of formula is regulation equation, which provides regulatory links between genes. According to mathematical forms of expression and regulation function, the dynamic networks can be categorized into linear and nonlinear state-space models.

2.2.1 Linear State-Space Models

In these models, the expression and regulation functions are depicted with the simplest linear function as follows.

$$
\begin{aligned}
x_i^*(t) &= x_i(t) + u_i(t), \quad i = 1, 2, \ldots m, \quad t = 1, 2, \ldots k \\
x_i(t+1) &= \sum_{j=1}^{m} a_{i,j} x_j(t) + v_i(t)
\end{aligned}
\tag{8}
$$

where m is the number of total genes (these genes are measured at k discrete time points); $x_i(t)$ is the actual expression value of the ith gene at time t, while $x_i^*(t)$ is corresponding measure value of the ith gene at time t; $u_i(t)$ and $v_i(t)$ are the measurement and system noise of the ith gene at time t; and $a_{i,j}$ presents relationship between genes i and j. Through an expectation maximization algorithm, both the model parameters (i.e., the A matrix and U, V vectors) and actual expression value X can be estimated effectively [25].

2.2.2 Nonlinear State-Space Models

It is imperative to inflect nonlinear effects on the expression and regulation function for dynamic GRNs since there is complex regulatory relationship between genes. Particularly, the sigmoid function is employed to capture these complex regulations between gene pairs.

$$x_i^*(t) = x_i(t) + u_i(t), \quad i = 1, 2, \ldots m, \quad t = 1, 2, \ldots k$$
$$x_i(t+1) = \sum_{j=1}^{m} b_{i,j} f(x_j(t)) + v_i(t), f(x_j(t)) = \frac{1}{1 + e^{-x_j}} \tag{9}$$

where $b_{i,j}$ presents nonlinear relationship between genes i and j and the other parameters are the same with those of the linear state-space models. One popular way of solving these nonlinear state-space models is the extended Kalman filter (EKF) method, which gives an approximating solution to the nonlinear system through first-order linear approximation. In addition, the variants of Kalman filter algorithm such as the cubature Kalman filter (CKF), unscented Kalman filter (UKF), and particle filter algorithm are proposed for solving the problem of the nonlinear state-space models [25].

2.2.3 Time-Delayed Dependency Models

In biological system, there exist time-delayed phenomena of regulation from regulators to downstream targets due to time requirement of the process of transcription and translation. Therefore, the time-delayed dependency models are developed to update regulation functions of the linear and nonlinear state-space models [26].

$$x_i(t+1) = \sum_{j=1}^{m} a_{i,j} x_j(t) + \sum_{p=1}^{q} c_{i,p} x_p(t-\tau) + v_i(t)$$
$$x_i(t+1) = \sum_{j=1}^{m} b_{i,j} f(x_j(t)) + \sum_{p=1}^{q} c_{i,p} x_p(t-\tau) + v_i(t), f(x_j(t)) = \frac{1}{1 + e^{-x_j}} \tag{10}$$

where q is the total numbers of regulators in the system and τ is the time-delayed value. In practice, the value of τ can be determined by the Akaike's information criterion (AIC) so as to avoid over-fitting the models.

3 Analysis Approaches of Network Topology Structure

After building up a gene regulatory network, the topology analysis is conducted to reveal structure of the network and thus give more insights into property of a biological system, which can help biologist explore function of interested genes and investigate mechanism of the system [27, 28]. Here, we will give some basic concept of network topology, and some structure analysis based on these concepts will be introduced in the following sections.

In terms of mathematics, the network is a graph consisting of nodes and edges. The former are molecules of a biological system, while the later are links between these molecules. The most elementary concept of nodes in network is degree (n), which describes the number of links a node connected to others. Specially, there are incoming (n_{in}) and outgoing (n_{out}) degrees for a node in a directed network. The incoming degree gives the number of links pointed to the target node. The outgoing degree presents the number of connections from the target node to others. The degree distribution ($p(n)$) gives probability of a node having n links. For many gene regulatory networks, they have modular structure, and genes of these networks take effect in a cluster manner [16, 29]. For these networks, they are regarded as scale-free networks in mathematics, and their degree distributions are in accordance with a power-law distribution, $p(n) \sim n^{-\gamma}$, where γ the is the degree exponent. The smaller the value of γ, the higher probability of modular structure the network has. Another elementary concept of nodes in network is path length, which gives the number of links needed to pass through for traveling between two nodes. The shortest path length (l) presents the minimum value of all possible paths between two nodes. The value of shortest path length from node A to B is equivalent to that of B to A in an undirected network. However, the value of shortest path length from node A to B may be different to that of B to A in a directed network. In general, the clustering coefficient is adopted to characterize the tendency of nodes in network to form clusters and modules [16]. For an undirected network, assuming a node A has n links to other nodes, then its clustering coefficient is defined as $c_a = 2m_a/n(n-1)$, where m_a is the number of links connecting n neighbors of node A to each other. The mean value of clustering coefficient for nodes with n links, termed as $c(n)$, depicts clustering property of the network.

3.1 Identification of Hub Genes in Networks

In many GRNs, they present scale-free characteristics, and most links are connected to a few nodes, which determine structure property of the network in a certain degree [30]. These genes are regarded as central nodes or hubs. It is imperative to identify hub genes in GRNs since these genes play important roles in regulatory patterns of networks. In practice, centrality of a node is defined by the betweenness measure, which gives the number of shortest paths passing through the node. The higher the value of betweenness, the more important the node is. As term of GRNs, the hub genes encode essential regulators, like transcription factors, which regulate expression level of many downstream target genes.

3.2 Identification of Gene Modules and Motifs in Networks

In most GRNs, genes work together in a modular manner to achieve a distinct function for internal and external stimuli. In a module or cluster, there are high connections between nodes and can be reduced to many triangle sets [16, 31, 32]. A high triangle density can be expressed by a big value of clustering coefficient. Therefore, biologist can estimate modular level of a network through calculating the mean value of clustering coefficient. After identifying hub genes in GRNs, the functional partners of these genes can be revealed through module analysis since they work together with hubs. In the past 20 years, some module mining approaches, such as MCL [33], NEMO [34], and MINE [35], have been proposed to identify highly interconnected gene sets. After obtaining modules of GRNs, an in-depth analysis is carried out to find functional motifs, which are overrepresented patterns when compared to random networks with equivalent size. Functional motifs are elementary unit of modules, and they are closely correlated to regulatory framework of GRNs, for example, bi-fan regulating motif, feed-forward loop regulatory unit, and dense overlapping regulons [31, 36]. In the past decade, some motif detection tools, such as FANMOD [37], MFINDER [31], and NetMODE [38], have been developed in the research field of computational biology. After identifying modules and motifs, gene ontology (GO) and pathway enrichment analysis are conducted to help biologist understand organization of genes within modules and motifs and their combinatory functional effect, which promotes mechanism investigation of cellular process in biological system.

3.3 Interpretation of Hierarchical Structure in Networks

For some GRNs, they have hierarchical structure, and they can be divided into topological frameworks in multilevels [29]. As for a network, topology modules, containing functional motifs, are elementary structure units, which are assembled into larger module frameworks. Next, these frameworks are combined in a hierarchical fashion to form the complete network. For these networks, they have not only scale-free property but also hierarchical characteristics, in which the mean value distribution of clustering coefficient is proportional to the reciprocal value of links n ($c(n) \sim n^{-1}$), while in random and scale-free networks, the mean value distribution of clustering coefficient is independent to links. According to the hierarchical characteristics of gene regulatory networks, behavior of the whole biological system can be predicted de novo through interpreting topological structures in multilevel of GRNs and explaining interactions of molecules within a hierarchical scale over the full range of cellular compartments [39].

3.4 Comparative Investigation of Networks

After reconstructing the gene regulatory networks, comparison study of GRNs across different species is carried out to reveal conservative subnetworks in evolution, which is thought to provide more insights into evolutional mechanism than gene sequence comparison study [40, 41]. These conservative subnetworks are essential modules, which are involved in many significant cellular processes and thought to be response units of internal and external stimuli in living cell. On the other hand, comparative investigation of GRNs between different organs for an organism is conducted to reveal organ-specific subnetworks, which are regarded as key factors controlling morphological formation and playing important roles in the organ [42].

4 Identification of Disease Gene

One of the key applications of network analysis is to identify novel disease genes based on known disease genes on the network. There are many different methods that have been proposed. Many of them were based on the principle called guilt by association [43]. The basic assumption is that the genes have similar functions with their neighbors on the network. It is reasonable in most scenarios. Based on this idea, the interaction neighbors of known disease genes are very likely to be also disease genes. Actually, the regulatory modules on network confirm the guilt-by-association principle [44]. In practice, guilt-by-association-based neighbor counting [45] is widely used. But the disadvantage of guilt-by-association methods is very obvious when the number of reported disease genes is too small and they locate far from each other on network. At that time, the guilt-by-association methods will not be able to identify possible novel disease genes. Therefore, two new methods are introduced in the following sections.

4.1 Random Walk with Restart (RWR)

Random walk with restart (RWR) [46–51] algorithm simulates a walker who starts from the nodes of reported disease genes and moves to its randomly chosen neighbors on the network at each step [48]. After many steps of walks, the procedures will be steady. Based on the final probability of the walker's walks to each node on the network, the highly possible candidate disease genes are identified.

It works as follows:

For a gene regulatory network G = (V, E) comprised of a set of genes V and a set of interactions E, we represent it by an $n \times n$ adjacency matrix A, where n is the number of genes. The entry at row i and column j is set to 1 if gene i interacts with gene j ; otherwise it is set to 0.

First, adjacency matrix A is column-wise normalized as follows [52–54]:

$$A_{[i,j]}{}' = \frac{A_{[i,j]}}{\sum_{k=1}^{n} A_{[k,j]}} \tag{11}$$

Then, in each step, the state probabilities P_{t+1} at time $t+1$ are calculated as

$$P_{t+1} = (1 - r)A'P_t + rP_0 \tag{12}$$

where P_t is state probabilities at time t, r is the restart probability, A' is the normalized adjacency matrix, and P_0 is the initial state probabilities which is a column vector with $1/m$ for the m known disease genes and to 0 for other genes on the gene regulatory network.

This process is repeated until the difference between two states is small enough to stop.

At last, each gene on the network will be assigned with a probability of being possible disease gene.

Based on RWR algorithm, Kohler et al. [48] developed RWOAG, and Lee et al. [55] developed HumanNet.

4.2 Shortest Path Method

Shortest path-based method has been used for identifying the genetic determinants of longevity [56] and disease genes [57–59].

Dijkstra's algorithm [60] is most widely used to discover the shortest paths between reported disease genes. The genes on the shortest path between known disease genes can only reveal the possible mechanism of disease progression but also indicate possible novel key disease genes. The procedure of Dijkstra's algorithm is as follows [60–62]:

Let $G = (V, E, w)$ be a weighted graph, where V is the set of vertices, E is the set of edges, and w is a function from E to R^+. Suppose u_0 and v_0 are two vertices in G, the shortest path between them can be discovered using the following procedures:

1. Let $S = \{u_0\}$, $\bar{S} = V - \{u_0\}$, $l(u_0) = 0$, and $l(v) = \infty$ for any vertex $v \in S - \{u_0\}$.
2. For each vertex $v \in \bar{S}$ such that $u'v \in E$, where $u' \in S$. If $l(v) \leq l(u') + w(u'v)$, then continue; otherwise, $l(v) = l(u') + w(u'v)$ and Parent$(v) = u'$.
3. Find a vertex $v' \in \bar{S}$ such that $l(v') = \min\{l(v) | v \in \bar{S}\}$.
4. $S = S \cup \{v'\}$ and $\bar{S} = \bar{S} - \{v'\}$.
5. If $v_0 \in S$, then continue; otherwise, return to **step 2**.
6. The label Parent was used to find a shortest path from u_0 to v_0.

4.3 Kth Shortest Path Methods

The Dijkstra's algorithm can only identify the shortest path between two nodes. But sometimes, the second or third shortest path may also include curtail information for understanding disease, especially when the weights on the network are not very accurate. Therefore, finding *K*th shortest paths in the graph G between each pair of genes ($K > =1$) using A* search algorithm [63] is very useful.

Given a weighted graph $G = (V, E, w)$, where V is the set of vertices, E is the set of edges, and w is a function from E to $\mathbf{R}^+$, the *K*th shortest path problem is to find the k shortest paths between two nodes s and t in a weighted graph. A* search algorithm works similarly with the Dijkstra's algorithm but adds an evaluation function $\widehat{f}(v) = \widehat{g}(v) + \widehat{h}(v)$, where $\widehat{g}(v)$ is the cost of the path from s to v with minimum cost so far found by A* and $\widehat{h}(v)$ is the estimate of the cost of an optimal path from v to t [64], to guide the search.

The evaluation function reduces the searching time, and if $\widehat{h}(v)$ is any lower bound on the cost of an optimal path from v to t, A* can find the optimal path from s to t [64]. Therefore, if $\widehat{h}(v)$ is set to be the cost of an optimal path from v to t, which can be found in advance by the Dijkstra's algorithm, A* search algorithm can still determine the optimal path from s to t.

By combining the Dijkstra's algorithm and A* search algorithm, the k shortest paths between s and t can be enumerated as follows:

1. The shortest path from t to other nodes in the graph is discovered using the Dijkstra's algorithm, and the cost of each path is stored.
2. A* search algorithm is used to search paths from s, with $\widehat{h}(v)$ in the evaluation function to be the value calculated in (1) for each node v in the graph. But, during this step, each node can be visited more than once. When t is visited the j-th time, the corresponding path is the j-th shortest path from s to t.

5 Network Reconstruction and Analysis Software

We introduced many network reconstruction and analysis methods in previous sections. To apply these methods, the easiest way for biological scientists is to use software that has been developed based on various algorithms. Therefore, some representative software for network reconstruction and analysis will be described.

5.1 ARACNE

ARACNE [12, 65] is the most widely used network reconstruction software based on information theory. It is proposed by Butte and Kohane [66] and uses mutual information to measure genes' expression pattern similarity. It requires each sample to be independent from the others. Therefore, it is not suitable for analyzing time-course gene expression profiles. The steady-state gene expression profile is the ideal input data. This software is easy to use and has graphical user interface (GUI). It has been integrated into geWorkbench.

5.2 WGCNA

WGCNA is a popular co-expression network reconstruction software. It calculates the co-expression coefficient between genes and uses the scale-free property as criteria to determine the cutoff of correlation coefficient. It not only constructs the co-expression network but also identifies the co-expressed modules which may be corresponding to certain pathways or biological processes. Therefore, usually, researchers do functional enrichment on each module and try to investigate the biological functions of the genes within the module. The module with biological significance will be selected for further studies.

5.3 bnlearn

Both the ARACNE and WGCNA software can only build up networks without directions. Such networks can be fast built and reveal some global network features, such as scale-free or densely connected modules, but they will not be able to tell which genes are the upstream regulators and which genes are the downstream targets. Therefore, after the genome-wide network is reconstructed by ARACNE or WGCNA, important modules will be selected to build refined regulatory networks with directions. Bayesian method and software, such as bnlearn [67], are able to infer the small-scale causal network in which the regulators and targets can be clearly seen. But the computational complexity is too high; usually only a dozen of genes can be handled. Therefore, the Bayesian method is not suitable for genome-wide network reconstruction, despite its network that is more informative. Usually, the genome-wide network methods, such as ARACNE or WGCNA, are used first to get the whole picture, and then the selected modules are further investigated using Bayesian method to get the refined regulatory picture [68].

5.4 GeneReg

GeneReg [69] is an R package for time delay gene regulatory network construction from short-time course gene expression profiles. The basis is time delay linear regression model. Different from ordinary linear regression model, this model has two parameters: both time delay and the regulation coefficient. Time delay is the time that the change of regulator's gene expression is transmitted and causes the change of target gene expression. It is difficult to measure but extremely important for understanding basic biological processes, such as cell cycle and signal cascade.

5.5 MCL

MCL [33] is a cluster extraction software for graph analysis, which can be utilized to identify modules of networks. As for the software, it is an implementation of the Markov Cluster Algorithm, which is based on simulation of stochastic flow in graphs. In practice, the interaction matrix between genes of a network is used as input data of the software, while modules of the network are produced as output. In addition, some properties of a network, such as degree, cluster coefficient, and betweenness of nodes, are presented also by the software.

6 The Network Databases

The network databases store either experimentally validated regulations/interactions or predicted regulations/interactions. They are important for evaluating network reconstruction methods and applying network analysis when there is not enough data to construct the disease- or condition-specific network. Several widely used network databases will be introduced.

6.1 STRING

STRING (Search Tool for the Retrieval of Interacting Genes/Proteins) [70] is technically a protein functional association network. But since it stores the networks of most organisms, covers most proteins, and has the largest number of functional associations, it has been widely used. What's more is that each interaction in STRING has a weight, called confidence score, which ranges from 150 to 900. A higher score indicates proteins in this association are more likely to have an actual association. The sources of STRING include genomic context, high-throughput experiments, conserved co-expression, and previous knowledge, such as database or literacy. Such diverse sources of both direct physical and indirect functional linkages between proteins make STRING the most comprehensive network database.

6.2 KEGG

KEGG (Kyoto Encyclopedia of Genes and Genomes) [71] stores high-quality manually curated pathways. The regulations in KEGG not only have directions but also have effects, such as activation or deactivation. Due to its high quality, the biologists use it to generate hypothesis of certain genes and try to add them as the upstream regulators or downstream targets of the known pathway. If their findings are verified, these genes may be included in KEGG pathway. Although the pathways stored in KEGG database are evolving and the number of pathways is growing, KEGG only covers a very small fraction of genes and their regulations. This limits its usage for finding novel mechanism of disease. Advanced network analysis, such as shortest path analysis and RWR, is difficult to be applied on the KEGG network due to its sparseness of regulations.

6.3 ConsensusPathDB

The coverage and quality of network are difficult to balance. To find the best trade-off of these contradict goals, many efforts have been done. For example, ConsensusPathDB [72] collects 12 pathway databases and finds the consensus interactions. It will certainly increase the coverage, but such ensemble approaches still need a lot of calculations and may introduce new errors. It is an open question and needs more efforts, such as high-throughput interaction screening technologies, to generate genome-wide network for different tissues and diseases. The complete and accurate dynamic condition-specific network is the ultimate goal for network studies.

7 Conclusions

The network is a great way to integrate complex omics big data and decipher the underlying mechanisms of many multigene diseases, such as cancers and diabetes. Here, we introduced the popular network reconstruction and analysis methods and software. With these tools, the regulatory pathways can be characterized, the key driver genes or hub genes can be identified, and novel disease genes can be inferred. Overall, the methods in this chapter are wonderful tools for studying complex diseases and biological processes.

References

1. Strogatz SH (2001) Exploring complex networks. Nature 410(6825):268–276. https://doi.org/10.1038/35065725
2. Bray D (2003) Molecular networks: the top-down view. Science 301 (5641):1864–1865. https://doi.org/10.1126/science.1089118
3. Noor A, Serpedin E, Nounou M, Nounou H (2013) Reverse engineering sparse gene regulatory networks using cubature Kalman filter and compressed sensing. Adv Bioinforma 205763. https://doi.org/10.1155/2013/205763
4. Barrett T, Troup DB, Wilhite SE, Ledoux P, Rudnev D, Evangelista C, Kim IF, Soboleva A, Tomashevsky M, Marshall KA, Phillippy KH, Sherman PM, Muertter RN, Edgar R (2009) NCBI GEO: archive for high-throughput functional genomic data. Nucleic Acids Res 37(Database issue):D885–D890. https://doi.org/10.1093/nar/gkn764
5. Zhu Y, Stephens RM, Meltzer PS, Davis SR SRAdb: query and use public next-generation sequencing data from within R. BMC Bioinformatics 14:19. https://doi.org/10.1186/1471-2105-14-19
6. Petryszak R, Keays M, Tang YA, Fonseca NA, Barrera E, Burdett T, Fullgrabe A, Fuentes AM, Jupp S, Koskinen S, Mannion O, Huerta L, Megy K, Snow C, Williams E, Barzine M, Hastings E, Weisser H, Wright J, Jaiswal P, Huber W, Choudhary J, Parkinson HE, Brazma A Expression Atlas update--an integrated database of gene and protein expression in humans, animals and plants. Nucleic Acids Res 44(D1):D746–D752. https://doi.org/10.1093/nar/gkv1045
7. Zhao W, Serpedin E, Dougherty ER (2008) Inferring connectivity of genetic regulatory networks using information-theoretic criteria. IEEE/ACM Trans Comput Biol Bioinform 5 (2):262–274. https://doi.org/10.1109/TCBB.2007.1067
8. Noor A, Serpedin E, Nounou M, Nounou H, Mohamed N, Chouchane L (2013) An overview of the statistical methods used for inferring gene regulatory networks and protein-protein interaction networks. Adv Bioinforma:953814. https://doi.org/10.1155/2013/953814
9. Usadel B, Obayashi T, Mutwil M, Giorgi FM, Bassel GW, Tanimoto M, Chow A, Steinhauser D, Persson S, Provart NJ (2009) Co-expression tools for plant biology: opportunities for hypothesis generation and caveats. Plant Cell Environ 32(12):1633–1651. https://doi.org/10.1111/j.1365-3040.2009.02040.x
10. Nounou M, Nounou H, Serpedin E, Datta A, Huang Y (2013) Computational and statistical approaches for modeling of proteomic and genomic networks. Adv Bioinforma:561968. https://doi.org/10.1155/2013/561968
11. Ma C, Wang X Application of the Gini correlation coefficient to infer regulatory relationships in transcriptome analysis. 160(1):192–Plant Physiol, 203. https://doi.org/10.1104/pp.112.201962
12. Margolin AA, Nemenman I, Basso K, Wiggins C, Stolovitzky G, Dalla Favera R, Califano A (2006) ARACNE: an algorithm for the reconstruction of gene regulatory networks in a mammalian cellular context. BMC Bioinformatics 7(Suppl 1):S7. https://doi.org/10.1186/1471-2105-7-S1-S7
13. Zhang X, Zhao XM, He K, Lu L, Cao Y, Liu J, Hao JK, Liu ZP, Chen L Inferring gene regulatory networks from gene expression data by path consistency algorithm based on conditional mutual information. Bioinformatics 28 (1):98–104. https://doi.org/10.1093/bioinformatics/btr626
14. Zheng G, Xu Y, Zhang X, Liu ZP, Wang Z, Chen L, Zhu XG CMIP: a software package capable of reconstructing genome-wide regulatory networks using gene expression data. BMC Bioinformatics 17(Suppl 17):535. https://doi.org/10.1186/s12859-016-1324-y
15. Zhao J, Zhou Y, Zhang X, Chen L Part mutual information for quantifying direct associations in networks. Proc Natl Acad Sci USA 113 (18):5130–5135. https://doi.org/10.1073/pnas.1522586113
16. Barabasi AL, Oltvai ZN (2004) Network biology: understanding the cell's functional organization. Nat Rev Genet 5(2):101–113. https://doi.org/10.1038/nrg1272
17. Needham CJ, Bradford JR, Bulpitt AJ, Westhead DR (2006) Inference in Bayesian

networks. Nat Biotechnol 24(1):51–53. https://doi.org/10.1038/nbt0106-51

18. Cooper GF (1990) The computational complexity of probabilistic inference using Bayesian belief networks. Artif Intell 42:393–405
19. Pedro Larranag HK, Bielza C, Santana R (2013) A review on evolutionary algorithms in Bayesian network learning and inference tasks. Inf Sci 233:109–125
20. Friedman N (2004) Inferring cellular networks using probabilistic graphical models. Science 303(5659):799–805. https://doi.org/10.1126/science.1094068
21. Menendez P, Kourmpetis YA, ter Braak CJ, van Eeuwijk FA Gene regulatory networks from multifactorial perturbations using graphical Lasso: application to the DREAM4 challenge. PLoS One 5(12):e14147. https://doi.org/10.1371/journal.pone.0014147
22. Kramer N, Schafer J, Boulesteix AL (2009) Regularized estimation of large-scale gene association networks using graphical Gaussian models. BMC Bioinformatics 10:384. https://doi.org/10.1186/1471-2105-10-384
23. Marbach D, Prill RJ, Schaffter T, Mattiussi C, Floreano D, Stolovitzky G Revealing strengths and weaknesses of methods for gene network inference. Proc Natl Acad Sci USA 107(14):6286–6291. https://doi.org/10.1073/pnas.0913357107
24. Noor A, Serpedin E, Nounou M, Nounou HN Inferring gene regulatory networks via nonlinear state-space models and exploiting sparsity. IEEE/ACM Trans Comput Biol Bioinform 9(4):1203–1211. https://doi.org/10.1109/TCBB.2012.32
25. Wang Z, Yang F, Ho DW, Swift S, Tucker A, Liu X (2008) Stochastic dynamic modeling of short gene expression time-series data. IEEE Trans Nanobioscience 7(1):44–55. https://doi.org/10.1109/TNB.2008.2000149
26. Koh C, Wu F-X, Selvaraj G, Kusalik AJ (2009) Using a State-Space Model and Location Analysis to Infer Time-Delayed Regulatory Networks. EURASIP Journal on Bioinformatics and Systems Biology 2009(1):484601
27. Califano A, Butte AJ, Friend S, Ideker T, Schadt E Leveraging models of cell regulation and GWAS data in integrative network-based association studies. Nat Genet 44(8):841–847. https://doi.org/10.1038/ng.2355
28. Marbach D, Costello JC, Kuffner R, Vega NM, Prill RJ, Camacho DM, Allison KR, Kellis M, Collins JJ, Stolovitzky G Wisdom of crowds for robust gene network inference. Nat Methods 9(8):796–804. https://doi.org/10.1038/nmeth.2016
29. Ravasz E, Barabasi AL (2003) Hierarchical organization in complex networks. Phys Rev E Stat Nonlinear Soft Matter Phys 67(2 Pt 2):026112. https://doi.org/10.1103/PhysRevE.67.026112
30. Barrat A, Barthelemy M, Pastor-Satorras R, Vespignani A (2004) The architecture of complex weighted networks. Proc Natl Acad Sci USA 101(11):3747–3752. https://doi.org/10.1073/pnas.0400087101
31. Kashtan N, Itzkovitz S, Milo R, Alon U (2004) Efficient sampling algorithm for estimating subgraph concentrations and detecting network motifs. Bioinformatics 20(11):1746–1758. https://doi.org/10.1093/bioinformatics/bth163
32. Karlebach G, Shamir R (2008) Modelling and analysis of gene regulatory networks. Nat Rev Mol Cell Biol 9(10):770–780. https://doi.org/10.1038/nrm2503
33. Enright AJ, Van Dongen S, Ouzounis CA (2002) An efficient algorithm for large-scale detection of protein families. Nucleic Acids Res 30(7):1575–1584
34. Rivera CG, Vakil R, Bader JS NeMo: Network Module identification in Cytoscape. BMC Bioinformatics 11(Suppl 1):S61. https://doi.org/10.1186/1471-2105-11-S1-S61
35. Rhrissorrakrai K, Gunsalus KCMINE Module identification in networks. BMC Bioinformatics 12:192. https://doi.org/10.1186/1471-2105-12-192
36. Shen-Orr SS, Milo R, Mangan S, Alon U (2002) Network motifs in the transcriptional regulation network of Escherichia Coli. Nat Genet 31(1):64–68. https://doi.org/10.1038/ng881
37. Wernicke S, Rasche F (2006) FANMOD: a tool for fast network motif detection. Bioinformatics 22(9):1152–1153. https://doi.org/10.1093/bioinformatics/btl038
38. Li X, Stones DS, Wang H, Deng H, Liu X, Wang G NetMODE: network motif detection without Nauty. PLoS One 7(12):e50093. https://doi.org/10.1371/journal.pone.0050093
39. Li Y, Pearl SA, Jackson SA Gene networks in plant biology: approaches in reconstruction and analysis. Trends Plant Sci 20(10):664–675. https://doi.org/10.1016/j.tplants.2015.06.013
40. Lynch M (2007) The evolution of genetic networks by non-adaptive processes. Nat Rev Genet 8(10):803–813. https://doi.org/10.1038/nrg2192
41. Crombach A, Hogeweg P (2008) Evolution of evolvability in gene regulatory networks. PLoS

Comput Biol 4(7):e1000112. https://doi.org/10.1371/journal.pcbi.1000112

42. Movahedi S, Van de Peer Y, Vandepoele K Comparative network analysis reveals that tissue specificity and gene function are important factors influencing the mode of expression evolution in Arabidopsis and rice. Plant Physiol 156(3):1316–1330. https://doi.org/10.1104/pp.111.177865
43. Oliver S (2000) Guilt-by-association goes global. Nature 403(6770):601–603. https://doi.org/10.1038/35001165
44. Barabasi AL, Gulbahce N, Loscalzo J (2011) Network medicine: a network-based approach to human disease. Nat Rev Genet 12 (1):56–68. https://doi.org/10.1038/nrg2918
45. Schwikowski B, Uetz P, Fields S (2000) A network of protein-protein interactions in yeast. Nat Biotechnol 18(12):1257–1261. https://doi.org/10.1038/82360
46. Macropol K, Can T, Singh AK (2009) RRW: repeated random walks on genome-scale protein networks for local cluster discovery. BMC bioinformatics 10:283. https://doi.org/10.1186/1471-2105-10-283
47. Li Y, Patra JC (2010) Genome-wide inferring gene-phenotype relationship by walking on the heterogeneous network. Bioinformatics 26 (9):1219–1224. https://doi.org/10.1093/bioinformatics/btq108
48. Kohler S, Bauer S, Horn D, Robinson PN (2008) Walking the interactome for prioritization of candidate disease genes. Am J Hum Genet 82(4):949–958. https://doi.org/10.1016/j.ajhg.2008.02.013
49. Jiang R, Gan M, He P (2011) Constructing a gene semantic similarity network for the inference of disease genes. BMC Syst Biol 5(Suppl 2):S2. https://doi.org/10.1186/1752-0509-5-S2-S2
50. Chen X, Liu MX, Yan GY (2012) Drug-target interaction prediction by random walk on the heterogeneous network. Mol BioSyst 8 (7):1970–1978. https://doi.org/10.1039/c2mb00002d
51. Shi H, Xu J, Zhang G, Xu L, Li C, Wang L, Zhao Z, Jiang W, Guo Z, Li X (2013) Walking the interactome to identify human miRNA-disease associations through the functional link between miRNA targets and disease genes. BMC Syst Biol 7:101. https://doi.org/10.1186/1752-0509-7-101
52. Huang T, Liu C-L, Li L-L, Cai M-H, Chen W-Z, Y-F X, O'Reilly PF, Cai L, He L (2016) A new method for identifying causal genes of schizophrenia and anti-tuberculosis drug-induced hepatotoxicity. Sci Rep 6:32571. https://doi.org/10.1038/srep32571
53. Chen L, Yang J, Xing Z, Yuan F, Shu Y, Zhang Y, Kong X, Huang T, Li H, Cai Y-D (2017) An integrated method for the identification of novel genes related to oral cancer. PLoS One 12(4):e0175185
54. Chen L, Chu C, Kong X, Huang G, Huang T, Cai YD (2015) A hybrid computational method for the discovery of novel reproduction-related genes. PLoS One 10(3): e0117090. https://doi.org/10.1371/journal.pone.0117090
55. Lee I, Blom UM, Wang PI, Shim JE, Marcotte EM (2011) Prioritizing candidate disease genes by network-based boosting of genome-wide association data. Genome Res 21 (7):1109–1121. https://doi.org/10.1101/gr.118992.110
56. Managbanag JR, Witten TM, Bonchev D, Fox LA, Tsuchiya M, Kennedy BK, Kaeberlein M (2008) Shortest-path network analysis is a useful approach toward identifying genetic determinants of longevity. PLoS One 3(11):e3802. https://doi.org/10.1371/journal.pone.0003802
57. Zhang J, Jiang M, Yuan F, Feng KY, Cai YD, Xu X, Chen L (2013) Identification of age-related macular degeneration related genes by applying shortest path algorithm in protein-protein interaction network. Biomed Res Int 2013:523415
58. Li B-Q, You J, Chen L, Zhang J, Zhang N, Li H-P, Huang T, Kong X-Y, Cai Y-D (2013) Identification of lung-cancer-related genes with the shortest path approach in a protein-protein interaction network. Biomed Res Int 2013:267375. https://doi.org/10.1155/2013/267375
59. Jiang M, Chen Y, Zhang Y, Chen L, Zhang N, Huang T, Cai Y-D, Kong X (2013) Identification of hepatocellular carcinoma related genes with k-th shortest paths in a protein–protein interaction network. Mol BioSyst 9 (11):2720–2728
60. Dijkstra EW (1959) A note on two problems in connexion with graphs. Numer Math 1:269–271
61. Chartrand G, Oellermann OR (1992) Applied and algorithmic graph theory. Mcgraw-Hill College, Pennsylvania NY
62. Cormen TH, Leiserson CE, R RL, Stein C (2001) Introduction to algorithms, second edn. MIT press and Mcgraw-Hill, Cambridge MA
63. Hart PENN, Raphael B (1968) A formal basis for the heuristic determination of minimum

cost paths. IEEE Transactions on Systems Science and Cybernetics 4:100–107

64. EW D (1959) A note on two problems in connection with graphs. Numer Math 1:269–271

65. Basso K, Margolin AA, Stolovitzky G, Klein U, Dalla-Favera R, Califano A (2005) Reverse engineering of regulatory networks in human B cells. Nat Genet 37(4):382–390. https://doi.org/10.1038/ng1532

66. Butte AJ, Kohane IS (2000) Mutual information relevance networks: functional genomic clustering using pairwise entropy measurements. Pac Symp Biocomput:418–429

67. Scutari M (2010) Learning Bayesian networks with the bnlearn R package. J Stat Softw 35 (3):1–22

68. Huang T, Yang J, Cai Y-D (2015) Novel candidate key drivers in the integrative network of genes, MicroRNAs, methylations, and copy number variations in squamous cell lung carcinoma. Biomed Res Int 2015:358125. https://doi.org/10.1155/2015/358125

69. Huang T, Liu L, Qian Z, Tu K, Li Y, Xie L (2010) Using GeneReg to construct time delay gene regulatory networks. BMC Res Notes 3 (1):142. https://doi.org/10.1186/1756-0500-3-142

70. Franceschini A, Szklarczyk D, Frankild S, Kuhn M, Simonovic M, Roth A, Lin J, Minguez P, Bork P, von Mering C, Jensen LJ (2013) STRING v9.1: protein-protein interaction networks, with increased coverage and integration. Nucleic Acids Res 41(Database issue):D808–D815. https://doi.org/10.1093/nar/gks1094

71. Ogata H, Goto S, Sato K, Fujibuchi W, Bono H, Kanehisa M (1999) KEGG: Kyoto Encyclopedia of genes and genomes. Nucleic Acids Res 27(1):29–34

72. Kamburov A, Wierling C, Lehrach H, Herwig R (2009) ConsensusPathDB--a database for integrating human functional interaction networks. Nucleic Acids Res 37(Database issue): D623–D628. https://doi.org/10.1093/nar/gkn698

Chapter 9

Differential Coexpression Network Analysis for Gene Expression Data

Bao-Hong Liu

Abstract

Gene expression profiling by microarray has been used to uncover molecular variations in many areas. The traditional analysis method to gene expression profiling just focuses on the individual genes, and the interactions among genes are ignored, while genes play their roles not by isolations but by interactions with each other. Consequently, gene-to-gene coexpression analysis emerged as a powerful approach to solve the above problems. Then complementary to the conventional differential expression analysis, the differential coexpression analysis can identify gene markers from the systematic level. There are three aspects for differential coexpression network analysis including the network global topological comparison, differential coexpression module identification, and differential coexpression genes and gene pairs identification. To date, the coexpression network and differential coexpression analysis are widely used in a variety of areas in response to environmental stresses, genetic differences, or disease changes. In this chapter, we reviewed the existing methods for differential coexpression network analysis and discussed the applications to cancer research.

Key words Coexpression, Differential coexpression network

1 Introduction

In biological systems, distinct groups of molecules that are functionally coordinated, physically interacting or co-regulated, drive complex biological processes. To dissect the complexity of biological systems, a complete map of intermolecular interactions is required. Networks provide a straightforward representation of interactions between the nodes, and there are multiple types of network including physical attachments underlying protein-protein interaction network, kinase-substrate interaction network, protein-DNA interaction network, and metabolic reaction network, as well as functional associations such as epistasis, synthetic lethality relationships, and correlated expression between genes [1, 2]. These various molecular networks have been successfully applied to address different biological questions, such as identification of

Tao Huang (ed.), *Computational Systems Biology: Methods and Protocols*, Methods in Molecular Biology, vol. 1754,
https://doi.org/10.1007/978-1-4939-7717-8_9,

disease genes [2] and drug discovery [3, 4]. Microarray is a form of high-throughput genomic data providing relative measurements of mRNA levels for thousands of genes in a biological sample. Besides the gene expression research, there are gene-to-gene interaction information inferred from microarray. And coexpression networks are typically constructed from gene expression data using correlation-based inference methods which have been commonly used to reveal gene functions and investigate gene regulatory systems [5–7]. Based on the coexpression network, numerous methods emerged to identify the differential coexpression modules, genes, or gene pairs which can further indicate the biological mechanisms underlining the data [8–10]. In this chapter, we will in detail introduce the coexpression network including the construction of coexpression network and the differential coexpression network analysis.

2 From Gene Coexpression Network to Differential Coexpression Analysis

Microarray technology has provided a powerful approach for analyzing the genome-wide gene expression profiling [11]. In this section, the construction of gene coexpression network will be described. Then, the conception of gene differential coexpression in coexpression network was defined and described in detail. Figure 1 showed the workflow from gene coexpression network construction to differential network analysis for expression data.

2.1 Gene Coexpression Network Analysis

A gene coexpression network is an undirected graph, where the graph nodes correspond to genes and edges between genes represent significant coexpression relationships. The network is usually constructed by measuring the gene expression similarity, which represents the coexpression relationships between genes. Gene coexpression takes into account the gene-to-gene interactions and makes it possible to investigate the whole-genome architecture under a certain condition. And the Pearson correlation coefficient is the most popular method to construct the gene coexpression network. When construct the coexpression network, the pairwise correlation should first be calculated. And then a correlation cutoff should be given to filter the low-correlation pairs. There are two thresholding strategies: the hard-thresholding method and the soft-thresholding method. The first method includes the correlation-based method [12], the q-value-based method [12], the percent-based method [12], the rank-based method [13], and other systematic threshold-finding methods [14]. The second method needs a power value (β) to scale the correlation coefficient and widen the difference between the low and high correlation values [9].

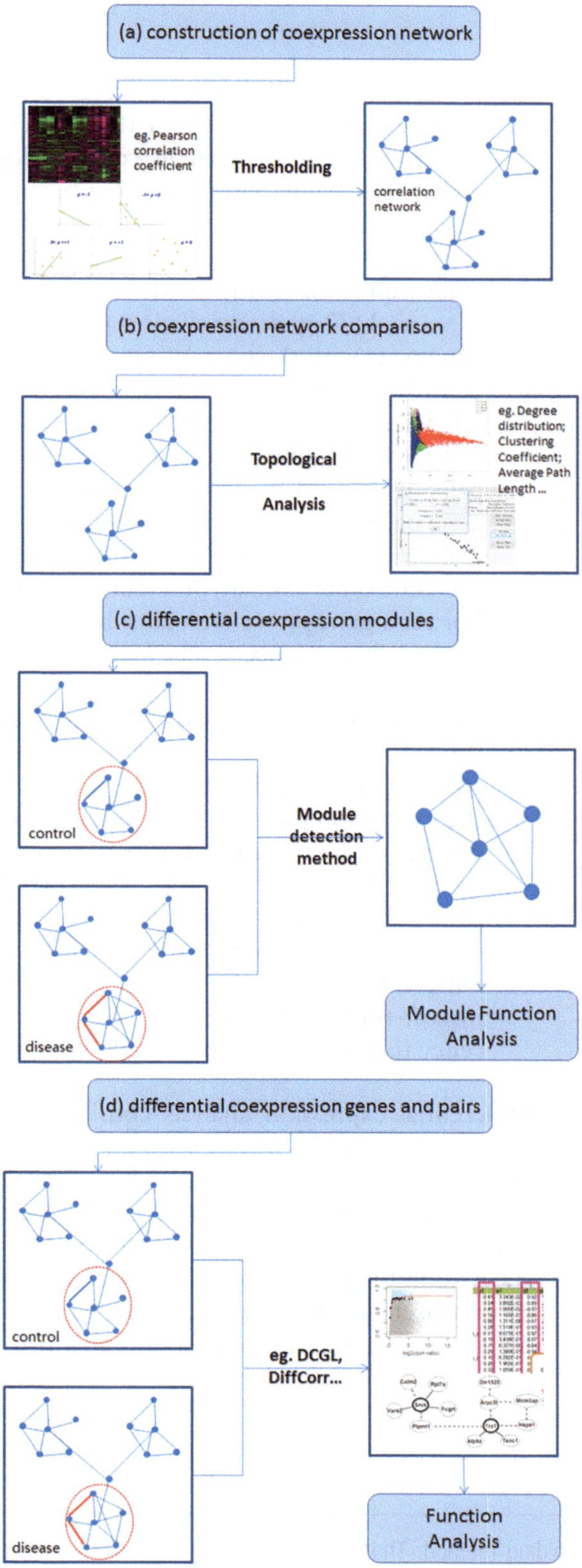

Fig. 1 Workflow of gene differential coexpression network analysis. First, the gene coexpression network was constructed by calculating the gene-to-gene pairwise correlation and filtering lower-correlation edges using a

1. *Hard-Thresholding-Based Coexpression Network.*

   Value based There are mainly three types of value-based methods to construct the gene coexpression network including correlation value-based, q value-based, and percent value-based methods. Pearson correlation coefficients (PCCs) are calculated based on the gene expression profiles. For a given gene expression matrix with n genes and m samples, all the $n \times (n-1)/2$ gene correlation pair number is generated. The gene pairs with the correlation coefficient higher than the Rth or with the adjusted p-value for the correlation lower than the qth will be left to form the coexpression network. On the other hand, the absolute values for all $n \times (n-1)/2$ gene correlation pairs are sorted in decreasing order. A fraction (percent) of gene pairs with the absolute correlation values will be retained.
   Rank based First calculate the Pearson correlation coefficient (or some other similarity measure) between every pair of genes. For each gene g_i, we rank all other genes by their similarity to g_i. And then connect every gene to the d genes that are most similar to it.
2. *Soft-Thresholding-Based Coexpression Network.*
   WGCNA (weighted gene coexpression network analysis) adopted the soft-thresholding strategy: x_{ij} and y_{ij} represent the expression profile of gene x and gene y. First, calculate the coexpression similarity measure of two genes by Pearson correlation coefficient $s_{ij} = \mathrm{cor}(x_{ij}, y_{ij})$. Then the coexpression similarity is transformed into the adjacency by raising the coexpression similarity to a power β: $a_{ij} = s_{ij}^{\beta}$, with $\beta \geq 1$. Different from the hard-thresholding method, the network by soft-thresholding method is weighted and allows the adjacency to take on continuous values between 0 and 1.

2.2 Coexpression Network Comparison and Differential Coexpression Network Analysis for Gene Expression Data

Differential expression analysis considers each gene individually, while their potential interactions are ignored. However, genes or their protein products do not act in isolation; instead, they are interacted with each other and act in close coordination. So differential coexpression analysis emerged to address this problem which is based on the gene coexpression network analysis [15]. There are three aspects for differential coexpression network analysis including the topological characteristic comparison, differential coexpression gene module identification, and differential coexpression genes and gene pair identification.

Fig. 1 (continued) thresholding method. Then for two network comparison, the network topological characteristics were calculated. Next, the differential coexpression modules, differential coexpression genes, and gene pairs were identified, and their function will be enriched by GO and KEGG analysis

2.2.1 Network Topological Characteristic Comparison

There are many parameters for network topology measurement which are shown in Table 1. The most robust measures are node degree distribution, network clustering coefficient, and average path length. Degree is the number of edges connected to a vertex, and the highest-degree nodes are often called hubs. For biological networks, the degree distribution always has a power-law distribution, that is, a few nodes own a very high number of degree and lots of nodes are connected to a few nodes. These networks have no characteristic scales for the degrees; hence, they are called scale-free networks. And the parameter of clustering coefficient is a measure of the degree to which nodes in a graph tend to cluster together. The networks tend to be modulated when it has a high clustering coefficient. Average path length is defined as the average number of steps along the shortest paths for all possible pairs of network nodes. Most real networks have a very short average path length leading to the concept of a small world where everyone is

Table 1
Network topological parameters for coexpression network comparison

Parameters	Definition	Formula
Degree	The number of edges connected to a vertex	
Power-law degree distribution	A network is said to have a power-law degree distribution when for degree k, the probability distribution of k follows a power law	$p(k) \propto k^{-\gamma}$, where p(.) indicates the probability mass function and $\gamma \geq 1$ is the parameter of the power-law distribution
Clustering coefficient	A measure of the degree to which nodes in a graph tend to cluster together	$C = \frac{1}{n}\sum_{i} C_i$
Average path length	Average number of steps along the shortest paths for all possible pairs of network nodes	$l = \frac{1}{n(n-1)}\sum_{i \neq j} d(v_i, v_j)$, where n is the number of vertices. $v_1, v_2 \in V$ denote the shortest distance between v_1 and v_2
Network diameter	The diameter of a network is the length (in number of edges) of the longest geodesic path between any two vertices	
Betweenness centrality	Number of shortest paths between all pairs of vertices that go through the vertex	$g(k) = \sum_{s \neq k \neq t} \frac{\sigma_{st}(k)}{\sigma_{st}}$ where σ_{st} is the total number of shortest paths from node s to node t and $\sigma_{st}(k)$ is the number of those paths that pass through k
Closeness centrality	A measure of centrality in a network, calculated as the sum of the length of the shortest paths between the node and all other nodes in the graph	$C(v) = \frac{1}{\sum_u d(v,u)}$, where $d(v, u)$ is the distance between vertices v and u
Network density	A ratio expressing the number of actual edges between vertices to the number of possible edges	

connected to everyone else through a very short path. And there are some other parameters to measure the structure of the network including the network diameter, the betweenness centrality, the average path length, the closeness centrality, and the network density. Network diameter measures the network size which is the length (in number of edges) of the longest geodesic path between any two vertices. The betweenness centrality is a measure of centrality in a graph based on shortest paths, and a node with higher betweenness centrality would have more control over the network. The closeness centrality is a measure of centrality in a network, calculated as the sum of the length of the shortest paths between the node and all other nodes in the graph. Thus the more central a node is, the closer it is to all other nodes. The network density is the ratio expressing the number of actual edges between vertices to the number of possible edges.

2.2.2 Differential Coexpression Gene Sets (Module) Identification

Numerous methods and tools were developed to measure the differential coexpression of genes including methods to identify differentially coexpressed gene modules which can be gene sets newly detected [9, 16–18] or a set of genes predefined [10, 19, 20] (Table 2). WGCNA is the most commonly used tool to detect the coexpressed modules. It took the soft-thresholding strategy in constructing the coexpression network and defined a dissimilarity measure for gene nodes. Then the average linkage hierarchical clustering coupled with the dissimilarity was applied to identify the coexpression modules. Then a preservation analysis can be applied to test whether the modules detected in one condition preserved in another condition. For the method coXpress, genes are clustered by the Pearson correlation coefficient (r), and the gene-to-gene distance can be calculated by $1-r$. Then the resulting tree will be cut by a predefined cutoff value getting a series of gene

Table 2
Methods for differential coexpression module identification

Method/tool name	Gene sets predefined	Extending to multiple conditions	References
WGCNA	No	No	[9]
coXpress	No	No	[17]
DiffCoEx	No	Yes	[18]
DICER	No	No	[21]
GSCA	Yes	Yes	[19]
GSNCA	Yes	No	[20]
CoGA	Yes	No	[10]

sets. The next step is to test the coexpression difference for the above gene sets in different experiments or conditions by calculating the pairwise correlation for gene set (size n) and summarizing them by t-statistics and sampling n genes randomly from the expression data and repeating the above step for m times to form the distribution of t-statistics. This will be repeated in the other condition. If the t-statistics is significant in condition 1 and not significant in condition 2, the gene set will be identified as differentially coexpressed.

For the method DiffCoEx, it provided two types for differential coexpression that is within-module differential coexpression and module-to-module differential coexpression. First, build adjacency matrix by Pearson correlation coefficient, and then compute the matrix of adjacency coefficient.

$$d_{ij} = \left(\sqrt{\frac{1}{2}}\left|\text{sign}\left(c_{ij}^{|1|}\right) * \left(c_{ij}^{|1|}\right)^2 - \text{sign}\left(c_{ij}^{|2|}\right) * \left(c_{ij}^{|2|}\right)^2\right|\right)^{\beta}$$

Then calculate the topological overlap measure to identify genes that share similar neighbors.

$$t_{ij} = 1 - \frac{\sum_k \left(d_{ik} d_{kj}\right) + d_{ij}}{\min\left(\sum_k d_{ik}, \sum_k d_{jk}\right) + 1 - d_{ij}}$$

And the modules will be identified by the dissimilarity values' formed matrix. The statistical significance of differential coexpression can be assessed using a measure of the statistics. This method can be extended to the study of differential coexpression over more than two conditions.

DICER detects differentially coexpressed gene sets using a probabilistic score. First a DC score is defined.

$$R_{D_i}^{u,v} - R_{D_j}^{u,v} \sim N\left(\mu_i - \mu_j, \sqrt{\sigma_i^2 + \sigma_j^2}\right)$$

Then a probabilistic framework will be applied to test the significance of the difference.

Besides the methods to detect new modules differentially coexpressed, there are methods for analyzing modules predefined. GSCA used the Euclidean distance to measure the difference for the pairwise correlation coefficient from the given pathway genes under different conditions. And test the significance of the distance using a permutation process. The method can be extended to multiple conditions. Then the GSNCA estimates net correlation changes by introducing for each gene a weight factor that characterizes its cross-correlations in the coexpression networks and tests the hypothesis that for a gene set there is no difference in the gene weight vectors between two conditions.

2.2.3 Differential Coexpression Genes and Gene Pair Identification

Complementary to traditional differential expression analysis, the differential coexpression genes and gene pairs can further help to explain the underlying molecular mechanisms for a biological problem from the systematic level. Differential coexpression genes are defined as genes whose correlated expression pattern differs between classes. Table 3 gave the widely used algorithms adopted by DCG and DCL identification tools. Yu et al. [12] reported that the traditional differential expression analysis just identified a part of T2D-associated genes, and a considerable amount of genes were identified by differential coexpression analysis as DCGs. Besides the DCGs, the DCLs can also be identified and validated to be very important. For example, RNA levels of prostate cancer biomarker gene *AMACR* have found to have positive with tumor suppressor gene *PTEN* in normal tissue but not in prostate cancer tissue samples [22]. So many differential coexpression genes and gene pair identification tools have been made based on different algorithms [8, 12, 16, 22–29]. Among them, DCGL is a very commonly used tool to identify differential coexpression genes and gene pairs simultaneously. And there are two methods of DCp and DCe to identify DCGs in which DCp used the length-normalized Euclidean distance to measure the difference of gene interaction with its neighbors and then calculate the significant *p*-value using a permutation test, while the method DCe adopted the hypermetric model to test whether the test gene enriched significantly more differentially coexpressed gene pairs.

Lai et al. extended the traditional *F*-statistic to ECF-statistics to identify differential gene-gene coexpression pattern. Choi et al. and Yoon et al. adopted *z*-score and Fisher's z-transformed score to measure the difference of gene pairs under two states and then applied methods to cancer research.

DiffCorr calculates correlations in each condition and uses the difference in z-transformed correlation coefficients to calculate *p*-values. EBcoexpress uses an empirical Bayesian approach and a nested expectation-maximization algorithm to estimate the posterior probability of differential correlation between gene pairs. Discordant fits a mixture distribution of correlation classes in each

Table 3
Methods for differential coexpression genes and gene pair identification

Methods	Identify genes	Identify gene pairs	References
Z-score	No	Yes	[22, 23, 26, 27]
F-statistics	No	Yes	[22]
Euclidean distance and hypergeometric model	Yes	Yes	[8, 12, 16]
Nested expectation-maximization algorithm	No	Yes	[28, 29]

condition and uses an expectation-maximization algorithm to estimate the posterior probability of each differential correlation category.

Like DiffCorr, DGCA transforms correlation coefficients to *z*-scores and uses differences in *z*-scores to calculate *p*-values of differential correlation between genes. Like Discordant, DGCA classifies differentially correlated gene pairs into the nine possible categories. However, DGCA differs from the existing differential correlation approaches. The simulation study indicated that the DGCA performs better than the above DiffCorr, Discordant, and EBcoexpress.

3 Applications of Differential Coexpression Network Analysis in Cancer Research

Diseases caused by single gene's variation can be detected by the traditional differential expression analysis method and then the molecular basis can be understood to discover disease biomarkers. However, cancer is a complex disease caused by multiple genes' aberration and can't be caught by the above methods while differential coexpression network analysis takes full account of the interactions of multiple genes and specific differential network of genes and gene pairs which can be identified as dysfunctional in cancer by comparing the difference of the coexpression networks. Deng et al. constructed coexpression networks at the bladder cancer and normal state and found there were great differences between the two networks in the network topological characteristics [30]. Since cancer is caused by aberrations of multiple genes which possess diverse functions and genes with similar functions are likely to be coexpressed, Jia et al. identified lung cancer related modules in coexpression networks using WGCNA and applied to facilitate cancer research and clinical diagnosis [31]. Four modules of ovarian cancer from a coexpression network were distinguished to be significantly associated with biological processes such as cell cycle and DNA replication [32]. Alexander et al. [33] explored gene networks in nine major human cancer types using a compendium of publicly available data. The analysis resulted in a large collection of high-resolution robust gene coexpression modules which offer insight in cancer biology.

Besides the global and local changes for the coexpression network under different cancer status, genes and gene pairs are also differentially coexpressed [23, 33–37]. Li et al. identified 204 differential coexpressed genes associated with cholangiocarcinoma which provides a set of targets useful for future investigations into molecular biomarker studies [37]. Fu et al. identified 37,094 differentially coexpressed links (DCLs) and 251 DCGs and then constructed the regulatory network which enhanced the understanding of disease mechanisms and leads to an improved diagnosis of lung cancer. Cao et al. designed two quantitative methods to prioritize differentially regulated genes (DRGs) and gene pairs or links (DRLs) for gastric carcinogenesis and generated testable hypotheses on the roles of GATA6, ESRRG, and their signaling pathways in gastric carcinogenesis [38].

4 Conclusions

Coexpression analysis has become a very useful tool to mining the cancer-related markers. With the accumulation of whole-genome expression data, and the improvement of computational algorithms, it is time to decipher the dysfunctional regulators and their relevant signaling pathway through efficient differential network analysis which will support the wet biological experiment and even further promote the prevention, treatment, diagnosis, and cure of cancer in the future. However, the above methods are all data-driven and need to predefine a cutoff. As we know, there are many validated networks such as protein-protein interaction networks, gene regulatory networks and so on, so it will be a good way to solve the thresholding problem by integrating the existing networks information.

References

1. Mitra K et al (2013) Integrative approaches for finding modular structure in biological networks. Nat Rev Genet 14(10):719–732
2. Vidal M, Cusick ME, Barabasi AL (2011) Interactome networks and human disease. Cell 144(6):986–998
3. Harrold JM, Ramanathan M, Mager DE (2013) Network-based approaches in drug discovery and early development. Clin Pharmacol Ther 94(6):651–658
4. Robin X et al (2013) Personalized network-based treatments in oncology. Clin Pharmacol Ther 94(6):646–650
5. Prieto C et al (2008) Human gene coexpression landscape: confident network derived from tissue transcriptomic profiles. PLoS One 3(12):e3911
6. Stanley D et al (2013) Genetic architecture of gene expression in the chicken. BMC Genomics 14:13
7. van Noort V, Snel B, Huynen MA (2004) The yeast coexpression network has a small-world, scale-free architecture and can be explained by a simple model. EMBO Rep 5(3): 280–284
8. Liu BH et al (2010) DCGL: an R package for identifying differentially coexpressed genes and links from gene expression microarray data. Bioinformatics 26(20):2637–2638
9. Langfelder P, Horvath S (2008) WGCNA: an R package for weighted correlation network analysis. BMC Bioinformatics 9:559
10. Santos Sde S et al (2015) CoGA: an R package to identify differentially co-expressed gene sets by analyzing the graph spectra. PLoS One 10 (8):e0135831
11. Jiang Z et al (2016) Differential coexpression analysis reveals extensive rewiring of Arabidopsis gene coexpression in response to pseudomonas syringae infection. Sci Rep 6:35064
12. Yu H et al (2011) Link-based quantitative methods to identify differentially coexpressed genes and gene pairs. BMC Bioinformatics 12:315
13. Ruan J, Dean AK, Zhang W (2010) A general co-expression network-based approach to gene expression analysis: comparison and applications. BMC Syst Biol 4:8
14. Elo LL et al (2007) Systematic construction of gene coexpression networks with applications

to human T helper cell differentiation process. Bioinformatics 23(16):2096–2103

15. Jiang X, Zhang H, Quan X (2016) Differentially Coexpressed disease gene identification based on gene Coexpression network. Biomed Res Int 2016:3962761
16. Yang J et al (2013) DCGL v2.0: an R package for unveiling differential regulation from differential co-expression. PLoS One 8(11):e79729
17. Watson M (2006) CoXpress: differential co-expression in gene expression data. BMC Bioinformatics 7:509
18. Tesson BM, Breitling R, Jansen RC (2010) DiffCoEx: a simple and sensitive method to find differentially coexpressed gene modules. BMC Bioinformatics 11:497
19. Choi Y, Kendziorski C (2009) Statistical methods for gene set co-expression analysis. Bioinformatics 25(21):2780–2786
20. Rahmatallah Y, Emmert-Streib F, Glazko G (2014) Gene sets net correlations analysis (GSNCA): a multivariate differential coexpression test for gene sets. Bioinformatics 30 (3):360–368
21. Amar D, Safer H, Shamir R (2013) Dissection of regulatory networks that are altered in disease via differential co-expression. PLoS Comput Biol 9(3):e1002955
22. Lai Y et al (2004) A statistical method for identifying differential gene-gene co-expression patterns. Bioinformatics 20(17):3146–3155
23. Choi JK et al (2005) Differential coexpression analysis using microarray data and its application to human cancer. Bioinformatics 21 (24):4348–4355
24. Yoon SH, Kim JS, Song HH (2003) Statistical inference methods for detecting altered gene associations. Genome Inform 14:54–63
25. Li KC (2002) Genome-wide coexpression dynamics: theory and application. Proc Natl Acad Sci USA 99(26):16875–16880
26. McKenzie AT et al (2016) DGCA: a comprehensive R package for differential gene correlation analysis. BMC Syst Biol 10(1):106
27. Fukushima A (2013) DiffCorr: an R package to analyze and visualize differential correlations in biological networks. Gene 518(1):209–214
28. Dawson JA, Ye S, Kendziorski C (2012) R/EBcoexpress: an empirical Bayesian framework for discovering differential co-expression. Bioinformatics 28(14):1939–1940
29. Siska C, Bowler R, Kechris K (2016) The discordant method: a novel approach for differential correlation. Bioinformatics 32 (5):690–696
30. Deng SP, Zhu L, Huang DS (2015) Mining the bladder cancer-associated genes by an integrated strategy for the construction and analysis of differential co-expression networks. BMC Genomics 16(Suppl 3):S4
31. Jia X et al (2014) Cancer-risk module identification and module-based disease risk evaluation: a case study on lung cancer. PLoS One 9 (3):e92395
32. Hong S et al (2011) Gene co-expression network and functional module analysis of ovarian cancer. Int J Comput Biol Drug Des 4 (2):147–164
33. Ivliev AE et al (2016) Drug repositioning through systematic Mining of Gene Coexpression Networks in cancer. PLoS One 11(11): e0165059
34. Giulietti M et al (2016) Weighted gene co-expression network analysis reveals key genes involved in pancreatic ductal adenocarcinoma development. Cell Oncol (Dordr) 39 (4):379–388
35. Gu Y et al (2017) Identification of prognostic genes in kidney renal clear cell carcinoma by RNAseq data analysis. Mol Med Rep 15 (4):1661–1667
36. Oros Klein K et al (2016) Gene Coexpression analyses differentiate networks associated with diverse cancers Harboring TP53 missense or null mutations. Front Genet 7:137
37. Li C et al (2013) Gene expression patterns combined with bioinformatics analysis identify genes associated with cholangiocarcinoma. Comput Biol Chem 47:192–197
38. Cao MS et al (2015) Differential network analysis reveals dysfunctional regulatory networks in gastric carcinogenesis. Am J Cancer Res 5 (9):2605–2625

Chapter 10

iSeq: Web-Based RNA-seq Data Analysis and Visualization

Chao Zhang, Caoqi Fan, Jingbo Gan, Ping Zhu, Lei Kong, and Cheng Li

Abstract

Transcriptome sequencing (RNA-seq) is becoming a standard experimental methodology for genome-wide characterization and quantification of transcripts at single base-pair resolution. However, downstream analysis of massive amount of sequencing data can be prohibitively technical for wet-lab researchers. A functionally integrated and user-friendly platform is required to meet this demand. Here, we present iSeq, an R-based Web server, for RNA-seq data analysis and visualization. iSeq is a streamlined Web-based R application under the Shiny framework, featuring a simple user interface and multiple data analysis modules. Users without programming and statistical skills can analyze their RNA-seq data and construct publication-level graphs through a standardized yet customizable analytical pipeline. iSeq is accessible via Web browsers on any operating system at http://iseq.cbi.pku.edu.cn.

Key words RNA-seq, R-Shiny, Gene expression analysis, Gene ontology enrichment, Data visualization

1 Introduction

Next-generation sequencing (NGS) technologies have been playing an essential role in the studies on genomics, transcriptomics, and epigenomics in the recent years. Their ability of sequencing multiple nucleic acid molecules in parallel makes it possible to generate large datasets and thus offer new insights to many biological questions [1]. Transcriptome sequencing (RNA-seq) utilizes NGS technologies for determination and quantification of RNA molecules in a biological sample. It provides higher coverage and improved sensitivity for genome-wide expression profiling compared to previous methods represented by microarray assays and has gained immense popularity due to ever-increasing through-

Chao Zhang and Caoqi Fan contributed equally to this work.

Tao Huang (ed.), *Computational Systems Biology: Methods and Protocols*, Methods in Molecular Biology, vol. 1754,
https://doi.org/10.1007/978-1-4939-7717-8_10, © Springer Science+Business Media, LLC, part of Springer Nature 2018

put and decreasing cost. The rise of RNA-seq methodologies has greatly deepened our understandings of embryonic development [2], carcinogenesis [3], cell differentiation [4], and many other research areas.

1.1 An Overview of RNA-seq Workflow

A complete RNA-seq procedure consists of both experimental stage and analysis stage. Although several sequencing protocols exist for RNA-seq, general steps and outputs in the experimental stage are similar. Briefly, RNA molecules with poly-A tails are first isolated by oligo-dT priming [5]. Alternatively, non-rRNAs are enriched by rRNA depletion [6]. The resulting RNAs are fragmented and then reverse-transcribed into short (200–1000 bp) cDNA fragments, which are then attached with sequencing adaptors and sequenced from one end or both ends. Several NGS technologies, including Illumina [7] and SOLiD [8], can be used for RNA-seq to generate millions or billions of short reads representing DNA segments.

The analysis stage of RNA-seq begins with mapping reads to the reference genome. Because eukaryotic genomes contain introns, RNA-seq reads often have gaps with varying lengths up to hundreds of thousands of base pairs, which make DNA sequence mapping tools generally unsuitable for direct use in RNA-seq. Widely used RNA-seq mapping tools include Tophat [9], SOAP [10], and GSNAP [11]. There are also programs that map reads onto a reference transcriptome, rather than reference genome, to circumvent the gap problem and to reduce computation time, such as Sailfish [12] and Kallisto [13]. Following read mapping is the quantification of each RNA species that are either provided by the reference transcriptome or de novo assembled from reads. Most mapping software packages also perform the quantification step. Generally, the final output of mapping and quantification steps can be described as a matrix with each column being a sample and each row being a gene or a splicing isoform of a transcript. This matrix, often called the expression profile, is the starting point of downstream analysis of RNA-seq datasets.

The expression profile contains rich transcriptomic information regarding the tested samples. How to draw biological meanings from it, however, is highly contingent on the specific research background. A one-size-fits-all analytical workflow does not exist. For example, different normalization methods have been proposed to alleviate technical variations and batch effects among samples, each with its strengths and drawbacks [14]. Theoretically predicting which method will give best results is a challenging or sometimes impossible task [14]. RNA-seq data are often analyzed by clustering methods to discover co-expressed gene groups or sample subclasses that share expression patterns. Commonly used clustering algorithms, including hierarchical clustering, principal component analysis (PCA) [15], and t-distributed stochastic neighbor embedding (t-SNE) [16], are each applicable for different data types, sample numbers, and research questions.

1.2 Web-Based RNA-seq Data Analysis

There is a notable contrast of skill requirement between the experimental stage and analysis stage of RNA-seq assays. The former needs experimental operation skills in cell culturing, RNA isolation, and library preparation and can be readily done in most traditional wet laboratories. However, the required computational and programming skills in the analysis stage are often less familiar to most biologists. There are commercial companies providing NGS data analysis services, but their high costs, inefficient information sharing, and communication delays are often not satisfying.

Web-based bioinformatics tools are now emerging to alleviate the situation. Some of them, exemplified by the Galaxy [17] and Seven Bridges (www.sbgenomics.com), generally aim at next-generation sequencing (NGS) data processing and are not specialized for RNA-seq analysis. Galaxy utilizes a graphical workflow editor to allow users to conduct their genomic data analysis workflow with interactivity and extensibility. However, it requires users to be familiar with each tool to achieve expected results and has many limitations including storage space, data transfer speed, and maximum job submissions. The Seven Bridges contains a suite of more than 200 pipelines and applications to help interpret bioinformatics data, emphasizing the security of data and analysis results. However, it is a commercial Web site that charges storage and computation costs and also requests high level of bioinformatics skills for proper usage.

Web-based tools specifically designed for RNA-seq include START [18], RAP [19], and CANEapp [20]. These tools provide more targeted solutions to extract information from RNA-seq datasets. RAP is a free cloud computing application with a fully automated and standardized pipeline dedicated mainly to read mapping, quantification, alternative splicing, and RNA editing detection. CANEapp shares many features with RAP except that it mainly focuses on detection of differential gene expression and novel noncoding RNA. START is an open-source application which can be run both locally and on the server side. It is user-friendly to wet-lab researchers and provides data visualization. However, this tool only provides basic visual interpretations of input datasets, such as heatmap, box plot, and volcano plot. Many important tasks within RNA-seq analysis, including normalization, differential expression detection, and functional enrichment, are lacking in this tool.

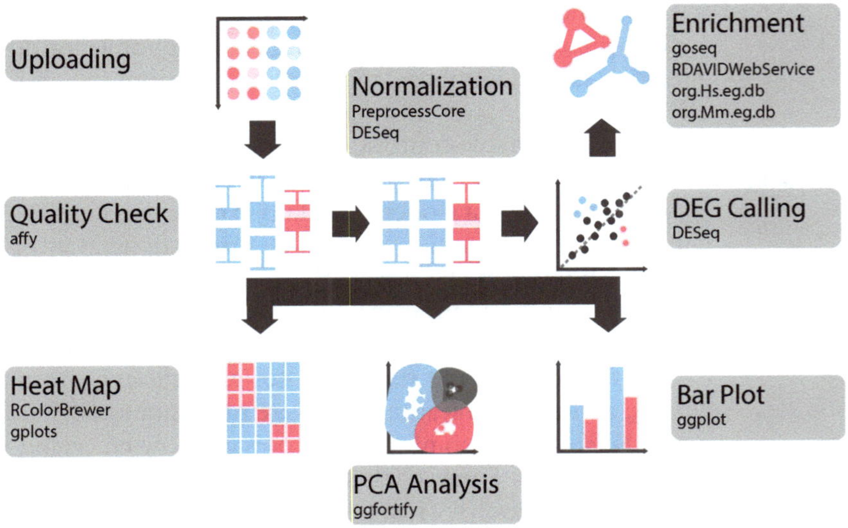

Fig. 1 An overview of iSeq analysis workflow

Table 1
Comparison between iSeq and other Web-based RNA-seq analysis tools

	iSeq	Galaxy	Seven Bridges	START	RAP	CANEapp
Local installation	Yes	Yes	No	Yes	No	No
Skill requirements	Low	High	High	Low	Middle	Middle
Mapping	No	Yes	Yes	No	Yes	Yes
RNA editing	No	Yes	Yes	No	Yes	No
Quality check[a]	Yes	No	No	No	No	No
Normalization	Yes	Yes	Yes	No	Yes	Yes
DEG calling	Yes	Yes	Yes	No	Yes	Yes
Visualization	Rich	Basic	Basic	Rich	Basic	No

[a]Quality check here refers to checking the comparability among samples which is susceptible to batch effects. Not to be confused with checking read quality as is often done by FastQC and similar tools

1.3 iSeq: A Web-Based Server for RNA-seq Data Analysis and Visualization

In this article, we introduce iSeq, a Web-based analysis suite specially designed for RNA-seq data analysis, which integrates robust analysis methods, data visualization schemes, and functional analysis tools to analyze and visualize RNA-seq data (Fig. 1). Users can explore their data through an intuitive graphical interface and focus on deciphering expressional patterns and functional relevance central to their experimental questions, without having to master complex computational and programming skills.

A comparison between iSeq and other Web-based RNA-seq analysis tools is summarized in Table 1. The greatest distinction of iSeq lies in that it focuses on a complete analytical workflow downstream of gene-expression profile. RNA-seq workflow involves a wide range of bioinformatics tools and requires a high level of computational skills to accomplish manually. Moreover, the interaction among different analysis steps brings additional workload and complexity. For example, the choice of normalization method will have impact on nearly all downstream results and may need to be re-performed after downstream quality-checking and clustering analysis. However, these tasks consume a very small amount of time, which makes them particularly suitable to be integrated in an interactive graphical user interface. Unlike most other tools (except START), iSeq is a lightweight application that makes it possible to complete all tasks within an hour (*see* **Note 1**).

2 Materials

In this section, we describe the software, packages, and methods used in building the iSeq Web server. The implementation of iSeq was based on Shiny, an open-source R package for turning R analyses into interactive Web applications that are easy to use. There are five modules in iSeq—data uploading, normalization, differentially expressed genes (DEGs) calling, functional enrichment, and plots. These modules realize a complete analysis pipeline starting from the gene-expression profile. Each module integrates a set of R packages that are key to its functioning, as listed below.

There are two available methods in the normalization module to normalize the input dataset. The size factor method is implemented in DESeq, an R package for differential RNA-seq analysis [21]. This package is also used in the DEG calling module to detect differentially expressed genes. The quantile method was initially developed to normalize microarray datasets but has also been shown effective to normalize RNA-seq data. This method is incorporated into the R package "preprocessCore" (https://github.com/bmbolstad/preprocessCore).

The functional enrichment module integrates multiple gene functional enrichment methods to facilitate a comprehensive functional analysis, revealing the biological meaning behind a select group of genes. DAVID [22, 23] is a widely used online Web server that provides functional enrichment analysis of a list of genes using gene ontology (GO) and pathway information. GOSeq [24], an R package that performs GO analysis, is also available in this module.

iSeq leverages powerful graphing packages in R to construct high-quality figures for visualization and publication. Most figure outputs in iSeq are produced by ggplot2 [25], an R package for providing beautiful plots while taking care of plotting details to meet individualized requirements. Several statistical plots including principal component analysis (PCA) are supported by the "ggfortify" R package [26]. Specialized color schemes in heatmaps and other plots are realized by the "RColorBrewer" R package (https://cran.r-project.org/web/packages/RColorBrewer/RColorBrewer.pdf).

3 Methods

In this section, we will show how to analyze RNA-seq expression data using iSeq. We reanalyzed a public RNA-seq data from Gene Expression Omnibus (GEO accession: GSE39866), which was published on *Nature Neuroscience* [27]. This dataset reported the mRNA expression differences between embryonic and adult mouse cerebral cortex and identified several genes (*Mobp*, *Igf2bp* et al.) involved in important pathways (ion transport, cell cycle) with altered expression during the cerebral cortex developmental processes. Using iSeq, we could easily reproduce the same results as in the paper without using command line analysis or programming.

3.1 Prepare Data

In the *Nature Neuroscience* paper, the authors generated seven RNA-seq samples for two biological conditions, consisting of four replicate samples for embryonic and three replicate samples for adult mouse cerebral cortex.

1. Expression File of Genes

   Download the expression data file from http://202.205.131.33:3838/expression.csv. In this file, each row represents a certain gene, and each column represents a certain sample (Fig. 2). The value of each entity represents the expression level as measured by raw sequence read counts; the FPKM and TPM are also acceptable. The first column and row list the names of genes and samples, respectively.
2. Condition File of Samples

   Download the sample description file from http://202.205.131.33:3838/condition.csv. This file allows classifying samples into biological conditions. It has two rows, with the first row listing sample names and the second row listing condition names (Fig. 2). Make sure to use the same sample names in the expression file and condition file.

3.2 Upload Data

There are two ways to access iSeq: either visit the online version (http://iseq.cbi.pku.edu.cn) or install a local version of iSeq on your own computer (https://github.com/ChengLiLab/iSeq). Figure 2 shows the interface to upload the expression table (expression.csv) and the sample condition table (condition.csv).

3.3 Normalization

Some systematic variations have been reported, such as the library size, gene length, and GC content [28]. Normalization is an import step for RNA-seq to remove these systematic biases. After

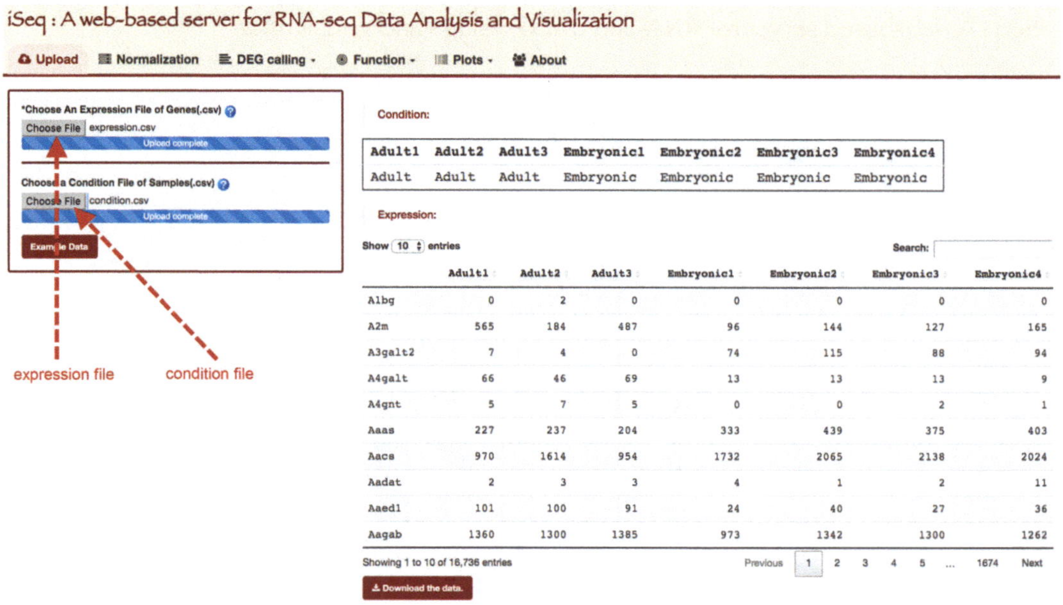

Fig. 2 The iSeq data uploading page. The arrows show the uploading buttons for expression file and sample condition file. The main page on the right shows the previews of the uploading files

normalizing the data, the effectiveness can be evaluated by examining the comparability of the gene expression distribution across samples (box plot) and the similarity among samples within each condition (hierarchical clustering).

iSeq provides two widely used normalization methods, quantile normalization and size factor. In this example, we choose size factor normalization method (Fig. 3). The plots on this page will automatically update, when user chose a new normalization method.

1. Box Plot

   In this plot, each box represents the distribution of gene expression levels of a sample (Fig. 3). Well-normalized expression profiles have expression patterns with similar distribution among samples.

2. Hierarchical Clustering

   Hierarchical clustering outputs a tree structure to visualize similarity relationships among samples (Fig. 4). Here we use the genes whose average expression values are higher than one to calculate the distance among samples. The height of a branching point stands for the similarity among samples in the subtree below it, with more similar samples having lower branching points connecting them. As expected, the sample clustering corresponds well with the partitioning by biological conditions. In this example, the four embryonic samples and three adult samples are clustered together, respectively (Fig. 4).

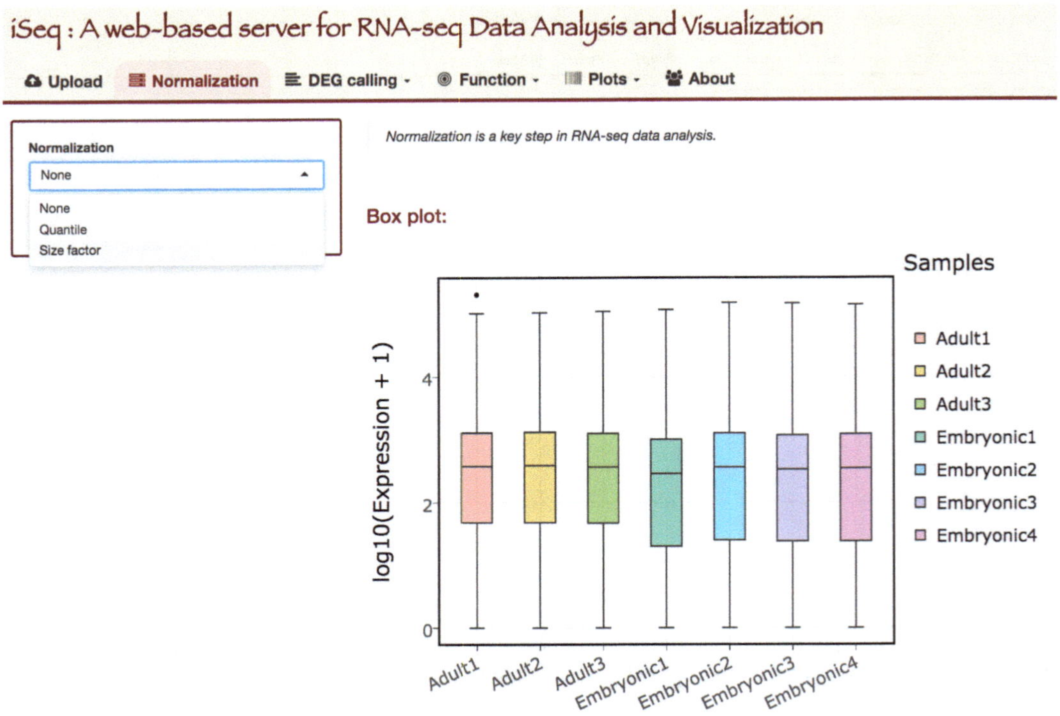

Fig. 3 Size factor normalization for the RNA-seq data. The box plot on the right shows the distribution of gene expression values in each sample. The box plot is a standardized way of displaying the distribution of a set of data points. The central rectangle spans from the first quartile to the third quartile (the 25th to 75th percentiles), and this distance is termed interquartile range (IQR). A segment inside the rectangle shows the median, and whiskers stretching outside the box show the 1.5*IQR from the lower and upper quartiles. Data points beyond the end of the whiskers are called "outlier points" and are plotted individually

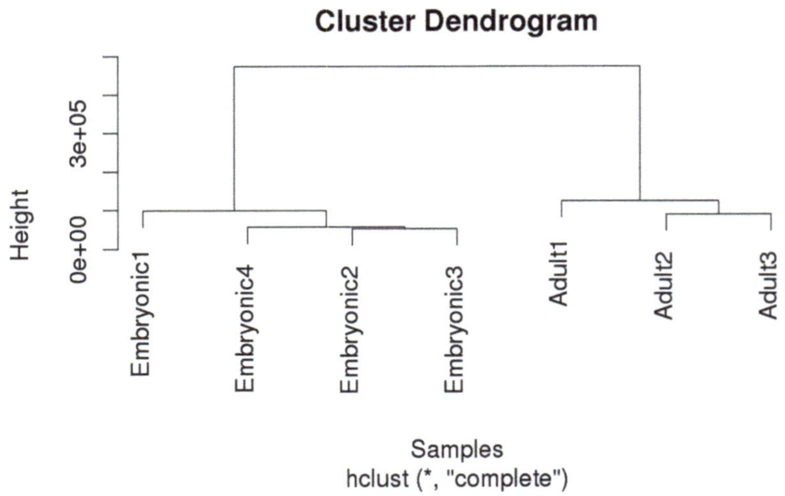

Fig. 4 Hierarchical clustering for all samples. The replicate embryonic and adult samples cluster together, respectively

3.4 Detect Differentially Expressed Genes (DEGs)

1. We use the R package, DESeq, to detect differentially expressed genes between two biological conditions. DESeq has great consistency in terms of the output gene list when the number of replicates is as small as 2–5. When the number of replicates increases to beyond 10, it also produces low false-positive rates [29].

   There are several parameters to run the DEG module:

   (a) padj cutoff: The individual *p*-value for each gene after being adjusted for multiple testing with the Benjamini-Hochberg procedure. Genes with smaller padj are regarded to be differential expressed with higher statistical significance. Setting a smaller cutoff value will result in a more stringent test and fewer DEG genes. The default padj is 0.05.

   (b) Fold-change cutoff: The fold change is defined as the ratio of mean gene expression values under two conditions. The greater the relative difference, the further fold change departs from 1. Setting a larger cutoff value will result in a more stringent test and fewer DEG genes and vice versa. Here we chose "less than 0.25 or greater than 4," which is used in the original paper of this dataset.

   (c) Base mean cutoff: The mean expression value of a gene among all samples under both conditions. This filter is intended to remove genes with very low expression, which often leads to unreliable large fold-change values. Here, we set the value to 10, which means, if a gene covers less than 10 reads on average, it will be not be called as a DEG.

2. Click the "Run to detect" button to start running. DEG calling is a time-consuming step in the RNA-seq data analysis pipeline. This step takes about 3 min. When it is finished, we will get a differentially expressed gene list (Fig. 5) and a volcano plot (Fig. 6), which is widely used in RNA-seq analysis to identify DEGs (upper-left and upper-right areas in the plot). This DEG list can be downloaded as a CSV file to be viewed or analyzed by other software.

3.5 Reveal the Biological Meaning Behind DEGs

1. Click the Function menu and select the Online servers (Fig. 7).
2. Copy the upregulated or downregulated gene list to clipboard.
3. Click the "David," and go to the official site of The Database for Annotation, Visualization and Integrated Discovery (DAVID, http://david.ncifcrf.gov).
4. Gene ontology and pathway enrichment analysis using DAVID.

   (a) Paste the upregulated or downregulated gene list.

   (b) Select gene identifier; the gene in the example list is "OFFICAL_GENE_SYMBOL". Make sure you select

Show 10 entries Search:

	baseMean	baseMeanA	baseMeanB	foldChange	log2FoldChange	pval	padj
Cd24a	12143.4285714286	222.666666666667	21084	94.688622754491	6.56511918497664	0	0
Clu	8015.71428571429	17869.3333333333	625.5	0.0350041038650948	-4.83633211689594	0	0
Cntnap1	2996.42857142857	6888.66666666667	77.25	0.0112140714216588	-6.47854602640193	0	0
Cx3cl1	7407.42857142857	16568.3333333333	536.75	0.0323961372095363	-4.94803438780291	0	0
Ddn	12609	28969	339	0.0117021643826159	-6.41708080054019	0	0
Dpysl3	34964.1428571429	1347.33333333333	60176.75	44.6635947550717	5.48102746621359	0	0
Dpysl5	17893.1428571429	1020.33333333333	30547.75	29.9389905259719	4.90395367273905	0	0
Draxin	2923.85714285714	37.6666666666667	5088.5	135.092920353982	7.07780826090149	0	0
Enpp2	5468.42857142857	12486.3333333333	205	0.0164179502923196	-5.92858216544151	0	0
Faim2	6308.71428571429	13873.6666666667	635	0.0457701640998534	-4.4494487253297	0	0

Showing 1 to 10 of 2,556 entries Previous 1 2 3 4 5 ... 256 Next

Fig. 5 The differentially expressed genes list between embryonic and adult samples. Each row represents a DEG, and columns contain the mean expression (base mean), the expression fold change, and the adjusted *p*-value (padj)

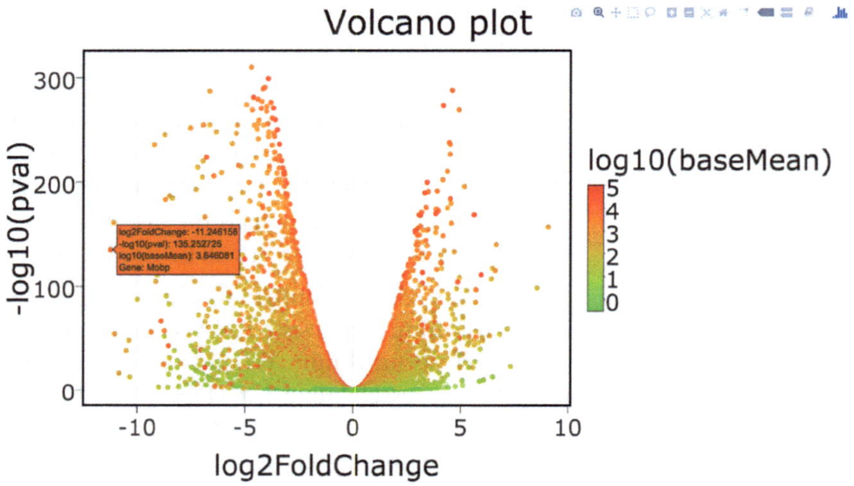

Fig. 6 The volcano plot shows the log *p*-value against the logarithm of fold change. Each dot represents a gene, with the color indicating the logarithm of mean expression of the gene. Pointing the mouse to a dot displays the supplementary information for the gene. The example is the supplementary information for *Mobp*

the right gene identifier. Take the gene "myelin-associated oligodendrocyte basic protein" as an example. The official gene symbol is *Mobp,* the HGNC gene id is "7189," the Entrez gene id is "4336," and the Ensemble gene id is "ENSG00000168314."

(c) Select gene list, which means that we upload a gene list.

(d) Submit the list.

(e) Select species.

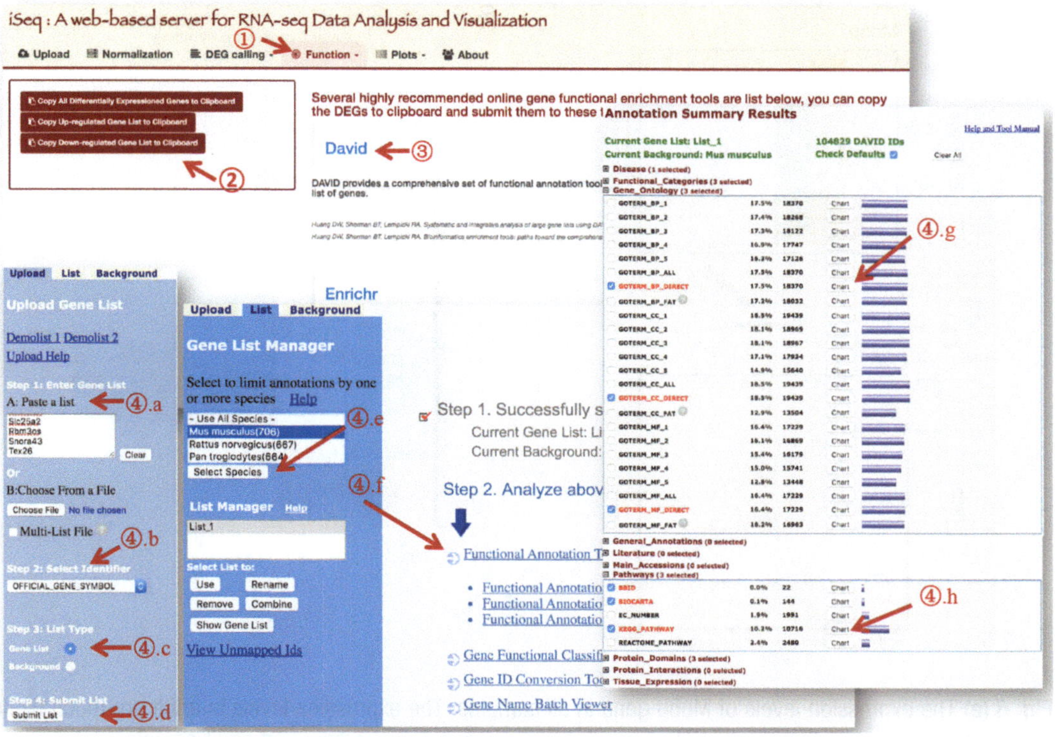

Fig. 7 The pipeline for gene enrichment analysis using iSeq and DAVID. The numbers and letters show the order of the enrichment analysis steps

(f) Analyze the uploaded gene list with "function annotation tools."

(g) Open the "Gene_Ontology" expanding menu and select the chart of "GOTERM_BP_DIRECT." This chart shows the enrichments of gene ontology terms in biological process. The results are similar to the original paper of the example data, with upregulated genes enriched in "cell cycle" and "cell division" and downregulated genes enriched in "ion transport" and "potassium ion transport."

(h) Open the "Pathways" expanding menu and select the chart of "KEGG_PATHWAY." This chart shows the enrichment KEGG pathways in the input gene list.

3.6 Frequently Used Plots in RNA-seq Analysis

We integrate some frequently used plots in the "plots" module, including the bar plot, heatmap, PCA and t-SNE.

1. Gene expression bar plot. The bar plot shows the expression level of a given gene in all samples, grouped by conditions. As showed in Fig. 8a, gene *Mobp* has much higher expression in adult samples than in embryonic samples.

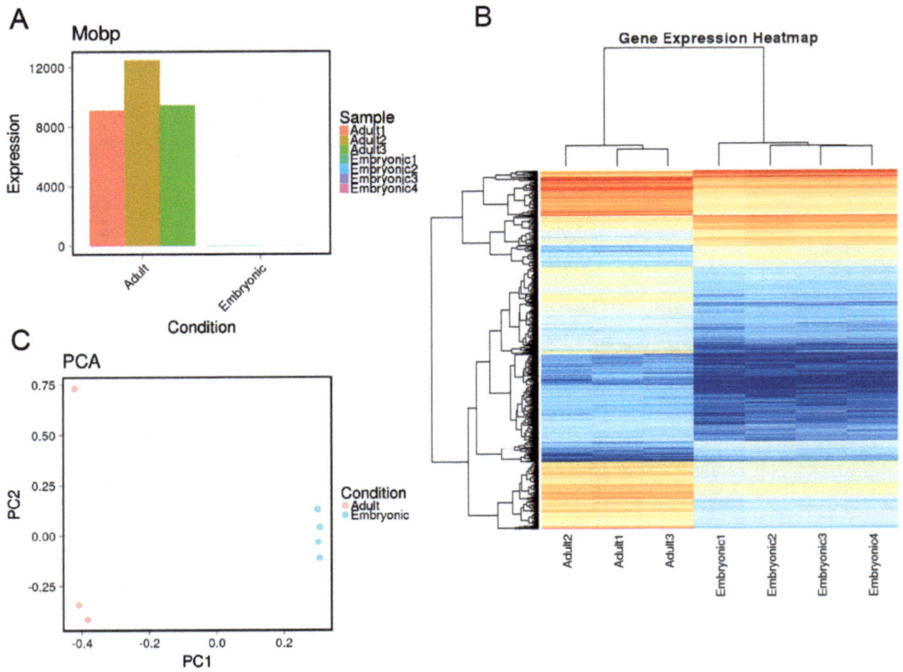

Fig. 8 (**a**) The expression levels of Mobp gene in all samples. The expression levels in adult are much higher than that in embryonic samples. (**b**) Heatmap for DEGs between embryonic and adult samples. (**c**) Principal component analysis shows that the adult and embryonic samples are separated into two clusters, indicating they are in two different conditions

2. Heatmap. In the gene expression heatmap (Fig. 8b), each row represents a gene, and each column represents a sample. Rows and columns are hierarchically clustered. Gene sets with specific expression patterns can be identified from the heatmap. The default genes used in the heatmap are the DEGs called by the "DEG calling" module; user can also upload other genes through the left menu to plot the heatmap.
3. Principal component analysis (PCA). PCA projects high-dimensional data points onto a low-dimensional space for visualization. The orthogonal axes of the space are names PC1 (principal component 1), PC2, and so on. They are chosen in such a way that the projected data points have the largest variance in the direction of PC1 and the second largest in PC2. The overall relationship or clustering of data points in the original high dimension can be visualized and identified in the low dimension more easily. In this example, the embryonic and adult samples are separated into two clusters, which are similar to the result from hierarchical clustering (Fig. 8c).

4 Notes

1. iSeq provides great flexibility in usage in that it is not required to run each module sequentially. For example, the DEG calling step could be skipped if a gene set other than DEGs is used in the enrichment module. Although iSeq is designed for NGS data analysis, datasets from Sanger sequencing or microarray assays, if converted to the required input format, could also be analyzed on this platform.

   As a lightweight RNA-seq data analysis application, iSeq imposes short waiting period between user operation and result display. Users generally complete all analysis tasks within an hour, without needing the requirement of registration. Currently, iSeq does not perform read mapping because it consumes massive computational resources and lasts much longer than other tasks. Researchers may resort to mapping software (e.g., Tophat) or online tools (e.g., Galaxy) for read alignment and then seamlessly shift to the iSeq workflow.

   iSeq is an ongoing project on which further enhancement and extension will be our future effort. In addition to providing a broader range of choices in each module, we will also introduce a mapping functionality and a cloud storage system to realize a one-stop analytical pipeline. We also notice that biologists with output data from other sequencing technologies such as bisulfite sequencing (BS-seq) and whole genome sequencing (WGS) are facing similar difficulties as with RNA-seq and will possibly extend iSeq into these areas.

Acknowledgments

We thank Yifang Liu for advice on Web server construction and the PKU Bioinformatics Core Discussion Group (Yangchen Zheng, Yong Peng) for testing and suggestions. This work was supported by funding from Peking-Tsinghua Center for Life Sciences and School of Life Sciences of Peking University, Natural Science Foundation of China (Key Research Grant 71532001), and Chinese National Key Projects of Research and Development (2016YFA0100103).

References

1. Schuster SC (2008) Next-generation sequencing transforms today's biology. Nat Methods 5 (1):16–18. https://doi.org/10.1038/nmeth1156
2. Yan L, Yang M, Guo H, Yang L, Wu J, Li R, Liu P, Lian Y, Zheng X, Yan J, Huang J, Li M, Wu X, Wen L, Lao K, Li R, Qiao J, Tang F (2013) Single-cell RNA-Seq profiling of human preimplantation embryos and embryonic stem cells. Nat Struct Mol Biol 20 (9):1131–1139. https://doi.org/10.1038/nsmb.2660

3. Edgren H, Murumagi A, Kangaspeska S, Nicorici D, Hongisto V, Kleivi K, Rye IH, Nyberg S, Wolf M, Borresen-Dale AL, Kallioniemi O (2011) Identification of fusion genes in breast cancer by paired-end RNA-sequencing. Genome Biol 12(1):R6. https://doi.org/10.1186/gb-2011-12-1-r6
4. Treutlein B, Brownfield DG, Wu AR, Neff NF, Mantalas GL, Espinoza FH, Desai TJ, Krasnow MA, Quake SR (2014) Reconstructing lineage hierarchies of the distal lung epithelium using single-cell RNA-seq. Nature 509 (7500):371–375. https://doi.org/10.1038/nature13173
5. Pan Q, Shai O, Lee LJ, Frey BJ, Blencowe BJ (2008) Deep surveying of alternative splicing complexity in the human transcriptome by high-throughput sequencing. Nat Genet 40 (12):1413–1415. https://doi.org/10.1038/ng.259
6. Yang L, Duff MO, Graveley BR, Carmichael GG, Chen LL (2011) Genomewide characterization of non-polyadenylated RNAs. Genome Biol 12(2):R16. https://doi.org/10.1186/gb-2011-12-2-r16
7. Nagalakshmi U, Wang Z, Waern K, Shou C, Raha D, Gerstein M, Snyder M (2008) The transcriptional landscape of the yeast genome defined by RNA sequencing. Science 320 (5881):1344–1349. https://doi.org/10.1126/science.1158441
8. Cloonan N, Forrest AR, Kolle G, Gardiner BB, Faulkner GJ, Brown MK, Taylor DF, Steptoe AL, Wani S, Bethel G, Robertson AJ, Perkins AC, Bruce SJ, Lee CC, Ranade SS, Peckham HE, Manning JM, McKernan KJ, Grimmond SM (2008) Stem cell transcriptome profiling via massive-scale mRNA sequencing. Nat Methods 5(7):613–619. https://doi.org/10.1038/nmeth.1223
9. Trapnell C, Pachter L, Salzberg SL (2009) TopHat: discovering splice junctions with RNA-Seq. Bioinformatics 25(9):1105–1111. https://doi.org/10.1093/bioinformatics/btp120
10. Li R, Li Y, Kristiansen K, Wang J (2008) SOAP: short oligonucleotide alignment program. Bioinformatics 24(5):713–714. https://doi.org/10.1093/bioinformatics/btn025
11. Wu TD, Nacu S (2010) Fast and SNP-tolerant detection of complex variants and splicing in short reads. Bioinformatics 26(7):873–881. https://doi.org/10.1093/bioinformatics/btq057
12. Patro R, Mount SM, Kingsford C (2014) Sailfish enables alignment-free isoform quantification from RNA-seq reads using lightweight algorithms. Nat Biotechnol 32(5):462–464. https://doi.org/10.1038/nbt.2862
13. Bray NL, Pimentel H, Melsted P, Pachter L (2016) Near-optimal probabilistic RNA-seq quantification. Nat Biotechnol 34 (5):525–527. https://doi.org/10.1038/nbt.3519
14. Zyprych-Walczak J, Szabelska A, Handschuh L, Gorczak K, Klamecka K, Figlerowicz M, Siatkowski I (2015) The impact of normalization methods on RNA-Seq data analysis. Biomed Res Int 2015:621690. https://doi.org/10.1155/2015/621690
15. Ringner M (2008) What is principal component analysis? Nat Biotechnol 26(3):303–304
16. van der Maaten L (2014) Accelerating t-SNE using Tree-Based Algorithms. J Mach Learn Res 15:3221–3245
17. Goecks J, Nekrutenko A, Taylor J, Galaxy T (2010) Galaxy: a comprehensive approach for supporting accessible, reproducible, and transparent computational research in the life sciences. Genome Biol 11(8):R86. https://doi.org/10.1186/gb-2010-11-8-r86
18. Nelson JW, Sklenar J, Barnes AP, Minnier J (2017) The START App: a web-based RNAseq analysis and visualization resource. Bioinformatics 33(3):447–449. https://doi.org/10.1093/bioinformatics/btw624
19. D'Antonio M, D'Onorio De Meo P, Pallocca M, Picardi E, D'Erchia AM, Calogero RA, Castrignano T, Pesole G (2015) RAP: RNA-Seq Analysis Pipeline, a new cloud-based NGS web application. BMC Genomics 16:S3. https://doi.org/10.1186/1471-2164-16-S6-S3
20. Velmeshev D, Lally P, Magistri M, Faghihi MA (2016) CANEapp: a user-friendly application for automated next generation transcriptomic data analysis. BMC Genomics 17:49. https://doi.org/10.1186/s12864-015-2346-y
21. Anders S, Huber W (2010) Differential expression analysis for sequence count data. Genome Biol 11(10):R106. https://doi.org/10.1186/gb-2010-11-10-r106
22. Dennis G Jr, Sherman BT, Hosack DA, Yang J, Gao W, Lane HC, Lempicki RA (2003) DAVID: Database for Annotation, Visualization, and Integrated Discovery. Genome Biol 4(5):P3
23. Huang DW, Sherman BT, Lempicki RA (2009) Systematic and integrative analysis of large gene lists using DAVID bioinformatics resources. Nat Protoc 4(1):44–57. https://doi.org/10.1038/nprot.2008.211
24. Young MD, Wakefield MJ, Smyth GK, Oshlack A (2010) Gene ontology analysis for RNA-seq:

accounting for selection bias. Genome Biol 11 (2):R14. https://doi.org/10.1186/gb-2010-11-2-r14
25. Ginestet C (2011) ggplot2: elegant graphics for data analysis. J R Stat Soc a Stat 174:245. https://doi.org/10.1111/j.1467-985X.2010.00676_9.x
26. Tang Y, Horikoshi M, Li WX (2016) ggfortify: unified interface to visualize statistical results of popular R packages. R J 8(2):474–485
27. Dillman AA, Hauser DN, Gibbs JR, Nalls MA, McCoy MK, Rudenko IN, Galter D, Cookson MR (2013) mRNA expression, splicing and editing in the embryonic and adult mouse cerebral cortex. Nat Neurosci 16(4):499–506. https://doi.org/10.1038/nn.3332
28. Dillies MA, Rau A, Aubert J, Hennequet-Antier C, Jeanmougin M, Servant N, Keime C, Marot G, Castel D, Estelle J, Guernec G, Jagla B, Jouneau L, Laloe D, Le Gall C, Schaeffer B, Le Crom S, Guedj M, Jaffrezic F, French StatOmique C (2013) A comprehensive evaluation of normalization methods for Illumina high-throughput RNA sequencing data analysis. Brief Bioinform 14 (6):671–683. https://doi.org/10.1093/bib/bbs046
29. Seyednasrollah F, Laiho A, Elo LL (2015) Comparison of software packages for detecting differential expression in RNA-seq studies. Brief Bioinform 16(1):59–70. https://doi.org/10.1093/bib/bbt086

Chapter 11

Revisit of Machine Learning Supported Biological and Biomedical Studies

Xiang-tian Yu, Lu Wang, and Tao Zeng

Abstract

Generally, machine learning includes many in silico methods to transform the principles underlying natural phenomenon to human understanding information, which aim to save human labor, to assist human judge, and to create human knowledge. It should have wide application potential in biological and biomedical studies, especially in the era of big biological data. To look through the application of machine learning along with biological development, this review provides wide cases to introduce the selection of machine learning methods in different practice scenarios involved in the whole biological and biomedical study cycle and further discusses the machine learning strategies for analyzing omics data in some cutting-edge biological studies. Finally, the notes on new challenges for machine learning due to small-sample high-dimension are summarized from the key points of sample unbalance, white box, and causality.

Key words Machine learning, Feature selection, Clustering, Classification, Omics big data, Association, Causality, Gut metagenomics, Precision medicine

1 Introduction

Machine learning includes many in silico methods to transform or abstract the natural principles to human knowledge. The main tasks have feature engineering or dimension reduction to extract the object-relevant elements in the study problem [1], clustering to determine the type or class of object observed in the study field [2], and classification/regression to predict the type or class for an unknown object [3].

The general purpose of machine learning is to save human labor, to assist human judge, and to create human knowledge. Thus, it should have wide application potential in biological and biomedical researches [4]. Actually, machine learning has an important branch involved in bioinformatics, which try to learn the data from biological technologies and transform such data to biological insights [5].

Tao Huang (ed.), *Computational Systems Biology: Methods and Protocols*, Methods in Molecular Biology, vol. 1754, https://doi.org/10.1007/978-1-4939-7717-8_11,

Previously, machine learning has been applied in many biological studies, such as (1) sequence analysis [6] to select ortholog genes, to identify binding motifs, or to predict the functional domains; (2) image analysis [7] to select image indicators, to identify object (e.g., cells) boundary in images, or to judge the object (e.g., molecules) type in images; (3) interaction analysis [8] to extract functional characteristics, to recognize functional modules, or to predict functional associations; (4) disease analysis [9] to select disease-associated genes, to identify disease subtypes, or to diagnose and know the prognoses of patients; and (5) annotation analysis [10] to select keywords in medical text, to recognize biological terms in literature, or to predict the person states by question survey.

Recently, along with the development of high-throughput technologies [11, 12], many novel machine learning technologies have been implemented to handle with these new big data [13], such as sequence assembly, modification pattern identification, confounding factor removal, heterogeneous data integration, and so on. This paper tries to provide wide cases to introduce the selection of machine learning methods in different practice scenarios involved in the whole biological and biomedical study cycle, rather than technical discussion on methodologies. After the brief introduction of several bioinformatic tools based on machine learning technologies, this review first demonstrates the categories of machine learning methods according to their biological application scenarios; next, developed on big biological data, machine learning strategies for analyzing omics data are discussed; and at last, the potential challenges on machine learning in the cutting-edge biological studies are deeply investigated and noted.

2 Materials

Generally, machine learning techniques aim to develop novel algorithms to improve the computers to assist human beings in the analysis of large, complex datasets [14]. For the big biomedical data, machine learning has entered into new and wide application fields [11, 15]. Many recent literature reviews have summarized the general and discriminative modeling approaches by applications of supervised, semi-supervised, and unsupervised machine learning methods [14], such as:

A survey on the machine learning applications for the annotation of sequence elements and epigenetic, proteomic, or metabolomic data [14].

A comprehensive review on the omics and clinical data integration techniques from a machine learning perspective [16].

A timely review on the supervised machine learning contributing to novel insights on aging and discussions on main findings and weaknesses [17].

A guideline to help non-specialists to notice the critical issues in machine learning, e.g., the large and diverse datasets, the overfitting reduction dependent on hidden parameters, and the novelty evaluation based on simple baseline strategies [18].

A perspective review of combination of machine learning and genomics for drug discovery in tuberculosis [19].

A summary about the interface between machine learning and big data technology to support basic research and biotechnology in the plant sciences [15].

A comprehensive overview and user-friendly taxonomy of machine learning tools to enable the plant community to correctly and easily apply the appropriate tools for various biotic and abiotic stress traits [20].

An extensive review of the existing models to predict protein solubility in *Escherichia coli* recombinant protein overexpression system before performing real laboratory experiments for saving labor, time, and cost [21].

An expert review of published approaches for predicting RNA-binding residues in proteins and a systematic comparison and critical assessment of protein-RNA interface residue predictors [22].

Different from such field-expert review of the application of machine learning, this paper tries to provide wide cases to introduce the selection of machine learning methods in different practical application scenarios involved in the whole biological and biomedical study cycle (Fig. 1), rather than technical discussion on methodologies.

Briefly, Table 1 lists the general key algorithms used in machine learning and their web available tools, such as *K*-means for sample clustering; C4.5, AdaBoost, KNN, and naive Bayes for sample classification; and PageRank, Apriori, and EM for feature extractions. And Table 2 supplies bioinformatic tools based on machine learning strategies to solve different biological problems. In follow-up, we will first introduce field-specific variants of machine learning methods according to their biological application scenarios. Then, focused on big biological data, the new developments of machine learning for analyzing omics data are discussed. Finally, we would like to summarize the potentials of machine learning in cutting-edge biological studies.

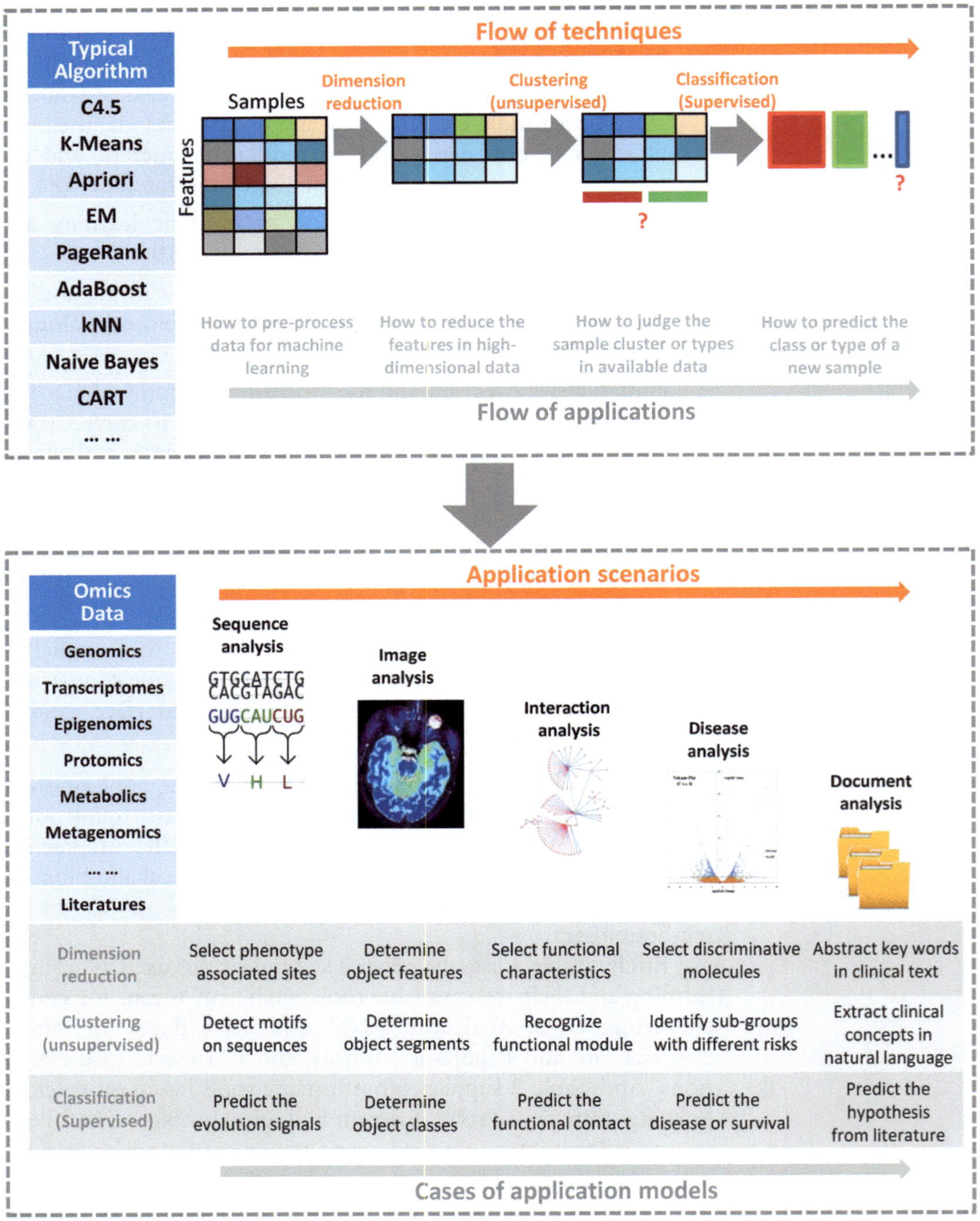

Fig. 1 The summary of machine learning workflow in omics data analysis

Table 1
The general key algorithms used in machine learning

Methods	Description	URL
C4.5	It is an algorithm used to generate a decision tree as an extension of earlier ID3 algorithm, and the decision trees can be used for classification [23]	http://www.cs.waikato.ac.nz/ml/weka/
PageRank	It is an algorithm used to rank websites by Google search engine results underlying an assumption that more important websites are likely to receive more links from other websites [24]	www.google.com
K-means	It aims to partition *N* samples into *K* groups where each sample belongs to the group with the nearest mean, serving as a prototype of the sample group [25]	https://github.com/mlpack/mlpack
Apriori	It identifies the frequent individual items in the database and extending them to larger and larger item sets as long as those item sets appear sufficiently often in the database [26]	http://www.borgelt.net/software.html
EM	The expectation–maximization (EM) algorithm is an iterative method to find maximum likelihood or maximum posteriori estimates of parameters in statistical models, where the model depends on unobserved latent variables [27]	http://wiki.stat.ucla.edu/socr/index.php/SOCR_EduMaterials_Activities_2D_PointSegmentation_EM_Mixture
AdaBoost	It is a machine learning meta-algorithm used in conjunction with many other types of learning algorithms to improve their performance. The individual learners can be weak, but as long as the performance of each one is slightly better than random guessing, the final model can be proven to converge to a strong learner [28]	http://luispedro.org/software/milk/
KNN	It is a nonparametric method used for classification and regression, where the input consists of the *k* closest training examples in the feature space and the output depends on whether KNN is used for classification or regression [29]	http://www.cs.waikato.ac.nz/ml/weka/
Naive Bayes	They are a family of simple probabilistic classifiers based on applying Bayes' theorem with strong (naive) independent assumptions between the features [30]	http://jbnc.sourceforge.net/

Table 2
The bioinformatic tools based on machine learning strategies

Methods	Description	URL
mlDNA	An R package for machine learning-based differential network analysis [31]	http://www.cmbb.arizona.edu/mlDNA/
Co-clustering	A co-clustering formulation to cluster the genes and the mesh elements, thereby identifying co-expressed embryonic domains and the associated genes simultaneously [32]	http://compbio.cs.odu.edu/fly/
ml2db	The software code (ml2db) uses InterPro signatures to predict enzyme mechanism [33]	http://sourceforge.net/projects/ml2db/
BetAware	Predict beta-barrels (which are poorly represented in the database of protein structures and difficult to identify with experimental approaches) in sets of proteins and the prediction of their topology [34]	http://betaware.biocomp.unibo.it/BetAware
PeakError	Supervised labeling method for quantitatively training and testing peak detection algorithms [35]	https://github.com/tdhock/PeakError
IDEPI	A domain-specific and extensible software library for supervised learning of models that relate genotype to phenotype for HIV-1 and other organisms [36]	https://github.com/veg/idepi
INTREPID	As an additional option for cases where sequence homologues are available, users can include evolutionary information from INTREPID for enhanced accuracy in site prediction [37]	http://www.pool.neu.edu
CoRAL	A machine learning package that can predict the precursor class of small RNAs present in a high-throughput RNA-sequencing dataset [38]	http://wanglab.pcbi.upenn.edu/coral/
SNooPer	A machine learning-based method for somatic variant identification from low-pass next-generation sequencing [39]	https://sourceforge.net/projects/snooper/
FingerID	A Matlab/Python package uses the predicted properties for matching against large molecule databases, such as PubChem, via machine learning [40]	http://www.sourceforge.net/p/fingerid
MFlux	A web-based platform predicts the bacterial central metabolism via machine learning, leveraging data from many papers on heterotrophic bacterial metabolisms [41]	http://mflux.org
apLCMS	The new peak detection approach based on the knowledge of known metabolites, as well as robust machine learning approaches can learn directly from various data features of the extracted ion chromatograms to differentiate between true peak regions from noise regions in the LC/MS profile [42]	http://web1.sph.emory.edu/apLCMS/
Taxonomic assignment	A package is implemented for efficient taxonomic assignment of metagenomic reads, which can be further improved by increasing the number of fragments sampled from reference genome or by increasing the *k*-mer size [43]	http://cbio.ensmp.fr/largescalemetagenomics

(continued)

Table 2 (continued)

Methods	Description	URL
Metaml	The software framework can analyze microbiome profiles and metadata for thousands of samples [44]	http://segatalab.cibio.unitn.it/tools/metaml
Hierarchical boosting	A machine learning classification framework can combine the selection tests to detect the features of polymorphism in hard sweeping with controls on population-specific demography [45]	http://hsb.upf.edu/

3 Methods

3.1 Variants of Machine Learning Methods in Biological and Biomedical Study

To apply machine learning methods in biological study, different variants have been proposed to address particular problem appeared in the biological questions. According to the biological background, there will be several application scenarios, e.g., sequence analysis, image analysis, interaction analysis, disease analysis, and annotation analysis. The following introduction will illustrate the wide strategy of applying machine learning in each scenario, rather than conventional survey on models of machine learning.

3.1.1 Application in Sequence-Focused Analysis

Based on general-purpose machine learning algorithms and libraries, many software packages have been designed to learn genotype-to-phenotype predictive models from sequences with known phenotypes [46, 47], which can computationally judge the genetic bases of phenotypes [36]. One is to recognize the regulatory elements on the biological sequence: to select appropriate features of promoters that distinguish them from non-promoters, the nonlinear time series descriptors along with nonlinear machine learning algorithms, such as support vector machine (SVM), are used to discriminate between promoter and non-promoter regions [48]; and a machine learning approach, MutPred Splice, has been developed to recognize coding region substitutions that disrupt pre-mRNA splicing, which can be applied to detect the splice site loss [49]; and a classifier is built and trained by using the enhancer set and identified related enhancers based on the presence or absence of known and putative TF binding sites, which combine the machine learning and evolutionary sequence analysis [50]; and an ortholog prediction meta-tool, WORMHOLE, is to integrate distinct ortholog prediction algorithms into meta-tools to identify novel least diverged orthologs (LDOs) with high confidence [51]. Two is to predict important proteins from sequences due to the difficulty on wet experiment: the top-performing methods based on machine learning approaches have been built to tackle both the detection of transmembrane beta-barrels in sets of proteins and the prediction of their topologies because most membrane proteins have poorly structure information in the available databases [34]; and a SVM classifier has been used to investigate the functional commonality and sequence homology of helical antimicrobial peptides, which can detect membrane activity in peptide sequences and penetrate the membranes of microbe [6]. Three is to select the functional key residues on sequences: to identify glycosylation sites requiring expensive/laborious experimental research, a bioinformatic tool called GlycoMine based on random forest algorithm is used for the systematic in silico identification of three types of glycosylation sites in the human proteome [52]; and to predict catalytic residues from 3-D structures, the partial order optimum likelihood (POOL) has used machine learning strategies to combine electrostatic and geometric information for enhancing the site prediction when sequence homologues are available, which is especially applicable to proteins with novel folds and engineered proteins [37]; and to identify residues that interact with ligands for designing small molecules interactive with target protein, a sequence-based method, called LIBRUS, combines homology-based transfer and direct prediction by support vector machine [53]; and similarly, the DISLOCATE, a two-step method based on machine learning models, is developed for predicting both the bonding state and the connectivity patterns of cysteine residues in a protein chain, which especially improves the overall performance when the features as protein subcellular localization are included [54].

3.1.2 Application in Image-Focused Analysis

Image analysis is an essential component in many biological experiments with multiple scales. For example, firstly on the molecule level, the StarryNite performs the automatic recognition of fluorescently labeled cells and traces their lineage, where a SVM classifier is assistant to decide whether StarryNite is correct or not to reduce the time required on correcting errors [55]; and two-dimensional gel electrophoresis (2-DE) is the protein separation method used in expression proteomics where 2-DE gel image analysis still remains a serious bottleneck, so that a hierarchical machine learning-based segmentation methodology has been proposed to improve the sensitivity and precision simultaneously [56]. Secondly on the molecular network level, the automated image analysis is able to effectively score many phenotypes, and a supervised machine learning approach can be used to iterative feedback to readily score phenotypes in high-throughput image-based screens rather than traditional screen by subjective visual inspection, which can speed up the discovery of biological pathways [7]. Thirdly on the cell level, previous live-cell imaging studies suggested that clathrin-mediated endocytosis (CME) is inefficient during cells internalizing molecules; a genome editing and machine learning method is applied to identify and analyze authentic CME sites by support vector machine classifiers [57]. Furthermore, on the organism level, compared to the morphological characteristics and molecular data, using image analysis and machine learning approaches (i.e., artificial neural network) is another invasive approach to investigate the house shrew, where an automated identification system is developed to reveal the shape characteristic features differentiating the specimens [58].

3.1.3 Application in Interaction-Focused Analysis

Protein-protein interactions (PPIs) may represent one of the next major classes of therapeutic targets [59], and such intricate biological systems cannot be cost-efficiently tackled using conventional high-throughput screening methods. To overcome the inherent problem of rigid approach on predicting the binding affinities when the modeling assumptions are not confirmed, a new RF-Score that circumvents the need for problematic modeling assumptions via nonparametric machine learning was used to implicitly capture binding effects that are hard to model explicitly [60]. And the protein-ligand interaction also requires predictive model for high-throughput screen: the machine learning-based models, PPI-HitProfiler, mainly decision trees, have been developed to determine a global physicochemical profile for putative PPI inhibitors, so that it can screen drug-like compound collection from any chemical library enriched in putative PPI inhibitors [61]; and to screen potential drug (or target) candidates for biochemical verification on drug-target interactions, the similarity-based machine learning-based approaches have been proposed to combine drug and target similarities to generate models for predicting new drug-target interactions [62]. Besides, to screen genome-wide targets of transcription factors (TFs) on regulatory level, the regulatory interaction predictor (RIP) with condition independent employs SVMs trained on a set of experimentally proven RIs from TRANSFAC, where the features of such RIs are extracted from the common TF (TF-module) of co-regulated genes by integrating the meta-analysis of gene expression correlation and in silico predictions of TF binding sites [63]. And on the epigenetic regulation, many computational methodologies for miRNA-mRNA target gene prediction have been developed based on cross-species sequence conservation of the seed segment of the miRNA and the region of the mRNA target [64]. Meanwhile, the methods that do not rely on conservation are increasing due to analyzing non-conserved genomic sequences. For example, the NBmiRTar adopts machine learning by a naive Bayes classifier and has shown higher sensitivity and specificity than algorithms that rely on conserved genomic regions [65]; and the TargetSpy can predict target sites regardless of the presence of a seed match, which is also based on machine learning and automatic feature selection using a wide spectrum of compositional, structural, and base-pairing features covering current biological knowledge [66].

3.1.4 Application in Disease-Focused Analysis

Machine learning techniques are becoming an alternative approach in medical diagnosis or prognosis. On one hand in the purpose of disease diagnosis, a molecular test is built to distinguish usual interstitial pneumonia from other interstitial lung diseases in surgical lung biopsy samples due to interstitial lung diseases having similar radiological and histopathological characteristics and also select the patient having to undergo surgery [67]; and to overcome a large imbalance of negative cases versus positive cases (*see* **Note 1**), the usage of an ensemble-based approach rather than a single classifier has been constructed with bagging, and a simple majority vote achieves a small positive effect on the accuracy rate depending on the studied diseases [68]; and to deal with redundant information and improve classification, a gene selection method, Recursive Feature Addition, is proposed to determine the final optimal gene set for disease prediction and classification [69]; and to assist physicians' subjective experience for the occurrence of skeletal-related events (SREs), the machine learning models (e.g., LR, DT, and SVM) ranked visual analog scale (VAS) as a key factor to assess the associations of clinical variables for predicting SREs risk groups [70]. On the other hand, in the purpose of disease prognosis, the ensemble classifier based on many logistic regression classifiers is applied to integrate mutation status with whole transcriptomes for high-performing prediction of NF1 inactivation in glioblastoma (GBM) with targeted therapies and personalized medicine [71]; and to assess response earlier in the treatment regimen avoiding no longer be surgically resectable, the Bayesian logistic regression is learned on the available clinical and quantitative MRI data to distinguish breast cancer responders from nonresponders after the first cycle of treatment [72]; and to tailor the prescription of prophylactic inguinal irradiation (PII) in deciding if deliver or not the PII in the treatment of anal cancer patients, the machine learning-based model (e.g., logistic regression, J48, random tree, and random forest) used a large set of clinical and therapeutic variables to obtain better performances [73]; and to capture deep molecular basis of clinical heterogeneity or specific therapeutic targets in clinical outcome models, the supervised learning prediction methods are required to delineate patients within specific risk categories who were likely to be cured or to die of their disease [9]. Besides, the computational methods in drug discovery are also accelerating drug-target prediction. Based on sequence-derived protein features, the most commonly used machine learning methods have been applied to predict whether a protein is druggable, and the feature selection procedures were used to provide the best performance of each classifier according to the optimum number of features [74]. And the identification of disease genes among the candidates remains time-consuming and expensive in the conventional way, so that ProDiGe, a new algorithm for Prioritization of Disease Genes, implements a new machine learning strategy based on learning from positive and unlabeled examples [75].

3.1.5 Application in Document-Focused Analysis

Machine learning taggers are greatly needed for clinical concept extraction from medical documents. For instance, BioTagger-GM is developed based on machine learning taggers for the automated detection of gene/protein names in the biological domain, which can be further improved by training on the datasets from multiple data sources [76]; and to extract clinical entities (e.g., medical problems, tests, and treatments, as well as their asserted status) from hospital discharge summaries, a hybrid clinical entity extraction system has been implemented for mining clinical text by conditional random fields [77]; and to collate coreferent chains of concepts from a corpus of clinical documents, a machine learning approach based on graphical models was employed to recognize and cluster coreferent concepts, which will be applicable for the assembly of problem and medication lists from clinical documents [78]. More advanced, the reported molecular mechanism can also be detected and collected from the clinical document mining, such as to search and identify gene-associated clinical trials by potential participants and clinicians, the automated methods are keys to extract genetic information from narrative trial documents, e.g., a two-stage machine learning-based approach as information retrieval tools targeting gene-associated clinical trials has been applied to identify genes and genetic lesion statuses in clinical trial documents held in the cancer clinical trial database [79]; and to identify drug side effects from free text being key for the development of up-to-date knowledge sources on drug adverse reactions, an intelligent system combining machine learning, rule- and knowledge-based approaches have been used to identify such drug side effects from the literatures [80].

3.2 New Development of Machine Learning in Omics Data Analysis

Along with the development of high-throughput technologies in different fields of biology [81, 82], the machine learning also enters a new development stage to address the new data and new problem [83–86], rather than the conventional methodology enhancement [87, 88].

3.2.1 Data Mining in Omics Data Analysis

In line with the application of high-throughput approach, firstly the sequencer like Illumina Genome Analyzer which can generate millions of short reads, many pre-procession packages (e.g., Ibis: Improved base identification system) have provided efficient base caller to increase the usable reads by reducing the error rate [89]. Next, the precise genome annotations are necessary to achieve the accurate definition of genomic segments, so that the genome annotation (e.g., to recognize exons and introns on the unspliced mRNA) is expected to be improved by using modern machine learning technology (e.g., support vector machines and label sequence learning) [90]; and to gain a detailed understanding of the dynamically composed transcription unit structures, the strand-specific RNA-seq datasets are collected to derive the genomic TU organization, and a machine learning approach is applied to predict the genomic boundaries of individual TUs [91]. Then, in addition to statistical model-based algorithms for in-depth investigation of next-generation sequencing of cancer genomes, the machine learning approaches (e.g., SNooPer based on random forest classification models) have been developed to accurately call somatic variants in low-depth, whole-exome, or whole-genome sequencing data [39]. With such accurate annotations, large-scale human genetic variation data can be obtained. For example, the single nucleotide polymorphisms (SNPs) are an important source of human genome variability and greatly contribute to human complex diseases, especially the amino acid mutations resulting from non-synonymous SNPs in coding regions; and the machine learning approach (e.g., support vector machine) has been used to predict cancer driver missense variants by training on cancer-causing variants and neutral polymorphisms with equal sample number [92]. To further detect the positive selection in those genomic regions as a natural population genetic study, a machine learning classification framework has been implemented to combine selection tests to detect the features of polymorphism in hard sweeping with controls on population-specific demography [45].

On the other hand, the high-throughput sequencing also allows researchers to examine more details on the transcriptome or other omics level than ever before (*see* **Note 2**), and a key of applying machine learning for such omics data is feature selection, i.e., to reduce the original high-dimensional omics data into a low-dimensional feature data. The CoRAL (Classification of RNAs by Analysis of Length) is a computational method for discriminating different classes of RNA, whose selected features are relevant to small RNA biogenesis pathways [38]. The RGIFE (Rule-guided Iterative Feature Elimination) is a heuristic method to select very small set of features by rule-based machine learning with balance on the objective of minimal features and high predictive power [93]. Based on the widely used *k*-top scoring pair (*k*-TSP) algorithm, the integration of *k*-TSP with other machine learning methods (e.g., multivariate classifiers such as SVM) would be a feature selector to tune certain data characteristics, i.e., correlations among informative genes [94]. More practical, the clinical application of omics data will ask for marker genes whose expression patterns will be sufficient to accurately predict the disease or not, such as the maximum difference subset algorithm has provided a coherent framework to combine the classical statistics and elements of machine learning [95]; and a supervised machine learning approach (radial and linear support vector machines) is designed to predict disease risk by genotypes incorporating gene expression data and rare variants [96]; and an extended computational method based on different machine learning approaches is used to identify the most confident candidate disease genes by integrating prior knowledge into the differential gene expressions between healthy and disease individuals [97].

3.2.2 Association Community in Gut Metagenomic Research

Almost all species in a microbial community could not be isolated and cultivated; the metagenomic methods have become one of the most important methods to analyze microbial community as a whole [44, 98]. With the fast accumulation of metagenomic samples and the advance of next-generation sequencing techniques [99], it is now possible to qualitatively and quantitatively assess all taxa (features) in a microbial community [99].

The number of gut microbiota is ten times more than the cells in the human body [100]. For understanding the interactions between human and human microbiome, three hypothesis are widely considered [101]: (1) the human genome may work as a part of larger sensorimotor organ, e.g., the human "metagenome," like our immune and nervous systems, responses to the environment change in real world; (2) the human body is an ecosystem composited with multiple ecological niches and habitats in which cellular species collaborate and compete; and (3) human beings are "super-organisms," which incorporate multiple symbiotic species into a massive individual. The complexity of human body and microbiome, especially gut microbiota, severally complicates the machine learning workflow.

Microbiome study has been growing with the advancement of machine learning in the past years. Similar to microarray data analysis, the sample-by-taxon abundance matrix is the most commonly used data structure in microbiome study. The machine learning methods are usually carried on such abundance data to determine which taxa differ between predefined groups of samples (e.g., diseased versus healthy) and build classification models by using these discriminatory taxa to predict the group of a new sample. The disadvantage at extracting salient features has limited the usage of traditional ecological assessment methods; thus, the classifying subject and body sites are the main goal for supervised classification. Usually, those available features include the taxon relative abundances, the α diversity and β diversity, and the general association between environment variable and operational taxonomy units (OTUs). The supervised classification can provide similar inspiration for microbiome study as those applied at microarray data [102]. Although random forest method does not provide clear important ranks on features, it has been widely applied and evaluated in many microbiome-end tasks. In a study to compare 18 major classification methods for microbiome studies [103], as a strongest performer, RFs are suitable for moderately sized microbial communities. And in another comparison study of 21 machine learning methods, random forests with optimized parameter are one of the best models at classifying cancer samples at gene expression microarray data [104], also efficient at microbiome data.

3.2.3 Deep Learning in Precision Medicine Study

Deep learning-based technologies have been successfully applied to learn the hidden representations of data with multiple levels of abstraction, which achieve great improvements in the conventional machine learning application fields, especially in domains such as drug discovery [105], regulatory genomics [106], computational biology [107], bioinformatics [108], human healthcare [109], and so on.

As the traditional data sources of biology, the genetic sequences can provide a large number of samples to feed the deep learning models [110]. For DNA sequences, a hybrid architecture combining a pre-trained, deep neural network and a hidden Markov model (DNN-HMM) has been built for the de novo identification of replication domains on DNA [111]; an open-source package Basset with deep convolutional neural networks is developed to learn the functional activity of DNA sequences from genomic data, e.g., DNase-seq, especially to annotate and interpret the noncoding genome [112]; and a deep learning-based hybrid architecture, BiRen, can predict enhancers using the DNA sequence alone [113]. For RNA sequences, a general and flexible deep learning framework for modeling structural binding preferences and predicting binding sites of RBPs takes (predicted) RNA tertiary structural information into account for the first time [114]; and DanQ, a hybrid convolutional and bidirectional long short-term memory recurrent neural network framework, is constructed to predict noncoding function de novo from sequence by learning regulatory "grammar" from the long-term dependencies between the sequence motifs [115]. Besides, for the protein sequences, a new deep learning method that predicts protein contact maps by integrating both evolutionary coupling and sequence conservation information is designed as an ultra-deep neural network to model contact occurrence patterns and complex sequence-structure relationship and has shown better quality than conventional template-based models [116]; and a computational program DeepConPred employed an effective scheme of two novel deep learning-based methods to identify optimal and important features for long-range residue contact prediction [117].

Recently, the precision medicine is developing rapidly, and many biological and biomedical images provide a new opportunity to introduce deep learning for enhancing the clinical practices [118]. Based on the high-content screening (HCS) technologies, large-scale imaging experiments are capable to study cell biology and for drug screening, and an approach combining deep convolutional neural networks (CNNs) with multiple-instance learning (MIL) is used to classify and segment such hundreds of thousands of microscopy images [119]. To overcome the pitfalls associated with conventional machine learning classifiers, a deep convolutional neural network (DeepLoc) is improving to analyze yeast cell images for automated classification of protein subcellular localization [120]. And a deep neural network is also applied to prospectively predict lineage choice in differentiating primary hematopoietic progenitors before conventional molecular markers are observable, by using image patches from bright-field microscopy and cellular movement [121]. Especially, a deep learning algorithm has been primarily used surface area information from magnetic resonance imaging of the brain of young individuals to efficiently predict the diagnosis of autism in old individual high-risk children [122]; and a single CNN, trained end to end from images directly, using only pixels and disease labels as inputs, can classify skin cancer with a level of competence comparable to dermatologists [123].

Although the deep learning has shown satisfied potential for analyzing omics data [124], the characteristic of biological high-throughput data as "small-sample high-dimension" is still a big challenge (*see* **Note 3**), and the "black box" of deep learning or other machine learning methods has usually hidden many useful readable information for biological or biomedical researches. Thus, it would be important to use multiple data resources to consistently improve collective health [125] in a discriminative and interpretative manner.

4 Notes

Totally, the machine learning plays an important role in current biological and biomedical researches. Especially, these computer-advanced technologies will be efficient to analyze the big biological data. However, different from conventional big social data, the big omics data are always "small-sample-high-dimension", which cause overwhelming application problems and also introduce new challenges.

1. The sample unbalance problem is usually discussed in the modeling of machine learning; some available solutions are resampling, one-class model or anomaly detection. But, in big biological data, the "extremely unbalance" problem exists, such as rare mutations or rare diseases, which is hard to obtain enough positive samples. *Thus, the prior-knowledge integrated methods are required to provide transferable learning methods to borrow (combine) multiple sources of data to assist the solution of single-sample analysis.*
2. A large number of machine learning models are "black box," which is enough to apply in social applications. However, in biological fields, the molecular mechanism underlying any

study is necessary to deepen our understanding, which means the "white-box" models are expected. *Although, some approaches have tried "gray-box" strategies to provide more biological details from the mathematical model, it is still urgent to develop new network-based theories and methods to balance the trade-off between accuracy and interpretability of machine learning in biological domains.*

3. The association is a "white clue" captured in conventional machine learning study and application, which is also a target of the conventional big data study. But, "causality" rather than association would be the key helpful for biologist, which can be used to determine the experimental target even the future research. *Therefore, how to obtain the causal relationship in biological community from the big biological data is asking for the new development of causality inference on small-sample high-dimension data.*

References

1. Ma S, Huang J (2008) Penalized feature selection and classification in bioinformatics. Brief Bioinform 9(5):392–403. https://doi.org/10.1093/bib/bbn027.bbn027[pii]
2. Rodriguez A, Laio A (2014) Machine learning. Clustering by fast search and find of density peaks. Science 344 (6191):1492–1496. https://doi.org/10.1126/science.1242072
3. Zeng T, Zhang W, Yu X, Liu X, Li M, Liu R, Chen L (2014) Edge biomarkers for classification and prediction of phenotypes. Sci China Life Sci 57(11):1103–1114. https://doi.org/10.1007/s11427-014-4757-4
4. Ghahramani Z (2015) Probabilistic machine learning and artificial intelligence. Nature 521 (7553):452–459. https://doi.org/10.1038/nature14541
5. Jordan MI, Mitchell TM (2015) Machine learning: trends, perspectives, and prospects. Science 349(6245):255–260. https://doi.org/10.1126/science.aaa8415
6. Lee EY, Fulan BM, Wong GC, Ferguson AL (2016) Mapping membrane activity in undiscovered peptide sequence space using machine learning. Proc Natl Acad Sci U S A 113(48):13588–13593. https://doi.org/10.1073/pnas.1609893113
7. Jones TR, Carpenter AE, Lamprecht MR, Moffat J, Silver SJ, Grenier JK, Castoreno AB, Eggert US, Root DE, Golland P, Sabatini DM (2009) Scoring diverse cellular morphologies in image-based screens with iterative feedback and machine learning. Proc Natl Acad Sci U S A 106(6):1826–1831. https://doi.org/10.1073/pnas.0808843106
8. King RD, Muggleton S, Lewis RA, Sternberg MJ (1992) Drug design by machine learning: the use of inductive logic programming to model the structure-activity relationships of trimethoprim analogues binding to dihydrofolate reductase. Proc Natl Acad Sci U S A 89 (23):11322–11326
9. Shipp MA, Ross KN, Tamayo P, Weng AP, Kutok JL, Aguiar RC, Gaasenbeek M, Angelo M, Reich M, Pinkus GS, Ray TS, Koval MA, Last KW, Norton A, Lister TA, Mesirov J, Neuberg DS, Lander ES, Aster JC, Golub TR (2002) Diffuse large B-cell lymphoma outcome prediction by gene-expression profiling and supervised machine learning. Nat Med 8(1):68–74. https://doi.org/10.1038/nm0102-68
10. Szarvas G, Farkas R, Busa-Fekete R (2007) State-of-the-art anonymization of medical records using an iterative machine learning framework. J Am Med Inform Assoc 14 (5):574–580. https://doi.org/10.1197/j.jamia.M2441
11. Obermeyer Z, Emanuel EJ (2016) Predicting the future - big data, machine learning, and clinical medicine. N Engl J Med 375 (13):1216–1219. https://doi.org/10.1056/NEJMp1606181
12. Passos IC, Mwangi B, Kapczinski F (2016) Big data analytics and machine learning:

2015 and beyond. Lancet Psychiatry 3 (1):13–15. https://doi.org/10.1016/S2215-0366(15)00549-0
13. Zeng T, Zhang W, Yu X, Liu X, Li M, Chen L (2016) Big-data-based edge biomarkers: study on dynamical drug sensitivity and resistance in individuals. Brief Bioinform 17 (4):576–592. https://doi.org/10.1093/bib/bbv078
14. Libbrecht MW, Noble WS (2015) Machine learning applications in genetics and genomics. Nat Rev Genet 16(6):321–332. https://doi.org/10.1038/nrg3920
15. Ma C, Zhang HH, Wang X (2014) Machine learning for Big Data analytics in plants. Trends Plant Sci 19(12):798–808. https://doi.org/10.1016/j.tplants.2014.08.004
16. Li Y, Wu FX, Ngom A (2016) A review on machine learning principles for multi-view biological data integration. Brief Bioinform. https://doi.org/10.1093/bib/bbw113
17. Fabris F, Magalhaes JP, Freitas AA (2017) A review of supervised machine learning applied to ageing research. Biogerontology 18 (2):171–188. https://doi.org/10.1007/s10522-017-9683-y
18. Walsh I, Pollastri G, Tosatto SC (2016) Correct machine learning on protein sequences: a peer-reviewing perspective. Brief Bioinform 17(5):831–840. https://doi.org/10.1093/bib/bbv082
19. Fu LM (2014) Machine learning and tubercular drug target recognition. Curr Pharm Des 20(27):4307–4318
20. Singh A, Ganapathysubramanian B, Singh AK, Sarkar S (2016) Machine learning for high-throughput stress phenotyping in plants. Trends Plant Sci 21(2):110–124. https://doi.org/10.1016/j.tplants.2015.10.015
21. Habibi N, Mohd Hashim SZ, Norouzi A, Samian MR (2014) A review of machine learning methods to predict the solubility of overexpressed recombinant proteins in Escherichia coli. BMC Bioinformatics 15:134. https://doi.org/10.1186/1471-2105-15-134
22. Walia RR, Caragea C, Lewis BA, Towfic F, Terribilini M, El-Manzalawy Y, Dobbs D, Honavar V (2012) Protein-RNA interface residue prediction using machine learning: an assessment of the state of the art. BMC Bioinformatics 13:89. https://doi.org/10.1186/1471-2105-13-89
23. Quinlan JR (1996) Improved use of continuous attributes in C4.5. J Artif Int Res 4 (1):77–90
24. Brin S, Page L (1998) The anatomy of a large-scale hypertextual Web search engine. Comput Netw ISDN Syst 30(1):107–117. https://doi.org/10.1016/S0169-7552(98)00110-X
25. Curtin RR, Cline JR, Slagle NP, March WB, Ram P, Mehta NA, Gray AG (2013) MLPACK: a scalable C++ machine learning library. J Mach Learn Res 14(1):801–805
26. Agrawal R, Srikant R (1994) Fast algorithms for mining association rules in large databases. In: Proceedings of the 20th International Conference on Very Large Data Bases, 1994. Morgan Kaufmann Publishers Inc., pp 487–499. doi:citeulike-article-id:217131
27. Dempster A, NM L, DB R (1977) Maximum likelihood from incomplete data via the EM algorithm. J R Stat Soc Series B Methodol 39 (1):1–38
28. Xie X, Wu S, Lam KM, Yan H (2006) PromoterExplorer: an effective promoter identification method based on the AdaBoost algorithm. Bioinformatics 22 (22):2722–2728. https://doi.org/10.1093/bioinformatics/btl482
29. Altman NS (1992) An introduction to Kernel and nearest-neighbor nonparametric regression. Am Stat 46(3):175–185
30. Rennie JDM (2003) Tackling the poor assumptions of naive Bayes text classifiers. In: Proceedings of the twentieth International Conference on Machine Learning (ICML-2003), Washington, DC, pp 616–623
31. Ma C, Xin M, Feldmann KA, Wang X (2014) Machine learning-based differential network analysis: a study of stress-responsive transcriptomes in Arabidopsis. Plant Cell 26 (2):520–537. https://doi.org/10.1105/tpc.113.121913
32. Zhang W, Feng D, Li R, Chernikov A, Chrisochoides N, Osgood C, Konikoff C, Newfeld S, Kumar S, Ji S (2013) A mesh generation and machine learning framework for Drosophila gene expression pattern image analysis. BMC Bioinformatics 14:372. https://doi.org/10.1186/1471-2105-14-372
33. De Ferrari L, Mitchell JB (2014) From sequence to enzyme mechanism using multi-label machine learning. BMC Bioinformatics 15:150. https://doi.org/10.1186/1471-2105-15-150
34. Savojardo C, Fariselli P, Casadio R (2013) BETAWARE: a machine-learning tool to detect and predict transmembrane beta-barrel proteins in prokaryotes. Bioinformatics 29

(4):504–505. https://doi.org/10.1093/bioinformatics/bts728

35. Hocking TD, Goerner-Potvin P, Morin A, Shao X, Pastinen T, Bourque G (2017) Optimizing ChIP-seq peak detectors using visual labels and supervised machine learning. Bioinformatics 33(4):491–499. https://doi.org/10.1093/bioinformatics/btw672
36. Hepler NL, Scheffler K, Weaver S, Murrell B, Richman DD, Burton DR, Poignard P, Smith DM, Kosakovsky Pond SL (2014) IDEPI: rapid prediction of HIV-1 antibody epitopes and other phenotypic features from sequence data using a flexible machine learning platform. PLoS Comput Biol 10(9):e1003842. https://doi.org/10.1371/journal.pcbi.1003842
37. Somarowthu S, Ondrechen MJ (2012) POOL server: machine learning application for functional site prediction in proteins. Bioinformatics 28(15):2078–2079. https://doi.org/10.1093/bioinformatics/bts321
38. Ryvkin P, Leung YY, Ungar LH, Gregory BD, Wang LS (2014) Using machine learning and high-throughput RNA sequencing to classify the precursors of small non-coding RNAs. Methods 67(1):28–35. https://doi.org/10.1016/j.ymeth.2013.10.002
39. Spinella JF, Mehanna P, Vidal R, Saillour V, Cassart P, Richer C, Ouimet M, Healy J, Sinnett D (2016) SNooPer: a machine learning-based method for somatic variant identification from low-pass next-generation sequencing. BMC Genomics 17(1):912. https://doi.org/10.1186/s12864-016-3281-2
40. Heinonen M, Shen H, Zamboni N, Rousu J (2012) Metabolite identification and molecular fingerprint prediction through machine learning. Bioinformatics 28(18):2333–2341. https://doi.org/10.1093/bioinformatics/bts437
41. Wu SG, Wang Y, Jiang W, Oyetunde T, Yao R, Zhang X, Shimizu K, Tang YJ, Bao FS (2016) Rapid prediction of bacterial heterotrophic fluxomics using machine learning and constraint programming. PLoS Comput Biol 12(4):e1004838. https://doi.org/10.1371/journal.pcbi.1004838
42. Yu T, Jones DP (2014) Improving peak detection in high-resolution LC/MS metabolomics data using preexisting knowledge and machine learning approach. Bioinformatics 30(20):2941–2948. https://doi.org/10.1093/bioinformatics/btu430
43. Vervier K, Mahe P, Tournoud M, Veyrieras JB, Vert JP (2016) Large-scale machine learning for metagenomics sequence classification. Bioinformatics 32(7):1023–1032. https://doi.org/10.1093/bioinformatics/btv683
44. Pasolli E, Truong DT, Malik F, Waldron L, Segata N (2016) Machine learning meta-analysis of large metagenomic datasets: tools and biological insights. PLoS Comput Biol 12(7):e1004977. https://doi.org/10.1371/journal.pcbi.1004977
45. Pybus M, Luisi P, Dall'Olio GM, Uzkudun M, Laayouni H, Bertranpetit J, Engelken J (2015) Hierarchical boosting: a machine-learning framework to detect and classify hard selective sweeps in human populations. Bioinformatics 31(24):3946–3952. https://doi.org/10.1093/bioinformatics/btv493
46. Magnan CN, Baldi P (2014) SSpro/ACCpro 5: almost perfect prediction of protein secondary structure and relative solvent accessibility using profiles, machine learning and structural similarity. Bioinformatics 30(18):2592–2597. https://doi.org/10.1093/bioinformatics/btu352
47. Cao R, Adhikari B, Bhattacharya D, Sun M, Hou J, Cheng J (2017) QAcon: single model quality assessment using protein structural and contact information with machine learning techniques. Bioinformatics 33(4):586–588. https://doi.org/10.1093/bioinformatics/btw694
48. Gangal R, Sharma P (2005) Human pol II promoter prediction: time series descriptors and machine learning. Nucleic Acids Res 33(4):1332–1336. https://doi.org/10.1093/nar/gki271
49. Mort M, Sterne-Weiler T, Li B, Ball EV, Cooper DN, Radivojac P, Sanford JR, Mooney SD (2014) MutPred Splice: machine learning-based prediction of exonic variants that disrupt splicing. Genome Biol 15(1):R19. https://doi.org/10.1186/gb-2014-15-1-r19
50. Busser BW, Taher L, Kim Y, Tansey T, Bloom MJ, Ovcharenko I, Michelson AM (2012) A machine learning approach for identifying novel cell type-specific transcriptional regulators of myogenesis. PLoS Genet 8(3):e1002531. https://doi.org/10.1371/journal.pgen.1002531
51. Sutphin GL, Mahoney JM, Sheppard K, Walton DO, Korstanje R (2016) WORMHOLE: novel least diverged ortholog prediction through machine learning. PLoS Comput Biol 12(11):e1005182. https://doi.org/10.1371/journal.pcbi.1005182
52. Li F, Li C, Wang M, Webb GI, Zhang Y, Whisstock JC, Song J (2015) GlycoMine: a machine learning-based approach for

predicting N-, C- and O-linked glycosylation in the human proteome. Bioinformatics 31 (9):1411–1419. https://doi.org/10.1093/bioinformatics/btu852

53. Kauffman C, Karypis G (2009) LIBRUS: combined machine learning and homology information for sequence-based ligand-binding residue prediction. Bioinformatics 25 (23):3099–3107. https://doi.org/10.1093/bioinformatics/btp561
54. Savojardo C, Fariselli P, Alhamdoosh M, Martelli PL, Pierleoni A, Casadio R (2011) Improving the prediction of disulfide bonds in Eukaryotes with machine learning methods and protein subcellular localization. Bioinformatics 27(16):2224–2230. https://doi.org/10.1093/bioinformatics/btr387
55. Aydin Z, Murray JI, Waterston RH, Noble WS (2010) Using machine learning to speed up manual image annotation: application to a 3D imaging protocol for measuring single cell gene expression in the developing C. elegans embryo. BMC Bioinformatics 11:84. https://doi.org/10.1186/1471-2105-11-84
56. Tsakanikas P, Manolakos ES (2011) Protein spot detection and quantification in 2-DE gel images using machine-learning methods. Proteomics 11(10):2038–2050. https://doi.org/10.1002/pmic.201000601
57. Hong SH, Cortesio CL, Drubin DG (2015) Machine-learning-based analysis in genome-edited cells reveals the efficiency of clathrin-mediated endocytosis. Cell Rep 12 (12):2121–2130. https://doi.org/10.1016/j.celrep.2015.08.048
58. Abu A, Leow LK, Ramli R, Omar H (2016) Classification of Suncus murinus species complex (Soricidae: Crocidurinae) in Peninsular Malaysia using image analysis and machine learning approaches. BMC Bioinformatics 17 (Suppl 19):505. https://doi.org/10.1186/s12859-016-1362-5
59. Hamp T, Rost B (2015) More challenges for machine-learning protein interactions. Bioinformatics 31(10):1521–1525. https://doi.org/10.1093/bioinformatics/btu857
60. Ballester PJ, Mitchell JB (2010) A machine learning approach to predicting protein-ligand binding affinity with applications to molecular docking. Bioinformatics 26 (9):1169–1175. https://doi.org/10.1093/bioinformatics/btq112
61. Reynes C, Host H, Camproux AC, Laconde G, Leroux F, Mazars A, Deprez B, Fahraeus R, Villoutreix BO, Sperandio O (2010) Designing focused chemical libraries enriched in protein-protein interaction inhibitors using machine-learning methods. PLoS Comput Biol 6(3):e1000695. https://doi.org/10.1371/journal.pcbi.1000695
62. Ding H, Takigawa I, Mamitsuka H, Zhu S (2014) Similarity-based machine learning methods for predicting drug-target interactions: a brief review. Brief Bioinform 15 (5):734–747. https://doi.org/10.1093/bib/bbt056
63. Bauer T, Eils R, Konig R (2011) RIP: the regulatory interaction predictor—a machine learning-based approach for predicting target genes of transcription factors. Bioinformatics 27(16):2239–2247. https://doi.org/10.1093/bioinformatics/btr366
64. Jha A, Shankar R (2011) Employing machine learning for reliable miRNA target identification in plants. BMC Genomics 12:636. https://doi.org/10.1186/1471-2164-12-636
65. Yousef M, Jung S, Kossenkov AV, Showe LC, Showe MK (2007) Naive Bayes for microRNA target predictions—machine learning for microRNA targets. Bioinformatics 23 (22):2987–2992. https://doi.org/10.1093/bioinformatics/btm484
66. Sturm M, Hackenberg M, Langenberger D, Frishman D (2010) TargetSpy: a supervised machine learning approach for microRNA target prediction. BMC Bioinformatics 11:292. https://doi.org/10.1186/1471-2105-11-292
67. Kim SY, Diggans J, Pankratz D, Huang J, Pagan M, Sindy N, Tom E, Anderson J, Choi Y, Lynch DA, Steele MP, Flaherty KR, Brown KK, Farah H, Bukstein MJ, Pardo A, Selman M, Wolters PJ, Nathan SD, Colby TV, Myers JL, Katzenstein AL, Raghu G, Kennedy GC (2015) Classification of usual interstitial pneumonia in patients with interstitial lung disease: assessment of a machine learning approach using high-dimensional transcriptional data. Lancet Respir Med 3 (6):473–482. https://doi.org/10.1016/S2213-2600(15)00140-X
68. Richardson AM, Lidbury BA (2013) Infection status outcome, machine learning method and virus type interact to affect the optimised prediction of hepatitis virus immunoassay results from routine pathology laboratory assays in unbalanced data. BMC Bioinformatics 14:206. https://doi.org/10.1186/1471-2105-14-206
69. Liu Q, Sung AH, Chen Z, Liu J, Chen L, Qiao M, Wang Z, Huang X, Deng Y (2011) Gene selection and classification for cancer microarray data based on machine learning

and similarity measures. BMC Genomics 12 (Suppl 5):S1. https://doi.org/10.1186/1471-2164-12-S5-S1

70. Wang Z, Wen X, Lu Y, Yao Y, Zhao H (2016) Exploiting machine learning for predicting skeletal-related events in cancer patients with bone metastases. Oncotarget 7 (11):12612–12622. https://doi.org/10.18632/oncotarget.7278
71. Way GP, Allaway RJ, Bouley SJ, Fadul CE, Sanchez Y, Greene CS (2017) A machine learning classifier trained on cancer transcriptomes detects NF1 inactivation signal in glioblastoma. BMC Genomics 18(1):127. https://doi.org/10.1186/s12864-017-3519-7
72. Mani S, Chen Y, Li X, Arlinghaus L, Chakravarthy AB, Abramson V, Bhave SR, Levy MA, Xu H, Yankeelov TE (2013) Machine learning for predicting the response of breast cancer to neoadjuvant chemotherapy. J Am Med Inform Assoc 20(4):688–695. https://doi.org/10.1136/amiajnl-2012-001332
73. De Bari B, Vallati M, Gatta R, Lestrade L, Manfrida S, Carrie C, Valentini V (2016) Development and validation of a machine learning-based predictive model to improve the prediction of inguinal status of anal cancer patients: a preliminary report. Oncotarget. https://doi.org/10.18632/oncotarget.10749
74. Lee BJ, Shin MS, YJ O, Oh HS, Ryu KH (2009) Identification of protein functions using a machine-learning approach based on sequence-derived properties. Proteome Sci 7:27. https://doi.org/10.1186/1477-5956-7-27
75. Mordelet F, Vert JP (2011) ProDiGe: Prioritization Of Disease Genes with multitask machine learning from positive and unlabeled examples. BMC Bioinformatics 12:389. https://doi.org/10.1186/1471-2105-12-389
76. Torii M, Wagholikar K, Liu H (2011) Using machine learning for concept extraction on clinical documents from multiple data sources. J Am Med Inform Assoc 18 (5):580–587. https://doi.org/10.1136/amiajnl-2011-000155
77. Jiang M, Chen Y, Liu M, Rosenbloom ST, Mani S, Denny JC, Xu H (2011) A study of machine-learning-based approaches to extract clinical entities and their assertions from discharge summaries. J Am Med Inform Assoc 18(5):601–606. https://doi.org/10.1136/amiajnl-2011-000163
78. Ware H, Mullett CJ, Jagannathan V, El-Rawas O (2012) Machine learning-based coreference resolution of concepts in clinical documents. J Am Med Inform Assoc 19 (5):883–887. https://doi.org/10.1136/amiajnl-2011-000774
79. Wu Y, Levy MA, Micheel CM, Yeh P, Tang B, Cantrell MJ, Cooreman SM, Xu H (2012) Identifying the status of genetic lesions in cancer clinical trial documents using machine learning. BMC Genomics 13(Suppl 8):S21. https://doi.org/10.1186/1471-2164-13-S8-S21
80. Bravo A, Li TS, Su AI, Good BM, Furlong LI (2016) Combining machine learning, crowdsourcing and expert knowledge to detect chemical-induced diseases in text. Database. https://doi.org/10.1093/database/baw094
81. Yip KY, Cheng C, Gerstein M (2013) Machine learning and genome annotation: a match meant to be? Genome Biol 14(5):205. https://doi.org/10.1186/gb-2013-14-5-205
82. Xu X, Hoang S, Mayo MW, Bekiranov S (2010) Application of machine learning methods to histone methylation ChIP-Seq data reveals H4R3me2 globally represses gene expression. BMC Bioinformatics 11:396. https://doi.org/10.1186/1471-2105-11-396
83. De Santis M, Rinaldi F, Falcone E, Lucidi S, Piaggio G, Gurtner A, Farina L (2014) Combining optimization and machine learning techniques for genome-wide prediction of human cell cycle-regulated genes. Bioinformatics 30(2):228–233. https://doi.org/10.1093/bioinformatics/btt671
84. Kangas LJ, Metz TO, Isaac G, Schrom BT, Ginovska-Pangovska B, Wang L, Tan L, Lewis RR, Miller JH (2012) In silico identification software (ISIS): a machine learning approach to tandem mass spectral identification of lipids. Bioinformatics 28(13):1705–1713. https://doi.org/10.1093/bioinformatics/bts194
85. Acharjee A, Ament Z, West JA, Stanley E, Griffin JL (2016) Integration of metabolomics, lipidomics and clinical data using a machine learning method. BMC Bioinformatics 17(Suppl 15):440. https://doi.org/10.1186/s12859-016-1292-2
86. Gorkin DU, Lee D, Reed X, Fletez-Brant C, Bessling SL, Loftus SK, Beer MA, Pavan WJ, McCallion AS (2012) Integration of ChIP-seq and machine learning reveals enhancers and a predictive regulatory sequence

vocabulary in melanocytes. Genome Res 22 (11):2290–2301. https://doi.org/10.1101/gr.139360.112

87. Ding Y, Tang S, Liao SG, Jia J, Oesterreich S, Lin Y, Tseng GC (2014) Bias correction for selecting the minimal-error classifier from many machine learning models. Bioinformatics 30(22):3152–3158. https://doi.org/10.1093/bioinformatics/btu520
88. Yang Y, Wu QM (2016) Extreme learning machine with subnetwork hidden nodes for regression and classification. IEEE Trans Cybernetics 46(12):2885–2898. https://doi.org/10.1109/TCYB.2015.2492468
89. Kircher M, Stenzel U, Kelso J (2009) Improved base calling for the Illumina Genome Analyzer using machine learning strategies. Genome Biol 10(8):R83. https://doi.org/10.1186/gb-2009-10-8-r83
90. Ratsch G, Sonnenburg S, Srinivasan J, Witte H, Muller KR, Sommer RJ, Scholkopf B (2007) Improving the Caenorhabditis elegans genome annotation using machine learning. PLoS Comput Biol 3(2):e20. https://doi.org/10.1371/journal.pcbi.0030020
91. Chou WC, Ma Q, Yang S, Cao S, Klingeman DM, Brown SD, Xu Y (2015) Analysis of strand-specific RNA-seq data using machine learning reveals the structures of transcription units in Clostridium thermocellum. Nucleic Acids Res 43(10):e67. https://doi.org/10.1093/nar/gkv177
92. Capriotti E, Altman RB (2011) A new disease-specific machine learning approach for the prediction of cancer-causing missense variants. Genomics 98(4):310–317. https://doi.org/10.1016/j.ygeno.2011.06.010
93. Swan AL, Stekel DJ, Hodgman C, Allaway D, Alqahtani MH, Mobasheri A, Bacardit J (2015) A machine learning heuristic to identify biologically relevant and minimal biomarker panels from omics data. BMC Genomics 16(Suppl 1):S2. https://doi.org/10.1186/1471-2164-16-S1-S2
94. Shi P, Ray S, Zhu Q, Kon MA (2011) Top scoring pairs for feature selection in machine learning and applications to cancer outcome prediction. BMC Bioinformatics 12:375. https://doi.org/10.1186/1471-2105-12-375
95. Lyons-Weiler J, Patel S, Bhattacharya S (2003) A classification-based machine learning approach for the analysis of genome-wide expression data. Genome Res 13(3):503–512. https://doi.org/10.1101/gr.104003
96. Held E, Cape J, Tintle N (2016) Comparing machine learning and logistic regression methods for predicting hypertension using a combination of gene expression and next-generation sequencing data. BMC Proc 10 (Suppl 7):141–145. https://doi.org/10.1186/s12919-016-0020-2
97. Jamal S, Goyal S, Shanker A, Grover A (2016) Integrating network, sequence and functional features using machine learning approaches towards identification of novel Alzheimer genes. BMC Genomics 17(1):807. https://doi.org/10.1186/s12864-016-3108-1
98. Ma B, Charkowski AO, Glasner JD, Perna NT (2014) Identification of host-microbe interaction factors in the genomes of soft rot-associated pathogens Dickeya dadantii 3937 and Pectobacterium carotovorum WPP14 with supervised machine learning. BMC Genomics 15:508. https://doi.org/10.1186/1471-2164-15-508
99. Wang X, Su X, Cui X, Ning K (2015) MetaBoot: a machine learning framework of taxonomical biomarker discovery for different microbial communities based on metagenomic data. PeerJ 3:e993. https://doi.org/10.7717/peerj.993
100. Khanna S, Tosh PK (2014) A clinician's primer on the role of the microbiome in human health and disease. Mayo Clin Proc 89(1):107–114. https://doi.org/10.1016/j.mayocp.2013.10.011
101. Drenthen MAM, Jozef KFW, Proctor J (2009) New visions of nature. Springer, New York
102. Knights D, Costello EK, Knight R (2011) Supervised classification of human microbiota. FEMS Microbiol Rev 35(2):343–359. https://doi.org/10.1111/j.1574-6976.2010.00251.x
103. Statnikov A, Henaff M, Narendra V, Konganti K, Li Z, Yang L, Pei Z, Blaser MJ, Aliferis CF, Alekseyenko AV (2013) A comprehensive evaluation of multicategory classification methods for microbiomic data. Microbiome 1(1):11. https://doi.org/10.1186/2049-2618-1-11
104. Statnikov A, Aliferis CF, Tsamardinos I, Hardin D, Levy S (2005) A comprehensive evaluation of multicategory classification methods for microarray gene expression cancer diagnosis. Bioinformatics 21(5):631–643. https://doi.org/10.1093/bioinformatics/bti033
105. LeCun Y, Bengio Y, Hinton G (2015) Deep learning. Nature 521(7553):436–444. https://doi.org/10.1038/nature14539

106. Park Y, Kellis M (2015) Deep learning for regulatory genomics. Nat Biotechnol 33 (8):825–826. https://doi.org/10.1038/nbt.3313
107. Angermueller C, Parnamaa T, Parts L, Stegle O (2016) Deep learning for computational biology. Mol Syst Biol 12(7):878. https://doi.org/10.15252/msb.20156651
108. Min S, Lee B, Yoon S (2016) Deep learning in bioinformatics. Brief Bioinform. https://doi.org/10.1093/bib/bbw068
109. Miotto R, Wang F, Wang S, Jiang X, Dudley JT (2017) Deep learning for healthcare: review, opportunities and challenges. Brief Bioinform. https://doi.org/10.1093/bib/bbx044
110. Uziela K, Menendez Hurtado D, Shu N, Wallner B, Elofsson A (2017) ProQ3D: improved model quality assessments using deep learning. Bioinformatics 33 (10):1578–1580. https://doi.org/10.1093/bioinformatics/btw819
111. Liu F, Ren C, Li H, Zhou P, Bo X, Shu W (2016) De novo identification of replication-timing domains in the human genome by deep learning. Bioinformatics 32 (5):641–649. https://doi.org/10.1093/bioinformatics/btv643
112. Kelley DR, Snoek J, Rinn JL (2016) Basset: learning the regulatory code of the accessible genome with deep convolutional neural networks. Genome Res 26(7):990–999. https://doi.org/10.1101/gr.200535.115
113. Yang B, Liu F, Ren C, Ouyang Z, Xie Z, Bo X, Shu W (2017) BiRen: predicting enhancers with a deep-learning-based model using the DNA sequence alone. Bioinformatics. https://doi.org/10.1093/bioinformatics/btx105
114. Zhang S, Zhou J, Hu H, Gong H, Chen L, Cheng C, Zeng J (2016) A deep learning framework for modeling structural features of RNA-binding protein targets. Nucleic Acids Res 44(4):e32. https://doi.org/10.1093/nar/gkv1025
115. Quang D, Xie X (2016) DanQ: a hybrid convolutional and recurrent deep neural network for quantifying the function of DNA sequences. Nucleic Acids Res 44(11):e107. https://doi.org/10.1093/nar/gkw226
116. Wang S, Sun S, Li Z, Zhang R, Xu J (2017) Accurate de novo prediction of protein contact map by ultra-deep learning model. PLoS Comput Biol 13(1):e1005324. https://doi.org/10.1371/journal.pcbi.1005324
117. Xiong D, Zeng J, Gong H (2017) A deep learning framework for improving long-range residue-residue contact prediction using a hierarchical strategy. Bioinformatics. https://doi.org/10.1093/bioinformatics/btx296
118. Yuan Y, Shi Y, Li C, Kim J, Cai W, Han Z, Feng DD (2016) DeepGene: an advanced cancer type classifier based on deep learning and somatic point mutations. BMC Bioinformatics 17(Suppl 17):476. https://doi.org/10.1186/s12859-016-1334-9
119. Kraus OZ, Ba JL, Frey BJ (2016) Classifying and segmenting microscopy images with deep multiple instance learning. Bioinformatics 32 (12):i52–i59. https://doi.org/10.1093/bioinformatics/btw252
120. Kraus OZ, Grys BT, Ba J, Chong Y, Frey BJ, Boone C, Andrews BJ (2017) Automated analysis of high-content microscopy data with deep learning. Mol Syst Biol 13(4):924. https://doi.org/10.15252/msb.20177551
121. Buggenthin F, Buettner F, Hoppe PS, Endele M, Kroiss M, Strasser M, Schwarzfischer M, Loeffler D, Kokkaliaris KD, Hilsenbeck O, Schroeder T, Theis FJ, Marr C (2017) Prospective identification of hematopoietic lineage choice by deep learning. Nat Methods 14(4):403–406. https://doi.org/10.1038/nmeth.4182
122. Hazlett HC, Gu H, Munsell BC, Kim SH, Styner M, Wolff JJ, Elison JT, Swanson MR, Zhu H, Botteron KN, Collins DL, Constantino JN, Dager SR, Estes AM, Evans AC, Fonov VS, Gerig G, Kostopoulos P, McKinstry RC, Pandey J, Paterson S, Pruett JR, Schultz RT, Shaw DW, Zwaigenbaum L, Piven J, IBIS Network; Clinical Sites; Data Coordinating Center; Image Processing Core; Statistical Analysis (2017) Early brain development in infants at high risk for autism spectrum disorder. Nature 542 (7641):348–351. https://doi.org/10.1038/nature21369
123. Esteva A, Kuprel B, Novoa RA, Ko J, Swetter SM, Blau HM, Thrun S (2017) Dermatologist-level classification of skin cancer with deep neural networks. Nature 542 (7639):115–118. https://doi.org/10.1038/nature21056
124. Chen Y, Li Y, Narayan R, Subramanian A, Xie X (2016) Gene expression inference with deep learning. Bioinformatics 32 (12):1832–1839. https://doi.org/10.1093/bioinformatics/btw074
125. Chen JH, Asch SM (2017) Machine learning and prediction in medicine - beyond the peak of inflated expectations. N Engl J Med 376 (26):2507–2509. https://doi.org/10.1056/NEJMp1702071

Identifying Interactions Between Long Noncoding RNAs and Diseases Based on Computational Methods

Wei Lan, Liyu Huang, Dehuan Lai, and Qingfeng Chen

Abstract

With the development and improvement of next-generation sequencing technology, a great number of noncoding RNAs have been discovered. Long noncoding RNAs (lncRNAs) are the biggest kind of noncoding RNAs with more than 200 nt nucleotides in length. There are increasing evidences showing that lncRNAs play key roles in many biological processes. Therefore, the mutation and dysregulation of lncRNAs have close association with a number of complex human diseases. Identifying the most likely interaction between lncRNAs and diseases becomes a fundamental challenge in human health. A common view is that lncRNAs with similar function tend to be related to phenotypic similar diseases. In this chapter, we firstly introduce the concept of lncRNA, their biological features, and available data resources. Further, the recent computational approaches are explored to identify interactions between long noncoding RNAs and diseases, including their advantages and disadvantages. The key issues and potential future works of predicting interactions between long noncoding RNAs and diseases are also discussed.

Key words Long noncoding RNA, Human disease, Biological networks, Heterogeneous data fusion, Machine learning

1 Introduction

With the completion of human genome project, only 2% in human genome were identified [1–3]. It means that about 98% of human genome does not encode protein sequences and these genome regions usually transcribe as noncoding RNA [4–6]. For a long time, noncoding RNAs (ncRNAs) have been viewed as transcriptional noise in biology [7–10]. However, more and more researches have shown that noncoding RNAs perform critical roles in many biological processes [11–14]. There are different kinds of functionally important RNAs such as small nucleolar RNA (snoRNA), microRNA (miRNA), small interfering RNA (siRNA), extracellular RNA (exRNA), piwi-interacting RNA (piRNA), small Cajal body RNA (scaRNA), and long noncoding RNA (lncRNA) [15, 16].

Tao Huang (ed.), *Computational Systems Biology: Methods and Protocols*, Methods in Molecular Biology, vol. 1754, https://doi.org/10.1007/978-1-4939-7717-8_12, © Springer Science+Business Media, LLC, part of Springer Nature 2018

Long noncoding RNAs are the group of noncoding RNA with the lengths more than 200 nucleotides. Recently, lncRNAs have attracted increasing attentions from biological researchers [17]. A number of evidences have demonstrated that lncRNA is essential in many biological processes such as RNA modification chromosome dosage compensation, genomic imprinting, epigenetic regulation, nuclear and cytoplasmic trafficking, cell proliferation, cell differentiation, cell growth, cell metabolism, cell apoptosis, etc. [18–20]. Furthermore, accumulating studies have proved that the mutations and dysregulations of lncRNAs have close association with many complex human diseases, such as breast cancer, cervix cancer, lung cancer, esophagus cancer, ovarian cancer, parotid cancer, tongue cancer, renal disease, rhabdomyosarcoma, cardiomyopathy, leukemia, dyskeratosis congenital, pancreaticobiliary maljunction, squamous carcinoma, Klinefelter's syndrome, autoimmune thyroid disease, ductal carcinoma, etc. [21, 22]. For example, MALAT1 (also known as NEAT2) was found to be highly expressed in lung cancer, and it was used as early prognostic marker for poor patient survival rates [23]. Figure 1 shows the abnormity of lncRNA-disease network. The triangle and rectangle denote lncRNA and disease, respectively. It can be observed that the mutation and dysregulation of lncRNA can cause disease.

Despite the results of lncRNA-disease association, the determination of the most likely lncRNA with disease is still a big challenge for molecular biologists and medical geneticists [24, 25]. Due to limitations of experimental approaches such as time and labor, it is appealing to develop efficient computational methods to tackle this obstacle. Recently, several computational approaches have been proposed to predict the interactions between lncRNAs and diseases [26]. This chapter aims at offering the state of arts of algorithms and tools used to prioritize candidate lncRNAs related to disease, by which to assist readers in catching up with recent and important developments in this filed. The paper is organized as follows: In Subheading 2, some available data resources are presented. Recent computational approaches for lncRNA-disease association identification are provided in Subheading 3. Subheading 4 highlights the key issues and the future works.

2 Biological Data Resources

Recently, with the rapid increase of biological data, some specific databases have been built to store and manage the data. In this section, we describe the public databases of disease and lncRNA. The overviews of disease and lncRNA databases are showed in Tables 1 and 2, respectively.

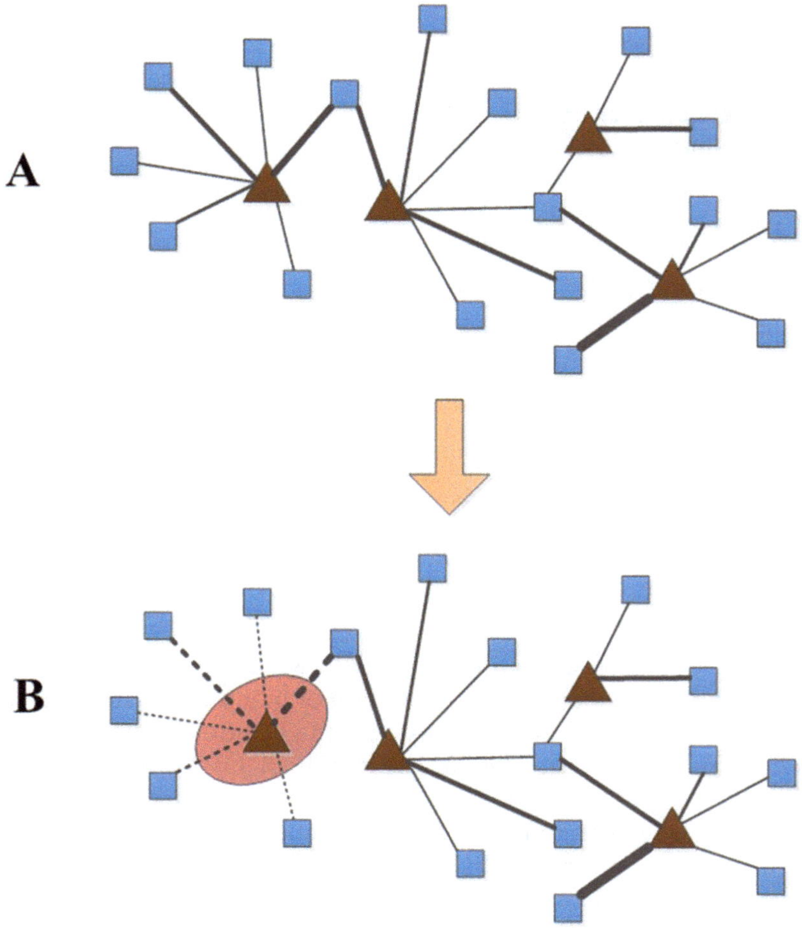

Fig. 1 The abnormity of lncRNA-disease network. The triangle and rectangle denote lncRNA and disease, respectively. (**a**) The normal lncRNA-disease interaction network. (**b**) The lncRNA mutation or perturbation in the lncRNA-disease network

Table 1
An overview of disease databases

Database	Records	URL	References
Disease Ontology	8043	http://disease-ontology.org	[27]
Medical Subject Headings	87,000	https://www.ncbi.nlm.nih.gov/mesh	
OMIM	24,065	http://www.omim.org/	[28]
GAD	167,130	http://geneticassociationdb.nih.gov/	[29]
DisGeNET	429,036	http://ibi.imim.es/DisGeNET/web/v02/home/	[30]

Table 2
An overview of lncRNA databases

Database	Number of lncRNA entries	URL	Reference
LNCipedia	113,513	http://www.lncipedia.org	[31]
NONCODE	527,336	http://www.bioinfo.org/noncode	[32]
LncRBase	216,562	http://bicresources.jcbose.ac.in/zhumur/lncrbase	[33]
lncRNAWiki	105,255	http://lncrna.big.ac.cn	[34]
lncRNAdb	About 300	http://www.lncrnadb.org	[35]
LncRNADisease	More than 1000	http://www.cuilab.cn/lncrnadisease	[36]
Lnc2Cancer	666	http://www.bio-bigdata.net/lnc2cancer	[37]
MNDR	807	http://www.rna-society.org/mndr	[38]

2.1 Disease Database

Disease Ontology (http://disease-ontology.org) [27] is an open database of the integration of biomedical data related with human diseases which is hosted at the Institute for Genome Sciences at the University Of Maryland School Of Medicine. It contains 8043 inherited, developmental, and acquired human diseases and aims to design to link disparate datasets through disease concepts.

Medical Subject Headings (https://www.ncbi.nlm.nih.gov/mesh) is a comprehensive controlled vocabulary thesaurus used for indexing articles in the life sciences. There are 27,883 descriptors and over 87,000 entry terms that are arranged in a hierarchy tree.

Online Mendelian Inheritance in Man (OMIM) [28] is an online knowledgebase of human genes and genetic disorders, which is maintained by McKusick-Nathans Institute of Genetic Medicine, Johns Hopkins University School of Medicine. Up to May 26th, 2017, there are 24,065 entries: 15,598 for gene description, 78 for the combination of genes and phenotypes, 5002 for molecular basis known phenotype description, 1602 for molecular basis unknown phenotype description, and 1785 for phenotypes with suspected Mendelian basis.

The Genetic Association Database (GAD) [29] is a comprehensive database of genetic association data from complex diseases and disorders. There are 5526 diseases classified into 19 categories such as cancer, aging, immune, and so on. It contains 167,130 records, out of which 82,285 records are given a description of whether they are reported to be associated or not with the disease phenotype for that specific record or not.

The DisGeNET database [30] is an integrated database of human gene-disease associations from various expert-curated databases including UniProt, Comparative Toxicogenomics Database, ClinVar, Orphanet, NHGRI-EBI GWAS Catalog, etc. In current version, it contains 429,036 associations, between 17,381 genes and 15,093 diseases and phenotypes.

2.2 lncRNA Database

LNCipedia (http://www.lncipedia.org/) [31] is a comprehensive database of human lncRNA database which contains 113,513 annotated human lncRNA transcripts by integrating different databases. In addition, it provides tools to predict protein-coding potential.

NONCODE (http://www.bioinfo.org/noncode/) [32] is an integrated knowledgebase including almost all noncoding RNA. It contains 16 species (human, mouse, cow, rat, chimpanzee, gorilla, orangutan, rhesus macaque, opossum, platypus, chicken, zebra fish, fruit fly, *Caenorhabditis elegans*, yeast, and *Arabidopsis*) and 527,336 lncRNA transcripts in the latest version.

LncRBase (http://bicresources.jcbose.ac.in/zhumur/lncrbase) [33] is a repository of human and mouse lncRNA which contains 216,562 lncRNA transcript entries in total. In this database, it collected basic lncRNA transcript features, genomic location, overlapping small noncoding RNAs, associated Repeat Elements, associated imprinted genes, and lncRNA promoter information.

lncRNAWiki (http://lncrna.big.ac.cn) [34] is a comprehensive resource of human long noncoding RNAs knowledge. 105,255 non-redundant lncRNA transcripts are obtained by integrating three databases: GENCODE, NONCODE, and LNCipedia. In addition, based on lncRNA genomic location in respect to protein-coding genes, lncRNAs are classified into seven groups (intergenic, intronic (S), intronic (AS), overlapping (S), overlapping (AS), sense, and antisense) in the database.

lncRNAdb (http://www.lncrnadb.org) [35] is a knowledge resource which aims to provide comprehensive annotations of eukaryotic long noncoding RNAs. In this database, every entry contains sequence, structure, genomic context, gene expression, subcellular localization, conservation, and function information of lncRNA. Users can search and download lncRNA information and submit new entries.

LncRNADisease (http://www.cuilab.cn/lncrnadisease) [36] is a manually curated database of experimentally supported lncRNA-disease associations. It includes more than 1000 lncRNA-disease entries and 475 lncRNA interaction entries, including 321 lncRNAs and 221 diseases from ~500 publications.

Lnc2Cancer (http://www.bio-bigdata.net/lnc2cancer) [37] is a resource of cancer-related lncRNAs which contains 1488 entries of associations between 666 human lncRNAs and 97 human cancers through review of more than 2000 published papers. It also provides the lncRNA expression pattern, experimental techniques, a brief functional description, the original reference, and additional annotation information. Further, it allows users to search, browse, and download the data or submit new data to the database.

MNDR (http://www.rna-society.org/mndr) [38] is an online knowledgebase of mammal ncRNA-disease relationships in mammals that aims to provide a platform to globally view the ncRNA-mediated disease network. In the present version, it contains 807 - lncRNA-associated, 229 miRNA-associated, 13 piRNA-associated, and 100 snoRNA-associated.

3 Predicting lncRNA-Disease Associations Based on Computational Methods

In the following, we review computational approaches for identification of lncRNA-disease interaction. The core assumption of lncRNA-disease interaction prediction is that functional similar lncRNAs with similar function are likely to relate with phenotypic similar diseases.

3.1 Information Propagation-Based Methods

It has been demonstrated that phenotypically similar diseases often share a set of functional similar lncRNAs [39–43]. According to this observation, several information propagation-based methods have been proposed to predict disease-related lncRNA.

Several studies have been developed to predict disease-related lncRNA based on lncRNA similarity network. This is made by integrating different biological data resources. Then, the information propagation method is employed to predict novel disease-related lncRNAs.

Sun et al. [44] presented a network-based method, RWRlncD, to infer potential human lncRNA-disease associations based on lncRNA functional similarity network. They constructed lncRNA similarity network by integrating lncRNA-disease association and Disease Ontology information. The random walk with restart method was employed to prioritize disease-related lncRNA on the function similarity network. The RWRlncD was robust to different parameter selections. Similar work has been developed by Cheng et al. [45]; they developed an integrative framework, IntNetLncSim, for lncRNA-disease interaction inference. They constructed lncRNA function similarity network by integrating lncRNA-regulatory network, mRNA-mRNA interaction network, and miRNA-mRNA interaction network. ITM Probe was applied for assigning a weight to each mRNA and miRNA for lncRNA and the cosine similarity was implemented for calculating disease similarity. The random walk with restart was used to predict disease-related lncRNAs. The performance of IntNetLncSim is superior to RWRlncD methods. Considering the limitations of traditional random walk with restart (RWR), Chen et al. [46] developed a model,

IRWRLDA, for lncRNA-disease interaction prediction. The work constructed an lncRNA similarity network by integrating known lncRNA-disease associations, disease semantic similarity, and various lncRNA similarity measures. The novelty of IRWRLDA lies in the incorporation of lncRNA expression similarity and disease semantic similarity to set the initial probability vector of the RWR. Therefore, IRWRLDA could be applied to diseases without any known related lncRNAs. In addition, Zhou et al. [47] constructed an lncRNA-lncRNA crosstalk network by examining the significant co-occurrence of shared miRNA response elements on lncRNA transcripts from the competing endogenous RNAs viewpoint. As expected, functional analysis showed that lncRNAs sharing significantly enriched interacting miRNAs tend to be involved in similar diseases and have more functionally related flanking gene sets. They further proposed a novel rank-based method, RWRHLD, to prioritize candidate lncRNA-disease associations by integrating three networks (miRNA-associated lncRNA-lncRNA crosstalk network, disease-disease similarity network, and known lncRNA-disease association network) into a heterogeneous network and implementing a random walk with restart on this heterogeneous network. Based on knowledge that lncRNAs have relationship with disease by regulating the expression of disease gene, Alaimo et al. [48] presented an information propagation method, ncPred, for novel ncRNA-disease association inference. In the first step, they constructed a tripartite network based on two levels of interaction: ncRNA-target and target-disease. Then, the network-based inference method was employed to predict potential lncRNA-disease associations.

Considering the difficulty of lncRNA similarity network construction, some researchers try to construct multi-level network of lncRNA-disease. Then, new disease-related lncRNAs are identified based on multi-level network.

Liu et al. [49] developed a computational method to identify candidate cancer-related lncRNAs based on interactions between protein-coding genes (PCGs) and lncRNAs. They constructed the lncRNA-PCG bipartite network of prostate cancer by combining expression profiles of lncRNAs and PCGs and protein-protein interactions. Six prostate cancer-related lncRNAs were included in this bipartite network. Based on this network, the random walk method was utilized to identify lncRNA related with prostate cancer. This method found that lncRNA ENSG00000261777 shares an intron with DDX19 and interacts with IGF2 P1, indicating its involvement in prostate cancer. In consideration that phenotype data directly reflected disease association, Yao et al. [50] proposed a novel algorithm, LncPriCNet, to prioritize candidate lncRNAs associated with diseases based on a multi-level composite network. They constructed a composite network by combining phenotype-phenotype interactions, lncRNA-lncRNA interactions, and gene-

gene interactions with disease-lncRNA relationships. Then, they prioritized disease-related lncRNA by using the random walking with restart (RWR) algorithm. Notably, LncPriCNet still performs well when information on known disease lncRNAs is insufficient. The reason may be that it considered the global functional interactions of the multi-level composite network. It is well known that special disease related with special tissue. Based on this knowledge, Ganegoda et al. [51] presented a computational method, KRWRH, to predict disease-lincRNA interactions based on phenotype information and lincRNA tissue expression details. They used Gaussian interaction profile kernel to calculate the similarity of diseases and lincRNAs, respectively. Then, the random walk with restart method is utilized to infer lincRNA-disease interactions.

3.2 Machine Learning Methods

Machine learning is a useful tool to prioritize candidate lncRNAs by training classifiers with features of known disease-related lncRNAs and unknown lncRNAs. Supervised machine learning prioritizes candidate lncRNAs based on the differences between disease-related lncRNAs and unknown lncRNAs of biological features [52–55].

Zhao et al. [56] proposed a computational model for cancer-related lncRNA identification by integrating genome, regulome, and transcriptome data. The naive Bayesian classifier was employed to classify lncRNA, and Database for Annotation, Visualization and Integrated Discovery (DAVID) was used for enrichment analysis. The results showed integration of multi-omic data can improve the performance of cancer-related lncRNA prediction. In addition, they predicted 707 potential cancer-related lncRNAs and found that these lncRNAs tend to exhibit significant differential expression and differential DNA methylation in multiple cancer types and prognosis effects in prostate cancer.

Considering the imbalance between known and unknown lncRNA-disease interactions, Lan et al. [57] presented a positive-unlabeled (PU) learning for discovering lncRNA-disease associations based on multiple data resources. Two lncRNA similarity and five disease similarity methods were employed to calculate similarities between lncRNA and lncRNA and between disease and disease, respectively. They used the geometric mean of matrix to fuse lncRNA and disease similarities, respectively. The bagging SVM is employed to identify potential lncRNA-disease associations. Figure 2 shows the flowchart of LDAP. Finally, this method is implemented as a web server (http://bioinformatics.csu.edu.cn/ldap) for new lncRNA-disease prediction. The LDAP took the input lncRNA sequence in FASTA format, either a pasted sequence or a file with multiple sequences (size limit <50 kb). The user can paste the sequence with FASTA format into textbox (1) and click the "submit" button (2) to submit single sequence. Figure 3 shows the usage guideline of LDAP by pasting a sequence. When the user

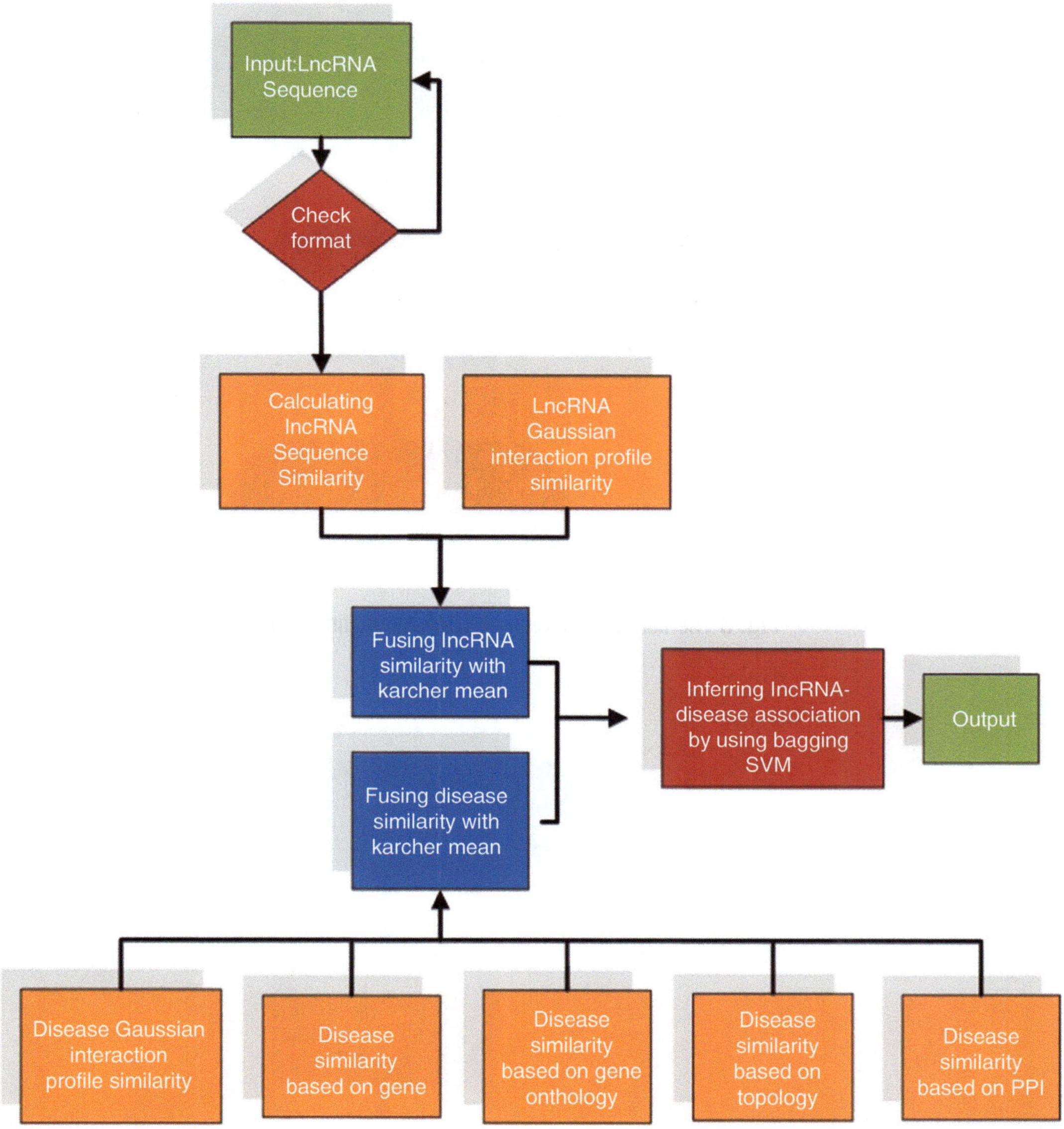

Fig. 2 The flow chart of LDAP

submits text file with multiple sequences, the user uploads the text with multiple sequences (1) and provides email address (2) and then clicks the "submit" button (3); the user can close the page when submitting successfully. The notification will be sent to the user in term of email when task is completed. The user can check their result according to job id and email (4). In addition, the user can download the results (5) and search the result with disease name (6). Figure 4 shows the usage guideline of LDAP by uploading multiple sequences.

Supervised learning methods based on hypothesis that the separation of disease-related lncRNAs and non-disease-related

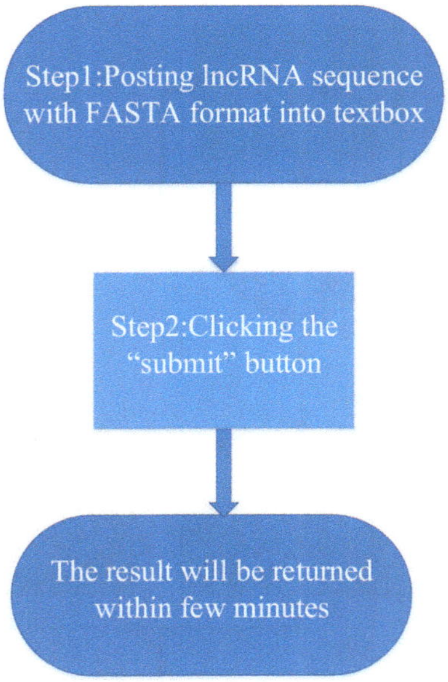

Fig. 3 The usage guanidine of LDAP by pasting one sequence

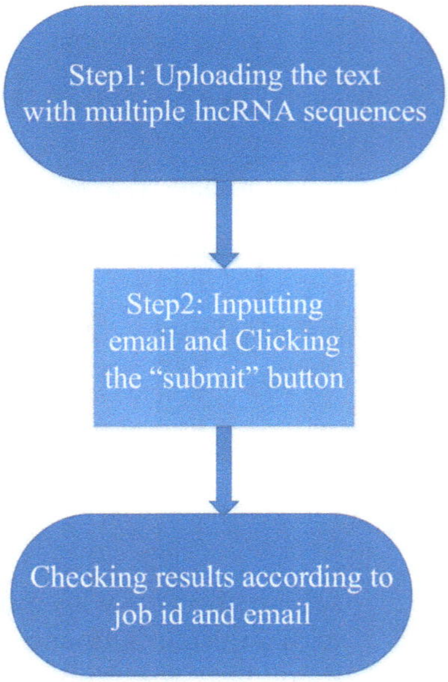

Fig. 4 The usage guanidine of LDAP by uploading multiple sequences

lncRNAs has been used to prioritize candidate lncRNAs. However, a limited number of lncRNAs are proved to relate with diseases. There are no experimentally verified negative samples. Some semi-supervised learning methods have been developed to prioritize disease-related lncRNAs.

Chen et al. [58] proposed a semi-supervised learning framework to predict lncRNA-disease associations. It integrated lncRNA expression similarity and Gaussian interaction profile kernel similarity for lncRNA. The Laplacian Regularized Least Squares method was employed for lncRNA-disease inference. In consideration of the important of lncRNA similarity, Chen et al. [59] proposed two novel lncRNA functional similarity calculation models based on disease information. The Laplacian Regularized Least Squares was used for lncRNA-disease association prediction. As a result, new predictive models improved the performance of LRLSLDA in the leave-one-out cross validation of various known lncRNA-disease association datasets. Similar works have been developed by Huang et al. [60] and Chen et al. [61]. Huang et al. [60] presented an improved lncRNA functional similarity calculation method (ILNCSIM) based on information content method. The main improvement was that it combines hierarchical structure of disease directed acyclic graphs and information content for disease similarity calculation. The Laplacian Regularized Least Squares was employed for lncRNA-disease association identification. The results showed that it can enhance the accuracy of prediction than previous method. Chen et al. [61] developed a fuzzy measure-based lncRNA functional similarity calculation model (FMLNCSIM) to identify interactions between lncRNAs and diseases. In this work, Sugeno λ-measure was used to calculate disease semantic similarity based on Medical Subject Headings (MeSH). Then, the Laplacian Regularized Least Squares was employed to evaluate the effectiveness of FMLNCSIM. The results demonstrated that FMLNCSIM could be used for searching functionally similar lncRNAs.

In addition, Biswas et al. [62] proposed a computational framework for lncRNA-disease interaction prediction based on matrix factorization. This work integrated lncRNA-disease association, experimentally validated gene-disease association, gene-gene interaction data, and expression profiles of both lncRNAs and genes. The integrative non-negative matrix factorization method was used to infer lncRNA-disease interactions, and bi-cluster was employed to identify the module of lncRNA. Experimental results showed the superiority of our proposed method over two state-of-the-art clustering algorithms—*k*-means and hierarchical clustering.

The challenge of machine learning method is how to select useful biological features to train classifier. Therefore, integrating multiple data resource is an effective method to improve performance. However, the redundant or irrelevant biological information may be futile even degrade the performance. How to select useful features from various biological resources would be another research focus [63, 64]. In addition, the different classification algorithms may be suit for different data resources. Hence, utilizing multiple learning algorithms can obtain better predictive performance.

3.3 Other Methods

Based on the fact that special disease has close association with special tissue, Liu et al. [65] proposed a computational framework to predict lncRNA-disease interactions by integrating lncRNA expression profiles, gene expression profiles, and human disease-associated gene data. The lncRNA expression profiles were used to divide lncRNA into tissue-specific lncRNAs and non-tissue-specific lncRNAs. For tissue-specific lncRNA, they considered this lncRNA might be associated with diseases that are related to this specific tissue. For non-tissue-specific lncRNA, the Spearman rank correlation coefficients between the integral expression level of this lncRNA and the corresponding genes in the common human tissue or cell types were calculated. Then, a certain cutoff score was used to select a set of genes co-expressed with this lncRNA. Finally, the hypergeometric distribution test was employed to find significantly enriched diseases in the corresponding co-expressed gene set for each lncRNA.

In addition, statistics methods are also used to predict lncRNA-disease interactions. Chen et al. [66] proposed a novel model, lncRNA-disease association inference (HGLDA), to predict lncRNA-disease associations based on hypergeometric distribution. The lncRNA functional similarity was calculated by integrating disease semantic similarity, miRNA-disease associations, and miRNA-lncRNA interactions. The hypergeometric distribution test was employed for each lncRNA-disease pair by examining whether this lncRNA and disease significantly shared common miRNAs which can interact with both of them. Li et al. [67] developed a computational method to identify the relationship between lncRNAs and gastric cancers. They constructed lncRNA and protein-coding gene co-expressed network by integrating Spearman correlation coefficient and Pearson correlation coefficient between the expression profiles of lncRNAs and coding genes. In addition, they utilized Wilcoxon signed-rank test to detect differentially expressed genes and lncRNAs in gastric cancer versus normal tissue. Finally, the hypergeometric distribution was used to find statistical significance of the differentially expressed genes among the coding gene set co-expressed with each differentially expressed lncRNA.

Based on assumption that functional similar lncRNAs tend to compose module in network, some researchers try to mine similar module of network to identify disease-related lncRNAs. Cogill et al. [68] proposed a computational method for lncRNA-cancer association prediction based on gene expression profiles and lncRNA expression profiles. In the first step, they constructed a gene co-expression network by computing Pearson product-moment correlation. Then, the co-expression modules were discovered and analyzed by using 1-Topological Overlap Matrix (TOM) and Database for Annotation, Visualization and Integrated Discovery (DAVID). Finally, 37 co-expression modules were found, and two

modules showed relatively high expression in blood and brain tissue. Zhou et al. [69] used computational method to analysis of lncRNA expression profiles across 473 breast cancer patients who did or did not develop recurrence by repurposing existing microarray datasets from the Gene Expression Omnibus database. The Pearson correlation coefficients were calculated to identify associations between lncRNAs and breast cancer. They identified 12 differentially expressed lncRNAs that associated with tumor recurrence of breast cancer patients. Furthermore, they used DAVID tool to perform functional enrichment analysis. The results showed that the predicted relapse-related lncRNAs may be involved in known breast cancer-related biological processes and pathways.

Network-based rank methods are also used to disease-related lncRNA prediction. Wang et al. [70] proposed a computational framework to lncRNA-disease interactions identification by using competing endogenous RNA (ceRNA) information. They calculated function similarity score between two genes based on the shortest path of two genes in biological molecular networks. Then, the lncRNA similarity of two lncRNAs was computed based on function score, and the lncRNA list was ranked according to lncRNA similarity. The experimental result demonstrated that DisLncPri had the highest rank enrichment score and AUC value in comparison to several other methods for case studies of Alzheimer's disease, ovarian cancer, pancreatic cancer, and gastric cancer.

4 Conclusion and Future Research

lncRNAs are arbitrarily considered to be more than 200 nucleotides without the ability of coding protein. It can be translated from intergenic, intronic, or overlapping regions and have 3′ polyadenylation, 5′ cap characteristic. Comparing with message RNAs (mRNAs) and lncRNAs, the average size of lncRNA is smaller than mRNAs. In addition, lncRNAs also have fewer exons number than mRNAs [71–74]. Furthermore, lncRNAs show less conservation than mRNA cross the species. It is interesting that many lncRNAs, just like mRNA, contain "K4-K36" domain that consist of histone 3 Lys 4 trimethylation at the promoter followed by histone 3 Lys 36 trimethylation along the transcribed region [75, 76]. In addition, most known lncRNAs are restricted to specific tissue types [77]. Increasing evidences have demonstrated that long noncoding RNAs (lncRNAs) play important roles in many human diseases. Therefore, predicting novel lncRNA-disease association would assist geneticists and molecular biologists to elucidate the genetic basis of human diseases. Further, it also contributes to the diagnosis, prevention, and treatment of human diseases. Some computational methods have been developed to infer lncRNA-disease association. In this article, we have reviewed the recent computational approaches for lncRNA-disease interaction prediction. Although these methods have achieved a great success in the past few years, there are some further researches that are needed to improve these methods.

1. It cannot be ignored that current available data contain large amounts of false-positive and false-negative noises. It has been showed that integrating other types of biological data can improve performance of disease-related lncRNA prioritization. However, it is still difficult to combine multiple data resources appropriate for lncRNA-disease association prediction. Therefore, methods for effectively integrating various data resources are needed to develop to improve network quality for lncRNA-disease interaction identification.
2. Different methods make use of different biological data to predict lncRNA-disease association. The criteria of data selection and data integration may be different. These differences give rise to biases among different approaches. Therefore, it is still challenging to design suitable strategy for assessing the performance of different methods.
3. Few computational tools have been developed to lncRNA-disease prediction. Although these tools have achieved great successes, some improvements are still necessary. For example, some tools only provide the ranking of lncRNA in the final report. It would be better to provide more supplemental information such as *P* values of ranking and other features of lncRNAs to enhance confidence of the results.

Acknowledgments

This work is supported in part by the National Natural Science Foundation of China under Grant No. 61702122, 61751314 and 61363025; a key project of Natural Science Foundation of Guangxi 2017GXNSFDA198033 and a key research and development plan of Guangxi AB17195055; the Director Open Fund of Qinzhou City Key Laboratory of Advanced Technology of Internet of Things IOT2017A04.

References

1. Altshuler DL, Durbin RM, Abecasis GR, Bentley DR, Chakravarti A, Clark AG, Collins FS, Vega FMDL, Donnelly P, Egholm M (2010) Consortium, T.G.P.: a map of human genome variation from population-scale sequencing. Nature 467(7319):1061–1073
2. Chen Q, Luo H, Zhang C, Chen P (2015) Bioinformatics in protein kinases regulatory network and drug discovery. Math Biosci 262:147–156
3. Green ED, Guyer MS (2011) Charting a course for genomic medicine from base pairs to bedside. Nature 470(7333):204
4. Yan C, Wang J, Lan W, Wu F, Pan Y (2017) SDTRLS: Predicting Drug-Target Interactions for Complex Diseases Based on Chemical

Substructures. Complexity 2017 (2017), Article ID 2713280

5. Chen Q, Chen P, Zhang C (2016) Effective classification of conserved RNA secondary structures using interval based similarity. IEEE Intell Syst 31(3):78–85
6. Chen Q, Lan W, Wang J (2013) Mining Featured Patterns of MiRNA Interaction based on Sequence and Structure Similarity. IEEE Trans Comput Biol Bioinformatics 10(2):415–422
7. Guttman M, Amit I, Garber M, French C, Lin M, Feldser D, Huarte M, Zuk O, Carey B, Cassady J, Cabili M, Jaenisch R, Mikkelsen T, Jacks T, Hacohen N, Bernstein B, Kellis M, Regev A, Rinn J, Lander E (2009) Chromatin signature reveals over a thousand highly conserved large non-coding RNAs in mammals. Nature 458 (7235):223–227
8. Louro R, Smirnova AS, Verjovski-Almeida S (2009) Long intronic noncoding RNA transcription: expression noise or expression choice. Genomics 93(4):291–298
9. Lan W, Chen Q, Li T, Yuan C, Mann S, Chen B (2014) Identification of important positions within miRNAs by integrating sequential and structural features. Curr Protein Pept Sci 15 (6):591–597
10. Guttman M, Rinn JL (2012) Modular regulatory principles of large non-coding RNAs. Nature 482(7385):339–346
11. Mercer T, Dinger M, Mattick J (2009) Long non-coding RNAs: insights into functions. Nat Rev Genet 10(3):155–159
12. Chen Q, Li G, Chen YP (2011) Interval-based distance function for identifying RNA structure candidates. J Theor Biol 269(1):280–286
13. Yang L, Lin C, Jin C, Yang JC, Tanasa B, Li W, Merkurjev D, Ohgi KA, Meng D, Zhang J (2013) LncRNA-dependent mechanisms of androgen receptor-regulated gene activation programs. Nature 500(7464):598–602
14. Li H, Yu B, Li J, Su L, Yan M, Zhu Z, Liu B (2014) Overexpression of lncRNA H19 enhances carcinogenesis and metastasis of gastric cancer. Oncotarget 5(8):2318–2329
15. Laurent GS, Wahlestedt C, Kapranov P (2015) The Landscape of long noncoding RNA classification. Trends Genet 31(5):239
16. Reddy TB, Thomas AD, Stamatis D, Bertsch J, Isbandi M, Jansson J, Mallajosyula J, Pagani I, Lobos EA, Kyrpides NC (2014) The Genomes OnLine Database (GOLD) v.5: a metadata management system based on a four level (meta)genome project classification. Nucleic Acids Res 43(Database issue):D1099
17. Bond CS, Fox AH (2009) Paraspeckles: nuclear bodies built on long noncoding RNA. J Cell Biol 186(5):637
18. Geisler S, Coller J (2013) RNA in unexpected places: long non-coding RNA functions in diverse cellular contexts. Nat Rev Mol Cell Biol 14(11):699–712
19. Ponting C, Oliver P, Reik W (2009) Evolution and functions of long noncoding RNAs. Cell 136(4):629–641
20. Rinn JL, Chang HY (2012) Genome regulation by long noncoding RNAs. Annu Rev Biochem 81(1):145
21. Tsai MC, Manor O, Wan Y, Mosammaparast N, Wang JK, Lan F, Shi Y, Segal E, Chang HY (2010) Long noncoding RNA as modular scaffold of histone modification complexes. Science 329(5992):689
22. Batista P, Chang H (2013) Long noncoding RNAs: cellular address codes in development and disease. Cell 152(6):1298
23. Gutschner T, Hämmerle M, Eissmann M, Hsu J, Kim Y, Hung G, Revenko A, Arun G, Stentrup M, Gross M, Zörnig M, MacLeod A, Spector D, Diederichs S (2013) The noncoding RNA MALAT1 is a critical regulator of the metastasis phenotype of lung cancer cells. Cancer Res 73(3):1180–1189
24. Bhartiya D, Kapoor S, Jalali S, Sati S, Kaushik K, Sachidanandan C, Sivasubbu S, Scaria V (2012) Conceptual approaches for lncRNA drug discovery and future strategies. Expert Opin Drug Discovery 7(6):503–513
25. Yang G, Lu X, Yuan L (2014) LncRNA: a link between RNA and cancer. Biochim Biophys Acta 1839(11):1097
26. Xing C, Yan CC, Xu Z, You ZH (2017) Long non-coding RNAs and complex diseases: from experimental results to computational models. Brief Bioinform 18:558
27. Schriml LM, Arze C, Nadendla S, Chang YW, Mazaitis M, Felix V, Feng G, Kibbe WA (2012) Disease ontology: a backbone for disease semantic integration. Nucleic Acids Res 40 (Database issue):940–946
28. Amberger JS, Bocchini CA, Schiettecatte F, Scott AF, Hamosh A (2014) OMIM.org: online mendelian inheritance in man (OMIM), an online catalog of human genes and genetic disorders. Nucleic Acids Res 43 (Database issue):D789
29. Becker KG, Barnes KC, Bright TJ, Wang SA (2004) The genetic association database. Nat Genet 36(5):431–432
30. Piñero J, Queraltrosinach N, Bravo À, Deupons J, Bauermehren A, Baron M, Sanz F, Furlong LI (2015) DisGeNET: a

discovery platform for the dynamical exploration of human diseases and their genes. Database 2015(3):bav028

31. Volders PJ, Helsens K, Wang X, Menten B, Martens L, Gevaert K, Vandesompele J, Mestdagh P (2013) LNCipedia: a database for annotated human lncRNA transcript sequences and structures. Nucleic Acids Res 41(Database issue):246–251
32. Zhao Y, Li H, Fang S, Kang Y, Wu W, Hao Y, Li Z, Bu D, Sun N, Zhang MQ (2016) NONCODE 2016: an informative and valuable data source of long non-coding RNAs. Nucleic Acids Res 44(Database issue):D203–D208
33. Chakraborty S, Deb A, Maji RK, Saha S, Ghosh Z (2013) LncRBase: an enriched resource for lncRNA information. PLoS One 9(9):e108010
34. Ma L, Li A, Zou D, Xu X, Xia L, Yu J, Bajic VB, Zhang Z (2015) LncRNAWiki: harnessing community knowledge in collaborative curation of human long non-coding RNAs. Nucleic Acids Res 43(Database issue):187–192
35. Quek XC, Thomson DW, Maag JL, Bartonicek N, Signal B, Clark MB, Gloss BS, Dinger ME (2015) lncRNAdb v2.0: expanding the reference database for functional long noncoding RNAs. Nucleic Acids Res 43(Database issue):168–173
36. Chen G, Wang Z, Wang D, Qiu C, Liu M, Chen X, Zhang Q, Yan G, Cui Q (2013) LncRNADisease: a database for long-non-coding RNA-associated diseases. Nucleic Acids Res 41(Database issue):D983
37. Ning S, Zhang J, Wang P, Zhi H, Wang J, Liu Y, Gao Y, Guo M, Yue M, Wang L (2016) Lnc2Cancer: a manually curated database of experimentally supported lncRNAs associated with various human cancers. Nucleic Acids Res 44(D1):D980
38. Wang Y, Chen L, Chen B, Li X, Kang J, Fan K, Hu Y, Xu J, Yi L, Yang J (2013) Mammalian ncRNA-disease repository: a global view of ncRNA-mediated disease network. Cell Death Dis 4(8):e765
39. Lan W, Wang J, Li M, Peng W, Wu F (2015) Computational approaches for prioritizing candidate disease genes based on PPI networks. Tsinghua Sci Technol 20(5):500–512
40. Guo X, Gao L, Wang Y, Chiu DK, Wang T, Deng Y (2015) Advances in long noncoding RNAs: identification, structure prediction and function annotation. Brief Funct Genomics 15(1):38
41. Chen LL, Carmichael GG (2010) Decoding the function of nuclear long noncoding RNAs. Curr Opin Cell Biol 22(3):357–364
42. Peng W, Lan W, Zhong J, Wang J, Pan Y (2017) A novel method of predicting microRNA-disease associations based on microRNA, disease, gene and environment factor networks. Methods 124:69. https://doi.org/10.1016/j.ymeth.2017.05.024
43. Yan C, Wang J, Ni P, Lan W, Wu F, Pan Y (2017) DNRLMF-MDA: Predicting microRNA-disease associations based on similarities of microRNAs and diseases. IEEE Transa Comput Biol Bioinformatics 99:1–1
44. Sun J, Shi H, Wang Z, Zhang C, Liu L, Wang L, He W, Hao D, Liu S, Zhou M (2014) Inferring novel lncRNA-disease associations based on a random walk model of a lncRNA functional similarity network. Mol Biosyst 10(8):2074–2081
45. Cheng L, Shi H, Wang Z, Hu Y, Yang H, Zhou C, Sun J, Zhou M (2016) IntNetLncSim: an integrative network analysis method to infer human lncRNA functional similarity. Oncotarget 7(30):47864–47874
46. Chen X, You ZH, Yan GY, Gong DW (2016) IRWRLDA: improved random walk with restart for lncRNA-disease association prediction. Oncotarget 7(36):57919
47. Zhou M, Wang X, Li J, Hao D, Wang Z, Shi H, Han L, Zhou H, Sun J (2015) Prioritizing candidate disease-related long non-coding RNAs by walking on the heterogeneous lncRNA and disease network. Mol Biosyst 11(3):760–769
48. Alaimo S, Giugno R, Pulvirenti A (2014) ncPred: ncRNA-disease association prediction through tripartite network-based inference. Front Bioeng Biotechnol 2:71
49. Liu Y, Zhang R, Qiu F, Li K, Zhou Y, Shang D, Xu Y (2015) Construction of a lncRNA-PCG bipartite network and identification of cancer-related lncRNAs: a case study in prostate cancer. Mol Biosyst 11(2):384
50. Yao Q, Wu L, Li J, Yang LG, Sun Y, Li Z, He S, Feng F, Li H, Li Y (2017) Global prioritizing disease candidate lncRNAs via a multi-level composite network. Sci Rep 7:39516
51. Ganegoda GU, Li M, Wang W, Feng Q (2015) Heterogeneous network model to infer human disease-long intergenic non coding RNA associations. IEEE Trans Nanobioscience 14(2):175–183
52. Lan W, Wang J, Min L, Jin L, Wu FX, Pan Y (2016) Predicting microRNA-disease associations based on improved microRNA and disease similarities. IEEE Transa Comput Biol Bioinformatics 99:1–1
53. Lan W, Wang J, Li M, Liu J, Li Y, Wu FX, Pan Y (2016) Predicting drug-target interaction

using positive-unlabeled learning. Neurocomputing 206(C):50–57

54. Liu J, Li M, Lan W, Wu FX, Pan Y, Wang J (2016) Classification of Alzheimer's disease using whole brain hierarchical network. IEEE Transa Comput Biol Bioinformatics 99:1–1
55. Peng W, Lan W, Zeng Y, Wang J, Yi P (2016) A Framework for integrating multiple biological networks to predict microRNA-disease associations. IEEE Trans Nanobiosci 99:1–1
56. Zhao T, Xu J, Liu L, Bai J, Xu C, Xiao Y, Li X, Zhang L (2015) Identification of cancer-related lncRNAs through integrating genome, regulome and transcriptome features. Mol Biosyst 11(1):126
57. Lan W, Li M, Zhao K, Jin L, Wu FX, Yi P, Wang J (2017) LDAP: a web server for lncRNA-disease association prediction. Bioinformatics 33:458
58. Chen X, Yan GY (2013) Novel human lncRNA-disease association inference based on lncRNA expression profiles. Bioinformatics 29(20):2617
59. Chen X, Yan CC, Luo C, Ji W, Zhang Y, Dai Q (2015) Constructing lncRNA functional similarity network based on lncRNA-disease associations and disease semantic similarity. Sci Rep 5:11338
60. Huang YA, Chen X, You ZH, Huang DS, Chan KC (2016) ILNCSIM: improved lncRNA functional similarity calculation model. Oncotarget 7(18):25902–25914
61. Chen X, Huang YA, Wang XS, You ZH, Chan KCC (2016) FMLNCSIM: fuzzy measure-based lncRNA functional similarity calculation model. Oncotarget 7(29):45948–45958
62. Biswas AK, Kang M, Kim DC, Ding CHQ, Zhang B, Wu X, Gao JX (2015) Inferring disease associations of the long non-coding RNAs through non-negative matrix factorization. Network Modeling Analysis in Health Informatics and. Bioinformatics 4(1):1–17
63. Li M, Zheng R, Zhang H, Wang J, Pan Y (2014) Effective identification of essential proteins based on priori knowledge, network topology and gene expressions. Methods 67 (3):325–333
64. Li M, Zheng R, Li Q, Wang J, Wu FX, Zhang Z (2016) Prioritizing disease genes by using search engine algorithm. Curr Bioinformatics 11:195
65. Liu MX, Chen X, Chen G, Cui QH, Yan GY (2014) A computational framework to infer human disease-associated long noncoding RNAs. PLoS One 9(1):e84408
66. Chen X (2015) Predicting lncRNA-disease associations and constructing lncRNA functional similarity network based on the information of miRNA. Sci Rep 5:13186
67. Ying L, Ye H, Han S, Liang Y (2017) Identification and functional inference for tumor-associated long non-coding RNA. IEEE Transa Comput Biol Bioinformatics 99:1–1
68. Cogill SB, Wang L (2014) Co-expression network analysis of human lncRNAs and cancer genes. Cancer Informatics 13(Suppl 5):49–59
69. Zhou M, Zhong L, Xu W, Sun Y, Zhang Z, Zhao H, Yang L, Sun J (2016) Discovery of potential prognostic long non-coding RNA biomarkers for predicting the risk of tumor recurrence of breast cancer patients. Sci Rep 6:31038
70. Peng W, Guo Q, Yue G, Hui Z, Yan Z, Yue L, Zhang J, Ming Y, Guo M, Ning S (2017) Improved method for prioritization of disease associated lncRNAs based on ceRNA theory and functional genomics data. Oncotarget 8 (3):4642–4655
71. Wery M, Descrimes M, Vogt N, Dallongeville AS, Gautheret D, Morillon A (2016) Nonsense-mediated decay restricts LncRNA levels in yeast unless blocked by double-stranded RNA structure. Mol Cell 61 (3):379–392
72. Chodroff RA, Goodstadt L, Sirey TM, Oliver PL, Davies KE, Green ED, Molnár Z, Ponting CP (2010) Long noncoding RNA genes: conservation of sequence and brain expression among diverse amniotes. Genome Biol 11(7): R72
73. Mitra SA, Mitra AP, Triche TJ (2012) A central role for long non-coding RNA in cancer. Front Genet 3:17
74. Sanbonmatsu KY (2015) Towards structural classification of long non-coding RNAs. Biochim Biophys Acta 1859(1):41–45
75. Liu J, Wang J, Hu B, Wu FX, Pan Y (2017) Alzheimer's disease classification based on individual hierarchical networks constructed with 3D texture features. IEEE Trans Nanobioscience 16:428. https://doi.org/10.1109/TNB.2017.2707139
76. Da SL, Baldassarre A, Masotti A (2012) Bioinformatics tools and novel challenges in long non-coding RNAs (lncRNAs) functional analysis. Int J Mol Sci 13(1):97–114
77. Grote P, Wittler L, Hendrix D, Koch F, Währisch S, Beisaw A, Macura K, Bläss G, Kellis M, Werber M (2013) The tissue-specific lncRNA Fendrr is an essential regulator of heart and body wall development in the mouse. Dev Cell 24(2):206

Chapter 13

Survey of Computational Approaches for Prediction of DNA-Binding Residues on Protein Surfaces

Yi Xiong, Xiaolei Zhu, Hao Dai, and Dong-Qing Wei

Abstract

The increasing number of protein structures with uncharacterized function necessitates the development of in silico prediction methods for functional annotations on proteins. In this chapter, different kinds of computational approaches are briefly introduced to predict DNA-binding residues on surface of DNA-binding proteins, and the merits and limitations of these methods are mainly discussed. This chapter focuses on the structure-based approaches and mainly discusses the framework of machine learning methods in application to DNA-binding prediction task.

Key words Structure-based function prediction, Functional annotation, DNA-binding residue, Machine learning method

1 Introduction

Protein-DNA interactions play vital roles in various biological activities such as gene regulation, transcription, DNA repair, and DNA packaging. It has been estimated that DNA-binding proteins represent 2–3% and 6–7% of all proteins encoded in prokaryotic and eukaryotic genomes, respectively [1–3]. As of early May 2017, a total of 3915 protein-DNA complex structures have already been deposited in the PDB (http://www.rcsb.org/). Due to the important role of DNA-binding proteins, a variety of computational approaches have been proposed for prediction of DNA-binding function from protein sequences or structures in the past decades. The first category of methods utilizes a comparative approach to infer protein function by global/local sequence or structural similarity [4–7]. While sequence comparison methods are powerful and widely adopted for function inference, structure comparison methods are more sensitive to detect remote homologs with low or no sequence similarity. However, significant sequence or structural similarity does not necessarily dictate identical function, since many proteins have functional divergence during the course of

Tao Huang (ed.), *Computational Systems Biology: Methods and Protocols*, Methods in Molecular Biology, vol. 1754,
https://doi.org/10.1007/978-1-4939-7717-8_13, © Springer Science+Business Media, LLC, part of Springer Nature 2018

evolution [8]. The second category of methods employs various machine learning techniques such as logistic regression, neural network, and support vector machines to train classification models with physicochemical, evolutionary, electrostatic, and structural features to distinguish DNA-binding proteins from other proteins [9–11]. If a protein is identified as DNA-binding protein, we will further develop the methods to detect which are DNA-binding residues on the given protein. The computational methods for prediction of DNA-binding residues can be categorized into two groups: (1) methods based on sequences and (2) methods based on structures. The first group of methods includes the sequence comparison methods and machine learning methods based on sequence-derived features. The sequence-based methods have the wide scope of application, since they require only sequence information as query input, rather than structures which have not been experimentally determined for most of the proteins encoded by genomes. However, these sequence-based predictors have at least two major limitations. One problem with sequence-based predictors is that amino acids that are sequential neighbors are not necessarily close in space to confer DNA-binding function. The other problem is that sequence information provides few clues to the interaction sites and is not sufficient for accurate prediction of DNA-binding residues. In fact, the information derived from protein structures is helpful for predicting DNA-binding function. In recent years, an increasing number of protein with unknown function are solved due to the efforts of structural genomics projects. Functional annotations of these targets are particularly challenging since many targets in structural genomics have low sequence identity to the proteins with known function. Therefore, it is urgent to develop computational approaches that utilize not only sequence but also structural information for function prediction. Since proteins always interact with other proteins or DNA/RNA molecules through their surface, we will focus on the review of computational approaches for prediction of DNA-binding residues on protein surface on the assumption that the given protein structure interacts with DNA.

2 Definition of Surface and DNA-Binding Residues

For any prediction model, the first step is to construct a reliable or benchmark data set of DNA-binding residues and nonbinding residues on the representative set of DNA-binding protein chains. It is relatively straightforward to determine DNA-binding residues if the three-dimensional (3D) structure of a protein-DNA complex is already solved. A residue is taken as a surface residue if its solvent-accessible surface area (SASA) is at least 10% of maximum values in a tripeptide state. The SASA of residues were calculated in each protein multimer in the absence of DNA. A surface residue is labeled as a binding residue if it satisfies one of the three definition approaches as follows. The most frequently used method to assign DNA-binding residues is based on a minimum distance cutoff of atoms between amino acids in a protein and nucleotides in DNA. However, different distance cutoffs lead to accuracy variations, while a single cutoff biases certain prediction programs [12]. Most studies used a cutoff distance (i.e., 3.5–6 Å) between atoms of amino acids and nucleotides to assign DNA-binding residues on proteins. The second approach to assign binding residues is based on the difference of the solvent-accessible surface areas when the structure of DNA-binding protein transforms from the isolated (the protein without DNA present) to the complexed state (the protein with DNA present). The third definition is based on the scoring function using AMBER potential to calculate the interaction free-energy between atoms in protein and DNA molecules [13]. The residues with the energy score less than −1 kcal/mol are identified as DNA-binding residues. The scoring function-based approach can quantitatively measure the interaction strength, in comparison to the distance-based approach in which the residue-nucleotide pairs with different distances have been treated in the same manner.

3 Structure-Based Methods for Prediction of DNA-Binding Residues

For prediction of DNA-binding residues, the structure-based methods can be categorized into three main types. The first type is the template-based methods based on the structural alignment [4, 14] or dynamic alignment [15]. The second type is based on the physical principles that ultimately govern protein-DNA interactions, such as knowledge-based [5] and docking-based methods [16]. The third type is feature-based methods using various machine learning technologies, which are elaborated in detail in the next section.

4 Machine Learning Methods for Prediction of DNA-Binding Residues Using Structure-Based Features

4.1 Representation of Environment of DNA-Binding Residues

As an input vector for training or testing by machine learning technologies, the sample of DNA-binding residue is commonly represented by the properties of the target residue and its neighbor residues to include the environmental information of the target residue. Similar to the sequence window used by sequence-based methods, structure-based methods utilize different types of structural windows or patches to incorporate the neighbor information of the target residue in 3D space. The common type of spatial window or patch is constructed as follows: for each surface residue, its distances by their alpha C atoms with all other surface residues in the same protein chain are calculated and sorted in ascending order, and then the *L* spatially nearest surface residues constitute a surface patch/window for including the environmental information. The size of the surface patch is a parameter to be optimized in the training stage. A topological patch or window is similarly defined by the *n* vertices with the smallest geodesic distances (shortest paths) to the center vertex. In this case, protein structures are recast as topological graphs based on protein residue contact maps, where each vertex of the graph represents the alpha C atom of an amino acid and edges connect vertices within a distance cutoff of 8 Å [17, 18].

4.2 Structure-Based Features of DNA-Binding Residues

For an effective classification model, the selected features should be highly related to the class of DNA-binding residues and have discrimination power to distinguish DNA-binding residues from nonbinding residues on surface of DNA-binding residues. A large number of studies have identified various sequence features [19–34], such as amino acid composition, physiochemical properties, and predicted structure features, and evolutionary features based on position-specific score matrix (PSSM) generated by PSI--BLAST [35]. However, sequence-based features and evolutionary features are not sufficient for prediction of DNA-binding residues, since the functions of proteins are more directly affected by their structural features. Thus, an increasing number of prediction methods have incorporated structural features, such as secondary structure, solvent-accessible surface area, spatial neighbors, *B*-factor, the empirical preference of electrostatic potential [36–39], and the shape of molecular surfaces [40–42].

4.2.1 Relative Solvent Accessibility (RSA)

Relative solvent accessibility of a residue was calculated as the ratio of its SASA to the nominal maximum area of its residue type in a tripeptide state. The results in previous studies show that positively charged residues Arg and Lys were more exposed in the binding group than in the nonbinding group, giving resultant more binding propensity, whereas for negatively charged residues Asp and Glu, it was opposite [43, 44].

4.2.2 B-Factor

B-factors are highly related to the flexibility of atoms and residues in a protein and are determined by X-ray crystallographic experiments. *B*-factor of alpha C atom was used to represent its residue flexibility and obtained from its PDB file. For each protein chain, the *B*-factor of each alpha C atom was normalized as follows:

$$\mathrm{NB} = \frac{B - \mu(B)}{\sigma(B)} \tag{1}$$

where B is the *B*-factor value of a given residue and $\mu(B)$ and $\sigma(B)$ are the average value and the standard deviation of the

B-factors for the selected chain, respectively. In our previous studies, the average values of the *B*-factors of 20 types of residues in DNA-binding group were also significantly lower than the non-binding group. This result can be explained by the fact that the atoms of DNA-binding residues are less exposed to solvent and experience less fluctuation, resulting in relatively lower *B*-factors. Generally, the *B*-factor is a distinguished feature to characterize the binding residues of biological macromolecules in their bound states. However, the *B*-factors of binding residues in apoproteins (without DNA present) were ranging from relatively lower values (rigidity) to higher values (flexibility), which suggests that the binding sites have dual character about mobility.

4.2.3 Betweenness Centrality

Protein structures are recast as topological graphs based on protein residue contact maps, where each vertex of the graph represents the alpha C atom of an amino acid and edges connect vertices within a distance cutoff of 8 Å. Once the graph is constructed, a variety of topological metrics can be used to describe functional residues.

Betweenness centrality (BC) measures how frequently a vertex occurs on the shortest path between all other vertex pairs within the contact graph (undirected graph) of a protein chain of length n. Since the chains vary in length, the measure is normalized by dividing through the number of pairs of vertices not including v, which is $(n - 1)(n - 2)/2$.

$$\mathrm{BC}(v) = \frac{2}{(n-1)(n-2)} \sum_{s \neq v \neq t \in V} \frac{\sigma_{st}(v)}{\sigma_{st}} \tag{2}$$

where V is the set of vertices, σ_{st} is the number of shortest paths from s to t, and $\sigma_{st}(v)$ is the number of shortest paths from s to t that pass through vertex v.

The weighted average betweenness centrality for a surface patch was calculated from the betweenness centrality values of component residues, weighted by the weighting factor w_i of residue i:

$$\mathrm{WBC} = \sum_{i=1}^{L} w_i \cdot \mathrm{BC}(i) \tag{3}$$

Our previous study found that residues located at protein-DNA interfaces exhibit the central role in the protein network with high betweenness centrality [18]. The predictive power of betweenness centrality was low (PR-AUC 0.208) for individual residues but was high (PR-AUC 0.228) when averaged over a patch of neighboring residues. This may suggest that a set of residues with higher betweenness centralities form a community so as to play an important role in protein-DNA interaction. It was indicated that DNA-binding residues are distinguishable from the nonbinding residues on protein surfaces by their higher weighted average betweenness centrality.

4.2.4 Electrostatic Potential

Electrostatic complementarity is shown to be important for protein-DNA interaction. DNA-binding sites have a large overlap with the surface patches which have the largest positive electrostatic potential [39, 45]. PBEQ-Solver can be used to calculate electrostatic potential of all atoms in a protein [36].

Protein surface was placed on a cubic grid. For each atom, the electrostatic potential of nearby grid points was averaged to construct an electrostatic feature at an atom-scale. The electrostatic potential values of grid points between the van der Waals and solvent-accessible surfaces were averaged, using a solvent probe at the radius of 1.4 Å. For each atom, values for the electrostatic potential at grid points were averaged at grid points outside the van der Waals surface of the given protein but within a distance that is the sum of the atom radius and the solvent radius. Three additional groups of features were derived by moving the shell slightly outward, by radius offsets 0.1 Å, 0.3 Å, and 0.5 Å. Note that the region of the shell maintains a width of 1.4 Å regardless of the size of the offset, but the regions move farther away from the van der Waals surface as the offset is varied. Mathematically, this is equivalent to adding the offset value to the radius of all the atoms and repeating the previous calculation described for the van der Waals and solvent-accessible surfaces. Figure 1 illustrates the details of this calculation of electrostatic feature at the atom-level.

Next, the local sums and averages of the residue-level electrostatic features were derived in the neighborhood of the target residue. More details can be found at [36].

4.3 Classification Algorithms

To our best knowledge, classification algorithms are mainly categorized into three classes: decision tree-based, artificial intelligence-based, and statistics-based methods. Decision tree algorithms provide an intuitive way for classifying a new sample based on a set of simple and easily interpretable rules. The artificial intelligence-based classification methods include the artificial neural network [46], deep learning algorithms, and evolutionary algorithms [47], which can be further classified into the genetic algorithm and swarm algorithm. The statistics-based methods include various algorithms such as support vector machine [48, 49], random forest [50], stochastic gradient boosting algorithm [51], and Bayesian classifier [52]. These diverse classification methods have already been explored to the prediction of DNA-binding residues on DNA-binding proteins. The comprehensive overview of the characteristics and specific application of these classification algorithms is out of scope of this review.

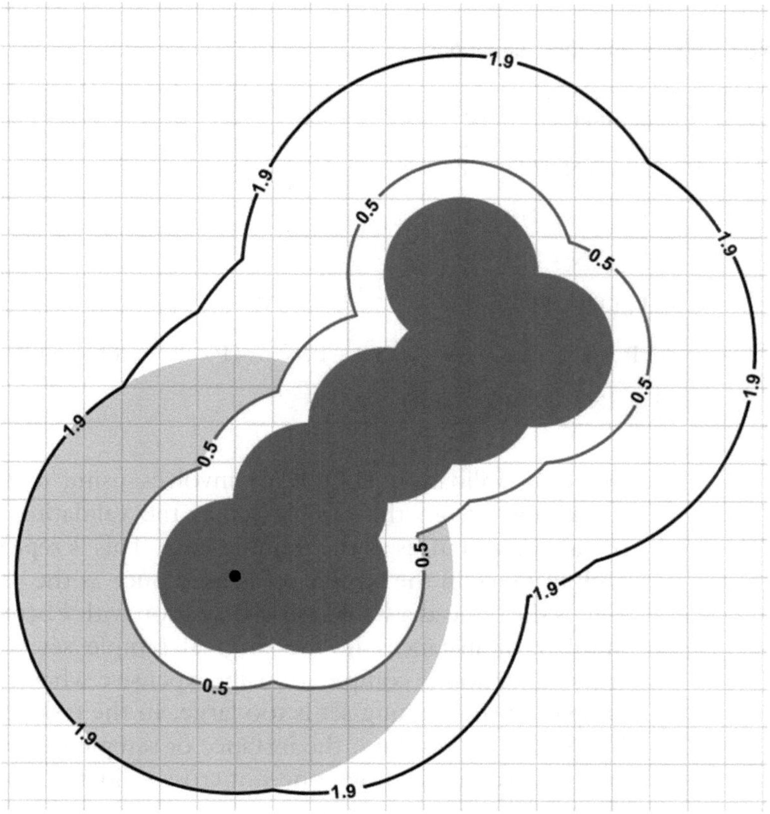

Fig. 1 The illustration of calculation of the atom-level electrostatic feature within a shell offset 0.5 Å from the van der Waals surface. The grid on which the electrostatic potential is calculated is shown relative to the molecule, shown in dark gray, and the atom at which the feature is calculated is marked using a black dot. Electrostatic potential values at grid points within the light gray annular region are those averaged to generate the feature for this atom. Grid points inside the 0.5 Å offset surface are excluded from the calculation. The light gray annular region is 1.4 Å in width regardless of the offset used to define the shell. The figure is extracted from [36]

4.4 Model Validation and Evaluation

When a prediction model is constructed by one type of classification algorithms, it requires a benchmark data set to validate and evaluate how the model works. The benchmark data set consists of well-labeled samples, which are divided into a training set and testing set. To evaluate the performance of classification models, the validation methods are mainly consisting of k-fold cross-validation, leave-one-out cross-validation, and independent tests. In k-fold cross-validation, the sample set is randomly partitioned into k subsets with equal size. Of the k subsets, one subset is selected as the validation data for testing the model, and the remaining $k - 1$ subsets are used as training data. The cross-validation process is then repeated k times (the folds), with each of the k subsets used exactly once as the validation data. The results from k folds are finally averaged to generate a single estimation metric. Leave-one-

Table 1
A list of common metrics and their equations

Metric	Equation
ACC	(TP + TN)/(TP + TN + FP + FN)
SN	TP/(TP + FN)
SP	TN/(TN + FP)
PR	TP/(TP + FP)
MCC	$(\mathrm{TP} \times \mathrm{TN} - \mathrm{FP} \times \mathrm{FN})/\sqrt{(\mathrm{TP}+\mathrm{FN}) \times (\mathrm{TP}+\mathrm{FP}) \times (\mathrm{TN}+\mathrm{FP}) \times (\mathrm{TN}+\mathrm{FN})}$
F_1	2 × SN × PR/(SN + PR)

out cross-validation (LOOCV) involves using a single instance (or sample) from the sample set as the validation data and the remaining instances as the training data. This is repeated such that each instance in the sample set is used once as the validation data. This is the same as a *k*-fold cross-validation with *k* being equal to the number of instances in the original sample set. Leave-one-out cross-validation is computationally expensive when the number of samples in the training set is too large. In the task of prediction of DNA-binding residues, the instance or sample can be a residue or protein chain in the leave-one-out cross-validation. In order to test the model in an unseen sample set, an independent test is usually adopted and conducted on a separate set which is independent of the training set. This type of test resembles a true prediction and reflects the generalization ability of a prediction model.

In order to assess the classification performance, various threshold-dependent metrics are utilized and defined as follows. They are accuracy (ACC), sensitivity (SN, also called recall), specificity (SP), precision (PR), Matthew's correlation coefficient (MCC), and F-measure (F_1). These metrics are calculated using the numbers of true positives (TP), false positives (FP), true negatives (TN) , and false negatives (FN) for each classifier. Their equations are defined in Table 1.

TP is the number of correctly predicted DNA-binding residues, TN is the number of correctly predicted nonbinding residues, FP is the number of nonbinding residues predicted as binding residues, and FN is the number of binding residues wrongly predicted as nonbinding.

The receiver operating characteristic (ROC) curve is a plot of the sensitivity versus (1-specificity) for a binary classifier at varying thresholds. The area under the curve (AUC) can be used as a threshold-independent measure of classification performance. It is a nontrivial task to assess the quality of prediction for heavily unbalanced data sets. On the unbalanced data sets, the accuracy and AUC of ROC curve can present overly optimistic assessments

Table 2
Web servers for prediction of DNA-binding residues using structure-based methods

Server	Website	Description/structural feature	Reference
DBS-Pred	http://www.netasa.org/dbs-pred	Solvent accessibility and secondary structure	[33]
PreDs	http://pre-s.protein.osaka-u.ac.jp/~preds	Electrostatic potential, the local curvature and the global curvature	[53]
DISPLAR	http://pipe.scs.fsu.edu/displar.html	Solvent accessibility	[44]
DBD-Hunter	http://cssb.biology.gatech.edu/skolnick/webservice/DBD-Hunter	Structural comparison and the evaluation of a statistical potential	[5]
DNABINDPROT	http://www.prc.boun.edu.tr/appserv/prc/dnabindprot	The fluctuations of residues in high-frequency modes	[54]
DNABind	http://mleg.cse.sc.edu/DNABind/	Combining machine learning- and template-based approaches	[14]
DBSI	http://dbsi.mitchell-lab.org	A banded electrostatic feature	[55]

of performance of an algorithm. Instead, the precision-recall (PR) curve is a plot of the recall versus precision for a binary classifier at varying thresholds.

4.5 Web Servers

To establish a useful structure-based predictor for a biological system, it is usual to develop a user-friendly web server for the predictor that is accessible to the public. Table 2 briefly presents a summary of publicly available methods that predict DNA-binding residues using structure-based features of DNA-binding proteins.

Acknowledgments

This work was supported by the grants from National Natural Science Foundation of China for Young Scholars (Grant No. 31601074 and 21403002), the funding from National Key Research Program (Contract No. 2016YFA0501703), and the Open Fund of Shanghai Key Laboratory of Intelligent Information Processing (Contract No. IIPL-2016-005).

References

1. Luscombe NM, Austin SE, Berman HM, Thornton JM (2000) An overview of the structures of protein-DNA complexes. Genome Biol 1(1):REVIEWS001
2. Biswas S, Guharoy M, Chakrabarti P (2009) Dissection, residue conservation, and structural classification of protein-DNA interfaces. Protein Struct Funct Bioinformatics 74 (3):643–654

3. Ahmad S, Keskin O, Sarai A, Nussinov R (2008) Protein-DNA interactions: structural, thermodynamic and clustering patterns of conserved residues in DNA-binding proteins. Nucleic Acids Res 36(18):5922–5932
4. Zhao H, Yang Y, Zhou Y (2010) Structure-based prediction of DNA-binding proteins by structural alignment and a volume-fraction corrected DFIRE-based energy function. Bioinformatics 26(15):1857–1863
5. Gao M, Skolnick J (2008) DBD-Hunter: a knowledge-based method for the prediction of DNA-protein interactions. Nucleic Acids Res 36(12):3978–3992
6. Jones S, Barker JA, Nobeli I, Thornton JM (2003) Using structural motif templates to identify proteins with DNA binding function. Nucleic Acids Res 31(11):2811–2823
7. Gao M, Skolnick J (2009) A threading-based method for the prediction of DNA-binding proteins with application to the human genome. PLoS Comput Biol 5(11):e1000567
8. Gherardini PF, Helmer-Citterich M (2008) Structure-based function prediction: approaches and applications. Brief Funct Genomic Proteomic 7(4):291–302
9. Nimrod G, Szilagyi A, Leslie C, Ben-Tal N (2009) Identification of DNA-binding proteins using structural, electrostatic and evolutionary features. J Mol Biol 387(4):1040–1053
10. Ahmad S, Sarai A (2004) Moment-based prediction of DNA-binding proteins. J Mol Biol 341(1):65–71
11. Liu B, Wang S, Wang X (2015) DNA binding protein identification by combining pseudo amino acid composition and profile-based protein representation. Sci Rep 5:15479
12. Miao Z, Westhof E (2015) A large-scale assessment of nucleic acids binding site prediction programs. PLoS Comput Biol 11(12): e1004639
13. Gromiha MM, Fukui K (2011) Scoring function based approach for locating binding sites and understanding recognition mechanism of protein-DNA complexes. J Chem Inf Model 51(3):721–729
14. Liu R, Hu J (2013) DNABind: a hybrid algorithm for structure-based prediction of DNA-binding residues by combining machine learning- and template-based approaches. Proteins 81(11):1885–1899
15. Zen A, de Chiara C, Pastore A, Micheletti C (2009) Using dynamics-based comparisons to predict nucleic acid binding sites in proteins: an application to OB-fold domains. Bioinformatics 25(15):1876–1883
16. Gao M, Skolnick J (2009) From nonspecific DNA-protein encounter complexes to the prediction of DNA-protein interactions. PLoS Comput Biol 5(3):e1000341
17. Maetschke SR, Yuan Z (2009) Exploiting structural and topological information to improve prediction of RNA-protein binding sites. BMC Bioinformatics 10:341
18. Xiong Y, Xia J, Zhang W, Liu J (2011) Exploiting a reduced set of weighted average features to improve prediction of DNA-binding residues from 3D structures. PLoS One 6(12): e28440
19. Zhou J, Xu R, He Y, Lu Q, Wang H, Kong B (2016) PDNAsite: identification of DNA-binding site from protein sequence by incorporating spatial and sequence context. Sci Rep 6:27653
20. Yan J, Friedrich S, Kurgan L (2016) A comprehensive comparative review of sequence-based predictors of DNA- and RNA-binding residues. Brief Bioinform 17(1):88–105
21. Peng Z, Kurgan L (2015) High-throughput prediction of RNA, DNA and protein binding regions mediated by intrinsic disorder. Nucleic Acids Res 43(18):e121
22. Si J, Zhang Z, Lin B, Schroeder M, Huang B (2011) MetaDBSite: a meta approach to improve protein DNA-binding sites prediction. BMC Syst Biol 5(Suppl 1):S7
23. Wang L, Huang C, Yang MQ, Yang JY (2010) BindN+ for accurate prediction of DNA and RNA-binding residues from protein sequence features. BMC Syst Biol 4(Suppl 1):S3
24. Cai Y, He Z, Shi X, Kong X, Gu L, Xie L (2010) A novel sequence-based method of predicting protein DNA-binding residues, using a machine learning approach. Mol Cells 30 (2):99–105
25. JS W, Liu HD, Duan XY, Ding Y, HT W, Bai YF, Sun X (2009) Prediction of DNA-binding residues in proteins from amino acid sequences using a random forest model with a hybrid feature. Bioinformatics 25(1):30–35
26. Wang L, Yang MQ, Yang JY (2009) Prediction of DNA-binding residues from protein sequence information using random forests. BMC Genomics 10(Suppl 1):S1
27. Badis G, Berger MF, Philippakis AA, Talukder S, Gehrke AR, Jaeger SA, Chan ET, Metzler G, Vedenko A, Chen X, Kuznetsov H, Wang CF, Coburn D, Newburger DE, Morris Q, Hughes TR, Bulyk ML (2009) Diversity and complexity in DNA recognition by transcription factors. Science 324 (5935):1720–1723

28. Ofran Y, Mysore V, Rost B (2007) Prediction of DNA-binding residues from sequence. Bioinformatics 23(13):I347–I353
29. Hwang S, Gou ZK, Kuznetsov IB (2007) DP-Bind: a Web server for sequence-based prediction of DNA-binding residues in DNA-binding proteins. Bioinformatics 23 (5):634–636
30. Ho SY, FC Y, Chang CY, Huang HL (2007) Design of accurate predictors for DNA-binding sites in proteins using hybrid SVM-PSSM method. Biosystems 90 (1):234–241
31. Wang LJ, Brown SJ (2006) BindN: a web-based tool for efficient prediction of DNA and RNA binding sites in amino acid sequences. Nucleic Acids Res 34:W243–W248
32. Wang L, Brown SJ (2006) Prediction of DNA-binding residues from sequence features. J Bioinform Comput Biol 4(6):1141–1158
33. Ahmad S, Gromiha MM, Sarai A (2004) Analysis and prediction of DNA-binding proteins and their binding residues based on composition, sequence and structural information. Bioinformatics 20(4):477–486
34. Yan J, Kurgan L (2017) DRNApred, fast sequence-based method that accurately predicts and discriminates DNA- and RNA-binding residues. Nucleic Acids Res 45 (10):e84
35. Ahmad S, Sarai A (2005) PSSM-based prediction of DNA binding sites in proteins. BMC Bioinformatics 6:33
36. Zhu X, Ericksen SS, Mitchell JC (2013) DBSI: DNA-binding site identifier. Nucleic Acids Res 41(16):e160
37. Tsuchiya Y, Kinoshita K, Nakamura H (2004) Structure-based prediction of DNA-binding sites on proteins using the empirical preference of electrostatic potential and the shape of molecular surfaces. Protein Struct Funct Bioinformatics 55(4):885–894
38. Chen YC, CY W, Lim C (2007) Predicting DNA-binding amino acid residues from electrostatic stabilization upon mutation to Asp/Glu and evolutionary conservation. Proteins 67(3):671–680
39. Bhardwaj N, Lu H (2007) Residue-level prediction of DNA-binding sites and its application on DNA-binding protein predictions. FEBS Lett 581(5):1058–1066
40. Zhou W, Yan H (2010) A discriminatory function for prediction of protein-DNA interactions based on alpha shape modeling. Bioinformatics 26(20):2541–2548
41. Zhou P, Tian F, Ren Y, Shang Z (2010) Systematic classification and analysis of themes in protein-DNA recognition. J Chem Inf Model 50(8):1476–1488
42. Sonavane S, Chakrabarti P (2009) Cavities in protein-DNA and protein-RNA interfaces. Nucleic Acids Res 37(14):4613–4620
43. Xiong Y, Liu J, Wei DQ (2011) An accurate feature-based method for identifying DNA-binding residues on protein surfaces. Proteins 79(2):509–517
44. Tjong H, Zhou HX (2007) DISPLAR: an accurate method for predicting DNA-binding sites on protein surfaces. Nucleic Acids Res 35 (5):1465–1477
45. Jones S, Shanahan HP, Berman HM, Thornton JM (2003) Using electrostatic potentials to predict DNA-binding sites on DNA-binding proteins. Nucleic Acids Res 31 (24):7189–7198
46. Dai H, Xu Q, Xiong Y, Liu WL, Wei DQ (2016) Improved prediction of michaelis constants in CYP450-mediated reactions by resilient back propagation algorithm. Curr Drug Metab 17(7):673–680
47. Yao Y, Zhang T, Xiong Y, Li L, Huo J, Wei DQ (2011) Mutation probability of cytochrome P450 based on a genetic algorithm and support vector machine. Biotechnol J 6 (11):1367–1376
48. Xiong Y, Liu J, Zhang W, Zeng T (2012) Prediction of heme binding residues from protein sequences with integrative sequence profiles. Proteome Sci 10(Suppl 1):S20
49. Li L, Xiong Y, Zhang ZY, Guo Q, Xu Q, Liow HH, Zhang YH, Wei DQ (2015) Improved feature-based prediction of SNPs in human cytochrome P450 enzymes. Interdiscip Sci Comput Life Sci 7(1):65–77
50. Zhang W, Xiong Y, Zhao M, Zou H, Ye X, Liu J (2011) Prediction of conformational B-cell epitopes from 3D structures by random forests with a distance-based feature. BMC Bioinformatics 12:341
51. Xu Q, Xiong Y, Dai H, Kumari KM, Xu Q, HY O, Wei DQ (2017) PDC-SGB: prediction of effective drug combinations using a stochastic gradient boosting algorithm. J Theor Biol 417:1–7
52. Sun Y, Xiong Y, Xu Q, Wei D (2014) A hadoop-based method to predict potential effective drug combination. Biomed Res Int 2014:196858
53. Tsuchiya Y, Kinoshita K, Nakamura H (2005) PreDs: a server for predicting dsDNA-binding

site on protein molecular surfaces. Bioinformatics 21(8):1721–1723

54. Ozbek P, Soner S, Erman B, Haliloglu T (2010) DNABINDPROT: fluctuation-based predictor of DNA-binding residues within a network of interacting residues. Nucleic Acids Res 38(Web Server issue):W417–W423

55. Sukumar S, Zhu X, Ericksen SS, Mitchell JC (2016) DBSI server: DNA binding site identifier. Bioinformatics 32(18):2853–2855

Chapter 14

Computational Prediction of Protein O-GlcNAc Modification

Cangzhi Jia and Yun Zuo

Abstract

Protein O-GlcNAcylation on serine and threonine residues is a significant posttranslational modification. Experimental techniques can uncover only a small portion of O-GlcNAcylation sites. Several computational algorithms have been proposed as necessary auxiliary tools to identify potential O-GlcNAcylation sites. This chapter discusses the metrics and procedures used to assess prediction tools and surveys six computational tools for the prediction of protein O-GlcNAcylation sites. Analyses of these tools using an independent test dataset indicated the advantages and disadvantages of the six existing prediction methods. We also discuss the challenges that may be faced while developing novel predictors in the future.

Key words O-GlcNAcylation, Prediction, Sequence information, Features

1 Introduction

O-linked *N*-acetylglucosamine (O-GlcNAc) was discovered in early 1984 by the famous biochemist Gerald W. Hart [1]. O-GlcNAcylation on serine and threonine residues of proteins is recognized as a significant extensive and dynamic posttranslational modification that occurs in animals, plants, microorganisms, and viruses. Protein O-GlcNAcylation is completed through the cooperation of O-GlcNAc transferase, which is responsible for the addition of O-GlcNAc, and O-GlcNAcase, which is also responsible for removal of O-GlcNAc to and from the hydroxyl groups of serine and threonine residues [2]. The chemical structure of O-GlcNAc glycosylation is illustrated in Fig. 1.

Recent studies have shown that protein O-GlcNAcylation is involved in multiple cellular processes, such as transcription, translation, cell signal transduction, and cell stress response. Abnormal O-GlcNAcylation can lead to diseases, such as cancer, diabetes, and a variety of neurodegenerative diseases [3, 4]. As a result, research of protein O-GlcNAcylation is attracting more and more attention.

Although experimental techniques to detect and accurately localize protein O-GlcNAcylation sites have improved significantly

Tao Huang (ed.), *Computational Systems Biology: Methods and Protocols*, Methods in Molecular Biology, vol. 1754, https://doi.org/10.1007/978-1-4939-7717-8_14, © Springer Science+Business Media, LLC, part of Springer Nature 2018

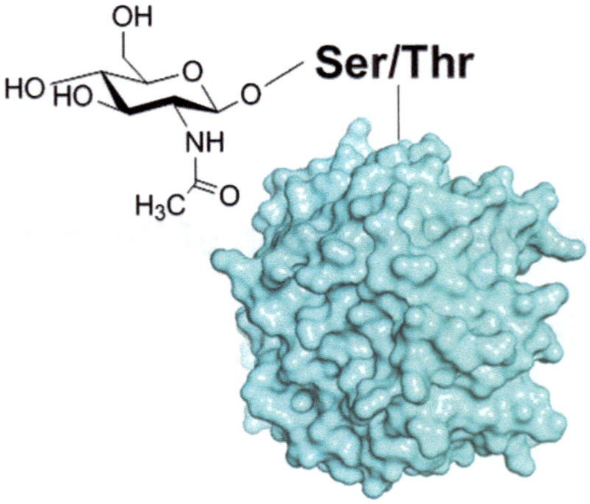

Fig. 1 The chemical structure of O-GlcNAc glycosylation

Table 1
A summary of O-GlcNAcylation sites prediction tools

Tool name	Data sources	Positive training data	Window size	Classify algorithms	Year
YingOYang	Swiss-Prot	40	–		2002
OGlcNAcScan	PubMed	373	11	SVM	2011
O-GlcNAcPRED	dbOGAP	373	23	SVM	2013
Wu et al. method	dbOGAP, UniProtKB, O-GlycBase, PhosphoSiteP	375	11	SVM	2015
Kao et al. method	dbOGAP, UniProtKB, O-GlycBase	410	11	HMM, SVM	2015
PGlcS	dbOGAP	373	15	SVM	2015

in the past several years, it remains likely that only a small portion of the O-GlcNAcylation proteins have been discovered. Even for a single protein, testing every serine/threonine residue for all possible modifications is typically prohibitively costly and laborious. Therefore, computational predictive techniques can serve as tools to generate hypotheses prior to carrying out experimental work [5]. Unlike the high number of available predictors of other protein posttranslational modifications, to our knowledge, there are only six computational predictors for O-GlcNAcylation sites identification. These predictors are listed in Table 1. In this chapter, we briefly review recent advances in the computational analysis of protein O-GlcNAcylation sites with respect to their datasets, feature extraction methods, and classifier algorithms. We also discuss the future challenges and outstanding questions.

2 Material and Methods

Computational prediction of protein O-GlcNAcylation sites can be formulated as a two-class classification problem. The systematic flowchart of the prediction method is summarized in Fig. 2. The method consists mainly of the following components: dataset construction and preprocessing, sequence feature representation and selection, and prediction algorithms. We discuss the six computational predictors for O-GlcNAcylation sites identification from the above three aspects that were provided by these studies [6–11].

2.1 Datasets Construction and Preprocessing

It is crucial to construct a high-quality benchmark dataset for unbiased performance evaluation. The datasets used to predict protein O-GlcNAcylation sites are generally constructed from the UniProtKB/Swiss-Prot Database [12], dbPTM [13], dbOGAP [6], O-GlycBase [14], PhosphoSitePlus [13], and the PubMed literature. For O-GlcNAcylation sites prediction, experimentally verified O-GlcNAcylation sites are defined as the positive dataset.

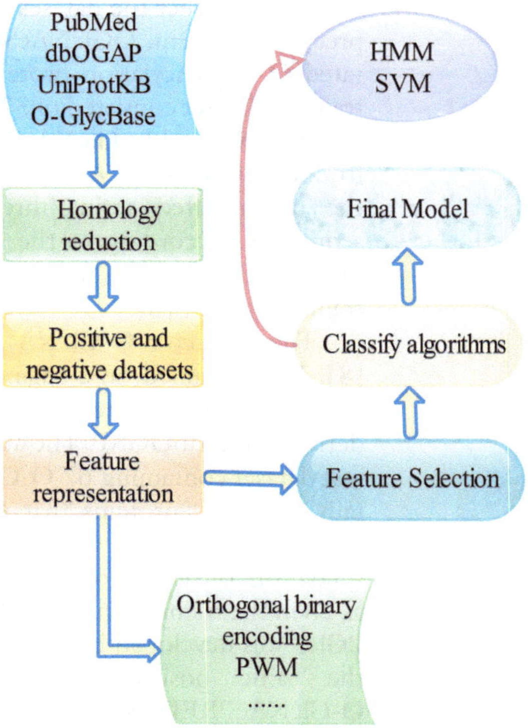

Fig. 2 The systematic flowchart of the O-GlcNAcylation prediction method

Because only neighboring residues can influence the status of centered serine or threonine residues, a sliding window strategy has been utilized to extract positive data from protein sequences as a training dataset, which will contain peptide sequences with O-GlcNAcylation sites symmetrically surrounded by flanking residues [5, 6, 16]. The number of residues considered in a window is important because too few may omit information useful for making predictions, while too many may introduce ineluctable redundancy and decrease the signal-to-noise ratio. The most appropriate window size is still unclear, and most researchers have preliminarily tested different fragment sizes and then have chosen the window size that gives the best predictive performance [5, 6, 8–11, 16, 18]. dbOGAP, which was created by Wang et al. [4], is a database of O-GlcNAcylated proteins and sites, primarily based on literature published before April 2010. The database currently contains approximately 800 proteins with experimental O-GlcNAcylation information; about 61% are proteins from human, and 172 of the proteins have a total of approximately 400 identified O-GlcNAcylation sites [6]. Most of the existing predictors have been trained on datasets constructed from dbOGAP.

The first predictor was constructed by Gupta and Brunak [7]. It is an artificial neural network system trained on sequence fragments of about 40 GlcNAcylation sites that were available at the time. However, as far as we know, the original proteins that were used to construct the dataset can no longer be found. The second prediction system, OGlcNAcScan, was developed based on annotated O-GlcNAcylation proteins collected in dbOGAP [6]. The training dataset consisted of 373 positive instances that were experimentally verified O-GlcNAcylation sites in 167 protein sequences from dbOGAP and 29,897 negative instances that were the rest of the unannotated serine/threonine sites in the same protein sequences. According to the prediction performance, the length of 11 of each protein peptide sequence is selected in Wang et al. training dataset [6].

The predictor O-GlcNAcPRED was developed by Jia et al. [8]. It was trained on a balanced dataset containing 339 positive O-GlcNAcylation sites and 339 non-O-GlcNAcylation sites derived from dbOGAP. The authors also provided an independent test dataset containing 67 O-GlcNAcylation sites in 38 experimentally identified proteins that were found by searching published literature and which were not included in dbOGAP. After cross-validation, the length of each protein peptide sequence was selected as 23 in the training and test datasets of Jia et al. [8]. The predictor PGlcS was developed by Zhao et al. [10] in 2015. It was based on the same positive training dataset and test dataset as O-GlcNAcPRED [8].

Lee's research group successively constructed two predictors in 2014 and 2015 [9, 11]. The most current training dataset for these predictors contains 410 O-GlcNAcylation sites and 410 non-O-GlcNAcylation sites from dbOGAP, OGlycBase, and UniProtKB, and the most current test dataset contains 956 O-GlcNAcylation sites and 60,976 non-O-GlcNAcylation sites from PhosphoSitePlus. Detailed information about the number of samples in the training datasets and the length of each sample for the six predictors is listed in Table 1.

It should be indicated that in the paper [19–30], a total of 1181 O-GlcNAc sites on 520 proteins were collected from publications since 2012, but these data have not been used in the recent predictors. Therefore, the construction of an up-to-date, comprehensive, reliable dataset is expected in the near future.

2.2 Feature Representation and Selection

To establish a really useful statistical predictor, the features of the protein or peptide samples need to be represented by an effective mathematical expression that can truly reflect their intrinsic correlation with the object to be predicted [9, 10, 31–41]. Here, ten different features, such as orthogonal binary coding and amino acid composition and so on, were used to represent peptides as described below.

2.2.1 Orthogonal Binary Coding

For this feature, each amino acid type is coded with 21 binary values; e.g., "100. . .0" (one followed by 20 zeros) for A (alanine), "010. . .0" for C (cysteine), . . ., "000. . .010" for Y (tyrosine), and "000. . .001" for X (pseudo amino acid) [6].

2.2.2 Amino Acid Composition and Amino Acid Pair Composition

Twenty elements (f_{20}) are used to specify the numbers of occurrences of 20 amino acids normalized with the total number of residues in a sequence fragment, and the 400 elements (f_{400}) specify the numbers of occurrences of 400 amino acid pairs normalized with the total number of amino acid pairs in a sequence fragment [9]. These two types of characteristics are calculated as follows:

$$f_{20}(i) = \frac{x_{20}(i)}{L}$$

$$f_{400}(j) = \frac{y_{400}(j)}{L-1}$$

where $x_{20}(i)$ and $y_{400}(j)$ represent the number of occurrences of residue i and the number of occurrences of amino acid j in a peptide sequence, respectively, and L is the length of each peptide sequence.

2.2.3 Positional Weighted Matrix (PWM)

The relative frequency of amino acids that surround O-GlcNAcylation sites and the fragment sequence are denoted and encoded by a PWM. A PWM containing $(2n + 1) \times \omega$ elements can profile the distribution of amino acids in the training dataset. Here, $2n + 1$ denotes the window size and ω represents 20 amino acids and one terminal signal [9].

2.2.4 Adapted Normal Distribution Bi-profile Bayes

The bi-profile Bayes feature extraction approach was first proposed by Xu et al. [31], in which each peptide in a training dataset can be encoded as $(p_1, p_2, \ldots, p_n, p_{n+1}, \ldots, p_{2n})$, where $(p_1, p_2, \ldots, p_n)$ represents the posterior probability of each amino acid at each position in the positive dataset and $(p_{n+1}, \ldots, p_{2n})$ represents the posterior probability of each amino acid at each position in the negative dataset. The posterior probability is calculated as the occurrence of each amino acid at each position in the training dataset. This feature extraction approach was improved by encoding the frequency of each amino acid at each position as random variables X_{ij}, where $i\,(i = 1,2,\ldots,20)$ represents the ith amino acid.

{A,C,D,E,F,G,H,I,K,L,M,N,P,Q,R,S,T,V,W,Y} and $j\,(j = 1,2,\ldots,L)$ represents the jth position. The random variables $X_{ij}\ (i = 1,2,\ldots,20;\ j = 1,2,\ldots,L)$ are independent and obey the same binomial distribution $b(n, p)$, where $n = 339$ is the number of peptide sequences in the positive/negative dataset and $p = 1/20$ is the probability of each amino acid occurring in each position. According to the De Moivre-Laplace theorem [42], the normal form variable $\frac{X_{ij}-np}{\sqrt{np(1-p)}}$ has a limiting cumulative distribution function which approximates a normal distribution $N(0,1)$. In O-GlcNAcPRED [8], the standard variable normalization method was modified to highlight and emphasize the distinct distribution of each amino acid at the same position. If V_j denotes the standard variance of $X_{ij}\ (i = 1,2,\ldots,20)$, i.e., the deviation of frequencies of each amino acid at the same jth position, then $X'_{ij} = \frac{X_{ij}-np}{\sqrt{V_j}}$ denotes the new normalization of X_{ij} and ensures it obeys the standard normal distribution. Thus, the posterior probability p_j $(j = 1,2,\ldots,2n)$ is coded by the adapted normal distribution as follows:

$p_j = P(X \leq X_{ij}) = \varphi(X'_{ij})$, where $\varphi(x)$ is the standard normal distribution function given by $\varphi(x) = \frac{1}{\sqrt{2\pi}} \int_{-\infty}^{x} e^{-\frac{t^2}{2}} \mathrm{d}t$.

2.2.5 Position-Specific Scoring Matrix

Evolutionary information is an important characteristic of proteins, because the conserved residues at specific sequence sites are under strong selective pressure and therefore are always functionally relevant [32–35]. In [9, 10], a position-specific scoring matrix (PSSM) was used to encode the evolutionary information of a protein sequence. The PSSM profiles were obtained by PSIBLAST searches [36] against nonredundant sequences at O-GlcNAcylated sites. The matrix of $(2n + 1) \times 20$ elements had rows centered on the substrate sites (serine/threonine) extracted from the PSSM profile, where $2n + 1$ represents the window size and 20 represents the position-specific scores for each type of amino acid. Then, the $(2n + 1) \times 20$ matrix was transformed into a 20×20 matrix by summing up the rows that were associated with the same type of amino acid. Finally, every element in the 20×20 matrix was divided by the window length $2n + 1$ and then normalized using the formula $\frac{1}{1+e^{-x}}$ in Wu et al. [9]. Zhao et al. [10] incorporated the output of the PSSM with amino acid physicochemical properties, as discussed in the following section.

2.2.6 Physicochemical Properties

The specificity and diversity of protein structure and function are largely attributed to the various properties of each of the 20 amino acids in the sequence [37]. The physicochemical properties of amino acids surrounding serine/threonine residues have been reported to influence their O-GlcNAcylation. Therefore, some predictors have applied the physicochemical and biochemical properties of amino acids to predict protein O-GlcNAcylation sites [8, 10]. The AAindex database (http://www.genome.jp/aaindex/) [43] includes amino acid mutation matrices and amino acid indexes. Version 9.0 contains 544 physicochemical properties. An amino acid index is a set of 20 numerical values that denote various physicochemical properties of amino acids. Physicochemical properties have been used successfully to predict several protein modifications [44–48]. Zhao et al. [10] selected 17 informative physicochemical properties to encode protein peptides in PGlcS method. They multiplied 20 PSSM values of each residue by the corresponding physicochemical property of the 20 amino acids and summed the properties of each residue according to the following equation:

$$F_{ki} = \sum_{j=1}^{20} P_{kj} M_{ij}$$

where P_{kj} represents the value of the kth physicochemical property for the jth amino acid type ($k = 1, 2,\ldots,L$) and M_{ij} represents the raw value of the ith position for the jth amino acid type in the PSSM ($i = 1, 2,\ldots, 15$).

2.2.7 Secondary Structure

To predict the secondary structure of all the amino acid residues, the full-length protein sequences with O-GlcNAcylated sites were submitted to PSIPRED [38]. PSIPRED outputs sequences composed of the characters "C," "H," and "E," where "C" represents coil, "H" represents helix, and "E" represents strand. The orthogonal binary coding scheme was used to transform these secondary structure terms into numeric vectors; for instance, a helix (H) is encoded as "001," a strand (E) is encoded as "010," and a coil (C) is encoded as "100" in PGlcS [10].

2.2.8 Disorder

DISOPRED2, which is one of the leading servers for predicting natively disordered regions in proteins [39], was applied to obtain disorder information for the amino acid sequences. A disordered structure is encoded as "0 1," while an ordered structure is encoded as "1 0" in PGlcS [10].

2.2.9 Solvent Accessible Surface Area

The solvent accessible surface area was predicted using RVP-Net [40, 41], which shows the proportion of solvent accessible area for each amino acid in a protein sequence. In PGlcS, the RVP-Net outputs were normalized to be approximately 0–1 for every amino acid close to the O-GlcNAcylation sites [10].

2.2.10 First- and Second-Order Composition Moment Vector (CMV)

Two encoding methods were used to obtain reduced amino acid sequences according to the physicochemical properties of the amino acids. Based on their acid–base properties, the 20 amino acid residues were classified into three groups: acidic (AA_{aci}), {D,E}; basic (AA_{bas}), {K,H,R}; and neutral (AA_{neu}), {A,C,F,G,I,L,M,N,P, Q,S,T,V,W,Y} amino acids. And based on their hydrophobicity, the 20 amino acid residues were classified into three other groups: internal (AA_{int}), {F,I,L,M,V}; external (AA_{ext}), {D,E,H,K,N,Q, R}; and ambivalent (AA_{amb}), {A,C,G,P,S,T,W,Y} amino acids.

To reflect the order of acidic and hydrophobic amino acids in the sequences surrounding candidate O-GlcNAcylation sites, CMVs were defined as

$$\mathrm{CMV}_i^k = \frac{\sum_{j=1}^{x_i} n_{ij}^k}{\Pi_{d=0}^k (N-d)}$$

where $k = 0, 1$ is the order of CMV; L is the sequence length; i is the ith amino acid with acid–base {AA_{aci}, AA_{bas}, AA_{neu}} and hydrophobicity {AA_{intl}, AA_{ext}, AA_{amb}} properties; x_j is the total number of i residues in the corresponding reduced amino acid sequence; and n_{ij} is the jth position of i residues in the corresponding reduced sequence. When $k = 0$, the CMV reduces to the content of the ith amino acid. Therefore, Acontent/Hcontent (the first order of CMV) and ACMV/HCMV (the second order of CMV) were defined to reflect the sequence order of acidic and hydrophobic amino acids, respectively [8].

2.3 Assessing Predictive Performance

The predictive performances of the O-GlcNAcylation sites predictors were assessed by calculating four commonly used measurements, sensitivity (Sn), specificity (Sp), accuracy (Acc), and Matthew's correlation coefficient (MCC), as follows:

$$\mathrm{Sn} = \frac{\mathrm{TP}}{\mathrm{TP}+\mathrm{FN}}$$

$$\mathrm{Sp} = \frac{\mathrm{TN}}{\mathrm{TN}+\mathrm{FP}}$$

$$\mathrm{Acc} = \frac{\mathrm{TP}+\mathrm{TN}}{\mathrm{TP}+\mathrm{TN}+\mathrm{FP}+\mathrm{FN}}$$

$$\mathrm{MCC} = \frac{\mathrm{TP} \times \mathrm{TN} - \mathrm{FP} \times \mathrm{FN}}{\sqrt{(\mathrm{TP} + \mathrm{FP})(\mathrm{TP} + \mathrm{FN})(\mathrm{TN} + \mathrm{FP})(\mathrm{TN} + \mathrm{FN})}}$$

where TP, TN, FP, and FN are the number of true positives (correctly predicted O-GlcNAcylated sites), true negatives (correctly predicted non-O-GlcNAcylated sites), false positives (falsely predicted O-GlcNAcylated sites), and false negatives (falsely predicted non-O-GlcNAcylated sites), respectively.

2.4 Classify Algorithms

All the prediction algorithms shown in Table 1 for identifying protein O-GlcNAcylated sites use machine learning methods, including support vector machines (SVM), artificial neural networks, and hidden Markov model (HMM). SVM is a set of related supervised learning methods used for classification and regression based on statistical learning theory, which has been shown to be a powerful tool in many fields of bioinformatics [8, 49–52]. Five of the six predictors (Table 1) were designed using a SVM classifier. The predictor designed by Kao et al. [10] is a two-layered machine learning method, incorporating a HMM profiles and SVM.

3 Performance of Online Tools for O-GlcNAcylation Sites Prediction

Each of the computational predictors for O-GlcNAcylation sites identification shown in Table 1 has self-reported their sensitivity and specificity values. However, these values can often deviate significantly from the actual sensitivity and specificity of the predictor for two reasons. First, developers can mistakenly include testing data when refining the predictor, leading to an “over-fit.” Second, when new O-GlcNAcylation modification data are published, the overall characteristics of the available data may change [53]. Three of the six predictors do not have an available web server. Therefore, to assess the performances of the six predictors, we selected the independent dataset used in O-GlcNAcPRED and PGlcS [8, 10]. The dataset is composed of 67 O-GlcNAcylation sites and 7244 non-O-GlcNAcylation sites extracted from 38 experimentally identified O-GlcNAcylated proteins that were not included in the original dbOGAP. The sequences of each of the 38 proteins were uploaded into each predictor, and the prediction results are shown in Table 2.

Among the six predictors, PGlcS achieved the best sensitivity of 64.62%, which obviously does not meet the requirements of experimental biologists. The predictor O-GlcNAcscan gave the best specificity of 92.45%, but also gave the worst sensitivity of 31.34%. The differences between sensitivity and specificity in O-GlcNAcscan were mainly due to the predictor being trained on

Table 2
Performances comparisons among six predictors

Tool name	Sn (%)	Sp (%)	Acc (%)	MCC
YinOYang	34.33	89.36	88.85	0.0725
O-GlcNAcscan	31.34	92.45	91.89	0.0847
O-GlcNAcPRED	56.72	64.77	64.70	0.0428
PGlcS	64.62	68.40	68.37	0.0697
Lee 2015	34.33	95.15	94.60	0.1275

an extremely unbalanced training dataset. The performance of O-GlcNAcylation sites predictors needs to be improved in the future work.

4 Conclusion and Perspectives

The development of computational tools for the accurate prediction of protein O-GlcNAcylation is a worthy goal, given the importance of these modifications in molecular biology and the incomplete state of our present knowledge. However, the findings presented in this chapter of six prediction tools suggest that the prediction performance of the majority of these tools is far from satisfactory, especially regarding their sensitivity. There is, therefore, great opportunity to update the accuracy of prediction tools by constructing optimized datasets, improving novel feature extraction methods, and developing ensemble classification algorithms. In the future, we suggest that researchers focus on these three aspects to develop powerful and efficient predictors for O-GlcNAcylation sites identification.

Acknowledgments

This work was supported by the Fundamental Research Funds for the Central Universities (3132016306, 3132017048).

References

1. Torres CR, Hart GW (1984) Topography and polypeptide distribution of terminal N-acetylglucosamine residues on the surfaces of intact lymphocytes. Evidence for O-linked GlcNAc. J Biol Chem 259:3308–3317
2. Comer FI, Hart GW (1999) O-GlcNAc and the control of gene expression. Biochim Biophys Acta 1473:161–171
3. McClain DA, Crook ED (1996) Hexosamines and insulin resistance. Diabetes 45:1003–1009
4. Liu F et al (2004) O-GlcNAcylation regulates phosphorylation of tau: a mechanism involved in Alzheimer's disease. Proc Natl Acad Sci U S A 101:10804–10809
5. Wang Z et al (2010) Enrichment and site mapping of O-linked N-acetylglucosamine by

a combination of chemical/enzymatic tagging, photochemical cleavage, and electron transfer dissociation mass spectrometry. Mol Cell Proteomics 9(1):153–160

6. Wang J, Torii M, Liu H et al (2011) dbOGAP - an integrated bioinformatics resource for protein O-GlcNAcylation. BMC Bioinformatics 12(1):91
7. Gupta R, Brunak S (2002) Prediction of glycosylation across the human proteome and the correlation to protein function. Pac Symp Biocomput 2002:310–322
8. Jia CZ, Liu T, Wang ZP (2013) O-GlcNAcPRED: a sensitive predictor to capture protein O-GlcNAcylation sites. Mol Biosyst 9(11):2909–2913
9. Wu HY et al (2014) Characterization and identification of protein O-GlcNAcylation sites with substrate specificity. BMC Bioinformatics 15(16):S1
10. Zhao XW et al (2015) PGlcS: prediction of protein O-GlcNAcylation sites with multiple features and analysis. J Theor Biol 380(3):524
11. Kao HJ et al (2015) A two-layered machine learning method to identify protein O-GlcNAcylation sites with O-GlcNAc transferase substrate motifs. BMC Bioinformatics 16(18):S10
12. Apweiler R, Bairoch A, Wu CH et al (2004) UniProt: the Universal Protein knowledgebase. Nucleic Acids Res 32(Database issue):115–119
13. Lee TY, Huang HD, Hung JH et al (2006) dbPTM: an information repository of protein post-translational modification. Nucleic Acids Res 34(Database issue):622–627
14. Hansen JE, Lund O, Nielsen JO et al (1999) O-GLYCBASE: a revised database of O-glycosylated proteins. Nucleic Acids Res 27 (1):370–372
15. Hornbeck PV, Kornhauser JM, Tkachev S et al (2012) PhosphoSitePlus: a comprehensive resource for investigating the structure and function of experimentally determined post-translational modifications in man and mouse. Nucleic Acids Res 40(Database issue):D261
16. Shi SP et al (2015) Progress and challenges in predicting protein methylation sites. Mol Biosyst 11:2610–2619
17. Chen SA, Lee TY, YY O (2010) Incorporating significant amino acid pairs to identify O-linked glycosylation sites on transmembrane proteins and non-transmembrane proteins. BMC Bioinformatics 11(1):1–13
18. Caragea C, Sinapov J, Silvescu A et al (2007) Glycosylation site prediction using ensembles of Support Vector Machine classifiers. BMC Bioinformatics 8(1):438
19. Trinidad JC et al (2012) Global identification and characterization of both O-GlcNAcylation and phosphorylation at the murine synapse. Mol Cell Proteomics 11(8):215–229
20. Jochmann R et al (2013) O-GlcNAc transferase inhibits KSHV propagation and modifies replication relevant viral proteins as detected by systematic O-GlcNAcylation analysis. Glycobiology 23(10):1114–1130
21. Hahne H, Gholami A, Kuster B (2012) Discovery of O-GlcNAc-modified proteins in published large-scale proteome data. Mol Cell Proteom 11(10):843
22. Hahne H et al (2013) Proteome wide purification and identification of O-GlcNAc-modified proteins using click chemistry and mass spectrometry. J Proteome Res 12(2):927–936
23. Allison DF et al (2012) Modification of RelA by O-linked N-acetylglucosamine links glucose metabolism to NF-κB acetylation and transcription. Proc Natl Acad Sci U S A 109 (42):16888–16893
24. Gawlowski T et al (2012) Modulation of dynamin-related protein 1 (DRP1) function by increased O-linked-β-N-acetylglucosamine modification (O-GlcNAc) in cardiac myocytes. J Biol Chem 287(35):30024–30034
25. Wang S et al (2012) Extensive crosstalk between O-GlcNAcylation and phosphorylation regulates Akt signaling. PLoS One 7(5): e37427
26. Floyd ZE, Stephens JM (2012) Controlling a master switch of adipocyte development and insulin sensitivity: covalent modifications of PPARγ. Biochim Biophys Acta 1822 (7):1090–1095
27. Ji S et al (2012) O-GlcNAc modification of PPARgamma reduces its transcriptional activity. Biochem Biophys Res Commun 417 (4):1158–1163
28. Alfaro JF et al (2012) Tandem mass spectrometry identifies many mouse brain O-GlcNAcylated proteins including EGF domain-specific O-GlcNAc transferase targets. Proc Natl Acad Sci U S A 109(19):7280–7285
29. Pathak S et al (2012) O-GlcNAcylation of TAB1 modulates TAK1-mediated cytokine release. EMBO J 31(6):1394–1404
30. Overath T et al (2012) Mapping of O-GlcNAc sites of 20 S proteasome subunits and Hsp90 by a novel biotin-cystamine tag. Mol Cell Proteom 11(8):467
31. Shao J et al (2009) Computational identification of protein methylation sites through

bi-profile bayes feature extraction. PLoS One 4 (3):e4920
32. Lee TY, Chen SA, Hung HY, YY O (2011) Incorporating distant sequence features and radial basis function networks to identify ubiquitin conjugation sites. PLoS One 6(3): e17331
33. Hsu BK et al (2012) Incorporating evolutionary information and functional domains for identifying RNA splicing factors in humans. PLoS One 6(11):e27567
34. Xie D et al (2005) LOCSVMPSI: a web server for subcellular localization of eukaryotic proteins using SVM and profile of PSI-BLAST. Nucleic Acids Res 33(Web Server issue):105–110
35. Jones DT (1999) Protein secondary structure prediction based on position-specific scoring matrices. J Mol Biol 292(2):195
36. Altschul SF et al (1997) Gapped BLAST and PSI-BLAST: a new generation of protein database search programs. Nucleic Acids Res 25 (17):3389–3402
37. Andersen MT, Packer NH (2014) Advances in LC–MS/MS-based glycoproteomics: getting closer to system-wide site-specific mapping of the N- and O-glycoproteome. Biochim Biophys Acta 1844:1437–1452
38. Mcguffin LJ, Bryson K, Jones DT (2000) The PSIPRED protein structure prediction server. Bioinformatics 16(4):404
39. Ward JJ et al (2004) Prediction and functional analysis of native disorder in proteins from the three kingdoms of life. J Mol Biol 337(3):635
40. Ahmad S, Gromiha MM, Sarai A (2003) RVP-net: online prediction of real valued accessible surface area of proteins from single sequences. Bioinformatics 19(14):1849–1851
41. Ahmad S, Gromiha MM, Sarai A (2003) Real value prediction of solvent accessibility from amino acid sequence. Proteins 50(4):629–635
42. Kenney JF, Mosak JL (1951), Mathematics of Statistics, Van Nostrand, Princeton, NJ, 2nd edn, pp. 36–41
43. Kawashima S et al (2008) AAindex: amino acid index database, progress report 2008. Nucleic Acids Res 36(Database issue):202–205
44. Tung CW, Ho SY (2008) Computational identification of ubiquitylation sites from protein sequences. BMC Bioinformatics 9:310
45. Cao DS, QS X, Liang YZ (2013) Propy: a tool to generate various modes of Chou's PseAAC. Bioinformatics 29:960–962
46. Du P, Gu S, Jiao Y (2014) PseAAC-General: fast building various modes of general form of Chou's pseudo-amino acid composition for large-scale protein datasets. Int J Mol Sci 15 (3):3495–3506
47. Qiu WR, Xiao X, Lin WZ (2014) iMethyl-PseAAC: identification of protein methylation sites via a pseudo amino acid composition approach. Biomed Res Int 2014(12):947416
48. Zhang Y, Liu B, Dong Q, Jin VX (2011) An improved profile-level domain linker propensity index for protein domain boundary prediction. Protein Pept Lett 18(1):7–16
49. Shao J et al (2012) Systematic analysis of human lysine acetylation proteins and accurate prediction of human lysine acetylation through bi-relative adapted binomial score Bayes feature representation. Mol Biosyst 8(11):2964–2973
50. Wee LJ et al (2010) SVM-based prediction of linear B-cell epitopes using Bayes feature extraction. BMC Genomics 11(4):S21
51. Song L et al (2014) nDNA-prot: identification of DNA-binding proteins based on unbalanced classification. BMC Bioinformatics 15:298
52. Li DP, Ju Y, Zou Q (2016) Protein folds prediction with hierarchical structured SVM. Curr Proteomics 13:79–85
53. Schwartz D (2012) Prediction of lysine post-translational modifications using bioinformatic tools. Essays Biochem 52:165–177

Chapter 15

Machine Learning-Based Modeling of Drug Toxicity

Jing Lu, Dong Lu, Zunyun Fu, Mingyue Zheng, and Xiaomin Luo

Abstract

Toxicity is an important reason for the failure of drug research and development (R&D). The traditional experimental testings for chemical toxicity profile are costly and time-consuming. Therefore, it is attractive to develop the effective and accurate alternatives, such as in silico prediction models. In this review, we discuss the practical use of some prediction models on three toxicity end points, including acute toxicity, carcinogenicity, and inhibition of the human ether-a-go-go-related gene ion channel (hERG). Special emphasis is put on the machine learning methods for developing in silico models, and their advantages and weaknesses are discussed. We conclude that machine learning methods are valuable for helping the process of designing new candidates with low toxicity in drug R&D studies. In the future, much still needs to be done to understand more completely the biological mechanisms for toxicity and to develop more accurate prediction models to screen compounds.

Key words Machine learning method, In silico model, Acute toxicity, Carcinogenicity, hERG

1 Introduction

Toxicity of drugs and candidates is an important issue for drug research and development (R&D), resulting in the increased attrition and cost, late-stage failures, and marked withdrawals. CMR International data demonstrated that in the period between 2006 and 2010, toxicity accounted for 22% and 54% of drug R&D failures in clinical development and preclinical development, respectively [1]. Moreover, Lasser et al. reported that of 548 drugs approved by FDA in the period 1975–2000, 10.2% had at least one black-box warning and 2.9% were withdrawn [2]. Thus, it is necessary to put an increased emphasis on the evaluation of toxicity profile of compounds in the early stage of drug R&D.

Nowadays, there are substantial toxicological data in many pharmaceutical companies and commercial databases. For example, almost 60,000 compounds had been screened by using in vitro hERG inhibition assay at Pfizer in 2005 [3]. The molecular clinical safety intelligence system developed by GlaxoSmithKline includes pharmacological, metabolic, and toxicological information of

Tao Huang (ed.), *Computational Systems Biology: Methods and Protocols*, Methods in Molecular Biology, vol. 1754, https://doi.org/10.1007/978-1-4939-7717-8_15,

approximately 80,000 drugs and drug-like compounds [4]. Accelrys Toxicity database collects six types of toxicological data for more than 0.17 million compounds from the open scientific literature, including acute toxicity, mutagenicity, tumorigenicity, skin and eye irritation, reproductive effects, and multiple dose effects [5]. In contrast, most of the open databases for chemical toxicities are much smaller, such as TOXNET [6] and ISSCAN [7, 8].

It is estimated that the number of possible compounds has already reached 10^{60} [9]. This poses a large challenge for toxicity evaluations of compounds using experimental screenings, which need to use compound entities and multiple biochemical assays. Therefore, particular interests have been raised to develop quick and effective in silico models based on the information available for recorded compounds, and supplement in vivo and in vitro testing for predicting the toxicities of new compounds. For example, the computational chemistry tools can use chemical structures and experimental toxicities of the compounds as input data, calculate molecular descriptors, and build prediction models using machine learning methods. In this review, we discuss some illustrative models employing machine learning methods against a series of toxicity end points, including acute toxicity, carcinogenicity, and inhibition of the human ether-a-go-go-related gene ion channel (hERG).

2 Acute Toxicity

Estimation of acute toxicity is one of the most common tasks in the safety assessment of drug R&D, which represents the adverse changes occurring immediately or within a short time after a single dose of a compound or multiple doses given within 24 h [10]. Median lethal dose (LD_{50}) or median lethal concentration (LC_{50}) is common criterion for evaluating acute toxicity of compounds in multiple species. The US EPA defined the toxicity categories based on LD_{50} or LC_{50} in 2014 (Table 1) [11]. The compounds in Category I are considered highly toxic and fatal if swallowed or inhaled. Category II means moderately toxic, and Category III indicates slightly toxic, while Category IV is practically nontoxic [12]. Due to ethical reasons for avoiding the use of animals, the alternatives, such as in silico models, are strongly recommended by FDA, NIH, and EMEA [13–16]. Some published models built by machine learning methods for acute toxicity are discussed in detail below.

2.1 Quantitative Structure-Toxicity Relationship (QSTR) Models for Acute Toxicity

For congeneric compounds with the same scaffold or mechanisms, most of the relationship between the toxicity values and structural descriptors approximates linear, and simple machine learning models, such as multiple linear regression (MLR) [17–19] and partial least squares regression (PLSR) [20, 21], can generally achieve good performance [22]. Furthermore, such linear regression

Table 1
Toxicity categories for acute toxicity defined by the US EPA

Acute toxicity	Category I (danger/poison)	Category II (warning)	Category III (caution)	Category IV (none required)
Oral (mg/kg)	≤ 50	>50 and ≤ 500	>500 and ≤ 5000	>5000
Dermal (mg/kg)	≤ 200	>200 and ≤ 2000	>2000 and ≤ 5000	>5000
Inhalation[a] (mg/l)	≤ 0.05	>0.05 and ≤ 0.5	>0.5 and ≤ 2	>2

[a]4 h exposure

models can provide explicit equations to explain which properties make positive or negative contributions to the toxicity. Toropov et al. developed a two-variable model using MLR coupled with the genetic algorithm (GA) [23] for 28 benzene derivatives [24]. The equation is shown in Eq. (1). X5Av characterizes the presence of heteroatoms, double and triple bonds in the compounds [25], while BELe1 represents the information associated with electronegativities, distances, and atom types [26, 27]. The negative contributions of these two variables indicated that the nitro groups had a greater impact on acute toxicity compared with the halogen atoms.

$$\log\left[{}^{1}/_{LD_{50}}\right] = -119.203 \times \text{X5Av} - 14.999 \times \text{BELe1} + 33.223 \quad (1)$$

The prediction models derived from congeneric compounds often cover a limited chemical space and therefore have limited applicability domain. Nowadays, a large number of compounds have been reported for their toxicological data, which involve multiple structural types and biochemical mechanisms. The increase of structural diversity and number of compounds in the data set makes it difficult to use linear methods for characterizing the structure-toxicity relationship. In contrast, nonlinear models, such as neutral network [28, 29] and support vector machine (SVM) [30, 31], tended to yield better performance than linear methods for such complex data sets [32–35]. SVM maps the features into a high-dimensional space to solve a linear function based on optimization theory, in which the calculations are simplified by introducing the kernel function [30]. Wang et al. developed a QSTR model based on a chemically diverse data set of 571 compounds for predicting acute toxicity to the fathead minnow [36]. The authors employed the GA to simultaneously select a descriptor subset and optimize the SVM parameters. Eight descriptors associated with acute toxicity, such as ALogP, ε_{HOMO}, ε_{LUMO}, and Fukui reactivity indices, were selected to build the prediction models. The GA-SVM model showed superiority to the linear models in the refs. 37, 38. However, the relationship between the toxicity values and the selected descriptors is not explicit because of the nature of the SVM model.

Lei et al. employed seven machine learning methods, including SVM, relevance vector machine (RVM), *k*-nearest neighbor (*k*NN), random forest (RF), local approximate Gaussian process [39, 40], multilayer perceptron ensemble [41], and eXtreme gradient boosting [42], to predict acute oral toxicity in rats based on 7314 diverse compounds [43]. RVM, which is a sparse Bayesian learning algorithm developed from the standard SVM [44, 45], showed better prediction ability than other models. Furthermore, the authors captured the important descriptors and fragments for acute toxicity by using multiple statistic methods. For example, one-dimensional sensitivity analysis indicated that descriptors associated with molecular polarity, molecular reactivity, and intramolecular interactions gave more contributions to acute toxicity than other descriptors. The R^2_{adj} change in the stepwise regression and Cramer's V coefficient demonstrated that nine fragments, such as trifluoromethyl and heterocyclic, made positive contributions to high pLD_{50} and four fragments, such as the count of nitrogen atoms and carbon-nitrogen double bond, had contrary effects. The analyses of descriptors and fragments based on such a large and structurally diverse data set can provide some instructions for designing drug candidates with lower toxicity.

An effective strategy to improve the prediction accuracy of the models for chemically diverse data sets is to divide the data set into some subsets based on structural features or mechanisms and build a local model for each subset. For example, *k*NN algorithm [46, 47], following the idea that "structurally similar chemicals are likely to have similar properties" [48], extracts *k* nearest neighbors from the training set for the query compound and explores local structure-activity relationships using these *k* neighbors instead of the global data set.

Zhu et al. employed multiple machine learning approaches, including *k*NN, RF [49], hierarchical clustering (HC) [50], nearest neighbor, and FDA MDL QSAR [51], to develop prediction models based on 7385 compounds. For eliminating the outliers, the distance-based methods [52–54] were used to define the applicability domain of the prediction models. The statistic results indicated that the *k*NN and RF models yielded good R^2 and low MAE, but at the expense of the low coverage of the test set (19%). Moreover, the authors built the consensus model, in which the predicted toxicity for each compound equals to the arithmetical average of all predicted values of the individual models, to reduce

Table 2
Performance of the GP model and the consensus model on the test set

	Using training set I (3472 compounds)		Using training set II (5743 compounds)	
Model	R^2	**MAE**	R^2	**MAE**
GP	0.413	0.550	0.587	0.436
Consensus model	0.466	0.510	0.619	0.422

the high variance of the individual models. The consensus model showed better performance than any individual models.

Lu et al. developed four kinds of local lazy learning (LLL) models, including local lazy regression (LLR), SA, SR, and GP, for LD_{50} prediction in rats [55]. SA, SR, and GP are directly based on the LD_{50} values of the query's neighbors, while LLR relies on the nearest neighbors as well as one selected descriptor used for building a linear model. Therefore, LLR has a higher risk of generating meaningless results compared with other models. For the training set I with 3472 compounds, the GP model achieved the best performance, yielding R^2 of 0.413 and MAE of 0.550 for the test set (Table 2). It is interesting that LLR produced better prediction ability for the query compounds outside the applicability domain. Therefore, it is hardly surprising that the consensus model obtained significantly higher R^2 and lower MAE as compared with those of any individual model, which indicated that different individual models could explain complementary portions of the variance in LD_{50} data. Moreover, the training set allows simple and fast upgrades when new data becomes available, and therefore 2271 compounds not in the training set I were added into the training set II. The results listed in Table 2 demonstrated that the performance of the individual and consensus models was significantly improved by extending the training set with diverse structures and broad activity distribution.

2.2 Structure-Toxicity Relationship (STR) Models for Acute Toxicity

In addition to multiple QSTR models, some STR models have been developed for the classification of toxic and nontoxic compounds.

Xue et al. compared five machine learning methods (SVM, *k*NN, logistic regression [56], C4.5 decision tree [57], and probabilistic neural network [58]) for predicting *Tetrahymena pyriformis* toxicity based on 1129 compounds with known IGC_{50} values [59]. The results indicated that the SVM model using 49 selected descriptors showed the best performance, which yielded overall accuracy of 96.8% and the Matthews correlation coefficient of 91.6% for the test set.

Li et al. developed multi-classification models for 12,204 compounds with rat LD_{50} values based on the US EPA toxicity categories [12]. Five machine learning methods, including SVM, RF, *k*NN, naive Bayes (NB), and C4.5 decision tree [60], were employed for model building. In this study, one-against-one (OAO) [61, 62] and binary tree (BT) [63, 64] strategies were used to extend SVM for multiclass classification, respectively. The results indicated that the OAO-SVM model not only achieved the best performance for external validation set I and II but also showed the good prediction ability for each class. Moreover, the privileged fragments in Categories I (Danger/Poison) and II (Warning) were extracted by substructure frequency analysis and information gain methods, which can help us to quickly identify the toxicity category of a new compound.

2.3 Software and Web Servers for Acute Toxicity

Nowadays, multiple popular software and web servers can predict acute toxicity following various administration routes in multiple species. For example, ACD/Labs ToxSuite predicted LD_{50} values of the compounds by considering expert knowledge of various physiological responses (e.g., inhibition of cholinesterase and ATP synthesis) and structure-activity relationship [65]. Toxicity Estimation Software Tool (TEST) employed the HC and *k*NN methods to build individual models and took an average of the predicted values as the final prediction [66]. Accelrys TOPKAT package developed 19 linear models for predicting LD_{50} values of multiple classes of compounds and therefore intuitively hinted the positive or negative contributions of descriptors to the toxicity values of the specific classes [67].

Gonella Diaza et al. employed five software programs, including ACD/ToxSuite, TEST, TOPKAT, ADMET Predictor [68], and TerraQSAR [69], to evaluate acute oral toxicity of 7417 compounds [70]. Among these five models, TEST showed the best performance, which yielded R^2 of 0.74 for the training set and 0.60 for the test set. TOPKAT had inferior prediction ability ($R^2 = 0.61$ for the training set and 0.34 for the test set).

ProTox is a web server for the prediction of oral LD_{50} values in rodents based on chemical similarity and toxic fragments [71]. ProTox produced better performance as compared with TOPKAT and TEST, especially the sensitivity and precision (Table 3). Moreover,

Table 3
Performance of ProTox, TOPKAT, and TEST on the external set

Model	ProTox (%)	TOPKAT (%)	TEST (%)
Sensitivity	73.1	44.8	46.3
Specificity	94.6	89.0	89.3
Precision	73.5	42.0	45.6
Coverage	91.8	89.4	78.6

the possible toxicity targets of the toxic compounds were identified based on protein-ligand-based toxicophores and the structures of similar reference compounds, which are valuable for studying the biochemical mechanisms of acute toxicity.

3 Carcinogenicity

Chemical carcinogenicity is a serious threat to human health. According to the regulatory authorities of European Union, Japan, and the USA, it is essential to perform the carcinogenicity studies before the marketing approval of medicines [72]. The conventional test for carcinogenicity is the 2-year rodent carcinogenicity assay, which is highly expensive, labor-intensive, and time-consuming [73]. The chemical carcinogens can be categorized as genotoxic and nongenotoxic/epigenetic carcinogens based on the mechanism of carcinogenesis [74]. Genotoxic carcinogens can cause damage directly to DNA and usually be detected by various short-term and less costly mutagenicity assays, such as Ames assay, gene mutation assay, chromosome aberration assay, DNA damage assay, and micronucleus assay [75]. However, these methods are not effective for those nongenotoxic carcinogens on account of their different and specific mechanisms of carcinogenesis [74].

As the highly material and time cost of bioassays, it is urgent to develop accurate computational models for predicting carcinogenicity based on the structures and properties of chemicals. Benfenati et al. concluded that the current in silico models for carcinogenicity were classified into structural alert (SA)-based models, local models, and global models [76].

3.1 SA-Based Models for Carcinogenicity

SAs are defined as the substructures that are considered to cause the potential toxicity [77]. The traditional SAs were often generated based on the expert opinion of toxicologists. For example, Ashby proposed a hypothetical structure with SAs that link to potential carcinogenicity in 1985 [78]. Thirty-three SAs were proposed by Bailey et al. [79] based on the Ashby's SAs and a related list compiled by Munro et al. [80]. Then, structurally diverse chemicals drive the development of the methods by using the machine learning methods to extract SAs. In 2005, a list of 29 SAs was automatically extracted based on the data mining analysis and produced a total classification error of 18% for 4337 chemicals [81]. In 2006, Kazius et al. adopted an elaborate chemical representation method (called hierarchical graphs) and a substructure mining method (called Gaston) and extracted six discriminative and nonredundant substructures with overall classification error of 21% [82]. Benigni and Bossa combined the previous work mentioned above and extracted a new list of SAs using Toxtree 1.50 [83]. Toxtree 1.50 showed higher accuracy (70%) for the same data set compared with the Ashby's SAs. Moreover, modulating factors can inhibit or cancel the toxic effects of SAs [83]. In 2012, Wang et al. developed a model by building and pruning a molecular fragments tree to select high-quality SAs for carcinogenicity [84]. Finally, 77 SAs and 4 modulating factors produced higher predictive ability than Benigni's model. As an effective measure in the evaluation of genotoxicity and carcinogenicity, the identification of SAs attracts much attention in the screening of drug candidates and toxicity testing [85].

3.2 Local Models

Most of the genotoxic carcinogens generally have the unifying feature that they are electrophiles or can be activated to electrophilic reactive intermediates [74]. Multiple QSTR models have been developed for numerous congeneric series of genotoxic carcinogens, such as aromatic amines, nitroaromatic compounds, *N*-nitroso compounds, quinolines, triazenes, polycyclic aromatic hydrocarbons, and halogenated aliphatics [74]. However, for non-genotoxic carcinogens, the QSTR predictions are still scarce due to their complex mechanisms of carcinogenesis [86].

3.3 Global Models

Global models are useful for predicting noncongeneric classes of chemicals that have diverse chemical scaffolds and complex mechanisms of carcinogenicity [76]. In general, the global models perform inferior to the local models because the global models consider several mechanisms of action at the same time [74]. According to the result of predictive toxicity challenge in 2000–2001, only 5 out of 111 models for classification performed better than random guessing [87]. In 2003, Contrera et al. constructed a MDL QSAR model using molecular structural similarity and E-state indices and had excellent coverage (93%) and good sensitivity (72%) and specificity (72%) for rodent carcinogenicity [51]. In 2004, Sun et al. developed a PLS-DA (partial least squares discriminant analysis) model for predicting carcinogenicity, showing $R^2 = 0.987$ and $Q^2 = 0.944$ for male mouse, $R^2 = 0.985$ and $Q^2 = 0.950$ for female mouse, $R^2 = 0.989$ and $Q^2 = 0.962$ for male rat, and $R^2 = 0.990$ and $Q^2 = 0.965$ for female rat [88]. Moreover, Tanabe et al. used ensemble learning technique to divide the data set into 20 subsets based on the contained substructures and built SVM models for each subset with an overall accuracy of approximately 80% [89]. In 2013, Singh et al. established the classification model using probabilistic neural network and the regression model using generalized regression neural network based on 834 structurally diverse chemicals from CPDB (Carcinogenic Potency Database) [90]. Both models exhibited excellent prediction ability, which are valuable for safety evaluations of chemicals. In 2015, Li et al. constructed the binary (carcinogen and non-carcinogen) and ternary (strong, weak carcinogen and non-carcinogen) classification models using six types of fingerprints and five machine learning methods (SVM, DT, RF, *k*NN, and NB) [91]. The best binary and ternary models were both developed by using MACCS keys and *k*NN, which yielded the prediction accuracy of 83.91% and 80.46%, respectively.

4 hERG

hERG (the human enter-a-go-go-related gene) encodes a voltage-dependent potassium channel that mediates a delayed rectifier potassium current (Ikr) in cardiomyocytes. Blockage of hERG channel is considered to be the primary factor for drug-induced prolongation of QT interval, which can cause sudden death in extreme situation. Several non-antiarrhythmic drugs, such as cisapride [92], caused death induced by the blockage of hERG channel and were withdrawn from the market. A variety of compounds covering a broad spectrum of therapeutic groups were also confirmed to block hERG [93]. Therefore, scientists should develop strategies to assess hERG blocking at the early stage of drug discovery process to avoid investing at risky lead series.

Several methods have been established to assess the potency of compounds to block hERG channel, including in vitro methods such as rubidium-efflux assay, radioligand binding assay, fluorescence-based assay, the whole cell patch-clamp assay, and in vivo methods such as electrocardiography (ECG) approaches. ECG and the patch-clamp technique are low-throughput and not suitable for screening lead compounds in the early phase of drug development. The rubidium-efflux assay, radioligand binding assay, and fluorescence-based assay have the advantages of high-throughput and low cost, but the correlation with the membrane patch-clamp and ECG is poor. It needs to mention that all in vitro testings are based on cells, and therefore the properties of cells would have an important impact on the experimental results. For example, the IC_{50} value that a compound blocks hERG may even have a deviation of 100-fold in *Xenopus* oocytes [94], making these data untrustable.

Compared with in vitro and in vivo experiments, computational models cost less time and expense. Hundreds of in silico models have been established, which can be divided into three categories, including homology modeling, QSTR model, and STR model [92].

4.1 Homology Modeling of hERG

Homology modeling is a comparative modeling procedure to construct a three-dimensional model for a protein sequence based on the structures of homologous proteins. The homology-derived models combined with docking and molecular dynamics simulation can be used to calculate the binding affinities and investigate the biochemical mechanisms of ligands.

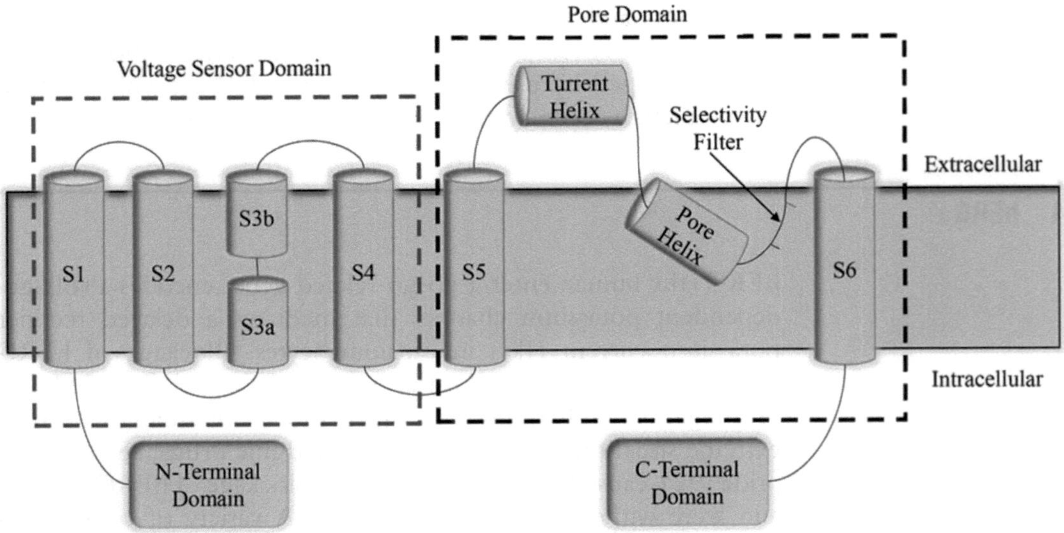

Fig. 1 Schematic diagram of the transmembrane topology of a hERG subunit

The hERG channel (also known as KCNH2 and Kv11.1) is a member of voltage-dependent potassium channel family and expressed in various tissues, especially in cardiomyocytes. Similar to other ion channels, hERG has three conformational states, including open, close, and inactivated states. However, hERG is activated slowly and inactivated rapidly compared with other K^+ channels [95]. hERG is composed of four identical α subunits, and each subunit contains six α-helical transmembrane segments known as S1–S6 (Fig. 1). Segments S1–S4 form the voltage sensor domain (VSD), which can sense the voltage changes of the membrane through the positively charged residues Lys and Arg in the S4 helix [96]. Segments S5–S6 form the pore domain (i.e., the central cavity), including the highly conserved K^+ channels features known as the pore helix and the selectivity filter [96]. In addition, there are two unique features in hERG: (a) aromatic residues Y652 and F656 in hERG instead of Ile and Val in other voltage-gated K^+ channels and (b) no Pro in the S6 domain [97].

Before the crystal structure of the hERG channel was available, a number of homology models had been primarily established based on the solved crystal structures of some bacterial K^+ channels, including KcsA (close form), KvaP (open form), and MthK (open form). However, the sequence identity between these templates and hERG is very low, and the widely used templates (KcsA and MthK) only have two segments S5–S6. Therefore, most of the homology models only contain the pore domain, which is the position that many drugs block hERG [96, 98].

To explore the key residues of hERG interacting with blockers, Mitcheson et al. used KcsA K^+ structure as a template, and established a tetrameric channel model composed of four S6 helices using MODELLER [97]. Alanine-scanning mutagenesis indicated that the residues (G648, Y652, V659, and F656) on the S6 transmembrane domain and the residues (T623 and V625) on pore helix had important effects on the interaction with MK-499 (a methanesulfonanilide antiarrhythmic drug). The docking analysis demonstrated that there was a π-stacking interaction between Y652, F656, and MK-499. V625 and G648 altered the size or the shape of the binding pocket and affected the binding with MK-499. Antihistamine terfenadine and cisapride interacted with Y652 and F656, but had no interaction with V625. Moreover, F656 engaged in π-π stacking interactions with the aromatic groups of most of the blockers, and Y652 produced π-cation interactions with the tertiary nitrogen in ligands [99].

Recently, the crystal structure of the hERG channel without ligands has been solved at 3.8 Å [100]. In this structure, the hERG is open, while the voltage sensors are in a depolarized conformation. The central cavity is surrounded by a unique environment, which may contribute to unusual properties of hERG blocking by many drugs. The subtle structural rearrangement of selectivity filter may correlate with hERG's fast inactivation. It is believed that the crystal structure of hERG is valuable for designing drugs without undesirable hERG-related cardiotoxicity.

4.2 Ligand-Based Models for hERG

Ligand-based approaches have been widely used to explore the structure-toxicity relationship of hERG blockers. The first hERG pharmacophore model based on 11 antipsychotic drugs and 15 compounds from the literature contained four hydrophobic features and one positive ionizable feature, producing the R^2 value of 0.9 [101]. Cavalli et al. developed a CoMFA model based on 31 QT-prolonging drugs, showing $R^2 = 0.952$ and $Q^2 = 0.767$ [102]. Inanobe et al. used HipHop algorithm to generate a 3D-QASR model, which contained three hydrophobic features and one positively ionizable feature [103]. The distance between hydrophobic features and the positively ionizable feature ranges from 5.5 to 8.9 Å, and the ionizable feature is 1.2 Å apart from the hydrophobic plane. Most of hERG blockers have a basic nitrogen center, which is protonated under physiological conditions, and other compounds lacking nitrogen center are referred to as "neutral" or "uncharged" hERG blockers. Aronov proposed two pharmacophore models based on 194 uncharged hERG blockers [104]. These two five-point pharmacophore models contain three hydrophobic or aromatic features and two hydrogen bond acceptors, but the location of the second hydrogen bond acceptor is different. The author combined them into a six-point pharmacophore model and concluded that ClogP < 1, decreasing lipophilicity, and introducing additional components could reduce hERG binding for neutral compounds. Moreover, Springer et al. indicated that increasing polarity, decreasing positive charge, and some changes in aromatic ring substituents could reduce the potency for hERG binding [105]. This conclusion is consistent with the observations of Braga et al. based on 4980 compounds, which indicated that removing carbons, changing the electronic environment around the basic nitrogen, and adding a hydroxyl group could reduce the potency of a compound inhibiting hERG [106].

A number of STR models have been developed by multiple machine learning methods, such as *k*NN, ANN, SVM, and RF, for the prediction of hERG blockage [98]. The first STR model was published by Roche et al., in which three classes were set with the cutoffs $IC_{50} = 1$ μM and $IC_{50} = 10$ μM [107]. The PLS, self-organizing maps, principal component analysis, and supervised neural networks were adopted to build classification models. Among them, the model using supervised neural networks showed the best performance, in which 93% of nonblockers and 71% of blockers were predicted correctly. Li et al. docked 495 compounds in a homology model of hERG based on the KvaP template and calculated pharmacophore-based GRIND descriptors, including hydrophobic interaction, hydrogen bond acceptor and donor, and molecular shape descriptors [108]. Then, the descriptors were applied into a SVM classifier to establish classification models at thresholds of 1, 5, 10, 20, 30, and 40 μM, respectively. The model was tested on an external set of 66 compounds and a large data set containing 1948 compounds and achieved the accuracy values of 72% and 73%, respectively. Wang et al. used NB and recursive partitioning (RP) to establish hERG classification model based on 806 compounds [109]. When the threshold was 1 μm, the Bayesian classifier based on 14 molecular properties and LCFP_8 fingerprint achieved the highest global accuracy of 91.5% for the training set and 88.3% for the test set.

5 Conclusions

Nowadays, a variety of in silico models for acute toxicity have been established with the aim of saving experimental resources in the early stage of drug development. However, the prediction accuracy is difficult to achieve a major breakthrough due to lack of sufficiently large data sets. Therefore, most of the previous prediction models improved the performance by limiting the model coverage. The future efforts will be devoted to enrich the data set with diverse structures and broad activity distribution.

Cancer is one of the leading causes of death, and it is necessary to identify chemical carcinogenicity as early as possible. The efficiency of machine learning models for carcinogenicity depends on the reliable and sufficient experimental data. In general, in silico models for nongenotoxic carcinogenicity performed inferior to those for genotoxic carcinogenicity. Moreover, global models generally performed less effective than local models because of the complex mechanisms of diverse chemical structures.

A lot of efforts have been made to uncover the mechanism of hERG blockage and predict the potency of compounds to inhibit hERG. For homology models, it can directly provide the information of compounds interacting with hERG, but the sequence identity between templates and hERG is low. Ligand-based models are not affected by the structure of hERG, but the quality of inhibition data has an important influence on the performance. Therefore, it is necessary to collect high-quality experimental data for the development of the prediction models.

References

1. CMR International 2010 Global R&D Performance Metrics Programme. http://cmr.thomsonreuters.com/services/programs/randd/
2. Lasser KE, Allen PD, Woolhandler SJ, Himmelstein DU, Wolfe SM, Bor DH (2002) Timing of new black box warnings and withdrawals for prescription medications. JAMA 287(17):2215–2220. https://doi.org/10.1001/jama.287.17.2215
3. O'Brien SE, de Groot MJ (2005) Greater than the sum of its parts: combining models for useful ADMET prediction. J Med Chem 48(4):1287
4. Vanderwall DE, Yuen N, Al-Ansari M, Bailey J, Fram D, Green DV, Pickett S, Vitulli G, Luengo JI, Almenoff JS (2011) Molecular clinical safety intelligence: a system for bridging clinically focused safety knowledge to early-stage drug discovery - the GSK experience. Drug Discov Today 16 (15–16):646–653. https://doi.org/10.1016/j.drudis.2011.05.001. S1359-6446(11)00143-7 [pii]
5. Accelrys Toxicity Database 2011.4. Accelrys Software Inc., San Diego, CA
6. TOXNET. http://toxnet.nlm.nih.gov/. Accessed 14 Oct 2011
7. SDF Download Page, U.S. EPA. http://www.epa.gov/ncct/dsstox/sdf_isscan_external.html. Accessed 8 July 2012
8. Istituto Superiore di Sanità Website. http://www.iss.it/ampp/dati/cont.php?id=233&lang=1&tipo=7. Accessed 8 July 2012
9. Dobson CM (2004) Chemical space and biology. Nature 432(7019):824–828. https://doi.org/10.1038/nature03192. nature03192 [pii]
10. Walum E (1998) Acute oral toxicity. Environ Health Perspect 106(Suppl 2):497–503
11. Label Review Manual (2014) Chapter7: Precautionary statements. U.S. EPA, Washington, DC. https://www.epa.gov/sites/production/files/2015-03/documents/chap-07-jul-2014.pdf. Accessed 20 Apr 2017
12. Li X, Chen L, Cheng F, Wu Z, Bian H, Xu C, Li W, Liu G, Shen X, Tang Y (2014) In silico prediction of chemical acute oral toxicity using multi-classification methods. J Chem Inform Model 54(4):1061–1069
13. Parasuraman S (2011) Toxicological screening. J Pharmacol Pharmacother 2(2):74–79. https://doi.org/10.4103/0976-500X.81895. JPP-2-74 [pii]
14. Nicolotti O, Benfenati E, Carotti A, Gadaleta D, Gissi A, Mangiatordi GF, Novellino E (2014) REACH and in silico methods: an attractive opportunity for medicinal chemists. Drug Discov Today 19 (11):1757–1768. https://doi.org/10.1016/j.drudis.2014.06.027. S1359-6446(14)00269-4 [pii]
15. Benz RD (2007) Toxicological and clinical computational analysis and the US FDA/CDER. Expert Opin Drug Metab Toxicol 3 (1):109–124. https://doi.org/10.1517/17425255.3.1.109
16. Creton S, Dewhurst IC, Earl LK, Gehen SC, Guest RL, Hotchkiss JA, Indans I, Woolhiser MR, Billington R (2010) Acute toxicity testing of chemicals-opportunities to avoid redundant testing and use alternative approaches. Crit Rev Toxicol 40(1):50–83. https://doi.org/10.3109/10408440903401511
17. Aiken LS, West SG, Pitts SC (2003) Multiple linear regression. In: Handbook of psychology. Wiley, New York. https://doi.org/10.1002/0471264385.wei0219

18. Slinker BK, Glantz SA (2008) Multiple linear regression. Accounting for multiple simultaneous determinants of a continuous dependent variable. Circulation 117 (13):1732–1737. https://doi.org/10.1161/circulationaha.106.654376
19. Tranmer M, Elliot M (2008) Multiple linear regression. The Cathie Marsh Centre for Census and Survey Research (CCSR), Oxford, UK
20. Helland I (2004) Partial least squares regression. In: Encyclopedia of statistical sciences. Wiley, New York. https://doi.org/10.1002/0471667196.ess6004.pub2
21. Geladi P, Kowalski BR (1986) Partial least squares regression: a tutorial. Anal Chim Acta 185:1–17. http://www.udel.edu/chem/analytical/cumes/text-partial%20least-squares%20regression.pdf
22. Le T, Epa VC, Burden FR, Winkler DA (2012) Quantitative structure-property relationship modeling of diverse materials properties. Chem Rev 112(5):2889–2919. https://doi.org/10.1021/cr200066h
23. Davis L (1991) Handbook of genetic algorithms. Van Nostrand Reinhold, New York
24. Toropov AA, Rasulev BF, Leszczynski J (2007) QSAR modeling of acute toxicity for nitrobenzene derivatives towards rats: comparative analysis by MLRA and optimal descriptors. QSAR Comb Sci 26 (5):686–693. https://doi.org/10.1002/qsar.200610135
25. Todeschini R, Consonni V (2008) Handbook of molecular descriptors, vol 11. Wiley, New York
26. Kubinyi H, Folkers G, Martin YC (1998) 3D QSAR in drug design, Ligand-protein interactions and molecular similarity, vol 2. Springer Science & Business Media, Dordrecht, Netherlands
27. Devillers J, Balaban AT (2000) Topological indices and related descriptors in QSAR and QSPAR. CRC Press, Boca Raton
28. Hecht-Nielsen R (1989) Theory of the backpropagation neural network. In: International 1989 Joint Conference on Neural Networks, 0–0, vol 591, pp 593–605. doi:10.1109/ijcnn.1989.118638
29. Patterson DW (ed) (1998) Artificial neural networks: theory and applications. Prentice Hall PTR, Upper Saddle River
30. Cortes C, Vapnik V (1995) Support-vector networks. Mach Learn 20(3):273–297
31. Wang L (2005) Support vector machines: theory and applications, vol 177. Springer Science & Business Media, New York
32. Eldred DV, Jurs PC (1999) Prediction of acute mammalian toxicity of organophosphorus pesticide compounds from molecular structure. SAR QSAR Environ Res 10 (2–3):75–99. https://doi.org/10.1080/10629369908039170
33. Lu J, Lu D, Zhang X, Bi Y, Cheng K, Zheng M, Luo X (2016) Estimation of elimination half-lives of organic chemicals in humans using gradient boosting machine. Biochim Biophys Acta 1860(11 Pt B):2664–2671. https://doi.org/10.1016/j.bbagen.2016.05.019
34. Peng J, Lu J, Shen Q, Zheng M, Luo X, Zhu W, Jiang H, Chen K (2014) In silico site of metabolism prediction for human UGT-catalyzed reactions. Bioinformatics 30 (3):398–405. https://doi.org/10.1093/bioinformatics/btt681
35. Kieslich CA, Smadbeck J, Khoury GA, Floudas CA (2016) conSSert: consensus SVM model for accurate prediction of ordered secondary structure. J Chem Inf Model 56 (3):455–461. https://doi.org/10.1021/acs.jcim.5b00566
36. Wang Y, Zheng M, Xiao J, Lu Y, Wang F, Lu J, Luo X, Zhu W, Jiang H, Chen K (2010) Using support vector regression coupled with the genetic algorithm for predicting acute toxicity to the fathead minnow. SAR QSAR Environ Res 21(5–6):559–570. https://doi.org/10.1080/1062936x.2010.502300
37. Papa E, Villa F, Gramatica P (2005) Statistically validated QSARs, based on theoretical descriptors, for modeling aquatic toxicity of organic chemicals in Pimephales promelas (fathead minnow). J Chem Inf Model 45 (5):1256–1266. https://doi.org/10.1021/ci050212l
38. Gini G, Craciun MV, König C, Benfenati E (2004) Combining unsupervised and supervised artificial neural networks to predict aquatic toxicity. J Chem Inf Comput Sci 44 (6):1897–1902. https://doi.org/10.1021/ci0401219
39. Obrezanova O, Csanyi G, Gola JM, Segall MD (2007) Gaussian processes: a method for automatic QSAR modeling of ADME properties. J Chem Inf Model 47 (5):1847–1857. https://doi.org/10.1021/ci7000633
40. Gramacy RB, Apley DW (2015) Local Gaussian process approximation for large computer experiments. J Comput Graph Stat 24 (2):561–578. https://doi.org/10.1080/10618600.2014.914442

41. González-Arjona D, López-Pérez G, Gustavo González A (2002) Non-linear QSAR modeling by using multilayer perceptron feedforward neural networks trained by back-propagation. Talanta 56(1):79–90. https://doi.org/10.1016/S0039-9140(01)00537-9
42. Friedman JH (2001) Greedy function approximation: a gradient boosting machine. Ann Stat 29(5):1189–1232
43. Lei T, Li Y, Song Y, Li D, Sun H, Hou T (2016) ADMET evaluation in drug discovery: 15. Accurate prediction of rat oral acute toxicity using relevance vector machine and consensus modeling. J Cheminform 8:6. https://doi.org/10.1186/s13321-016-0117-7
44. Tipping ME (2001) Sparse Bayesian learning and the relevance vector machine. J Mach Learn Res 1:211–244. https://doi.org/10.1162/15324430152748236
45. Burden FR, Winkler DA (2015) Relevance vector machines: sparse classification methods for QSAR. J Chem Inf Model 55 (8):1529–1534. https://doi.org/10.1021/acs.jcim.5b00261
46. Larose DT (2005) k-Nearest neighbor algorithm. In: Discovering knowledge in data. Wiley, New York, pp 90–106. https://doi.org/10.1002/0471687545.ch5
47. Johnson RA, Wichern DW (2002) Applied multivariate statistical analysis, vol 8. Prentice Hall, Upper Saddle River, NJ
48. Johnson MA, Maggiora GM (eds) (1990) Concepts and applications of molecular similarity. Wiley, New York
49. Breiman L (2001) Random forests. Machine Learning 45(1):5–32. citeulike-article-id:12416445. https://doi.org/10.1023/a%253a1010933404324
50. Romesburg CH (1984) Cluster analysis for researchers. Lifetime Learning publications, Belmont, CA
51. Contrera JF, Matthews EJ, Daniel Benz R (2003) Predicting the carcinogenic potential of pharmaceuticals in rodents using molecular structural similarity and E-state indices. Regul Toxicol Pharmacol 38(3):243–259. S0273230003000710 [pii]
52. Netzeva TI, Worth A, Aldenberg T, Benigni R, Cronin MT, Gramatica P, Jaworska JS, Kahn S, Klopman G, Marchant CA, Myatt G, Nikolova-Jeliazkova N, Patlewicz GY, Perkins R, Roberts D, Schultz T, Stanton DW, van de Sandt JJ, Tong W, Veith G, Yang C (2005) Current status of methods for defining the applicability domain of (quantitative) structure-activity relationships. The report and recommendations of ECVAM Workshop 52. Altern Lab Anim 33 (2):155–173
53. Jaworska J, Nikolova-Jeliazkova N, Aldenberg T (2005) QSAR applicabilty domain estimation by projection of the training set descriptor space: a review. Altern Lab Anim 33 (5):445–459
54. Tropsha A, Gramatica P, Gombar VK (2003) The importance of being earnest: validation is the absolute essential for successful application and interpretation of QSPR models. QSAR Comb Sci 22(1):69–77. https://doi.org/10.1002/qsar.200390007
55. Lu J, Peng J, Wang J, Shen Q, Bi Y, Gong L, Zheng M, Luo X, Zhu W, Jiang H, Chen K (2014) Estimation of acute oral toxicity in rat using local lazy learning. J Cheminform 6:26. https://doi.org/10.1186/1758-2946-6-26
56. Hosmer DW Jr, Lemeshow S, Sturdivant RX (2013) Applied logistic regression, vol 398. Wiley, New York
57. Quinlan JR (2014) C4. 5: programs for machine learning. Elsevier, San Francisco
58. Specht DF (1990) Probabilistic neural networks and the polynomial Adaline as complementary techniques for classification. IEEE Trans Neural Netw 1(1):111–121. https://doi.org/10.1109/72.80210
59. Xue Y, Li H, Ung CY, Yap CW, Chen YZ (2006) Classification of a diverse set of Tetrahymena pyriformis toxicity chemical compounds from molecular descriptors by statistical learning methods. Chem Res Toxicol 19(8):1030–1039. https://doi.org/10.1021/tx0600550
60. Watson P (2008) Naïve Bayes classification using 2D Pharmacophore feature triplet vectors. J Chem Inf Model 48(1):166–178. https://doi.org/10.1021/ci7003253
61. Hsu C-W, Lin C-J (2002) A comparison of methods for multiclass support vector machines. Trans Neur Netw 13(2):415–425. https://doi.org/10.1109/72.991427
62. Chang CC, Lin CJ. LIBSVM -- A library for support vector machines. http://www.csie.ntu.edu.tw/~cjlin/libsvm. Accessed 14 Feb 2013
63. Fei B, Liu J (2006) Binary tree of SVM: a new fast multiclass training and classification algorithm. Trans Neur Netw 17(3):696–704. https://doi.org/10.1109/tnn.2006.872343
64. Cheong S, Sang H, Lee SY (2004) Support vector machines with binary tree architecture for multi-class classification. Neural Inf Process 2:47–51

65. ACD/Labs ToxSuite. Advanced Chemistry Development Inc., Toronto, Canada. Software available at www.acdlabs.com

66. Martin T Toxicity Estimation Software Tool (TEST). Software available at www.epa.gov/nrmrl/std/qsar/qsar.html

67. TOxicity Prediction by Komputer Assisted Technology (TOPKAT). Accelrys Inc., San Diego, USA. Software available at http://accelrys.com

68. ADMET Predictor. Simulations Plus Inc., Lancaster, USA. Software available at www.simulationsplus.com

69. TerraQSAR. TerraBase Inc., Hamilton, Canada. Software available at www.terrabase-inc.com

70. Gonella Diaza R, Manganelli S, Esposito A, Roncaglioni A, Manganaro A, Benfenati E (2015) Comparison of in silico tools for evaluating rat oral acute toxicity. SAR QSAR Environ Res 26(1):1–27. https://doi.org/10.1080/1062936x.2014.977819

71. Drwal MN, Banerjee P, Dunkel M, Wettig MR, Preissner R (2014) ProTox: a web server for the in silico prediction of rodent oral toxicity. Nucleic Acids Res 42(Web Server issue): W53–W58. https://doi.org/10.1093/nar/gku401.nar/gku401

72. Muller L, Kikuchi Y, Probst G, Schechtman L, Shimada H, Sofuni T, Tweats D (1999) ICH-harmonised guidances on genotoxicity testing of pharmaceuticals: evolution, reasoning and impact. Mutat Res 436 (3):195–225. https://doi.org/10.1016/s1383-5742(99)00004-6

73. Lagunin AA, Dearden JC, Filimonov DA, Poroikov VV (2005) Computer-aided rodent carcinogenicity prediction. Mutat Res 586 (2):138–146. https://doi.org/10.1016/j.mrgentox.2005.06.005

74. Benigni R (2005) Structure-activity relationship studies of chemical mutagens and carcinogens: mechanistic investigations and prediction approaches. Chem Rev 105 (5):1767–1800. https://doi.org/10.1021/cr030049y

75. Sato S, Tomita I (2001) Short-term screening method for the prediction of carcinogenicity of chemical substances: current status and problem of an in vivo rodent micronucleus assay. J Health Sci 47(1):1–8. https://doi.org/10.1248/jhs.47.1

76. Benfenati E, Benigni R, Demarini DM, Helma C, Kirkland D, Martin TM, Mazzatorta P, Ouedraogo-Arras G, Richard AM, Schilter B, Schoonen WG, Snyder RD, Yang C (2009) Predictive models for carcinogenicity and mutagenicity: frameworks, state-of-the-art, and perspectives. J Environ Sci Health C Environ Carcinog Ecotoxicol Rev 27(2):57–90. https://doi.org/10.1080/10590500902885593

77. Kruhlak NL, Contrera JF, Benz RD, Matthews EJ (2007) Progress in QSAR toxicity screening of pharmaceutical impurities and other FDA regulated products. Adv Drug Deliv Rev 59(1):43–55. https://doi.org/10.1016/j.addr.2006.10.008

78. Ashby J (1985) Fundamental structural alerts to potential carcinogenicity or noncarcinogenicity. Environ Mutagen 7(6):919–921. https://doi.org/10.1002/em.2860070613

79. Bailey AB, Chanderbhan R, Collazo-Braier N, Cheeseman MA, Twaroski ML (2005) The use of structure-activity relationship analysis in the food contact notification program. Regul Toxicol Pharmacol 42(2):225–235. https://doi.org/10.1016/j.yrtph.2005.04.006

80. Munro IC, Ford RA, Kennepohl E, Sprenger JG (1996) Thresholds of toxicological concern based on structure-activity relationships. Drug Metab Rev 28(1–2):209–217. https://doi.org/10.3109/03602539608994000

81. Kazius J, McGuire R, Bursi R (2005) Derivation and validation of toxicophores for mutagenicity prediction. J Med Chem 48 (1):312–320. https://doi.org/10.1021/jm040835a

82. Kazius J, Nijssen S, Kok J, Back T, Ijzerman AP (2006) Substructure mining using elaborate chemical representation. J Chem Inf Model 46(2):597–605. https://doi.org/10.1021/ci0503715

83. Benigni R, Bossa C (2008) Structure alerts for carcinogenicity, and the Salmonella assay system: a novel insight through the chemical relational databases technology. Mutat Res 659(3):248–261. https://doi.org/10.1016/j.mrrev.2008.05.003

84. Wang Y, Lu J, Wang F, Shen Q, Zheng M, Luo X, Zhu W, Jiang H, Chen K (2012) Estimation of carcinogenicity using molecular fragments tree. J Chem Inf Model 52 (8):1994–2003. https://doi.org/10.1021/ci300266p

85. Kalgutkar AS, Didiuk MT (2009) Structural alerts, reactive metabolites, and protein covalent binding: how reliable are these attributes as predictors of drug toxicity? Chem Biodivers 6(11):2115–2137

86. Benigni R, Bossa C, Tcheremenskaia O (2013) Nongenotoxic carcinogenicity of chemicals: mechanisms of action and early

recognition through a new set of structural alerts. Chem Rev 113(5):2940–2957. https://doi.org/10.1021/cr300206t

87. Toivonen H, Srinivasan A, King RD, Kramer S, Helma C (2003) Statistical evaluation of the Predictive Toxicology Challenge 2000–2001. Bioinformatics 19 (10):1183–1193. https://doi.org/10.1093/bioinformatics/btg130

88. Sun HM (2004) Prediction of chemical carcinogenicity from molecular structure. J Chem Inf Comput Sci 44(4):1506–1514. https://doi.org/10.1021/ci049917y

89. Tanabe K, Lucic B, Amic D, Kurita T, Kaihara M, Onodera N, Suzuki T (2010) Prediction of carcinogenicity for diverse chemicals based on substructure grouping and SVM modeling. Mol Divers 14(4):789–802. https://doi.org/10.1007/s11030-010-9232-y

90. Singh KP, Gupta S, Rai P (2013) Predicting carcinogenicity of diverse chemicals using probabilistic neural network modeling approaches. Toxicol Appl Pharmacol 272 (2):465–475. https://doi.org/10.1016/j.taap.2013.06.029

91. Li X, Du Z, Wang J, Wu Z, Li W, Liu G, Shen X, Tang Y (2015) In silico estimation of chemical carcinogenicity with binary and ternary classification methods. Mol Informatics 34(4):228–235. https://doi.org/10.1002/minf.201400127

92. Aronov AM (2005) Predictive in silico modeling for hERG channel blockers. Drug Discov Today 10(2):149–155. https://doi.org/10.1016/s1359-6446(04)03278-7

93. Taboureau O, Jorgensen FS (2011) In silico predictions of hERG channel blockers in drug discovery: from ligand-based and target-based approaches to systems chemical biology. Comb Chem High Throughput Screen 14 (5):375–387

94. Cavero I, Mestre M, Guillon JM, Crumb W (2000) Drugs that prolong QT interval as an unwanted effect: assessing their likelihood of inducing hazardous cardiac dysrhythmias. Expert Opin Pharmacother 1(5):947–973. https://doi.org/10.1517/14656566.1.5.947

95. Elliott DJ, Dondas NY, Munsey TS, Sivaprasadarao A (2009) Movement of the S4 segment in the hERG potassium channel during membrane depolarization. Mol Membr Biol 26(8):435–447. https://doi.org/10.3109/09687680903321081

96. Stansfeld PJ, Gedeck P, Gosling M, Cox B, Mitcheson JS, Sutcliffe MJ (2007) Drug block of the hERG potassium channel: insight from modeling. Proteins 68(2):568–580. https://doi.org/10.1002/prot.21400

97. Mitcheson JS, Chen J, Lin M, Culberson C, Sanguinetti MC (2000) A structural basis for drug-induced long QT syndrome. Proc Natl Acad Sci U S A 97(22):12329–12333. https://doi.org/10.1073/pnas.210244497

98. Wang S, Li Y, Xu L, Li D, Hou T (2013) Recent developments in computational prediction of HERG blockage. Curr Top Med Chem 13(11):1317–1326. CTMC-EPUB-20130509-6 [pii]

99. Pearlstein RA, Vaz RJ, Kang J, Chen XL, Preobrazhenskaya M, Shchekotikhin AE, Korolev AM, Lysenkova LN, Miroshnikova OV, Hendrix J, Rampe D (2003) Characterization of HERG potassium channel inhibition using CoMSiA 3D QSAR and homology modeling approaches. Bioorg Med Chem Lett 13(10):1829–1835

100. Wang W, MacKinnon R (2017) Cryo-EM structure of the open human ether-a-go-go-Related K+ Channel hERG. Cell 169 (3):422–430 e410. S0092-8674(17)30410-5 [pii]. https://doi.org/10.1016/j.cell.2017.03.048

101. Ekins S, Crumb WJ, Sarazan RD, Wikel JH, Wrighton SA (2002) Three-dimensional quantitative structure-activity relationship for inhibition of human ether-a-go-go-related gene potassium channel. J Pharmacol Exp Ther 301(2):427–434

102. Cavalli A, Poluzzi E, De Ponti F, Recanatini M (2002) Toward a pharmacophore for drugs inducing the long QT syndrome: insights from a CoMFA study of HERG K(+) channel blockers. J Med Chem 45(18):3844–3853

103. Inanobe A, Kamiya N, Murakami S, Fukunishi Y, Nakamura H, Kurachi Y (2008) In silico prediction of the chemical block of human ether-a-go-go-related gene (hERG) K+ current. J Physiol Sci 58 (7):459–470. https://doi.org/10.2170/physiolsci.RV-0114-08-07-R1

104. Aronov AM (2006) Common pharmacophores for uncharged human ether-a-go-go-related gene (hERG) blockers. J Med Chem 49(23):6917–6921. https://doi.org/10.1021/jm060500o

105. Springer C, Sokolnicki KL (2013) A fingerprint pair analysis of hERG inhibition data. Chem Cent J 7(1):167. https://doi.org/10.1186/1752-153x-7-167

106. Braga RC, Alves VM, Silva MF, Muratov E, Fourches D, Tropsha A, Andrade CH (2014) Tuning HERG out: antitarget QSAR models

for drug development. Curr Top Med Chem 14(11):1399–1415

107. Roche O, Trube G, Zuegge J, Pflimlin P, Alanine A, Schneider G (2002) A virtual screening method for prediction of the HERG potassium channel liability of compound libraries. Chembiochem 3 (5):455–459. https://doi.org/10.1002/1439-7633(20020503)3:5<455::AID-CBIC455>3.0.CO;2-L

108. Li Q, Jorgensen FS, Oprea T, Brunak S, Taboureau O (2008) hERG classification model based on a combination of support vector machine method and GRIND descriptors. Mol Pharm 5(1):117–127. https://doi.org/10.1021/mp700124e

109. Wang S, Li Y, Wang J, Chen L, Zhang L, Yu H, Hou T (2012) ADMET evaluation in drug discovery. 12. Development of binary classification models for prediction of hERG potassium channel blockage. Mol Pharm 9 (4):996–1010. https://doi.org/10.1021/mp300023x

Chapter 16

Metabolomics: A High-Throughput Platform for Metabolite Profile Exploration

Jing Cheng, Wenxian Lan, Guangyong Zheng, and Xianfu Gao

Abstract

Metabolomics aims to quantitatively measure small-molecule metabolites in biological samples, such as bodily fluids (e.g., urine, blood, and saliva), tissues, and breathe exhalation, which reflects metabolic responses of a living system to pathophysiological stimuli or genetic modification. In the past decade, metabolomics has made notable progresses in providing useful systematic insights into the underlying mechanisms and offering potential biomarkers of many diseases. Metabolomics is a complementary manner of genomics and transcriptomics, and bridges the gap between genotype and phenotype, which reflects the functional output of a biological system interplaying with environmental factors. Recently, the technology of metabolomics study has been developed quickly. This review will discuss the whole pipeline of metabolomics study, including experimental design, sample collection and preparation, sample detection and data analysis, as well as mechanism interpretation, which can help understand metabolic effects and metabolite function for living organism in system level.

Key words Metabolomics, Metabolite profile, Metabolic response, Biomarker, Underlying mechanism

1 Introduction

In the past decade, metabolomics, as a vital component of systems biology, has made notable progresses in providing systematic insights into the underlying mechanisms of biological processes and presenting potential biomarkers of many diseases such as cancer, obesity, and type 2 diabetes. Metabolomics aims to quantitatively measure endogenous small molecular metabolites (typically <1500 Da) containing substrates, intermediates, and products (e.g., carbohydrates, fatty acids, organic acids, amino acids, nucleotides, steroids, and lipids) in a biological sample. Accurate measurement of variations of small molecular metabolites reflects multiparametric metabolic responses of a living system to

Jing Cheng and Wenxian Lan contributed equally to this work.

Tao Huang (ed.), *Computational Systems Biology: Methods and Protocols*, Methods in Molecular Biology, vol. 1754, https://doi.org/10.1007/978-1-4939-7717-8_16, © Springer Science+Business Media, LLC, part of Springer Nature 2018

pathophysiological stimuli or genetic modification [1, 2]. From the view of genetic information flow, small molecular metabolites locate in the downstream of genes and proteins and correlate to the phenotype closely. A minor change of gene expressions or protein modifications upstream might result in the readily amplification of metabolites downstream. And genomic study may not directly indicate what may happen, while metabolomics research may depict what has happened. Thus, metabolomics has the potential as a more reliable and sensitive measuring approach in early diagnosis of disorder for organism and plays an important role in bridging the gaps between genotype and phenotype. Metabolomics covers a wide range of metabolites with heterogeneous structures and broad dynamic concentration change of compounds in a biological system [3]. Nowadays, over 72,619 small-molecule metabolites and their concentration changes in biological systems are described in the Human Metabolites Database (HMDB: http://www.hmdb.ca/, version 3.6). Metabolic profile in a biological system covers a wide range of metabolites with heterogeneous structures and their dynamic range of concentrations across nine orders of magnitude in a biological system; thus a comprehensive detection of metabolic profile for a biological system is a great challenge [4]. The combination of multiple detecting platforms of metabolomics can enhance analysis of such a broad spectrum of metabolites. In practice, numerous detecting platforms have been utilized, for example, nuclear magnetic resonance (NMR), gas chromatography (GC) coupled with mass spectrometry (MS), and ultrahigh-performance liquid chromatography (UHPLC) coupled with mass spectrometry (MS) [5]. Compared with NMR methods, MS-based approaches exhibit a higher sensitivity for metabolite detection. Typically, GC and LC coupled to MS are employed widely since these methods combine an effective chromatographic separation with a sensitive mass detection. A classical research roadmap of metabolomics is illustrated in Fig. 1 for both untargeted and targeted strategies, which includes section of experimental design, sample collection, sample preparation, data acquisition, data preprocessing, data analysis, mechanism interpretation, and hypothesis generation. In this review we will discuss these different sections of metabolomics.

2 Experimental Design

Metabolomics is a powerful "omic" tool in the study of many research fields ranging from biomarker identification to disorder mechanism interpretation. One of the purposes for metabolomics study is to quantitatively measure small-molecule metabolites in biological samples, which contain complex compounds and have complicate bias. An appropriate experimental design should

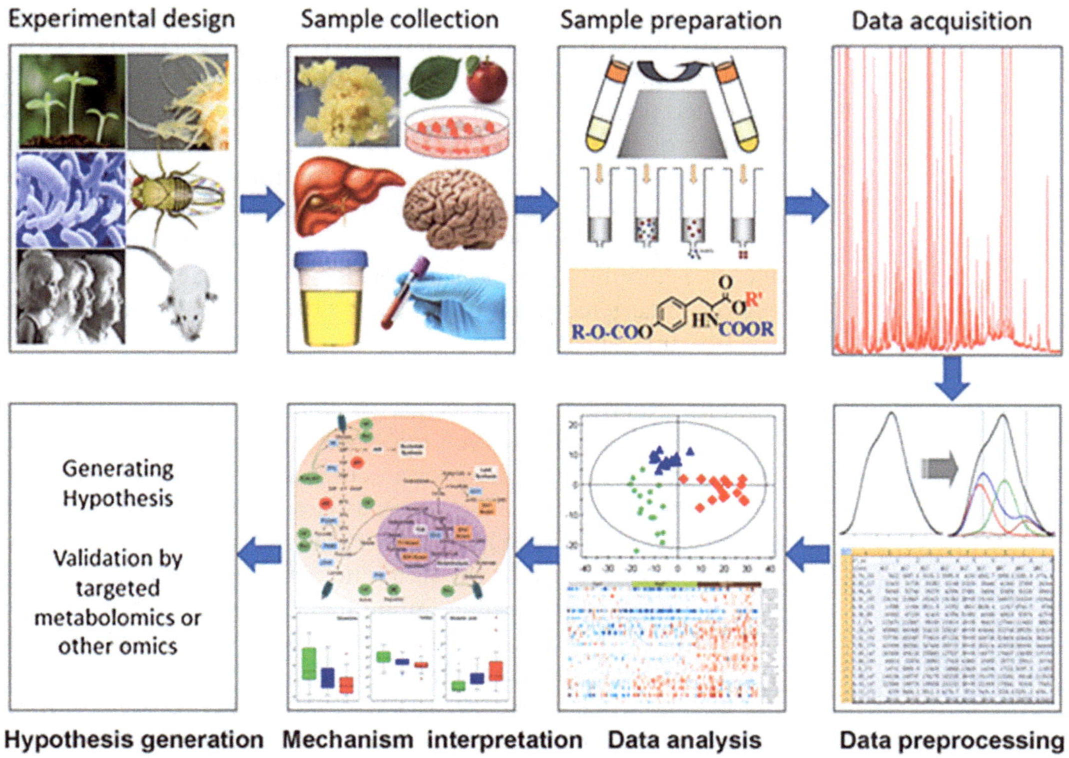

Fig. 1 Flow chart of a metabolomics study

minimize biological and technical bias impact on research results, give correct answers to the questions raised, and generate reasonable hypothesis. The experiment design is determined by the research goal (e.g., identification of diagnosis biomarker, influence of drug treatment, outcomes of nutrition intervention), which includes the following four important considerations: (1) experiment strategy (untargeted or targeted strategies), (2) experimental subjects (e.g., model animals or human samples, plant, cells, or microbes), (3) biological replicates, and (4) data analytical strategy (e.g., case-control analysis, paired analysis, time series analysis).

Untargeted and targeted strategies are two distinct experiment strategies, which are applied in metabolomics studies widely. Each strategy has its own inherent advantages and limitations. The untargeted metabolomics aims to detect metabolites as many as possible in a biological sample. Thus in untargeted strategy metabolomics, samples should be less preprocessed to keep more metabolites. This strategy can identify patterns or fingerprints of metabolites responding to genetic alterations and environmental stimuli, which often used for hypothesis-generating and hypothesis-refining study [6]. However, there are some difficulties of this strategy in comparing investigation results of different laboratories, combining research data in several time points, and characterizing unknown metabolites in multi-experiments. While for targeted metabolomics, it is a hypothesis-driven research and focuses on absolute quantification of targeted metabolites involved in some interested pathways. Appropriate sample preparation is required for accurate quantification of target metabolites. The strength of this strategy is high sensitivity and specificity; thus numerous metabolites of low abundance in biological samples might be quantified. However, an obvious weakness of this strategy is its narrow coverage of metabolite detection. According to characteristics of untargeted and targeted strategy, it is a powerful strategy to combine the initially untargeted metabolomics for exploration with the subsequently targeted metabolomics for validation.

The experimental subject is an important factor that could influence results of metabolomics. For example, Sprague-Dawley (SD) and Wistar rats are experimental animals commonly interchangeable in practice. However, a metabolomics study revealed that SD rats have higher individual metabolic variations than Wistar rats [7], suggesting that the difference of subjects should be considered thoroughly when designing metabolomics experiment.

Biological bias, existed in all organisms substantially, greatly affects the results of metabolomics study. Irrelevant biological bias induced by some impact factors, such as gender, dose, and time, should be eliminated by biological replicates (repetitive samples obtained from different individuals). Thus one important aspect of experimental design is to determine how many biological replicates are required in statistics. In metabolomics study, biological replicates are preferred over analytical replicates (repetitive analyses of the same sample obtained from the same individual), since biological bias almost exceeds analytical one. In practice, three biological replicates are required and five replicates are preferred [8]. Moreover, dozens of biological replicates are required for model animals, while for human population investigation, a larger scale of samples (usually hundreds to thousands samples) are necessary for early diagnosis in disease [9].

For some metabolomics studies, small treatments and effects can easily be overlaid by large variation between human subjects, and thus conventional data analytical strategy can't work effectively. A solution for these studies is the case-control analysis based on cross-over experiment design, where each subject in the study population acts as case specimen and the other subjects act as control in turn [10]. In addition, multivariate analysis methods for paired data remarkably improve the statistical power and give more reasonable interpretability to results of these metabolomics studies [11–13]. Therefore, the data analytical strategy should be chosen thoroughly according to the research goals.

3 Sample Collection and Preparation

Metabolomics usually requires a prior step of sample preparation. The aims of sample preparation are to extract analytes from a complicated biological matrix and to improve the detectability of analytes by chemical derivatization. Sample preparation procedure mainly includes metabolism quenching, sample storage, metabolite extraction, and chemical derivatization. When biological samples are collected from organism, metabolic reaction should be efficiently terminated to achieve a true snapshot of metabolome of biological samples. Thus quenching is the first step instantly to deactivate metabolic reaction by inhibiting endogenous enzymes. After that, most biological samples are stored until enough samples are collected. The storage factors, such as storage time, temperature, and freeze-thaw cycle, should be considered comprehensively since these factors affect metabolic stability significantly. Metabolite extraction is a key step of metabolomics study, which is coupled with protein precipitation, super-centrifugation, and liquid-liquid extraction [14]. Finally, some metabolites should be derivatized before detection. The whole sample preparation steps should not induce significant variations in chemical and physical properties as well as the concentration of metabolites. It is vital to build standard protocols for specific sample and instrument platform, in which the re-productivity across batches and laboratories should be considered thoroughly.

3.1 Sample Preparation for MS-Based Metabolomics

3.1.1 Sample Quenching and Storage

After collection, metabolic reaction in biological samples, such as biofluids, tissues, and cells, should be deactivated instantly by inhibiting endogenous enzymes. In this way, the snapshot of metabolome can be amenable to subsequent detection step. It has been found that the levels of lactic acid, citric acid, and lipids are changed in the plasma sample which is not quenched [15]. Some hydrolases and phosphatase tend to remain active even in methanol solution at ambient temperatures [16]. It is worth to note that metabolic reaction may last less than 1 s. Therefore an ideal quenching strategy should allow a rapid metabolism interruption in a sub-second time scale, which may not induce significant variations of metabolome. The general strategies for sample quenching is freezing coupled with pH regulation by adding acid or alkali [17]. Snap-freezing by liquid nitrogen (−196 °C) is an effective quenching approach especially for tissue samples, in which removing contaminants from the anesthetic drug in residual blood is of importance [18–20]. A pilot study of stability of urine metabolome based on GC-TOF-MS by Dunn et al. [21] indicated that the metabolome of urine sample stored at 4 °C for 24 h before being frozen at −80 °C was not significantly changed as compared to that of immediately frozen at −80 °C after collection. It is reported that the urine samples freezing at −80 °C instantly after collection is difficult [22]. The common strategy for urine samples freezing is to superinduce antibacterial preservatives (e.g., sodium azide (0.05–0.1% wt/vol)) and storage samples at 4 °C or −20 °C before being frozen at −80 °C [23, 24]. Methanol is a perfect organic solvent of quenching solutions, since it is miscible with water and has a low freezing point (much lower than ethanol or glycerol). The cold methanol (60%, vol/vol) allows a rapid termination of metabolism in sub-second time scale for mammalian cells and *Escherichia coli* [25]. Quenching in liquid nitrogen coupled with filtration has been developed as a widely used strategy of preparation for microorganisms, because the risk of cell leakage and damage in this strategy is considerably lower than that of the methanol-quenching strategy [26]. The metabolomics study of feces sample is significant to reveal metabolic interactions between host and gut flora. The instant preparation of fecal samples is important since living environment of gut flora is changed during sample collection, which may lead to metabolic profile distortion. Fecal samples frozen at −20 °C manifested remarkably higher amino acids than that of crude fecal control samples, suggesting a release of microbial intracellular metabolites [27, 28]. The recommendation for fecal samples is collected at 4 °C and processed within 1 h of collection [28]. After that, samples should be stored under low temperature (commonly at −80 °C). Freeze-thaw cycles should be avoided to prevent the variation of metabolites. However, urine sample is insensitive to freeze-thaw cycle [24]. A practicable approach is to divide the quenched sample into aliquots and utilize one aliquot sample in each time.

3.1.2 Metabolite Extraction

Metabolite extraction is a key step in the metabolomics analytical procedure, which directly affects the quality of the final data. To keep the nature of metabolome, minimal sample preparation is essential for metabolomics research. The basic step of sample preparation is to remove interferes (e.g., salts and proteins) and make the extracts compatible with the detecting technique [17]. The extraction methods are determined by the physicochemical properties of samples and detective technique for samples. The conventional extraction methods for untargeted metabolomics include liquid-liquid extraction (LLE), solid-liquid extraction (SLE), solid phase extraction (SPE), microwave-assisted extraction (MAE), accelerated solvent extraction (ASE), microdialysis, and ultrafiltration [29]. During extraction step, it is preferable to spike analytical standards or isotope-labeled internal standards into samples for efficiency evaluation of the metabolite recoveries. In general, protein precipitation is required after metabolite extraction. For this purpose, organic solvent is added to the extractive complex followed by centrifugation or ultrafiltration [30]. Methanol and acetonitrile are the most usual organic solvents of protein precipitation

[31, 32]. In addition, the lipids in the supernatant may interfere with mass spectrometry due to ion suppression. Therefore, the combination of protein precipitation and lipid removal by chloroform is recommended to improve metabolite coverage and reproducibility [33]. The extraction methods of different samples are various because of their different natural characterization. Urine samples usually contain relative less protein than plasma or tissues. Thus, the sample preparation is performed through a simple centrifugation commonly. For the urine of renal disease, however, the injury kidney function usually causes the excessive excretion of urine proteins, and thus protein precipitation by organic solvent is an essential step [34]. Compared to biofluids (e.g., plasma and urine), the sample preparation of tissues is labor-intensive and time-consuming, because an additional step of homogenization (physical disruption of tissue sample) is prerequisite. The breakdown of the frozen tissue is performed through grinding in liquid nitrogen with mortar and pestle, in which a specifically designed tissue lyzer is utilized to homogenize multiple tissue samples simultaneously [19]. For specific tissues such as the lung and heart, a prior step of enzyme digestion before bead-beater treatment can effectively improve homogenization [35]. Methanol/water in different proportions (50–100%) is the common solvent for homogenization. After that, proteins in tissue homogenates are precipitated by a one-step solvent method (adding methanol/water mixture) [18] or a two-step solvent method (adding methanol/water (1:1, vol/vol) mixture followed by dichloromethane/methanol (3:1, vol/vol) mixture) [19]. The feces is a kind of complicated sample including undigested food residues, metabolites, and gut microbes. To avoid the disturbance from the intracellular contents of gut microbes especially amino acids, super-centrifugation (171,500 × *g* at 4 °C for 2 h) of feces samples within 24 h after collection is recommended [27, 28]. Short-chain fatty acids and bile acids are two kinds of targeted metabolites closely correlating to the interactions between host and gut microbes. Interestingly, our study found that these metabolites were almost not affected by the intracellular contents of gut microbes when the feces samples were stored at low temperature (e.g., −80 °C) [27].

3.1.3 Chemical Derivatization for MS-Based Metabolomics

Due to the complexity of metabolome and the diversity of physicochemical property of metabolites, it is practically impossible to simultaneously detect all compounds in a biological sample through MS-based detection methods. To detect more metabolites, derivatization is a prerequisite step in GC/MS based on methods for rendering highly polar compounds with sufficiently volatile and increasing ionization efficiency of these compounds. In derivatization, a chemical reagent is applied to react with one or more active groups (such as carboxylic, amide, amino, and hydroxyl groups) of compounds, and they are transformed into a structurally altered complexes, which have some features improving effectiveness of the subsequent chromatographic retention, separation, detection, and identification [36].

Currently, there are two common used derivation methods, i.e., silylation derivatization and alkylation derivatization. As for silylation derivatization, most GC-MS based metabolomics studies apply a two-step process, including the first oximation and subsequent trimethylsilylation (TMS). The aim of oximation is to prevent side derivatization of compounds containing carbonyl group (such as pyruvic acid, α-ketoglutaric acid, and reducing sugars), by reacting with methoxyamine hydrochloride [37]. *N*-Methyl-*N*-(trimethylsilyl)trifluoroacetamide (MSTFA) with 1% trimethylchlorosilane (TMCS) and *N*,*O*-bis(Trimethylsilyl)trifluoroacetamide (BSTFA) with 1% trimethylchlorosilane (TMCS) are two commonly used TMS reagents in GC-MS metabolomics [37–39]. The sequential oximation and TMS derivatization can improve effectiveness of metabolic analysis of compounds containing nitrogen (amino acids, amines, purine, pyrimidine, and nucleosides), carboxylic acids (fatty acids, the intermediates of citric acid cycle, hydroxyl acids), polyols, carbohydrates, sugar acids, cholesterol and free bile acids, and phenolics. These compounds cover the main metabolites in central carbon metabolism, which are very important in the study of biological science. In general, the silylation derivatization requires a non-aqueous environment, while the alkylation derivatization can be performed in an aqueous environment. For the alkyl chloroformate derivatization, proposed by Husek, its chemical reaction system is consists of alkyl chloroformate, water, ethanol, and pyridine [40, 41]. During the derivatization reaction, alkoxycarbonyl esters of amino and hydroxyl groups are stable, but the alkoxycarbonyl ester of carboxylic group is unstable, and it subsequently transformed into stable alkyl esters by an exchange reaction with ethanol [27]. Alkyl chloroformate derivatization allows simultaneous esterification of carboxylic group, amino group, and hydroxyl group linked to a benzene ring or the side chain, which remarkably improve peak intensity and chromatographic separation [42]. The alkyl chloroformate derivatization has higher reproducibility and stability than the TMS derivatization [27, 43]. However, carbohydrates and polyols existed in biofluids can be detected by the TMS derivatization but not the alkyl chloroformate derivatization. While alkyl chloroformate derivatization can help detect short-chain fatty acids, amino acids, carboxylic acids, hydroxyl acids, phenolic acids, benzoyl and phenyl derivatives, and indoles [27, 42].

Generally, metabolomics based on LC-MS just apply a simple protein precipitation step in sample preparation process, allowing hundreds to thousands of metabolites to be routinely detected and relatively quantitated. However, there are still numerous metabolites that can't be detected by the standard LC-MS techniques, due to the weak retention on the chromatographic column, lower responses of mass detection, wide concentration range, poor ionization, low abundance, matrix effects, and potential degradation of compounds. Chemical derivatization of compounds can help to improve the detection effectiveness of LC-MS methods, in which a charged or chargeable functional group is introduced into compounds. An ideal derivatization reaction should be conducted in a mild condition with high yields and less side reactions, and the resulting derivatives should have good chemical stability in the subsequent analytical procedures. The stable-isotope-labeled derivatives, presumed to have identical physicochemical properties to those of the corresponding unlabeled compounds, are utilized as internal standard (IS) to correct for the variability arising from sample preparations and instrumental analyses. Generally, the ^{13}C- or ^{15}N-labeled ISs are more preferable than deuterium (d-) labeled ISs because they avoid chromatographic isotope effect that could lead to low accuracy in quantitative LC-MS assays. For example, Guo and his colleagues synthesized a new reagent of $^{13}C_2$-dansyl chloride to derivatize standard compounds as ISs, and then 93 metabolites containing amino or phenol hydroxy group were absolutely quantified from human urine samples by the RPLC-ESI-MS method [44]. This approach was afterward extended for metabolomics studies of body fluids [45–48]. Another widely available derivatization reagent is ^{13}C-labeled benzoyl chloride (BzCl), which is used for primary and secondary amines, phenols, thiols, and ribose hydroxyls. Compared to $^{13}C_2$-dansyl chloride, ^{13}C-labeled benzoyl chloride has the following advantages in derivatization: (1) the derivative reaction of BzCl is a fast (seconds) and mild (room temperature) reaction at a wider pH range; (2) the reaction does not need to scavenge reagents; (3) benzoyl chloride and the benzoylated products are insensitive to light; and (4) ^{13}C-labeled benzoyl chloride is commercially available at relatively low cost. It was reported by using the ^{13}C-labeled benzoyl chloride as derivatization reagents, up to 70 neurochemicals, including catecholamines, indoleamines, amino acids, polyamines, trace amines, antioxidants, energy compounds, and their metabolic products were quantified from cerebrospinal fluid, plasma, tissue, and hemolymph with a 10 nM detection limits [49, 50].

3.2 Sample Preparation for NMR-Based Metabolomics

Compared with MS-based metabolomics, the sample preparation in NMR-based metabolomics is simpler. The alteration in pH value and salt concentration affect the chemical shift of metabolites, which lead to spurious clustering of compounds in metabolomics studies. Therefore, it is necessary to adjust the pH value of each sample through a phosphate buffer (pH 7.4). Various preparation methods are required for different samples. Serum or plasma samples should be collected by lithium heparin tubes, because EDTA could result in extra resonances. They are usually diluted with saline-buffer solution containing 10% D_2O. Similar to MS, proteins in serum or in plasma should be removed before NMR measurements by either organic solvent precipitation or by ultrafiltration using a 10 kDa molecular weight as cutoff [51]. Urine samples should be supplemented with NaN_3 (in a total concentration of sodium azide of min 0.05%, wt/vol) after sample collection and stored at −40 °C before analysis [52]. The supernatant should be mixed with potassium phosphate buffer (containing 10 mM sodium trimethylsilyl [2,2,3,3-D_4] propionate TMSP, in 100% D_2O). For the urine of subjects with injury kidney function, additional steps of protein precipitation by organic solvent or ultrafiltration should be conducted. For tissues and cells samples, it is necessary to quench metabolic reactions by flash-freezing in liquid nitrogen after collection. Tissues can be analyzed directly when NMR machine is equipped with HR MAS probe, which allows the sample to be spun with a high speed (more than 2 kHz) at the magic angle 54.7° [53]. However, HR MAS NMR spectra might be overcrowded by the overlapping lipid signals. Thus, it is of necessity to perform an extraction procedure prior to NMR detection. Inorganic acids (e.g., perchloric acid) can be applied to extract polar metabolites, but the risks of metabolite degradation and the ionic strength increasement of the final extracts are raised. The solvent of methanol/water (2:1, vol/vol) can be used to extract low molecular weight metabolites, while acetonitrile/water (2:1, v/v) is used for lipids and macromolecules. Solvents containing methanol, chloroform, and water can be utilized for extraction of polar and lipophilic metabolites [54]. For cell samples, it is recommended to disrupt these samples by repeated freeze-thaw procedure. Then the supernatant is collected after centrifugation in 2000 × *g* at 4 °C for 3 min [55]. The approaches of protein precipitation and metabolite extraction are the same as those of tissue samples. For fecal samples, multiple solvents, such as phosphate buffer, methanol, and alkalized solutions, can be utilized for metabolite extraction. The recommended extraction solvent is phosphate buffer (0.1 M, pH 7.4), which present maximum signal-to-noise ratio for subsequent analysis [55, 56].

4 Multiple Detecting Platforms for Data Acquisition

The ambition of metabolomics is to detect and quantify all small molecular metabolites in a biological sample, which is a huge challenge in practice because of the diverse physicochemical properties of metabolites (e.g., polarity, concentration, stability). Multiple analytical techniques are required to detect metabolites as many as possible. Currently, most metabolomics studies are performed based on mass spectrometry (MS) and proton nuclear magnetic resonance spectroscopy (NMR) platforms. However, none of these techniques alone can detect the entire metabolome of an organism. NMR is a global and nondestructive technique characterized by high-throughout, reproducible, and minimal sample pretreatment. The main limitation of NMR is low sensitivity compared to MS technology. Although MS can be successfully operated in flow infusion analysis mode, it is usually linked with an additional device, such as GC, LC, and capillary electrophoresis (CE), to increase the resolution of analytes and lower the matrix effects (ion suppression or enhancement when ionization in ion source). In general, the combination of MS- and NMR-based approaches can detect more metabolites.

4.1 GC-MS Platform

GC-MS is a widely used analytical platform for the identification and quantification of volatile and semivolatile organic compounds in complex matrix due to its high sensitivity, peak resolution, and excellent reproducibility [57]. For the GC-MS method, the structural information of compounds can be obtained through matching detecting data to the well-established metabolite spectrometry database, which facilitates the identification of metabolites and the elucidation of physiological or pathological mechanisms. The available databases include the commercial NIST, Wiley, Fiehn Metabolomics library, and the free Golm Metabolome Database. For GC-MS platform, the compound identification is quite straightforward and robust because of the extensively reproducible fragmentation pattern across mass spectrometers in electron impact (EI) ionization with full-scan mode. Moreover, EI ionization in GC-MS is less prone to matrix effect due to its use of the gaseous phase and the nature of EI ionization; therefore GC-MS in EI ionization has been the standard reference method for quantification of many compounds [58].

Time-of-flight (TOF) and quadrupole mass analyzers are most frequently applied in GC-MS metabolomics. TOF mass analyzer provides faster spectral acquisition (up to 500 Hz) and more scan points recorded for a peak; thus TOF mass analyzer can more accurate deconvolution of mass spectra for co-eluted analytes than other mass analyzer. The total run time of each GC-TOFMS analysis can be remarkably shorten since both chromatographic separation and spectral acquisition are faster than conventional GC-MS analysis. Therefore, the predominant advantage of GC-TOFMS is the higher throughout as compare to the conventional GC-MS based on quadrupole mass analyzer [59]. Currently, comprehensive two-dimensional gas chromatography (GC × GC) is applied for metabolomics study, which provides superior separation capacity, chemical selectivity, and sensitivity than conventional GC for complex sample analysis [60, 61]. All analytes in a complex sample can be separated by nonpolar column and subsequently a polar column

[62]. GC × GC-TOFMS or recently tested GC × GC-qMS has been applied to metabolomics studies for different types of samples (cell, biofluid, tissue, bacteria, yeast, plant) [62, 63].

GC-MS is a highly attractive metabolomics tool due to the inherent advantages: (1) the technology is easy to use and low cost, (2) has excellent chromatographic resolution power, and (3) has readily available mass spectral databases independent of instrument manufacturers [64]. However, GC-MS is limited to the analysis of small volatile molecules, and molecules can be volatilized [65].

4.2 LC-MS Platform

LC coupled with MS is a powerful platform for metabolomics study, which can be tailored for targeted or untargeted strategies. This platform is attributed to the remarkable development of ultrahigh-performance liquid chromatography (UPLC or UHPLC) based on sub-2 μm size of packing particles and high-resolution mass spectrometer. The UPLC or UHPLC significantly increase chromatographic resolution and lead to narrow peak width [39], which can detect more metabolites at less time compared with conventional HPLC [66]. Mostly published LC-MS-based untargeted metabolomics studies used reverse phase (RP, e.g., C18) silica-based columns with less than 2 μm particle sizes so as to obtain high resolution, sensitivity, and throughout. In general, RP stationary phase is biased in favor of the separation of nonpolar and polar metabolites rather than highly polar metabolites (e.g., glucose 6-phosphate and ATP). These highly polar metabolites are usually eluted without chromatographic retention, and the reproducibility of quantification is poor due to the matrix effects. Normal phase (NP) and hydrophilic interaction chromatography (HILIC) are the complementary separation technologies to RP. Recently, HILIC has been preferable technology in metabolomics studies for the separation of highly polar metabolites because it uses the same mobile phase and salt modifiers as RP chromatography [67, 68]. In general, the application of RP and HILIC technology together improves the chromatographic separation of different metabolites, which is vital to the qualification and quantification of metabolomics.

High-resolution mass spectrometry (HRMS, typically >5000 full width at half maximum) is a prerequisite in untargeted metabolomics for unknown metabolite identification. The application of HRMS will significantly increase the accuracy of mass-to-charge (m/z) as well as reduce the number of unknown molecule candidates. Time-of-flight (TOF), Fourier transform ion cyclotron (FTICR), and Orbitrap are the most commonly used HRMS analyzers. The application of hybrid mass spectrometers such as Q-TOF, FTICR, and Q-Orbitrap further improves the accuracy of identification of metabolites through integrating the profile of mass fragments. The narrow peak width by U(H)PLC requires faster acquisition rates or scanning speed. In general, the mass resolutions of FTICR and Orbitrap mass spectrometers are significantly higher than that of Q-TOF mass spectrometer at long scan time. It is reported that mass resolutions of FTICR and Orbitrap mass spectrometers are seriously decreased if faster scan speed is adopted. Therefore, it is important to coordinate the chromatographic gradient of U(H)PLC and the scan speed of Orbitrap or FTICR mass spectrometers. Electrospray ionization is the popular ionization mode in LC-MS, which include the positive and negative modes. It is recommended that both modes should be performed to ensure a wide panel of metabolites in a sample. For targeted metabolomics, LC usually is hyphenated to tandem mass spectrometers, such as triple quadrupole (QqQ) or hybrid triple quadrupole/linear ion trap (QTrap), which has high sensitivity and specificity of metabolite qualification. The multiple reaction monitoring (MRM) mode is mostly applied, where the precursor ion filtered in the first quadrupole is fragmented in the collision cell of the second quadrupole, and subsequently the characteristic fragments (product ions) are further selected in the third quadrupole. Compared to UPLC-Q-TOF-MS and UPLC-Q-Orbitrap-MS, LC-MS/MS can more effectively detect the low abundance metabolites.

LC-MS-based platform can detect more metabolites than GC-MS platform. However, the inherent matrix effects in ESI mode are the major limitation of LC-MS for the quantification of co-eluted metabolites [39, 64, 69]. Recently, submetabolome is introduced to improve quantitative metabolomics, where QC sample derivatized by ^{13}C-labeled chemical reagents is mixed with the samples derivatized by ^{12}C-labeled chemical reagents to eliminate the disturbances of matrix effects [44, 47, 48].

4.3 NMR-Based Detection Platforms

NMR technique is based on the energy absorption and reemission of the atom nuclei due to variations in an external magnetic field [70]. High-resolution NMR spectroscopy is a quantitative and nondestructive technique. The increase in field strength will tremendously improve the resolution and sensitivity, and more metabolites are observed. Generally, 500 or 600 MHz NMR instruments are used in metabolomics studies. Additionally, the introduction of cryo-cooled NMR probes reduces the level of thermal noise, and a four- or fivefold increase of the signal-to-noise (S/N) ratio is achieved as compared to a room temperature probe, which can detect metabolites with micromolar (μM) concentration [71].

High-resolution ^{1}H NMR spectroscopy is a powerful technology of metabolite detection, which can present comprehensive metabolic profile within a relatively short time scale. However, for the ^{1}H NMR spectroscopy, the dominant signal of the water protons could cover signals of low abundant metabolites and hinder spectral assignment and integration. In addition, the existence of macromolecule in analytical samples could produce broader resonances and cause difficulties in spectral assignment and integration. Thus the signals of water and macromolecule should be suppressed and filtered. Currently, NMR metabolomics are primarily conducted with 1D NOESY presat for water suppression and 1D Carr-Purcell-Meiboom-Gill (CPMG) presat as a T2 filter to remove macromolecule signals [72].

Recently, 2D NMR spectrum experiments are applied since signals of 1D ^{1}H NMR spectrum are seriously overlapped. One of the most commonly used 2D experiments is ^{1}H-^{1}H J-resolved spectrum that separates chemical shifts and J coupling. In this experiment, the peaks from macromolecules are attenuated, and then multiplicity information and coupling patterns of the remaining resonances are obtained. Correlation spectroscopy (COSY) shows correlations between scalar coupled hydrogen atoms over two to several (up to four in favorable cases) chemical bonds. Total correlation spectroscopy (TOCSY) provides ^{1}H-^{1}H spin-spin coupling connectivity with all hydrogen atoms within a spin system, forming an unbroken chain of couplings and thereby unraveling the number and nature of the molecular fragments. Heteronuclear single quantum coherence spectroscopy (HSQC) shows correlations between hydrogen atoms and carbon atom to which they are attached. Heteronuclear multiple bond correlation (HMBC) provides correlations between hydrogen and heteronucleus (generally two or three bonds away). These 2D sequences are used for signal assignments and identification of unknown metabolites commonly.

In addition, ^{31}P and ^{13}C NMR technologies are also applied in metabolomics studies recently. Nuclei ^{31}P has large gyromagnetic ratio; therefore ^{31}P NMR has high sensitive in metabolite detection, which can be used for investigation of cellular energy states [73]. ^{13}C NMR signals have better resolution than ^{1}H NMR signals for its large chemical shifts range. ^{13}C NMR technology is utilized on indirect detecting [74] or isotope labeling [75] scenarios commonly.

5 Data Preprocessing and Analysis

Data preprocessing is an important step before data analysis, which is crucial in metabolomics study. After acquiring data by MS or NMR platform, the subsequent step is to translate these raw data into available extracted data (e.g., peak tables) that can be applied in statistical analysis for metabolite identification easily.

5.1 Raw Data Preprocessing and Metabolite Identification of MS Platform

For GC-MS-based metabolomics, the main procedure of data preprocessing includes baseline filtering, peak finding, retention time correction, peak alignment, and co-eluted peaks deconvoluting. Currently, some popular tools are developed for GC-MS data processing, such as MetAlign [76], metaMS [77], TargetSearch [78], and web-based MetabolomeExpress [79] and TagFinder [80]. Specially, TagFinder is a free software package for mass spectral tags retrieving based on the tag intensity correlation within a group of time serial samples, which can be applied for retention index (RI) and calculation, mass spectra tag (MST) extraction, and tag structural identification. Thus, TagFinder supports both non-targeted fingerprinting analyses and targeted metabolite profiling.

For LC-MS-based metabolomics, the main procedure of data preprocessing is similar to that of GC-MS-based studies. Currently, a range of commercial software packages for LC-MS data preprocessing are developed by instrument vendors (e.g., Waters MarkerLynx and Progenesis QI, Agilent MassHunter, Sciex MarkerView, and LECO ChromaTOF), while some other packages are implemented by academic community (e.g., XCMS [81], MZmine [82, 83], MetAlign [76], TagFinder [84]). Specially, the XCMS software is a widely used R package of data preprocessing with command line mode. To facilitate application of the XCMS package, a graphic version has been developed, which allow users to upload and process LC-MS data online. XCMS online provides a whole workflow for untargeted metabolomics study, which includes feature detection, retention time correction, peak alignment, peak annotation, statistical analysis, and data visualization [85]. In addition, a free R package, IPO, is further designed as assistant tool of XCMS software, which can automatically select parameters for data processing of the LC-HRMS platform [86]. Specially, spectra of data-independent acquisition coupled with tandem mass spectrometry (DIA-MS/MS) platform contain total fragment ions of all precursors simultaneously. Therefore, DIA-MS/MS spectra are needed to be deconvoluted for fragment ion purification. Currently, tools of DIA-MS/MS data analysis include MS-DIAL, MetaboDIA, and MetDIA [87–89], and they can be utilized to quantify and identify metabolites through matching the MS peaks to spectra reference library, such as Metlin, MassBank, HMDB, NIST, ReSpect, and mzCloud, LIPID MAPS, and LipidBlast [90].

5.2 Raw Data Preprocessing and Metabolite Identification of NMR Platform

The preprocessing steps in NMR platform are described as follows: baseline correction, spectra alignment, binning, normalization, and scaling. Baseline correction is the first step of data preprocessing, since baseline distortion will affect the subsequent spectral analysis. Peak shifts between different spectra are a common phenomenon caused by instrumental factors as well as changes of the pH, salt concentration, temperature, and excessive dilution. In general, spectra are initially aligned to an internal standard to remove the global shifts. But this is not enough because there are still shifts of peaks in one spectrum, so a refine or local alignment should be conducted for peaks in one spectrum. For this purpose, various alignment algorithms, including interval correlated shifting (icoshift) [91], correlation optimized warping (COW) [92], fuzzy warping, and hierarchical cluster-based peak alignment (CluPA) [93], are proposed. The NMR spectra are divided into segments (so-called bins or buckets), and the spectral intensity within each bin is calculated. Binning attempts to minimize effects from variations in peak positions caused by the pH, ionic strength, and composition in a sample. Binning can reduce the dimensionality for multivariate statistical analysis but decreases the spectral resolution. The most common type of spectral binning is equidistant binning with 0.04 ppm. This indicates that every spectrum is divided into evenly spaced integral regions with spectral width of 0.04 ppm. Sometimes a peak is split between two bins, for equal size binning is the lack of flexibility of the boundaries. To prevent peak splitting, some new binning methods are proposed [94–96]. As for peak normalization, constant sum normalization (CSN) is the most commonly used method, in which the total spectral peak area is considered as a constant across. In addition, histogram matching (HM) normalization, volume normalization, and weight normalization (dry weight/wet weight) are also used in metabolomics studies [55]. After normalization, the scaling step is conducted to avoid the situation where the highest peaks are the most influential in the multivariate data models. Currently, several scaling approaches, such as auto scaling, Pareto scaling, mean centering, and range scaling, are proposed. Among these approaches, the Pareto scaling is the most commonly used approaches, where each variable is divided by the square root of the standard deviation of the column values.

Although a number of metabolites are detected by the 1D ^{1}H NMR methods, the identification of metabolites in biological samples is still a huge challenge because many peaks tend to overlap in 1D ^{1}H NMR spectra. This problem can be solved by the 2D ^{1}H-^{1}H TOCSY and 2D ^{1}H-^{13}C HSQC approaches. After spectral data preprocessing, the known metabolites is characterized by querying libraries of NMR spectra (1D and 2D). Now, many NMR spectra databases have been constructed, including the Human Metabolome Database (HMDB) [97, 98], the Biological Magnetic Resonance Data Bank (BMRB) [99], the Madison-Qingdao Metabolomics Consortium Database (MQMCD) [100], and the MetaboLights database [101]. Various forms of queries can be performed to extract metabolite information from these databases. Users can submit structure, chemical shift, or *J*-coupling constant of metabolites to these databases for query, and relevant biological pathway metabolite involved are retrieved as a result. The features of different databases are summarized by Ellinger et al. [102]. In addition, a new database named ^{13}C-TOCCATA customized database is recently developed [103], which contains complete ^{1}H and ^{13}C chemical shift information on individual spin systems and isomeric states of common metabolites. Unknown metabolites need to be separated temporarily from target metabolites, which can simplify the NMR spectra to a great extent. Then, the target metabolite can be determined combined 2D NMR (TOCSY, COSY, HMBC, HSQC) with MS according routine procedure [104]. The functional groups can be identified conclusively by their ^{1}H and ^{13}C chemical shifts. The skeletal connectivity is deduced by combining data from both homonuclear (COSY, TOCSY) and heteronuclear (HSQC and HMBC) correlation spectroscopy. The relative stereochemistry is inferred from NOE correlations between protons and coupling constants ($^{n}J_{H,H}$, $^{n}J_{C,H}$). Finally, the proposed structure is verified in light of all of the observed data [104]. With new NMR pulse sequence emerging, unknown metabolites can be identified in the complicated mixtures [105]. The methods are demonstrated for covariance ^{1}H-^{1}H TOCSY and ^{13}C-^{1}H HSQC-TOCSY spectra and triple-rank correlation spectra constructed from pairs of ^{13}C-^{1}H HSQC and ^{13}C-^{1}H HSQC-TOCSY spectra [105].

5.3 Signal Correction and Data Normalization

In metabolomics, samples of different batches span several months or even years commonly. The signal drift of metabolites over time and across batches is a major confounding factor in large-scale metabolomics studies. The unexpected variations of metabolite ion peaks in the measurements during data acquisition (intra- and inter-batch) are unavoidable, probably derived from sample handling and preparation, the degradation of column efficiency, MS instrument contamination, and nonlinear drift over long runs [106–108]. To correct signal drift, the quality control (QC) of samples are necessary over the entire time course of large-scale study. For this purpose, QC samples are needed, which have similar matrix composition to biological samples under study. Applying QC sample can help correct signal drift and remove peak with extreme change so as to eliminate batch effects [109–111]. Recently, the popular signal correction methods include the robust locally estimated scatter-plot smoothing (QC-RLSC) [39] and support vector regression (SVR) [112]. After signal correction, peak change should be checked, where peaks with variation over a threshold (e.g., 20% and 30% for UPLC-MS and GC-MS, respectively) are removed to make subsequent statistical analysis can be conducted correctly.

After signal correction, data normalization is an essential step before statistical analysis to remove unwanted variations (e.g., unwanted biological variations, experimental variations, and technical errors) in metabolomics study. Currently, numerous methods are applied in the normalization of metabolomics [113–115], and they are generally categorized into two groups: (1) method-driven normalization approaches that remove the unwanted variations case by case based on the internal standards or QC samples and (2) data-driven normalization approaches that normalize data based on biases among various metabolites [116].

5.4 Statistical Analysis

In metabolomics study, huge variables and high-dimension datasets are universal, which raise great challenge in statistics. Chemometric analysis has become an integral part of the metabolomic workflow because it provides interpretable models for complex intercorrelated data. Recently, MetaboAnalyst and XCMS online are developed as chemometric tools for data analysis, which are widely adopted by the metabolomics research community.

Prior to multivariate statistical analysis, mean centering and variance scaling are often conducted. The aim of mean centering is to make data centralization in multidimensional space implemented by subtracting the mean value to each variable. The variance scaling tends to normalize variable value through dividing each variable by the standard deviation of all variables. The mean centering and variance scaling are combined to form so-called auto scaling. Pareto scaling is an alternative popular scaling method, which is performed through dividing each variable by the square root of standard deviation of all variables; thus it minimizes the effects of undesired noises. In practice, auto-scaling is appropriate for the data scaling of GC-MS metabolomics, while Pareto scaling is more popular in the data scaling of LC-MS metabolomics.

When raw data is transformed into expression matrix table, statistical analysis is performed to characterize the correlation between groups (control group and case group) and identify candidate biomarkers for group classification. In metabolomics study, the number of variables is large and the sample size is small; therefore variable selection or reduction is a crucial step in statistical analysis. After variable selection, multivariable analysis is conducted to extract metabolites with different abundance between groups. Commonly, the multivariable analysis methods are divided into two categories, e.g., non-supervised approach and supervised approach. In the non-supervised approach, the data matrix of metabolite is divided into several parts based on inherent characterization of data, in which the relationship between parts can be understood by comparing different parts. Principal component analysis (PCA) [117] is a popular non-supervised approach, which attempts to uncover hidden internal structures through constructing components to distinguish group samples with maximum likelihood. Hierarchical cluster analysis (HCA) [118] is another non-supervised method used in metabolomics study commonly. HCA can cluster the metabolite data to divide the instance space.

Aside from the non-supervised approach, the supervised multivariate technique is utilized to identify interested metabolites directly. Partial Least Squares-Discriminant Analyses (PLS-DA) and Orthogonal Partial Least Squares-Discriminant Analyses (OPLS-DA) are regress methods with a binary "dummy" response variable, which are used to extract meaningful metabolites from expression matrix table directly.

6 Mechanism Interpretation and Hypothesis Generation

After interested metabolites are identified and quantified, the interpretation of complex interaction between genes, proteins, and metabolites is one important goal of metabolomics study. Functional annotations to metabolites can give meaningful biological information and provide mechanistic insights into a living system [119]. In general, an enrichment analysis is conducted to reveal biological pathways of target metabolites, which describe function of these metabolites and can help understand their physiological roles in a living system. Recently, some software packages and databases are developed for mechanism interpretation and hypothesis generation in metabolomics (*see* Table 1 for details). The KEGG, BioCyc, iPath, KaPPA-View, MapMan, MetPA, MetScape, MGV, Paintomics, Pathos, PathVisio, ProMeTra, Reactome, and VANTED resources provide mapping and visualizing tools of metabolites correlated to biological pathways, while the MSEA, MBRole, MPEA, and IMPaLA tools can carry out enrichment analysis [120, 121]. In detail, the KEGG's pathway browser is created and maintained by the Kanehisa Laboratories, which provides a list of pathways and compounds within these pathways. Biocyc.org resource is developed and managed by SRI International, which present mapping and visualizing tools for metabolites. The visualization tools of Reactome website can provide integrative pathway information retrieving from multi-data source (such as VANTED, iPath, and MetScape). Specially, experimental data can be mapped to correlated metabolism graphs quickly in VANTED website. iPath offers an "interactive" mode for biological pathways visualization. Users can visualize and operate biological networks with customized mode in MetScape website. In addition, a lot of tools of omics data visualization are presented recently, such as ProMeTra, Paintomics, PathVisio, and MGV. Specially, the KaPPA-View and MapMan are developed for plant metabolomics. Currently, some pathway visualization tools are integrated into general analytical software to facilitate data analysis of metabolomics, for example, the MetPA and the Pathos tool have been integrated in the Metabo and MeltDB platform, respectively. MSEA, MBRole, MPEA, and IMPaLA are enrichment analysis tools, which are utilized for human and mammalian metabolomics

Table 1
Software of mechanism interpretation for metabolites

Name	URL	Reference
MSEA	http://www.msea.ca	[122]
MBRole	http://csbg.cnb.csic.es/mbrole	[123]
MPEA	http://ekhidna.biocenter.helsinki.fi/poxo/mpea/	[124]
IMPaLA	http://impala.molgen.mpg.de	[125]
KEGG	http://www.genome.jp/kegg/pathway.html	[126]
BioCyc—Omics Viewer	http://biocyc.org	[127]
iPath	http://pathways.embl.de	[128]
KaPPA-View	http://kpv.kazusa.or.jp/en/	[129]
MapMan	http://mapman.gabipd.org/web/guest/mapman	[130]
MetPA	http://metpa.metabolomics.ca	[131]
Metscape	http://metscape.ncibi.org	[132]
MGV	http://www.microarray-analysis.org/mayday	[133]
Paintomics	http://www.paintomics.org	[134]
Pathos	http://motif.gla.ac.uk/Pathos/	[135]
PathVisio	http://www.pathvisio.org/	[136]
ProMeTra	http://www.cebitec.uni-bielefeld.de/groups/brf/software/prometra_info/Reactome	[137]
Reactome	http://www.reactome.org	[138]
VANTED	http://vanted.ipk-gatersleben.de	[139]

study. They provide integrative metabolite information retrieving from public databases, such as KEGG, SMPDB, HMDB, PubChem, and ChEBI. In short, these databases and tools can help understand the biological roles of metabolites play in living systems.

7 Conclusions

Metabolomics is an important "omic" technology that can discover clinic biomarker and provide deep mechanistic insights into biological process for a living system. Here we review the experimental design, sample collection and preparation methods, analytical platform of data acquisition (such as GC-MS, LC-MS, CE-MS, NMR), the data preprocessing and analyzing methods, as well as mechanism interpretation and hypothesis generation approaches for metabolomics study. The challenges in metabolomics are the generation of metabolite profile and the interpretation of mechanism underlying. In addition, combining with other omics technologies, such as transcriptomics and proteomics, to provide a comprehensive insight into the molecular mechanism of biological process in living system is a significant aim of metabolomics study. Nowadays, rapid developments of bioinformatics accelerate investigation of metabolomics and thus can make the investigators spend more time in interpreting metabolism result and deciphering molecular mechanism.

References

1. Fiehn O (2002) Metabolomics—the link between genotypes and phenotypes. Plant Mol Biol 48(1–2):155–171
2. Nicholson JK, Lindon JC, Holmes E (1999) 'Metabonomics': understanding the metabolic responses of living systems to pathophysiological stimuli via multivariate statistical analysis of biological NMR spectroscopic data. Xenobiotica 29(11):1181–1189. https://doi.org/10.1080/004982599238047
3. Cacciatore S, Loda M (2015) Innovation in metabolomics to improve personalized healthcare. Ann N Y Acad Sci 1346(1):57–62. https://doi.org/10.1111/nyas.12775
4. Naz S, Garcia A, Barbas C (2013) Multiplatform analytical methodology for metabolic fingerprinting of lung tissue. Anal Chem 85 (22):10941–10948
5. Gowda GA, Djukovic D (2014) Overview of mass spectrometry-based metabolomics: opportunities and challenges. Methods Mol Biol 1198:3–12. https://doi.org/10.1007/978-1-4939-1258-2_1
6. Naz S, Vallejo M, Garcia A, Barbas C (2014) Method validation strategies involved in non-targeted metabolomics. J Chromatogr A 1353:99–105. https://doi.org/10.1016/j.chroma.2014.04.071
7. Gao X, Zhao A, Zhou M, Lin J, Qiu Y, Su M, Jia W (2011) GC/MS-based urinary metabolomics reveals systematic differences in metabolism and ethanol response between Sprague–Dawley and Wistar rats. Metabolomics 7(3):363–374. https://doi.org/10.1007/s11306-010-0252-5
8. Sumner LW, Amberg A, Barrett D, Beale MH, Beger R, Daykin CA, Fan TW, Fiehn O, Goodacre R, Griffin JL, Hankemeier T, Hardy N, Harnly J, Higashi R, Kopka J, Lane AN, Lindon JC, Marriott P, Nicholls AW, Reily MD, Thaden JJ, Viant MR (2007) Proposed minimum reporting standards for chemical analysis Chemical Analysis Working Group (CAWG) Metabolomics Standards Initiative (MSI). Metabolomics 3(3):211–221. https://doi.org/10.1007/s11306-007-0082-2
9. Hart CD, Vignoli A, Tenori L, Uy GL, Van To T, Adebamowo C, Hossain SM, Biganzoli L, Risi E, Love RR, Luchinat C, Di Leo A (2017) Serum metabolomic profiles identify ER-positive early breast cancer patients at increased risk of disease recurrence in a multicenter population. Clin Cancer Res 23(6):1422–1431. https://doi.org/10.1158/1078-0432.CCR-16-1153
10. Westerhuis JA, van Velzen EJ, Hoefsloot HC, Smilde AK (2010) Multivariate paired data analysis: multilevel PLSDA versus OPLSDA. Metabolomics 6(1):119–128. https://doi.org/10.1007/s11306-009-0185-z
11. Gregory JF III, Park Y, Lamers Y, Bandyopadhyay N, Chi YY, Lee K, Kim S, da Silva V, Hove N, Ranka S, Kahveci T, Muller KE, Stevens RD, Newgard CB, Stacpoole PW, Jones DP (2013) Metabolomic analysis reveals extended metabolic consequences of marginal vitamin B-6 deficiency in healthy human subjects. PLoS One 8(6):e63544. https://doi.org/10.1371/journal.pone.0063544
12. Ghini V, Unger FT, Tenori L, Turano P, Juhl H, David KA (2015) Metabolomics profiling of pre- and post-anesthesia plasma samples of colorectal patients obtained via Ficoll separation. Metabolomics 11 (6):1769–1778. https://doi.org/10.1007/s11306-015-0832-5
13. van Velzen EJ, Westerhuis JA, van Duynhoven JP, van Dorsten FA, Hoefsloot HC, Jacobs DM, Smit S, Draijer R, Kroner CI, Smilde AK (2008) Multilevel data analysis of a cross-over designed human nutritional intervention study. J Proteome Res 7(10):4483–4491. https://doi.org/10.1021/pr800145j

14. Vuckovic D (2012) Current trends and challenges in sample preparation for global metabolomics using liquid chromatography-mass spectrometry. Anal Bioanal Chem 403 (6):1523–1548. https://doi.org/10.1007/s00216-012-6039-y
15. Hund E, Vander Heyden Y, Massart DL, Smeyers-Verbeke J (2002) Derivation of system suitability test limits from a robustness test on an LC assay with complex antibiotic samples. J Pharmaceut Biomed 30 (4):1197–1206
16. t'Kindt R, Morreel K, Deforce D, Boerjan W, Van Bocxlaer J (2009) Joint GC-MS and LC-MS platforms for comprehensive plant metabolomics: repeatability and sample pre-treatment. J Chromatogr B 877 (29):3572–3580
17. Álvarez-Sánchez B, Priego-Capote F, Luque de Castro MD (2010) Metabolomics analysis II. Preparation of biological samples prior to detection. TrAC Trends Anal Chem 29 (2):120–127. https://doi.org/10.1016/j.trac.2009.12.004
18. Naz S, Moreira dos Santos DC, García A, Barbas C (2014) Analytical protocols based on LC–MS, GC–MS and CE–MS for nontargeted metabolomics of biological tissues. Bioanalysis 6(12):1657–1677
19. Want EJ, Masson P, Michopoulos F, Wilson ID, Theodoridis G, Plumb RS, Shockcor J, Loftus N, Holmes E, Nicholson JK (2013) Global metabolic profiling of animal and human tissues via UPLC-MS. Nat Protoc 8 (1):17–32. https://doi.org/10.1038/nprot.2012.135
20. Ly-Verdu S, Schaefer A, Kahle M, Groeger T, Neschen S, Arteaga-Salas JM, Ueffing M, de Angelis MH, Zimmermann R (2014) The impact of blood on liver metabolite profiling - a combined metabolomic and proteomic approach. Biomed Chromatogr 28 (2):231–240. https://doi.org/10.1002/bmc.3010
21. Dunn WB, Broadhurst D, Ellis DI, Brown M, Halsall A, O'Hagan S, Spasic I, Tseng A, Kell DB (2008) A GC-TOF-MS study of the stability of serum and urine metabolomes during the UK Biobank sample collection and preparation protocols. Int J Epidemiol 37(Suppl 1): i23–i30. https://doi.org/10.1093/ije/dym281
22. Fernández-Peralbo MA, Luque de Castro MD (2012) Preparation of urine samples prior to targeted or untargeted metabolomics mass-spectrometry analysis. TrAC Trends Anal Chem 41:75–85. https://doi.org/10.1016/j.trac.2012.08.011
23. Want EJ, Wilson ID, Gika H, Theodoridis G, Plumb RS, Shockcor J, Holmes E, Nicholson JK (2010) Global metabolic profiling procedures for urine using UPLC-MS. Nat Protoc 5(6):1005–1018
24. Gika HG, Theodoridis GA, Wilson ID (2008) Liquid chromatography and ultra-performance liquid chromatography–mass spectrometry fingerprinting of human urine. J Chromatogr A 1189(1):314–322. https://doi.org/10.1016/j.chroma.2007.10.066
25. Winder CL, Dunn WB, Schuler S, Broadhurst D, Jarvis R, Stephens GM, Goodacre R (2008) Global metabolic profiling of Escherichia coli cultures: an evaluation of methods for quenching and extraction of intracellular metabolites. Anal Chem 80 (8):2939–2948. https://doi.org/10.1021/ac7023409
26. Meyer H, Weidmann H, Lalk M (2013) Methodological approaches to help unravel the intracellular metabolome of Bacillus subtilis. Microb Cell Factories 12:69. https://doi.org/10.1186/1475-2859-12-69
27. Gao X, Pujos-Guillot E, Martin JF, Galan P, Juste C, Jia W, Sebedio JL (2009) Metabolite analysis of human fecal water by gas chromatography/mass spectrometry with ethyl chloroformate derivatization. Anal Biochem 393(2):163–175. https://doi.org/10.1016/j.ab.2009.06.036
28. Gratton J, Phetcharaburanin J, Mullish BH, Williams HR, Thursz M, Nicholson JK, Holmes E, Marchesi JR, Li JV (2016) Optimized sample handling strategy for metabolic profiling of human feces. Anal Chem 88 (9):4661–4668. https://doi.org/10.1021/acs.analchem.5b04159
29. Zhou B, Xiao JF, Tuli L, Ressom HW (2012) LC-MS-based metabolomics. Mol BioSyst 8 (2):470–481. https://doi.org/10.1039/c1mb05350g
30. Raterink R-J, Lindenburg PW, Vreeken RJ, Ramautar R, Hankemeier T (2014) Recent developments in sample-pretreatment techniques for mass spectrometry-based metabolomics. TrAC Trends Anal Chem 61:157–167. https://doi.org/10.1016/j.trac.2014.06.003
31. Gowda GAN, Raftery D (2014) Quantitating metabolites in protein precipitated serum using NMR spectroscopy. Anal Chem 86 (11):5433–5440
32. Michopoulos F, Lai L, Gika H, Theodoridis G, Wilson I (2009) UPLC-MS-based analysis of human plasma for metabonomics using solvent precipitation or solid phase extraction. J Proteome Res 8 (4):2114–2121

33. Masson P, Alves AC, Ebbels TM, Nicholson JK, Want EJ (2010) Optimization and evaluation of metabolite extraction protocols for untargeted metabolic profiling of liver samples by UPLC-MS. Anal Chem 82 (18):7779–7786. https://doi.org/10.1021/ac101722e
34. Gao X, Chen W, Li R, Wang M, Chen C, Zeng R, Deng Y (2012) Systematic variations associated with renal disease uncovered by parallel metabolomics of urine and serum. BMC Syst Biol 6(Suppl 1):S14. https://doi.org/10.1186/1752-0509-6-S1-S14
35. Liang X, Ubhayakar S, Liederer BM, Dean B, Ran-Ran Qin A, Shahidi-Latham S, Deng Y (2011) Evaluation of homogenization techniques for the preparation of mouse tissue samples to support drug discovery. Bioanalysis 3 (17):1923–1933. https://doi.org/10.4155/bio.11.181
36. Han J, Lin K, Sequeira C, Borchers CH (2015) An isotope-labeled chemical derivatization method for the quantitation of short-chain fatty acids in human feces by liquid chromatography-tandem mass spectrometry. Anal Chim Acta 854:86–94. https://doi.org/10.1016/j.aca.2014.11.015
37. Gao X, Pujos-Guillot E, Sebedio JL (2010) Development of a quantitative metabolomic approach to study clinical human fecal water metabolome based on trimethylsilylation derivatization and GC/MS analysis. Anal Chem 82(15):6447–6456. https://doi.org/10.1021/ac1006552
38. A J, Trygg J, Gullberg J, Johansson AI, Jonsson P, Antti H, Marklund SL, Moritz T (2005) Extraction and GC/MS analysis of the human blood plasma metabolome. Anal Chem 77(24):8086–8094. https://doi.org/10.1021/ac051211v
39. Dunn WB, Broadhurst D, Begley P, Zelena E, Francis-McIntyre S, Anderson N, Brown M, Knowles JD, Halsall A, Haselden JN, Nicholls AW, Wilson ID, Kell DB, Goodacre R, Human Serum Metabolome Consortium (2011) Procedures for large-scale metabolic profiling of serum and plasma using gas chromatography and liquid chromatography coupled to mass spectrometry. Nat Protoc 6 (7):1060–1083. https://doi.org/10.1038/nprot.2011.335
40. Husek P (1991) Amino acid derivatization and analysis in five minutes. FEBS Lett 280 (2):354–356
41. Husek P (1998) Chloroformates in gas chromatography as general purpose derivatizing agents. J Chromatogr B Biomed Sci Appl 717(1–2):57–91
42. Zhao L, Ni Y, Su M, Li H, Dong F, Chen W, Wei R, Zhang L, Guiraud SP, Martin FP, Rajani C, Xie G, Jia W (2017) High throughput and quantitative measurement of microbial metabolome by gas chromatography/mass spectrometry using automated alkyl chloroformate derivatization. Anal Chem 89 (10):5565–5577. https://doi.org/10.1021/acs.analchem.7b00660
43. Villas-Boas SG, Smart KF, Sivakumaran S, Lane GA (2011) Alkylation or silylation for analysis of amino and non-amino organic acids by GC-MS? Metabolites 1(1):3–20. https://doi.org/10.3390/metabo1010003
44. Guo K, Li L (2009) Differential 12C-/13C-isotope dansylation labeling and fast liquid chromatography/mass spectrometry for absolute and relative quantification of the metabolome. Anal Chem 81 (10):3919–3932. https://doi.org/10.1021/ac900166a
45. Chen D, Su X, Wang N, Li Y, Yin H, Li L, Li L (2017) Chemical isotope labeling LC-MS for monitoring disease progression and treatment in animal models: plasma metabolomics study of osteoarthritis rat model. Sci Rep 7:40543. https://doi.org/10.1038/srep40543
46. Wu Y, Streijger F, Wang Y, Lin G, Christie S, Mac-Thiong JM, Parent S, Bailey CS, Paquette S, Boyd MC, Ailon T, Street J, Fisher CG, Dvorak MF, Kwon BK, Li L (2016) Parallel metabolomic profiling of cerebrospinal fluid and serum for identifying biomarkers of injury severity after acute human spinal cord injury. Sci Rep 6:38718. https://doi.org/10.1038/srep38718
47. Zhao S, Luo X, Li L (2016) Chemical isotope labeling LC-MS for high coverage and quantitative profiling of the hydroxyl submetabolome in metabolomics. Anal Chem 88 (21):10617–10623. https://doi.org/10.1021/acs.analchem.6b02967
48. Su X, Wang N, Chen D, Li Y, Lu Y, Huan T, Xu W, Li L, Li L (2016) Dansylation isotope labeling liquid chromatography mass spectrometry for parallel profiling of human urinary and fecal submetabolomes. Anal Chim Acta 903:100–109. https://doi.org/10.1016/j.aca.2015.11.027
49. Song P, Mabrouk OS, Hershey ND, Kennedy RT (2012) In vivo neurochemical monitoring using benzoyl chloride derivatization and liquid chromatography-mass spectrometry. Anal Chem 84(1):412–419. https://doi.org/10.1021/ac202794q
50. Wong JM, Malec PA, Mabrouk OS, Ro J, Dus M, Kennedy RT (2016) Benzoyl chloride derivatization with liquid chromatography-

mass spectrometry for targeted metabolomics of neurochemicals in biological samples. J Chromatogr A 1446:78–90. https://doi.org/10.1016/j.chroma.2016.04.006

51. Issaq HJ, Van QN, Waybright TJ, Muschik GM, Veenstra TD (2009) Analytical and statistical approaches to metabolomics research. J Sep Sci 32(13):2183–2199
52. Beckonert O, Keun HC, Ebbels TM, Bundy J, Holmes E, Lindon JC, Nicholson JK (2007) Metabolic profiling, metabolomic and metabonomic procedures for NMR spectroscopy of urine, plasma, serum and tissue extracts. Nat Protoc 2(11):2692–2703. https://doi.org/10.1038/nprot.2007.376
53. Sitter B, Bathen TF, Tessem M-B, Gribbestad IS (2009) High-resolution magic angle spinning (HR MAS) MR spectroscopy in metabolic characterization of human cancer. Prog Nucl Magn Reson Spectrosc 54 (3–4):239–254. https://doi.org/10.1016/j.pnmrs.2008.10.001
54. Beltran A, Suarez M, Rodriguez MA, Vinaixa M, Samino S, Arola L, Correig X, Yanes O (2012) Assessment of compatibility between extraction methods for NMR- and LC/MS-based metabolomics. Anal Chem 84 (14):5838–5844. https://doi.org/10.1021/ac3005567
55. Li N, Song Y, Tang H, Wang Y (2016) Recent developments in sample preparation and data pre-treatment in metabonomics research. Arch Biochem Biophys 589:4–9. https://doi.org/10.1016/j.abb.2015.08.024
56. Le Gall G (2015) Sample collection and preparation of biofluids and extracts for NMR spectroscopy. Methods Mol Biol 1277:15–28. https://doi.org/10.1007/978-1-4939-2377-9_2
57. Pasikanti KK, Ho PC, Chan EC (2008) Gas chromatography/mass spectrometry in metabolic profiling of biological fluids. J Chromatogr B Analyt Technol Biomed Life Sci 871 (2):202–211. https://doi.org/10.1016/j.jchromb.2008.04.033
58. Halket JM, Waterman D, Przyborowska AM, Patel RK, Fraser PD, Bramley PM (2005) Chemical derivatization and mass spectral libraries in metabolic profiling by GC/MS and LC/MS/MS. J Exp Bot 56 (410):219–243. https://doi.org/10.1093/jxb/eri069
59. Lisec J, Schauer N, Kopka J, Willmitzer L, Fernie AR (2006) Gas chromatography mass spectrometry-based metabolite profiling in plants. Nat Protoc 1(1):387–396. https://doi.org/10.1038/nprot.2006.59
60. Welthagen W, Shellie RA, Spranger J, Ristow M, Zimmermann R, Fiehn O (2005) Comprehensive two-dimensional gas chromatography–time-of-flight mass spectrometry (GC × GC-TOF) for high resolution metabolomics: biomarker discovery on spleen tissue extracts of obese NZO compared to lean C57BL/6 mice. Metabolomics 1(1):65–73. https://doi.org/10.1007/s11306-005-1108-2
61. Adahchour M, Beens J, Brinkman UA (2008) Recent developments in the application of comprehensive two-dimensional gas chromatography. J Chromatogr A 1186 (1–2):67–108. https://doi.org/10.1016/j.chroma.2008.01.002
62. Weinert CH, Egert B, Kulling SE (2015) On the applicability of comprehensive two-dimensional gas chromatography combined with a fast-scanning quadrupole mass spectrometer for untargeted large-scale metabolomics. J Chromatogr A 1405:156–167. https://doi.org/10.1016/j.chroma.2015.04.011
63. Almstetter MF, Oefner PJ, Dettmer K (2012) Comprehensive two-dimensional gas chromatography in metabolomics. Anal Bioanal Chem 402(6):1993–2013. https://doi.org/10.1007/s00216-011-5630-y
64. Khamis MM, Adamko DJ, El-Aneed A (2017) Mass spectrometric based approaches in urine metabolomics and biomarker discovery. Mass Spectrom Rev 36(2):115–134. https://doi.org/10.1002/mas.21455
65. Theodoridis GA, Gika HG, Want EJ, Wilson ID (2012) Liquid chromatography-mass spectrometry based global metabolite profiling: a review. Anal Chim Acta 711:7–16. https://doi.org/10.1016/j.aca.2011.09.042
66. Wilson ID, Nicholson JK, Castro-Perez J, Granger JH, Johnson KA, Smith BW, Plumb RS (2005) High resolution "ultra performance" liquid chromatography coupled to oa-TOF mass spectrometry as a tool for differential metabolic pathway profiling in functional genomic studies. J Proteome Res 4 (2):591–598. https://doi.org/10.1021/pr049769r
67. Tang DQ, Zou L, Yin XX, Ong CN (2016) HILIC-MS for metabolomics: an attractive and complementary approach to RPLC-MS. Mass Spectrom Rev 35(5):574–600. https://doi.org/10.1002/mas.21445
68. Chen J, Wang W, Lv S, Yin P, Zhao X, Lu X, Zhang F, Xu G (2009) Metabonomics study of liver cancer based on ultra performance liquid chromatography coupled to mass

spectrometry with HILIC and RPLC separations. Anal Chim Acta 650(1):3–9. https://doi.org/10.1016/j.aca.2009.03.039

69. Dunn WB, Ellis DI (2005) Metabolomics: current analytical platforms and methodologies. TrAC Trends Anal Chem 24 (4):285–294. https://doi.org/10.1016/j.trac.2004.11.021
70. Bothwell JHF, Griffin JL (2011) An introduction to biological nuclear magnetic resonance spectroscopy. Biol Rev 86(2):493–510. https://doi.org/10.1111/j.1469-185X.2010.00157.x
71. Grimes JH, O'Connell TM (2011) The application of micro-coil NMR probe technology to metabolomics of urine and serum. J Biomol NMR 49(3–4):297–305. https://doi.org/10.1007/s10858-011-9488-2
72. Le Guennec A, Tayyari F, Edison AS (2017) Alternatives to nuclear overhauser enhancement spectroscopy presat and Carr-Purcell-Meiboom-Gill presat for NMR-based metabolomics. Anal Chem. https://doi.org/10.1021/acs.analchem.7b02354
73. Chaudhry V, Bhatia A, Bharti SK, Mishra SK, Chauhan PS, Mishra A, Sidhu OP, Nautiyal CS (2015) Metabolite profiling reveals abiotic stress tolerance in Tn5 mutant of Pseudomonas putida. PLoS One 10(1):e0113487. https://doi.org/10.1371/journal.pone.0113487
74. Wan Q, Wang Y, Tang H (2017) Quantitative 13C traces of glucose fate in hepatitis B virus-infected hepatocytes. Anal Chem 89 (6):3293–3299. https://doi.org/10.1021/acs.analchem.6b03200
75. Hollinshead KE, Williams DS, Tennant DA, Ludwig C (2016) Probing cancer cell metabolism using NMR spectroscopy. Adv Exp Med Biol 899:89–111. https://doi.org/10.1007/978-3-319-26666-4_6
76. Lommen A (2009) MetAlign: interface-driven, versatile metabolomics tool for hyphenated full-scan mass spectrometry data preprocessing. Anal Chem 81(8):3079–3086. https://doi.org/10.1021/ac900036d
77. Wehrens R, Weingart G, Mattivi F (2014) metaMS: an open-source pipeline for GC-MS-based untargeted metabolomics. J Chromatogr B Analyt Technol Biomed Life Sci 966:109–116. https://doi.org/10.1016/j.jchromb.2014.02.051
78. Cuadros-Inostroza A, Caldana C, Redestig H, Kusano M, Lisec J, Pena-Cortes H, Willmitzer L, Hannah MA (2009) TargetSearch—a Bioconductor package for the efficient preprocessing of GC-MS metabolite profiling data. BMC Bioinformatics 10:428. https://doi.org/10.1186/1471-2105-10-428
79. Carroll AJ, Badger MR, Harvey Millar A (2010) The MetabolomeExpress Project: enabling web-based processing, analysis and transparent dissemination of GC/MS metabolomics datasets. BMC Bioinformatics 11:376. https://doi.org/10.1186/1471-2105-11-376
80. Luedemann A, von Malotky L, Erban A, Kopka J (2012) TagFinder: preprocessing software for the fingerprinting and the profiling of gas chromatography-mass spectrometry based metabolome analyses. Methods Mol Biol 860:255–286. https://doi.org/10.1007/978-1-61779-594-7_16
81. Smith CA, Want EJ, O'Maille G, Abagyan R, Siuzdak G (2006) XCMS: processing mass spectrometry data for metabolite profiling using nonlinear peak alignment, matching, and identification. Anal Chem 78 (3):779–787. https://doi.org/10.1021/ac051437y
82. Katajamaa M, Miettinen J, Oresic M (2006) MZmine: toolbox for processing and visualization of mass spectrometry based molecular profile data. Bioinformatics 22(5):634–636. https://doi.org/10.1093/bioinformatics/btk039
83. Pluskal T, Castillo S, Villar-Briones A, Oresic M (2010) MZmine 2: modular framework for processing, visualizing, and analyzing mass spectrometry-based molecular profile data. BMC Bioinformatics 11:395. https://doi.org/10.1186/1471-2105-11-395
84. Luedemann A, Strassburg K, Erban A, Kopka J (2008) TagFinder for the quantitative analysis of gas chromatography—mass spectrometry (GC-MS)-based metabolite profiling experiments. Bioinformatics 24(5):732–737. https://doi.org/10.1093/bioinformatics/btn023
85. Tautenhahn R, Patti GJ, Rinehart D, Siuzdak G (2012) XCMS Online: a web-based platform to process untargeted metabolomic data. Anal Chem 84(11):5035–5039. https://doi.org/10.1021/ac300698c
86. Libiseller G, Dvorzak M, Kleb U, Gander E, Eisenberg T, Madeo F, Neumann S, Trausinger G, Sinner F, Pieber T, Magnes C (2015) IPO: a tool for automated optimization of XCMS parameters. BMC Bioinformatics 16:118. https://doi.org/10.1186/s12859-015-0562-8
87. Tsugawa H, Cajka T, Kind T, Ma Y, Higgins B, Ikeda K, Kanazawa M, Van der

Gheynst J, Fiehn O, Arita M (2015) MS-DIAL: data-independent MS/MS deconvolution for comprehensive metabolome analysis. Nat Methods 12(6):523–526. https://doi.org/10.1038/nmeth.3393

88. Chen G, Walmsley S, Cheung GCM, Chen L, Cheng CY, Beuerman RW, Wong TY, Zhou L, Choi H (2017) Customized consensus spectral library building for untargeted quantitative metabolomics analysis with data independent acquisition mass spectrometry and MetaboDIA workflow. Anal Chem 89 (9):4897–4906. https://doi.org/10.1021/acs.analchem.6b05006
89. Li H, Cai Y, Guo Y, Chen F, Zhu ZJ (2016) MetDIA: targeted metabolite extraction of multiplexed MS/MS spectra generated by data-independent acquisition. Anal Chem 88 (17):8757–8764. https://doi.org/10.1021/acs.analchem.6b02122
90. Jansen BC, Reiding KR, Bondt A, Ederveen ALH, Palmblad M, Falck D, Wuhrer M (2015) MassyTools: a high-throughput targeted data processing tool for relative quantitation and quality control developed for glycomic and glycoproteomic MALDI-MS. J Proteome Res 14(12):5088–5098
91. Savorani F, Tomasi G, Engelsen SB (2010) icoshift: a versatile tool for the rapid alignment of 1D NMR spectra. J Magn Reson 202(2):190–202. https://doi.org/10.1016/j.jmr.2009.11.012
92. Christensen JH, Tomasi G, Hansen AB (2005) Chemical fingerprinting of petroleum biomarkers using time warping and PCA. Environ Sci Technol 39(1):255–260
93. Vu TN, Valkenborg D, Smets K, Verwaest KA, Dommisse R, Lemiere F, Verschoren A, Goethals B, Laukens K (2011) An integrated workflow for robust alignment and simplified quantitative analysis of NMR spectrometry data. BMC Bioinformatics 12:405. https://doi.org/10.1186/1471-2105-12-405
94. De Meyer T, Sinnaeve D, Van Gasse B, Tsiporkova E, Rietzschel ER, De Buyzere ML, Gillebert TC, Bekaert S, Martins JC, Van Criekinge W (2008) NMR-based characterization of metabolic alterations in hypertension using an adaptive, intelligent binning algorithm. Anal Chem 80(10):3783–3790. https://doi.org/10.1021/ac7025964
95. Jacob D, Deborde C, Moing A (2013) An efficient spectra processing method for metabolite identification from 1H-NMR metabolomics data. Anal Bioanal Chem 405 (15):5049–5061. https://doi.org/10.1007/s00216-013-6852-y
96. Worley B, Powers R (2015) Generalized adaptive intelligent binning of multiway data. Chemom Intell Lab Syst 146:42–46. https://doi.org/10.1016/j.chemolab.2015.05.005
97. Wishart DS, Jewison T, Guo AC, Wilson M, Knox C, Liu Y, Djoumbou Y, Mandal R, Aziat F, Dong E, Bouatra S, Sinelnikov I, Arndt D, Xia J, Liu P, Yallou F, Bjorndahl T, Perez-Pineiro R, Eisner R, Allen F, Neveu V, Greiner R, Scalbert A (2013) HMDB 3.0—the human metabolome database in 2013. Nucleic Acids Res 41(Database issue): D801–D807. https://doi.org/10.1093/nar/gks1065
98. Wishart DS, Tzur D, Knox C, Eisner R, Guo AC, Young N, Cheng D, Jewell K, Arndt D, Sawhney S, Fung C, Nikolai L, Lewis M, Coutouly MA, Forsythe I, Tang P, Shrivastava S, Jeroncic K, Stothard P, Amegbey G, Block D, Hau DD, Wagner J, Miniaci J, Clements M, Gebremedhin M, Guo N, Zhang Y, Duggan GE, Macinnis GD, Weljie AM, Dowlatabadi R, Bamforth F, Clive D, Greiner R, Li L, Marrie T, Sykes BD, Vogel HJ, Querengesser L (2007) HMDB: the human metabolome database. Nucleic Acids Res 35(Database issue):D521–D526. https://doi.org/10.1093/nar/gkl923
99. Markley JL, Anderson ME, Cui Q, Eghbalnia HR, Lewis IA, Hegeman AD, Li J, Schulte CF, Sussman MR, Westler WM, Ulrich EL, Zolnai Z (2007) New bioinformatics resources for metabolomics. Pacific Symposium on Biocomputing. Pac Symp Biocomput 12:157–168
100. Cui Q, Lewis IA, Hegeman AD, Anderson ME, Li J, Schulte CF, Westler WM, Eghbalnia HR, Sussman MR, Markley JL (2008) Metabolite identification via the Madison Metabolomics Consortium Database. Nat Biotechnol 26(2):162–164. https://doi.org/10.1038/nbt0208-162
101. Kale NS, Haug K, Conesa P, Jayseelan K, Moreno P, Rocca-Serra P, Nainala VC, Spicer RA, Williams M, Li X, Salek RM, Griffin JL, Steinbeck C (2016) MetaboLights: an open-access database repository for metabolomics data. Curr Protoc Bioinformatics 53:14.13.11–14.13.18. https://doi.org/10.1002/0471250953.bi1413s53
102. Ellinger JJ, Chylla RA, Ulrich EL, Markley JL (2013) Databases and software for NMR-based metabolomics. Curr Metabolomics 1(1). https://doi.org/10.2174/2213235X11301010028
103. Bingol K, Zhang F, Bruschweiler-Li L, Bruschweiler R (2012) TOCCATA: a

customized carbon total correlation spectroscopy NMR metabolomics database. Anal Chem 84(21):9395–9401. https://doi.org/10.1021/ac302197e

104. Kwan EE, Huang SG (2008) Structural elucidation with NMR spectroscopy: practical strategies for organic chemists. Eur J Org Chem 2008(16):2671–2688. https://doi.org/10.1002/ejoc.200700966

105. Bingol K, Bruschweiler R (2011) Deconvolution of chemical mixtures with high complexity by NMR consensus trace clustering. Anal Chem 83(19):7412–7417. https://doi.org/10.1021/ac201464y

106. Leek JT, Scharpf RB, Bravo HC, Simcha D, Langmead B, Johnson WE, Geman D, Baggerly K, Irizarry RA (2010) Tackling the widespread and critical impact of batch effects in high-throughput data. Nat Rev Genet 11 (10):733–739. https://doi.org/10.1038/nrg2825

107. Burton L, Ivosev G, Tate S, Impey G, Wingate J, Bonner R (2008) Instrumental and experimental effects in LC-MS-based metabolomics. J Chromatogr B Analyt Technol Biomed Life Sci 871(2):227–235. https://doi.org/10.1016/j.jchromb.2008.04.044

108. De Livera AM, Sysi-Aho M, Jacob L, Gagnon-Bartsch JA, Castillo S, Simpson JA, Speed TP (2015) Statistical methods for handling unwanted variation in metabolomics data. Anal Chem 87(7):3606–3615. https://doi.org/10.1021/ac502439y

109. Hendriks MMWB, van FA E, Jellema RH, Westerhuis JA, Reijmers TH, Hoefsloot HCJ, Smilde AK (2011) Data-processing strategies for metabolomics studies. TrAC Trends Anal Chem 30(10):1685–1698. https://doi.org/10.1016/j.trac.2011.04.019

110. Wehrens R, Hageman JA, van Eeuwijk F, Kooke R, Flood PJ, Wijnker E, Keurentjes JJ, Lommen A, van Eekelen HD, Hall RD, Mumm R, de Vos RC (2016) Improved batch correction in untargeted MS-based metabolomics. Metabolomics 12:88. https://doi.org/10.1007/s11306-016-1015-8

111. Brunius C, Shi L, Landberg R (2016) Large-scale untargeted LC-MS metabolomics data correction using between-batch feature alignment and cluster-based within-batch signal intensity drift correction. Metabolomics 12 (11):173. https://doi.org/10.1007/s11306-016-1124-4

112. Shen X, Gong X, Cai Y, Guo Y, Tu J, Li H, Zhang T, Wang J, Xue F, Zhu Z-J (2016) Normalization and integration of large-scale metabolomics data using support vector regression. Metabolomics 12(5):89. https://doi.org/10.1007/s11306-016-1026-5

113. Li B, Tang J, Yang Q, Li S, Cui X, Li Y, Chen Y, Xue W, Li X, Zhu F (2017) NOREVA: normalization and evaluation of MS-based metabolomics data. Nucleic Acids Res. https://doi.org/10.1093/nar/gkx449

114. Hochrein J, Zacharias HU, Taruttis F, Samol C, Engelmann JC, Spang R, Oefner PJ, Gronwald W (2015) Data normalization of 1H NMR metabolite fingerprinting data sets in the presence of unbalanced metabolite regulation. J Proteome Res 14 (8):3217–3228. https://doi.org/10.1021/acs.jproteome.5b00192

115. Chen J, Zhang P, Lv M, Guo H, Huang Y, Zhang Z, Xu F (2017) Influences of normalization method on biomarker discovery in gas chromatography-mass spectrometry-based untargeted metabolomics: what should be considered? Anal Chem 89(10):5342–5348. https://doi.org/10.1021/acs.analchem.6b05152

116. Li B, Tang J, Yang Q, Cui X, Li S, Chen S, Cao Q, Xue W, Chen N, Zhu F (2016) Performance evaluation and online realization of data-driven normalization methods used in LC/MS based untargeted metabolomics analysis. Sci Rep 6:38881. https://doi.org/10.1038/srep38881

117. Putri SP, Yamamoto S, Tsugawa H, Fukusaki E (2013) Current metabolomics: technological advances. J Biosci Bioeng 116(1):9–16. https://doi.org/10.1016/j.jbiosc.2013.01.004

118. Boccard J, Veuthey JL, Rudaz S (2010) Knowledge discovery in metabolomics: an overview of MS data handling. J Sep Sci 33 (3):290–304

119. Tagore S, Chowdhury N, De RK (2014) Analyzing methods for path mining with applications in metabolomics. Gene 534 (2):125–138

120. Chagoyen M, Pazos F (2013) Tools for the functional interpretation of metabolomic experiments. Brief Bioinform 14(6):737–744

121. Johnson CH, Ivanisevic J, Siuzdak G (2016) Metabolomics: beyond biomarkers and towards mechanisms. Nat Rev Mol Cell Biol 17(7):451–459

122. Xia JG, Wishart DS (2010) MSEA: a web-based tool to identify biologically meaningful patterns in quantitative metabolomic data. Nucleic Acids Res 38:W71–W77
123. Chagoyen M, Pazos F (2011) MBRole: enrichment analysis of metabolomic data. Bioinformatics 27(5):730–731
124. Kankainen M, Gopalacharyulu P, Holm L, Oresic M (2011) MPEA-metabolite pathway enrichment analysis. Bioinformatics 27 (13):1878–1879
125. Kamburov A, Cavill R, Ebbels TMD, Herwig R, Keun HC (2011) Integrated pathway-level analysis of transcriptomics and metabolomics data with IMPaLA. Bioinformatics 27(20):2917–2918
126. Okuda S, Yamada T, Hamajima M, Itoh M, Katayama T, Bork P, Goto S, Kanehisa M (2008) KEGG Atlas mapping for global analysis of metabolic pathways. Nucleic Acids Res 36:W423–W426
127. Paley SM, Karp PD (2006) The Pathway Tools cellular overview diagram and Omics Viewer. Nucleic Acids Res 34 (13):3771–3778
128. Letunic I, Yamada T, Kanehisa M, Bork P (2008) iPath: interactive exploration of biochemical pathways and networks. Trends Biochem Sci 33(3):101–103
129. Tokimatsu T, Sakurai N, Suzuki H, Ohta H, Nishitani K, Koyama T, Umezawa T, Misawa N, Saito K, Shibata D (2005) KaPPA-View. A web-based analysis tool for integration of transcript and metabolite data on plant metabolic pathway maps. Plant Physiol 138(3):1289–1300. https://doi.org/10.1104/pp.105.060525
130. Thimm O, Blasing O, Gibon Y, Nagel A, Meyer S, Kruger P, Selbig J, Muller LA, Rhee SY, Stitt M (2004) MAPMAN: a user-driven tool to display genomics data sets onto diagrams of metabolic pathways and other biological processes. Plant J 37(6):914–939. https://doi.org/10.1111/j.1365-313X.2004.02016.x
131. Xia JG, Wishart DS (2010) MetPA: a web-based metabolomics tool for pathway analysis and visualization. Bioinformatics 26 (18):2342–2344. https://doi.org/10.1093/bioinformatics/btq418
132. Gao J, Tarcea VG, Karnovsky A, Mirel BR, Weymouth TE, Beecher CW, Cavalcoli JD, Athey BD, Omenn GS, Burant CF, Jagadish HV (2010) Metscape: a Cytoscape plug-in for visualizing and interpreting metabolomic data in the context of human metabolic networks. Bioinformatics 26(7):971–973
133. Symons S, Nieselt K (2011) MGV: a generic graph viewer for comparative omics data. Bioinformatics 27(16):2248–2255. https://doi.org/10.1093/bioinformatics/btr351
134. Garcia-Alcalde F, Garcia-Lopez F, Dopazo J, Conesa A (2011) Paintomics: a web based tool for the joint visualization of transcriptomics and metabolomics data. Bioinformatics 27(1):137–139. https://doi.org/10.1093/bioinformatics/btq594
135. Leader DP, Burgess K, Creek D, Barrett MP (2011) Pathos: a web facility that uses metabolic maps to display experimental changes in metabolites identified by mass spectrometry. Rapid Commun Mass Spectrom 25 (22):3422–3426. https://doi.org/10.1002/rcm.5245
136. van Iersel MP, Kelder T, Pico AR, Hanspers K, Coort S, Conklin BR, Evelo C (2008) Presenting and exploring biological pathways with PathVisio. BMC Bioinformatics 9:399. https://doi.org/10.1186/1471-2105-9-399
137. Neuweger H, Persicke M, Albaum SP, Bekel T, Dondrup M, Huser AT, Winnebald J, Schneider J, Kalinowski J, Goesmann A (2009) Visualizing post genomics data-sets on customized pathway maps by ProMeTra-aeration-dependent gene expression and metabolism of Corynebacterium glutamicum as an example. BMC Syst Biol 3:82. https://doi.org/10.1186/1752-0509-3-82
138. Joshi-Tope G, Gillespie M, Vastrik I, D'Eustachio P, Schmidt E, de Bono B, Jassal B, Gopinath GR, Wu GR, Matthews L, Lewis S, Birney E, Stein L (2005) Reactome: a knowledgebase of biological pathways. Nucleic Acids Res 33: D428–D432
139. Junker BH, Klukas C, Schreiber F (2006) VANTED: a system for advanced data analysis and visualization in the context of biological networks. BMC Bioinformatics 7:109

Check for updates

Chapter 17

Single-Cell Protein Assays: A Review

Beiyuan Fan, Junbo Wang, Ying Xu, and Jian Chen

Abstract

Quantification of single-cell proteomics provides key insights in the field of cellular heterogeneity. This chapter discusses the emerging techniques that are being used to measure the protein copy numbers at the single-cell level, which includes flow cytometry, mass cytometry, droplet cytometry, microengraving, and single-cell barcoding microchip. The advantages and limitations of each technique are compared, and future research opportunities are highlighted.

Key words Single-cell protein assays, Flow cytometry, Mass cytometry, Droplet cytometry, Microengraving, Single-cell barcoding microchip

1 Introduction

Proteins (i.e., macromolecules composed of chains of amino acid residues) perform dominant functions within organisms, which function as key regulators in metabolic reactions, DNA replications, signal transductions, and molecule transportations [1]. Thus, the status of proteins has been regarded as the most important indicators of biological activities, which are closely related to cellular differentiation, neuron transmission, and disease progressions, to name a few [2].

A large number of characterization approaches (e.g., immunoassay and electrophoresis) have been proposed to estimate protein expressions [3], which significantly promote the developments in biological and medical sciences [4]. However, these conventional approaches can only quantify protein copy numbers in population and cannot be used to address the issue of cellular heterogeneity.

Within the last decades, we have witnessed huge developments in tools enabling single-cell protein analysis, providing key insights in the field of cellular heterogeneity (e.g., immune response variations and tumor heterogeneity) [5–8]. Currently, approaches in the field of single-cell protein analysis can be classified into five major types, which are (1) flow cytometry, (2) mass cytometry, (3) droplet

Tao Huang (ed.), *Computational Systems Biology: Methods and Protocols*, Methods in Molecular Biology, vol. 1754, https://doi.org/10.1007/978-1-4939-7717-8_17, © Springer Science+Business Media, LLC, part of Springer Nature 2018

cytometry, (4) microengraving, and (5) single-cell barcoding microchip. In this chapter, we examine the advantages and limitations of each technique (*see* Fig. 1) and summarize single-cell protein copy numbers enabled by these approaches (*see* Table 1). Future research opportunities are highlighted based on four measurement considerations, which are absolute quantification, detection resolution, multiplexing capability, and throughput.

2 Flow Cytometry

Currently, fluorescent flow cytometry is the golden standard in single-cell protein analysis where single cells stained with fluorescence-labeled antibodies are flushed in a capillary tube with fluorescent levels quantified by photomultiplier tubes [44, 45] (*see* Fig. 1a). After its initial invention, key technical developments include the detection of multiple fluorescence simultaneous and the inclusion of fluorescence-labeled calibration beads for absolute quantitative measurement [46]. Leveraging commercially available flow cytometry, quantitative evaluations of surface antigen expressions of single cells were realized, contributing dramatically to a variety of diseases.

Although well-established, fluorescent flow cytometry still has several key limitations, which severely compromise its functionalities in the field of single-cell protein analysis. Fluorescent flow cytometry can only provide an absolute quantitation of surface proteins of single cells leveraging calibration beads, while it cannot quantify intracellular proteins due to the lack of calibration approaches [47–50]. In addition, fluorescent flow cytometry cannot also effectively quantify secreted proteins at the single-cell level, and it cannot simultaneously quantify more than 20 types of proteins due to fluorescent spectrum overlap.

3 Mass Cytometry

Since flow cytometry has the limitation of simultaneously characterizing multiple fluorescent signals due to the emission spectral overlaps, mass spectroscopy was proposed where rare earth metal isotope tags were used to replace fluorescence labels and 100 of signals can be detected simultaneously [51–55]. When performing a mass cytometry experiment, cells of interest are first stained with antibodies previously conjugated to stable heavy-metal isotopes. Then cells are loaded into a nebulizer, which places cells into droplets for introduction into the mass cytometer. Upon entering the instrument, cells travel through an argon plasma, in which covalent bonds are broken to produce charged free atoms. The resulting ion cloud is passed through a quadrupole to discard

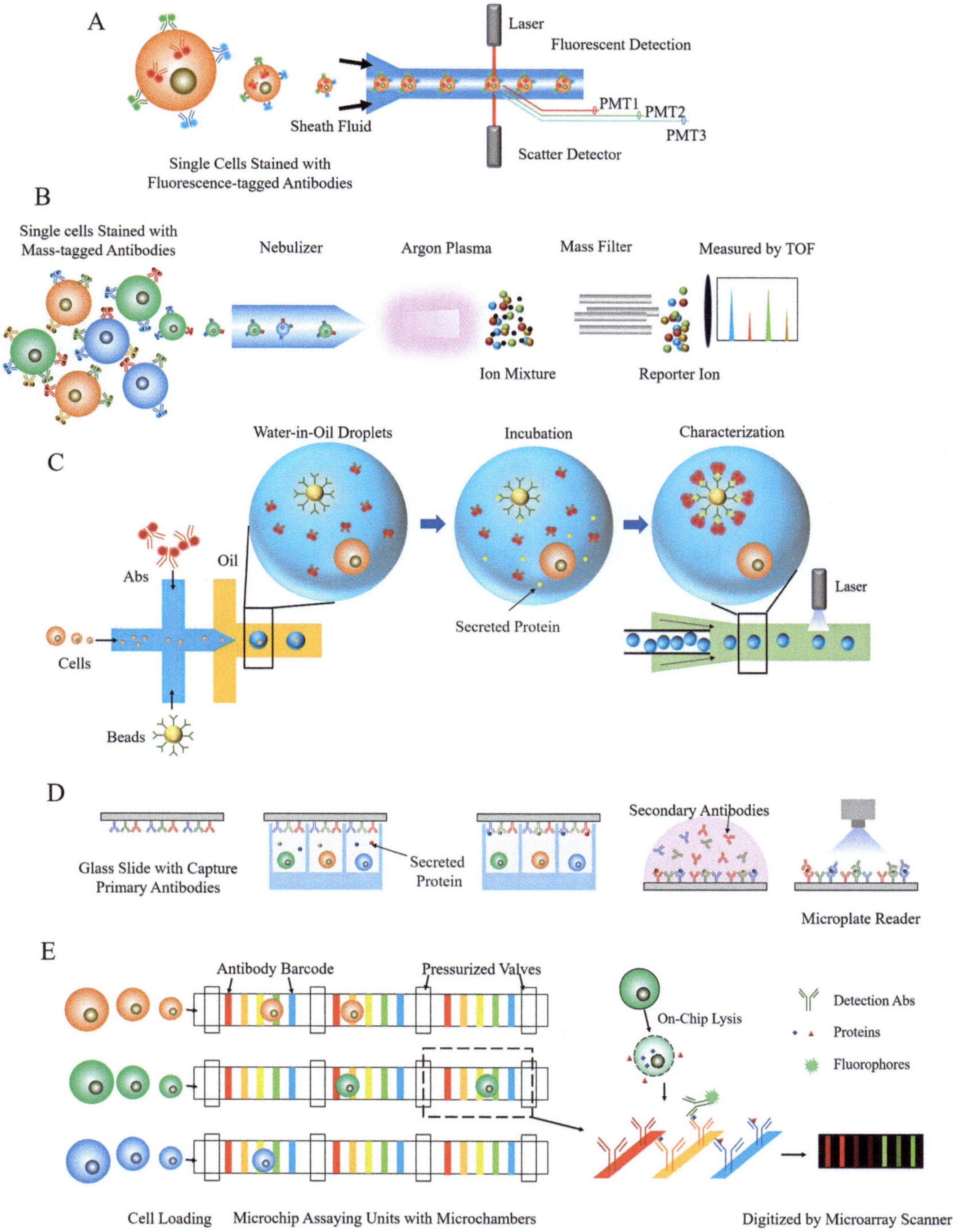

Fig. 1 Techniques enabling characterization of the protein expressions of single cells, which mainly include (**a**) flow cytometry, (**b**) mass cytometry, (**c**) droplet cytometry, (**d**) microengraving, and (**e**) single-cell barcoding microchips

Table 1
Quantitative copy numbers of single-cell proteins

Technique	Key results
Flow cytometry	Thy-1, Lyt-1, Lyt-2, and Lyt-3 of single positive T cells were quantified within the range of 10^5–10^6 per cell [9]
Flow cytometry	Fc receptors of positive lymphocytes (5–25%) were quantified as 4.2×10^4 per cell [10]
Flow cytometry	Fc receptors of small and large monocytes were quantified as $3.3 \pm 0.6 \times 10^5$ and $10 \pm 1 \times 10^5$ per cell, respectively [11]
Flow cytometry	Surface antigens of TI01, RFTI1, OKT3, OKT4, OKT8, OKT6, and 3AI of single cells of RPMI 8402, MOLT-4, CEM, and HSB-2 were quantified in the range of 10^3–10^5 copy numbers per cell [12]
Flow cytometry	Fc receptors of monocytes were quantified as $1.75 \pm 0.34 \times 10^4$ per cell [13]
Flow cytometry	PI-bound proteins of blood cells were quantified within the range of 10^4 and 10^5 per cells [14]
Flow cytometry	CD11a of CD8+ lymphocytes was quantified, locating the value of 4.3×10^3 per cell as the cutoff point between high and low expressions [15]
Flow cytometry	CD10 of normal lymphoid precursors was quantified as $3–12.5 \times 10^3$ per cell with an upper limit of 5×10^4 per cell. The median CD10 in acute B-lineage acute lymphoblastic leukemia were variable and ranged from undetectable to very high values ($>1.8 \times 10^5$) [16]
Flow cytometry	Normal TdT$^+$ precursors have a significantly higher number of TdT ($>100 \times 10^3$) and a lower number of CD10 ($<50 \times 10^3$) and CD19 ($<10 \times 10^3$) molecules per cell than B-lineage ALL blasts (<100, >50, $>10 \times 10^3$ molecules per cell, respectively) [17]
Flow cytometry	For T cells, CD2, CD3, CD4, CD5, CD7, and CD8 were quantified as 24 ± 7, 57 ± 7, 47 ± 14, 50 ± 11, 28 ± 7, and 145 ± 29 ($\times 10^3$/cell), while for B cells, CD19, CD20, CD21, CD22, CD24, and CD37 were quantified as 27 ± 3, 149 ± 10, 12 ± 2, 17 ± 2, 39 ± 2, and 99 ± 26 ($\times 10^3$/cell) [18]
Flow cytometry	CD3, CD4, CD8, CD16, and CD19 of lymphocytes and CD4 of monocytes were quantified as 134,34,945,905, 54,17,446,106, 268,868–190,622, 38,307,436, 25,252–11,689, and 9246–3094 per cell, respectively [19]
Flow cytometry	For subsets of lymphocytes, CD2, CD3, CD4, CD5, CD7, and CD8 were quantified in the ranges of 54.78–59.38, 20.71–35.56, 34.41–54.61, 81.85–94.33, 26.53–65.05, and 30.46–130.96 ($\times 10^3$ per cell) [20]
Flow cytometry	CD3 molecules were quantified as 124–25 (normal T cells), 30–21 (T lymphocytes from healthy donors), 38–31 (adult T-cell leukemia/lymphoma), 92–47 (T-prolymphocytic leukemia), and 95–21 (granular lymphocyte leukemia) ($\times 10^3$ per cell); CD7 molecules were quantified as 20–3 (normal T cells), 35–7 (T lymphocytes from healthy donors), 13–3 (adult T-cell leukemia/lymphoma), 29–12 (T-prolymphocytic leukemia), and 12–3 (granular lymphocyte leukemia) ($\times 10^3$ per cell) [21]

(continued)

Table 1
(continued)

Technique	Key results
Flow cytometry	CD19 molecules were quantified as 22 (normal B cells), 13 (chronic lymphocytic leukemia), 16 (B prolymphocytic leukemia), 15 (splenic lymphoma with villous lymphocytes), 10 (mantle cell lymphoma), and 38 (hairy cell leukemia) ($\times 10^3$ per cell); CD20 molecules were quantified as 22 (normal B cells), 13 (chronic lymphocytic leukemia), 16 (B prolymphocytic leukemia), 15 (splenic lymphoma with villous lymphocytes), 10 (mantle cell lymphoma), and 38 (hairy cell leukemia) ($\times 10^3$ per cell) [22]
Flow cytometry	CD19 molecules were quantified as 13 ± 1 (normal B cells), 4 ± 4 (chronic lymphocytic leukemia), 8 ± 7 (B-cell prolymphocytic leukemia), 4 ± 4 (mantle cell lymphoma), and 4 ± 5 splenic lymphoma with villous lymphocytes ($\times 10^3$ per cell); CD5 molecules were quantified as 11 ± 2 (normal B cells), 31 ± 17 (chronic lymphocytic leukemia), 16 ± 10 (B-cell prolymphocytic leukemia), 24 ± 14 (mantle cell lymphoma), and 18 ± 9 splenic lymphoma with villous lymphocytes ($\times 10^3$ per cell); CD79b molecules were quantified as 43 ± 8 (normal B cells), 2 ± 3 (chronic lymphocytic leukemia), 37 ± 18 (B-cell prolymphocytic leukemia), 17 ± 16 (mantle cell lymphoma), and 12 ± 13 splenic lymphoma with villous lymphocytes ($\times 10^3$ per cell) [23]
Flow cytometry	CD19, CD20, CD22, CD23, CD79b, and CD5 of typical chronic lymphocytic leukemias and atypical chronic lymphocytic leukemias were quantified within the ranges of 3844–38,347 vs. 2022–23,300, 1900–109,600 vs. 4000–70,640, 1689–20,941 vs. 2790–28,679, 1736–22,342 vs. 2011–3768, 1721–5087 vs. 2043–10,840, and 1790–59,258 vs. 2620–24,440, respectively [24]
Flow cytometry	CD23 molecules were quantified in the range of 0.6–167.3 (chronic lymphocytic leukemia), 1.1–85.2 (chronic lymphocytic leukemia with prolymphocytes), 0.6–42.7 (B-cell lymphoma), and 0.8–3.6 (control) ($\times 10^3$ per cell) [25]
Flow cytometry	CD10, TdT, CD34, CD19, and CD79a of normal and leukemic bone marrow precursors were quantified in the ranges of 8.1–23.4 vs. 17.1–211.7, 90.7–314.8 vs. 20.8–582, 99.7–280.8 vs. 1.9–423.5, 20.6–33.2 vs. 8.2–85.8, and 5.9–7.8 vs. 1.7–37.5 ($\times 10^3$/cell) [26]
Flow cytometry	Estrogen receptors were quantified as 9213 ± 1572 (MDA-MB-231), 5912 ± 1768 (SKBr3), 57,765 ± 12,998 (T47-D), 104,000 ± 16,683 (MCF-7), and 6267–1,390,000 (clinical breast tumor samples) (per cell) [27]
Flow cytometry	CD58 molecules in regenerating, normal, and leukemic bone marrows were quantified as 1649 ± 837, 1705 ± 669, and 11,245 ± 7672 (per cell) [28]
Flow cytometry	CD11b molecules of single polymorphonuclear cells were quantified as 79,440 of *Aspergillus* septicemia postrenal transplantation, 80,090 of septicemia *Streptococcus bovis* postvalvuloplasty infection, 37,330 of meningoencephalitis to *Herpesviridae*, 36,060 of aortic endocarditis *Streptococcus agalactiae*, 53,430 of septicemia *Streptococcus mitis*, 36,050 of pneumopathy to *Staphylococcus aureus*, 37,790 of pneumonia *Chlamydia psittaci*, 14,300 of HIV infection, 34,090 of pneumonia *Pseudomonas aeruginosa*, 42,550 of acute pyelonephritis to *Proteus mirabilis*, 45,680 of endocarditis to *Streptococcus viridans*, 85,360 of peritonitis secondary to appendicitis, 16,210 of hepatitis B infection, 57,920 of acute pancreatitis, and 69,540 of septicemia to *Staphylococcus aureus* (per cell) [29]

(continued)

Table 1 (continued)

Technique	Key results
Flow cytometry	CD51 molecules of Hela cells with and without irradiation were quantified as 403,700 ± 26,400 and 212,700 ± 12,000 per cell, respectively [30]
Flow cytometry	CD5, CD10, CD11c, CD19, CD20, CD22, CD23, and CD79b molecules of positive single B cells were quantified as 8.7 (normal), 13.4 (CLL), and 6.0 (B-ALL); 10.6 (normal), 11.4 (CLL), 21.7 (B-ALL), and 23.9 (HCL); 10.4 (normal), 5.4 (CLL), 2.3 (B-All), and 49.3 (HCL); 21.5 (normal), 13.3 (CLL), 12.5 (B-All), and 25.5 (HCL); 53.9 (normal), 10.5 (CLL), 14.3 (B-All), and 66.6 (HCL); 14.3 (normal), 2.4 (CLL), 4.0 (B-All), and 12.7 (HCL); 7.2 (normal), 4.5 (CLL), 0.2 (B-All), and 5.9 (HCL); and 13.9 (normal), 8.8 (CLL), 9.6 (B-All), and 14.8 (HCL) ($\times 10^3$) [31]
Flow cytometry	Glycophorin A molecules of red blood cells were quantified as in the range of 3–13 $\times 10^5$ per cell [32]
Flow cytometry	FY1, FY2, MNS3, and MNS4 molecules of single red blood cells were quantified as 11,332 (524–26,918), 7204 (4405–11,988), 10,387 (149–68,819), and 100,589 (2638–282,788) per cell, respectively [33]
Flow cytometry	CD20 molecules of chronic lymphocytic leukemia were quantified as 9341 (1039–123,045) for 17p depletion, 5886 (2248–26,221) for 11q depletion, 23,603 (1688–112,301) for trisomy, 8828 (2420–122,875) for negative FISH, and 10,781 (1711–98,361) for 13q depletion (per cell) [34]
Flow cytometry	Met1, Met2, Her2, and Her3 of cell lines including T47D, H441, Du145, HT29, and A549 were quantified as NA, ND, 80,677 ± 14,938, and 5080 ± 1641; 109,909 ± 36,334, 137,252 ± 42,899, 66,913 ± 11,847, and 8297 ± 1932; 7000 ± 1057, 11,211 ± 2009, 20,524 ± 2317, 2751 ± 429; 9440 ± 2219, 15,460 ± 1898, 45,029 ± 1683, and 4020 ± 86; and 37,224 ± 6878, 45,096 ± 6530, 12,219 ± 3157, and 1498 ± 266 (per cell) [35]
Flow cytometry	CD5 molecules of B cells of normal people, CLL, atypical CLL, MCL, MZL, HCL, LPL, and FL were quantified as 315 ± 142 (145–586), 8877 ± 5708 (169–21,017), 4475 ± 4141 (396–11,536), 5371 ± 3331 (126–10,254), 234 ± 253 (19–770), 178 ± 139 (96–338), 140 ± 139 (34–344), and 46 ± 17 (29–82) (per cell) [36]
Flow cytometry	Intracellular Bcl-2 molecules of tumor cell lines of HL-60, K562, DOHH2, Jurkat, MCF7, MDA435/LCC6, and PBL were quantified as 12.6 ± 6.5, 32.9 ± 18, 200 ± 58.3, 65.1 ± 4.6, 97.4 ± 23.1, 52.9 ± 7.3, and 38.5 ± 1.7 (10^3 per cell) [37]
Flow cytometry	Intracellular perforin molecules of natural killer cells of healthy controls and individuals with heterozygous and homozygous perforin deficiencies were quantified as 3561 ± 1175, 2260, and 212, respectively (per cell) [38]
Flow cytometry	Intracellular proteins of ZAP-70 were quantified as 12,621 (normal B cells) and 102,615 ± 119,252 and 233,259 ± 182,071 from CD19+/CD5+ and CD3+/CD45+ B cells from chronic lymphocytic leukemia (per cell) [39]
Mass cytometry	CD117, CD34, CD133, and CD45 of progenitors of normal bone marrows were quantified as 2.8×10^4, 6.1×10^4, and 4.1×10^4, 2.0×10^5 per cell, while CD95, CD45, CD38, CD15, and CD11 of granulocytes/monocytes of normal bone marrows were quantified as 3.1×10^5, 4.1×10^4, 7.0×10^4, 1.8×10^5, and 1.6×10^5 per cell [40]

(continued)

Table 1 (continued)

Technique	Key results
Mass cytometry	CD4 molecules of fixed whole blood, fixed cryopreserved peripheral blood mononuclear cells, and lyophilized peripheral blood mononuclear cells were quantified as 45, 46, and 14×10^3 per cells [41]
Microengraving	The secretion rates of IL-6 of PBMCs were quantified as 6.5 ± 3.2 molecules/s (3 h) and 10.6 ± 7.1 molecules/s (12 h) at the single-cell level [42]
Microengraving	The secretion rates off CXCL1, CXCL5, and CXCL8 of single tumor cells were quantified as 0.3–0.9, 0.6–3.1, and 1.9–4.8 molecules/s, respectively [43]

common biological elements, enriching for heavy-metal reporter ions, which are separated by their mass-to-charge ratio in a time-of-flight mass spectrometer (*see* Fig. 1b). Based on mass cytometry, deep phenotyping [40, 41, 56–64], signaling state characterization [59, 65–68], and cycle analysis [57, 69, 70] at the single-cell level have been demonstrated.

Although mass cytometry can simultaneously characterize roughly 100 types of proteins, it still has several drawbacks. In mass cytometry, cells are atomized and ionized, and thus it remains infeasible to recover living cells after analysis, and thus it cannot function as a cell sorter. In addition, due to the dynamics of ion flight in the mass spectrometer, the throughput of mass cytometry lags behind that of fluorescence-based instruments. Furthermore, both fluorescent flow cytometry and mass cytometry cannot produce absolute copy numbers of intracellular proteins and cannot be used to estimate secreted proteins at the single-cell level.

4 Droplet Cytometry

Since flow cytometry and mass cytometry cannot effectively estimate the secreted proteins from single cells, droplet microfluidics [71–77] was proposed where aqueous droplets encapsulate individual cells and confine secreted proteins, which are further characterized based on fluorescent detection [78–94] (*see* Fig. 1c). Based on droplet cytometry, secreted proteins at the single-cell level were characterized, which includes (1) yellow fluorescent protein mutant "Venus" of single *E. coli* [78]; (2) the activities of enzyme alkaline phosphatase secreted by single *E. coli* [79]; (3) cytokine (IL-10) secretion of single $CD4^+CD25^+$ regulatory T cells [80]; (4) intracellular proteins of HRas-mCitrine, expressed within single HEK-293 cells, and actin-EGFP expressed within single MCF-7 cells [81]; and (5) cytokine (IL-2, IFN-γ, TNF-α) secretion of single, activated T-cells [82].

Although powerful, droplet cytometry has two limitations, which need to be further addressed. (1) This technique is not capable of absolute quantification of secreted proteins due to the lack of calibration approaches. In droplet cytometry, cytokine-capture beads are usually confined in the same droplet for the capture of secreted proteins, which, however, leads to an uneven distribution of fluorescence among individual droplets (intensity peaks around individual beads), posing obstacles in calibration. (2) The cell loading rates in individual droplets need to be further optimized. In the process of forming droplets for single-cell encapsulation, as time goes by, there is a concern of cell sedimentation, leading to variations in the concentrations of cell suspensions, and thus the cell loading rates in the individual droplet cannot be finely tuned. Although cell-suspension solutions with enhanced densities (e.g., percoll) were used to address this issue, the addition of percoll may lead to enhanced auto-fluorescence, bringing another concern in fluorescence-based detections.

5 Microengraving

As an enabling technique, microengraving proposed by the Love group effectively isolates individual cells to realize the absolute quantification of secreted proteins in a time sequence [43, 95–107]. As shown in Fig. 1d, single cells, suspended in media, are deposited into a large array of polymeric microwells, which are then sealed by a glass slides coated with pre-coated with capture reagents, sequentially. The resulting glass slides are then interrogated with laser-based fluorescence scanners to quantify the protein copy numbers.

Based on microengraving, antigen-specific antibodies [95, 96, 105], cytokines [42, 96–98, 101, 102], and chemokines [43] secreted at the single-cell level were reported. Although microengraving can enable absolute quantification of multiple proteins secreted by individual cells, compared to flow cytometry, it requires a careful manipulation of glass slides without stimulating or even dislodging single cells within microwells. This process requires highly trained technicians and cannot be further increased in the throughput, easily translated to function in an automatic manner, limiting its widespread applications in single-cell proteomics.

6 Single-Cell Barcoding Microchip

Both flow cytometry and mass cytometry cannot realize the absolute quantification of intracellular proteins at the single-cell level due to the lacks of calibration approaches. To address this issue, Heath et al. proposed large-array microchambers (single-cell barcoding microchips), which enables absolute quantification of both cytosolic and membrane proteins at the single-cell level

[108–126]. As shown in Fig. 1e, the single-cell barcoding microchips include thousands of individually addressed microchambers to realize single-cell trapping, lysis, and protein capture by pre-printed antibodies on the surface of the chambers. Then the fluorescent immunosandwich approach is used for absolute quantification. This powerful approach has been used to quantify multiple key cytokines at the single-cell level, reporting cellular heterogeneities in both immunology [111] and tumor biology [114, 115].

As an approach capable of absolute quantification of both surface and cytosolic proteins of single cells in large arrays, there are two practical concerns. When running experiments with this approach, it is possible that large portions of single cells may stick to the bottom surfaces of cellular inlets, and thus this technique may not be directly used to characterize rare cells (e.g., circulating tumor cells). In addition, the potential of trapping multiple cells within individual chambers is a key concern. If a manual check of the number of cells within each chamber is conducted, it is definitely time-consuming. If a very low concentration of cell suspension is used in the cell loading step, a large number of empty chambers can be generated. Thus this approach suffers from limited throughput, although instrumentation can address this issue to an extent.

7 Discussion and Future Development

In this study, we summarized key technical approaches enabling the quantification of single-cell proteins. Although significant improvements have been made within the last decade, there are still several important issues that should be addressed in the near future.

The absolute quantification of intracellular proteins in a high-throughput manner is a key issue in the field of single-cell proteomics. Currently, both flow cytometry and mass cytometry cannot produce absolute copy numbers of interested intracellular proteins due to the lack of calibration beads. In addition, the step of intracellular staining in flow or mass cytometry also brings concerns that not all of the targeted intracellular proteins are bound with corresponding antibodies. As an enabling technique, single-cell barcoding microchip can quantify intracellular proteins, which, however, suffers from limited throughput. Thus, more efforts are suggested to be devoted in this field to realize the absolute quantification of single-cell intracellular proteins to construct single-cell data, functioning as a reference for future analysis.

Other issues in the field of single-cell proteomics may include detection multiplexity, resolution, and throughput. With the help of mass cytometry, 100 of proteins can be simultaneously characterized, which provide key insights in understanding cellular signal transductions. Further increases in the type of proteins that can be simultaneously characterized are definitely beneficial, which may help us obtain deep-view comprehensive information of biology at the cellular level. Currently the detection resolution of flow cytometry is several hundreds of molecules at the single-cell level, and further improvements in the detection resolution help locate new functionalities of proteins with low copy numbers. As to the throughput, mass cytometry still lags behind flow cytometry, and further work in this field is also suggested.

Acknowledgments

The authors would like to acknowledge the discussions with Xiufeng Li, Deyong Chen, and Dong Men and financial supports from the National Basic Research Program of China (973 Program, Grant No. 2014CB744600), National Natural Science Foundation of China (Grant No. 61431019, 61671430), Key Project of Chinese Academy of Sciences (QYZDB-SSW-JSC011), Natural Science Foundation of Beijing (4152056), Instrument Development Program of the Chinese Academy of Sciences, Beijing NOVA Program of Science and Technology, and Youth Innovation Promotion Association of Chinese Academy of Sciences.

References

1. Kim MS, Pinto SM, Getnet D, Nirujogi RS, Manda SS, Chaerkady R, Madugundu AK, Kelkar DS, Isserlin R, Jain S, Thomas JK, Muthusamy B, Leal-Rojas P, Kumar P, Sahasrabuddhe NA, Balakrishnan L, Advani J, George B, Renuse S, Selvan LD, Patil AH, Nanjappa V, Radhakrishnan A, Prasad S, Subbannayya T, Raju R, Kumar M, Sreenivasamurthy SK, Marimuthu A, Sathe GJ, Chavan S, Datta KK, Subbannayya Y, Sahu A, Yelamanchi SD, Jayaram S, Rajagopalan P, Sharma J, Murthy KR, Syed N, Goel R, Khan AA, Ahmad S, Dey G, Mudgal K, Chatterjee A, Huang TC, Zhong J, Wu X, Shaw PG, Freed D, Zahari MS, Mukherjee KK, Shankar S, Mahadevan A, Lam H, Mitchell CJ, Shankar SK, Satishchandra P, Schroeder JT, Sirdeshmukh R, Maitra A, Leach SD, Drake CG, Halushka MK, Prasad TS, Hruban RH, Kerr CL, Bader GD, Iacobuzio-Donahue CA, Gowda H, Pandey A (2014) A draft map of the human proteome. Nature 509 (7502):575–581. https://doi.org/10.1038/nature13302
2. Savage N (2015) Proteomics: high-protein research. Nature 527(7576):S6–S7. https://doi.org/10.1038/527S6a
3. Phizicky E, Bastiaens PI, Zhu H, Snyder M, Fields S (2003) Protein analysis on a proteomic scale. Nature 422(6928):208–215. https://doi.org/10.1038/nature01512
4. Hanash S (2003) Disease proteomics. Nature 422(6928):226–232. https://doi.org/10.1038/nature01514
5. Ma C, Fan R, Elitas M (2013) Single cell functional proteomics for assessing immune response in cancer therapy: technology, methods, and applications. Front Oncol 3:133. https://doi.org/10.3389/fonc.2013.00133
6. Wei W, Shin YS, Ma C, Wang J, Elitas M, Fan R, Heath JR (2013) Microchip platforms for multiplex single-cell functional proteomics with applications to immunology and cancer research. Genome Med 5(8):75. https://doi.org/10.1186/gm479
7. Yu J, Zhou J, Sutherland A, Wei W, Shin YS, Xue M, Heath JR (2014) Microfluidics-based single-cell functional proteomics for fundamental and applied biomedical applications. Annu Rev Anal Chem 7:275–295. https://doi.org/10.1146/annurev-anchem-071213-020323
8. Su Y, Shi Q, Wei W (2017) Single cell proteomics in biomedicine: High-dimensional data

acquisition, visualization, and analysis. Proteomics 17(3–4):1600267. https://doi.org/10.1002/pmic.201600267

9. Ledbetter JA, Rouse RV, Micklem HS, Herzenberg LA (1980) T cell subsets defined by expression of Lyt-1,2,3 and Thy-1 antigens. Two-parameter immunofluorescence and cytotoxicity analysis with monoclonal antibodies modifies current views. J Exp Med 152(2):280–295
10. Titus JA, Sharrow SO, Connolly JM, Segal DM (1981) Fc (IgG) receptor distributions in homogeneous and heterogeneous cell populations by flow microfluorometry. Proc Natl Acad Sci U S A 78(1):519–523
11. Kavai M, Bodolay E, Szollosi J (1982) Characterization of human monocyte subpopulations by flow cytometry. Immunology 47(2):255–262
12. Poncelet P, Carayon P (1985) Cytofluorometric quantification of cell-surface antigens by indirect immunofluorescence using monoclonal-antibodies. J Immunol Methods 85(1):65–74. https://doi.org/10.1016/0022-1759(85)90274-1
13. Christensen J, Leslie RG (1990) Quantitative measurement of Fc receptor activity on human peripheral blood monocytes and the monocyte-like cell line, U937, by laser flow cytometry. J Immunol Methods 132(2):211–219. https://doi.org/10.1016/0022-1759(90)90032-Q
14. Plesner T, Hansen NE, Carlsen K (1990) Estimation of PI-bound proteins on blood cells from PNH patients by quantitative flow cytometry. Br J Haematol 75(4):585–590
15. Pallis M, Robins A, Powell R (1993) Quantitative analysis of lymphocyte CD1 la using standardized flow cytometry. Scand J Immunol 38(6):559–564. https://doi.org/10.1111/j.1365-3083.1993.tb03241.x
16. Lavabre-Bertrand T, Janossy G, Ivory K, Peters R, Secker-Walker L, Porwit-MacDonald A (1994) Leukemia-associated changes identified by quantitative flow cytometry: I. CD10 expression. Cytometry 18(4):209–217. https://doi.org/10.1002/cyto.990180404
17. Farahat N, Lens D, Zomas A, Morilla R, Matutes E, Catovsky D (1995) Quantitative flow cytometry can distinguish between normal and leukaemic B-cell precursors. Br J Haematol 91(3):640–646
18. Bikoue A, George F, Poncelet P, Mutin M, Janossy G, Sampol J (1996) Quantitative analysis of leukocyte membrane antigen expression: normal adult values. Cytometry 26(2):137–147. https://doi.org/10.1002/(SICI)1097-0320(19960615)26:2<137::AID-CYTO7>3.0.CO;2-D
19. Denny TN, Stein D, Mui T, Scolpino A, Holland B (1996) Quantitative determination of surface antibody binding capacities of immune subsets present in peripheral blood of healthy adult donors. Cytometry 26(4):265–274. https://doi.org/10.1002/(SICI)1097-0320(19961215)26:4<265::AID-CYTO5>3.0.CO;2-B
20. Ginaldi L, Farahat N, Matutes E, De Martinis M, Morilla R, Catovsky D (1996) Differential expression of T cell antigens in normal peripheral blood lymphocytes: a quantitative analysis by flow cytometry. J Clin Pathol 49(7):539–544
21. Ginaldi L, Matutes E, Farahat N, De Martinis M, Morilla R, Catovsky D (1996) Differential expression of CD3 and CD7 in T-cell malignancies: a quantitative study by flow cytometry. Br J Haematol 93(4):921–927
22. Ginaldi L, De Martinis M, Matutes E, Farahat N, Morilla R, Catovsky D (1998) Levels of expression of CD19 and CD20 in chronic B cell leukaemias. J Clin Pathol 51(5):364–369
23. Cabezudo E, Carrara P, Morilla R, Matutes E (1999) Quantitative analysis of CD79b, CD5 and CD19 in mature B-cell lymphoproliferative disorders. Haematologica 84(5):413–418
24. D'Arena G, Musto P, Cascavilla N, Dell'Olio M, Di Renzo N, Carotenuto M (2000) Quantitative flow cytometry for the differential diagnosis of leukemic B-cell chronic lymphoproliferative disorders. Am J Hematol 64(4):275–281
25. Lopez-Matas M, Rodriguez-Justo M, Morilla R, Catovsky D, Matutes E (2000) Quantitative expression of CD23 and its ligand CD21 in chronic lymphocytic leukemia. Haematologica 85(11):1140–1145
26. Rego EM, Garcia AB, Carneiro JJ, Falcao RP (2001) Immunophenotype of normal and leukemic bone marrow B-precursors in a Brazilian population. A comparative analysis by quantitative fluorescence cytometry. Braz J Med Biol Res 34(2):183–194
27. Gritzapis AD, Baxevanis CN, Missitzis I, Katsanou ES, Alexis MN, Yotis J, Papamichail M (2003) Quantitative fluorescence cytometric measurement of estrogen and progesterone receptors: correlation with the hormone binding assay. Breast Cancer Res Treat 80(1):1–13. https://doi.org/10.1023/A:1024462416640

28. Veltroni M, De Zen L, Sanzari MC, Maglia O, Dworzak MN, Ratei R, Biondi A, Basso G, Gaipa G, Group IB-A-F-M-S (2003) Expression of CD58 in normal, regenerating and leukemic bone marrow B cells: implications for the detection of minimal residual disease in acute lymphocytic leukemia. Haematologica 88(11):1245–1252
29. Latger-Cannard V, Besson I, Doco-Lecompte T, Lecompte T (2004) A standardized procedure for quantitation of CD11b on polymorphonuclear neutrophil by flow cytometry: potential application in infectious diseases. Clin Lab Haematol 26(3):177–186. https://doi.org/10.1111/j.1365-2257.2004.00599.x
30. Smith RA, Giorgio TD (2004) Quantitation and kinetics of CD51 surface receptor expression: implications for targeted delivery. Ann Biomed Eng 32(5):635–644
31. Olejniczak SH, Stewart CC, Donohue K, Czuczman MS (2006) A quantitative exploration of surface antigen expression in common B-cell malignancies using flow cytometry. Immunol Investig 35(1):93–114. https://doi.org/10.1080/08820130500496878
32. Riquelme BD, de Isla NG, Valverde JR, Stoltz JF (2006) A simple method for quantifying high density antigens in erythrocyte membrane by flow cytometry. J Biochem Biophys Methods 68(1):31–42. https://doi.org/10.1016/j.jbbm.2005.12.009
33. Calonego SB, Barjas-Castro Mde L, Metze K, Pereira FG, Lorand-Metze I (2008) The influence of storage and leukocyte depletion on the antigen densities of FY1, FY2, MNS3 and MNS4 measured by flow cytometry. Transfus Apher Sci 38(2):101–107. https://doi.org/10.1016/j.transci.2007.11.001
34. Tam CS, Otero-Palacios J, Abruzzo LV, Jorgensen JL, Ferrajoli A, Wierda WG, Lerner S, O'Brien S, Keating MJ (2008) Chronic lymphocytic leukaemia CD20 expression is dependent on the genetic subtype: a study of quantitative flow cytometry and fluorescent in-situ hybridization in 510 patients. Br J Haematol 141(1):36–40. https://doi.org/10.1111/j.1365-2141.2008.07012.x
35. Panke C, Weininger D, Haas A, Schelter F, Schlothauer T, Bader S, Sircar R, Josel HP, Baer U, Burtscher H, Mundigl O, Grote M, Brinkmann U, Sustmann C (2013) Quantification of cell surface proteins with bispecific antibodies. Protein Eng Des Sel 26 (10):645–654. https://doi.org/10.1093/protein/gzt035
36. Challagundla P, Jorgensen JL, Kanagal-Shamanna R, Gurevich I, Pierson DM, Ferrajoli A, Reyes SR, Medeiros LJ, Miranda RN (2014) Utility of quantitative flow cytometry immunophenotypic analysis of CD5 expression in small B-cell neoplasms. Arch Pathol Lab Med 138(7):903–909. https://doi.org/10.5858/arpa.2013-0367-OA
37. Dragowska WH, Lopes de Menezes DE, Sartor J, Mayer LD (2000) Quantitative fluorescence cytometric analysis of Bcl-2 levels in tumor cells exhibiting a wide range of inherent Bcl-2 protein expression: correlation with Western blot analysis. Cytometry 40 (4):346–352
38. Maher KJ, Klimas NG, Hurwitz B, Schiff R, Fletcher MA (2002) Quantitative fluorescence measures for determination of intracellular perforin content. Clin Diagn Lab Immunol 9(6):1248–1252. https://doi.org/10.1128/cdli.9.6.1248-1252.2002
39. Kay S, Herishanu Y, Pick M, Rogowski O, Baron S, Naparstek E, Polliack A, Deutsch VR (2006) Quantitative flow cytometry of ZAP-70 levels in chronic lymphocytic leukemia using molecules of equivalent soluble fluorochrome. Cytometry B Clin Cytom 70 (4):218–226. https://doi.org/10.1002/cyto.b.20078
40. Ornatsky O, Bandura D, Baranov V, Nitz M, Winnik MA, Tanner S (2010) Highly multiparametric analysis by mass cytometry. J Immunol Methods 361(1–2):1–20. https://doi.org/10.1016/j.jim.2010.07.002
41. Wang L, Abbasi F, Ornatsky O, Cole KD, Misakian M, Gaigalas AK, He HJ, Marti GE, Tanner S, Stebbings R (2012) Human CD4+ lymphocytes for antigen quantification: characterization using conventional flow cytometry and mass cytometry. Cytometry A 81 (7):567–575. https://doi.org/10.1002/cyto.a.22060
42. Han Q, Bradshaw EM, Nilsson B, Hafler DA, Love JC (2010) Multidimensional analysis of the frequencies and rates of cytokine secretion from single cells by quantitative microengraving. Lab Chip 10(11):1391–1400. https://doi.org/10.1039/b926849a
43. Adalsteinsson VA, Tahirova N, Tallapragada N, Yao X, Campion L, Angelini A, Douce TB, Huang C, Bowman B, Williamson CA, Kwon DS, Wittrup KD, Love JC (2013) Single cells from human primary colorectal tumors exhibit polyfunctional heterogeneity in secretions of ELR+ CXC chemokines. Integr Biol 5 (10):1272–1281. https://doi.org/10.1039/c3ib40059j
44. De Rosa SC, Herzenberg LA, Herzenberg LA, Roederer M (2001) 11-color,

13-parameter flow cytometry: identification of human naive T cells by phenotype, function, and T-cell receptor diversity. Nat Med 7 (2):245–248. https://doi.org/10.1038/84701

45. Perfetto SP, Chattopadhyay PK, Roederer M (2004) Seventeen-colour flow cytometry: unravelling the immune system. Nat Rev Immunol 4(8):648–655. https://doi.org/10.1038/nri1416
46. Herzenberg LA, Parks D, Sahaf B, Perez O, Roederer M, Herzenberg LA (2002) The history and future of the fluorescence activated cell sorter and flow cytometry: a view from stanford. Clin Chem 48(10):1819–1827
47. Zenger VE, Vogt R, Mandy F, Schwartz A, Marti GE (1998) Quantitative flow cytometry: inter-laboratory variation. Cytometry 33 (2):138–145. https://doi.org/10.1002/(SICI)1097-0320(19981001)33:2<138::AID-CYTO8>3.0.CO;2-F
48. Schwartz A, Fernandez-Repollet E (2001) Quantitative flow cytometry. Clin Lab Med 21(4):743–761
49. Marti GE, Zenger VE, Vogt R, Gaigalas A (2002) Quantitative flow cytometry: history, practice, theory, consensus, inter-laboratory variation and present status. Cytotherapy 4 (1):97–98. https://doi.org/10.1080/146532402317251626
50. Maher KJ, Fletcher MA (2005) Quantitative flow cytometry in the clinical laboratory. Clin Appl Immunol Rev 5(6):353–372. https://doi.org/10.1016/j.cair.2005.10.001
51. Atkuri KR, Stevens JC, Neubert H (2015) Mass cytometry: a highly multiplexed single-cell technology for advancing drug development. Drug Metab Dispos 43(2):227–233. https://doi.org/10.1124/dmd.114.060798
52. Spitzer Matthew H, Nolan Garry P (2016) Mass cytometry: single cells, many features. Cell 165(4):780–791. https://doi.org/10.1016/j.cell.2016.04.019
53. Comi TJ, Do TD (2017) Categorizing cells on the basis of their chemical profiles: progress in single-cell mass spectrometry. J Am Chem Soc 139(11):3920–3929. https://doi.org/10.1021/jacs.6b12822
54. Bendall SC, Nolan GP, Roederer M, Chattopadhyay PK (2012) A deep profiler's guide to cytometry. Trends Immunol 33(7):323–332. https://doi.org/10.1016/j.it.2012.02.010
55. Bjornson ZB, Nolan GP, Fantl WJ (2013) Single-cell mass cytometry for analysis of immune system functional states. Curr Opin Immunol 25(4):484–494. https://doi.org/10.1016/j.coi.2013.07.004
56. Ornatsky OI, Kinach R, Bandura DR, Lou X, Tanner SD, Baranov VI, Nitz M, Winnik MA (2008) Development of analytical methods for multiplex bio-assay with inductively coupled plasma mass spectrometry. J Anal At Spectrom 23(4):463–469. https://doi.org/10.1039/b710510j
57. Ornatsky OI, Lou X, Nitz M, Schafer S, Sheldrick WS, Baranov VI, Bandura DR, Tanner SD (2008) Study of cell antigens and intracellular DNA by identification of element-containing labels and metallointercalators using inductively coupled plasma mass spectrometry. Anal Chem 80(7):2539–2547. https://doi.org/10.1021/ac702128m
58. Bandura DR, Baranov VI, Ornatsky OI, Antonov A, Kinach R, Lou X, Pavlov S, Vorobiev S, Dick JE, Tanner SD (2009) Mass cytometry: technique for real time single cell multitarget immunoassay based on inductively coupled plasma time-of-flight mass spectrometry. Anal Chem 81 (16):6813–6822. https://doi.org/10.1021/ac901049w
59. Bendall SC, Simonds EF, Qiu P, Amir el AD, Krutzik PO, Finck R, Bruggner RV, Melamed R, Trejo A, Ornatsky OI, Balderas RS, Plevritis SK, Sachs K, Pe'er D, Tanner SD, Nolan GP (2011) Single-cell mass cytometry of differential immune and drug responses across a human hematopoietic continuum. Science 332(6030):687–696. https://doi.org/10.1126/science.1198704
60. Newell Evan W, Sigal N, Bendall Sean C, Nolan Garry P, Davis Mark M (2012) Cytometry by time-of-flight shows combinatorial cytokine expression and virus-specific cell niches within a continuum of CD8+ T cell phenotypes. Immunity 36(1):142–152. https://doi.org/10.1016/j.immuni.2012.01.002
61. Horowitz A, Strauss-Albee DM, Leipold M, Kubo J, Nemat-Gorgani N, Dogan OC, Dekker CL, Mackey S, Maecker H, Swan GE, Davis MM, Norman PJ, Guethlein LA, Desai M, Parham P, Blish CA (2013) Genetic and environmental determinants of human NK cell diversity revealed by mass cytometry. Sci Transl Med 5(208):208ra145. https://doi.org/10.1126/scitranslmed.3006702
62. Newell EW, Sigal N, Nair N, Kidd BA, Greenberg HB, Davis MM (2013) Combinatorial tetramer staining and mass cytometry analysis facilitate T-cell epitope mapping and characterization. Nat Biotechnol 31(7):623–629. https://doi.org/10.1038/nbt.2593
63. Tanner SD, Baranov VI, Ornatsky OI, Bandura DR, George TC (2013) An introduction

to mass cytometry: fundamentals and applications. Cancer Immunol Immunother 62 (5):955–965. https://doi.org/10.1007/s00262-013-1416-8

64. Levine JH, Simonds EF, Bendall SC, Davis KL, Amir el AD, Tadmor MD, Litvin O, Fienberg HG, Jager A, Zunder ER, Finck R, Gedman AL, Radtke I, Downing JR, Pe'er D, Nolan GP (2015). Data-driven phenotypic dissection of AML reveals progenitor-like cells that correlate with prognosis. Cell 162 (1):184–197. https://doi.org/10.1016/j.cell.2015.05.047
65. Bodenmiller B, Zunder ER, Finck R, Chen TJ, Savig ES, Bruggner RV, Simonds EF, Bendall SC, Sachs K, Krutzik PO, Nolan GP (2012) Multiplexed mass cytometry profiling of cellular states perturbed by small-molecule regulators. Nat Biotechnol 30(9):858–867. https://doi.org/10.1038/nbt.2317
66. Mingueneau M, Krishnaswamy S, Spitzer MH, Bendall SC, Stone EL, Hedrick SM, Pe'er D, Mathis D, Nolan GP, Benoist C (2014) Single-cell mass cytometry of TCR signaling: amplification of small initial differences results in low ERK activation in NOD mice. Proc Natl Acad Sci U S A 111 (46):16466–16471. https://doi.org/10.1073/pnas.1419337111
67. Spitzer MH, Gherardini PF, Fragiadakis GK, Bhattacharya N, Yuan RT, Hotson AN, Finck R, Carmi Y, Zunder ER, Fantl WJ, Bendall SC, Engleman EG, Nolan GP (2015) Immunology. An interactive reference framework for modeling a dynamic immune system. Science 349(6244):1259425. https://doi.org/10.1126/science.1259425
68. Zunder ER, Lujan E, Goltsev Y, Wernig M, Nolan GP (2015) A continuous molecular roadmap to iPSC reprogramming through progression analysis of single-cell mass cytometry. Cell Stem Cell 16(3):323–337. https://doi.org/10.1016/j.stem.2015.01.015
69. Behbehani GK, Bendall SC, Clutter MR, Fantl WJ, Nolan GP (2012) Single-cell mass cytometry adapted to measurements of the cell cycle. Cytometry A 81(7):552–566. https://doi.org/10.1002/cyto.a.22075
70. Darzynkiewicz Z (2012) Cycling into future: mass cytometry for the cell-cycle analysis. Cytometry A 81(7):546–548. https://doi.org/10.1002/cyto.a.22080
71. Kang D-K, Monsur Ali M, Zhang K, Pone EJ, Zhao W (2014) Droplet microfluidics for single-molecule and single-cell analysis for cancer research, diagnosis and therapy. TrAC Trends Anal Chem 2014:145–153. https://doi.org/10.1016/j.trac.2014.03.006
72. Brouzes E (2012) Droplet microfluidics for single-cell analysis. Methods Mol Biol 853:105–139. https://doi.org/10.1007/978-1-61779-567-1_10
73. Joensson H, Andersson Svahn H (2012) Droplet microfluidics-a tool for single-cell analysis. Angew Chem Int Ed 51 (49):12176–12192
74. Ralf S, Martin B, Thomas P, Stephan H (2012) Droplet based microfluidics. Rep Prog Phys 75(1):016601
75. Lagus TP, Edd JF (2013) A review of the theory, methods and recent applications of high-throughput single-cell droplet microfluidics. J Phys D Appl Phys 46(11):114005
76. Basova EY, Foret F (2015) Droplet microfluidics in (bio)chemical analysis. Analyst 140 (1):22–38. https://doi.org/10.1039/c4an01209g
77. Collins DJ, Neild A, deMello A, Liu AQ, Ai Y (2015) The Poisson distribution and beyond: methods for microfluidic droplet production and single cell encapsulation. Lab Chip 15 (17):3439–3459. https://doi.org/10.1039/c5lc00614g
78. Huebner A, Srisa-Art M, Holt D, Abell C, Hollfelder F, deMello AJ, Edel JB (2007) Quantitative detection of protein expression in single cells using droplet microfluidics. Chem Commun 12:1218–1220. https://doi.org/10.1039/B618570C
79. Huebner A, Olguin LF, Bratton D, Whyte G, Huck WT, de Mello AJ, Edel JB, Abell C, Hollfelder F (2008) Development of quantitative cell-based enzyme assays in microdroplets. Anal Chem 80(10):3890–3896. https://doi.org/10.1021/ac800338z
80. Konry T, Dominguez-Villar M, Baecher-Allan C, Hafler DA, Yarmush ML (2011) Droplet-based microfluidic platforms for single T cell secretion analysis of IL-10 cytokine. Biosensors Bioelectron 26(5):2707–2710. https://doi.org/10.1016/j.bios.2010.09.006. S0956-5663(10)00608-1 [pii]
81. Martino C, Zagnoni M, Sandison ME, Chanasakulniyom M, Pitt AR, Cooper JM (2011) Intracellular protein determination using droplet-based immunoassays. Anal Chem 83(13):5361–5368. https://doi.org/10.1021/ac200876q
82. Chokkalingam V, Tel J, Wimmers F, Liu X, Semenov S, Thiele J, Figdor CG, Huck WT (2013) Probing cellular heterogeneity in cytokine-secreting immune cells using droplet-based microfluidics. Lab Chip 13

(24):4740–4744. https://doi.org/10.1039/c3lc50945a

83. Brouzes E, Medkova M, Savenelli N, Marran D, Twardowski M, Hutchison JB, Rothberg JM, Link DR, Perrimon N, Samuels ML (2009) Droplet microfluidic technology for single-cell high-throughput screening. Proc Natl Acad Sci U S A 106 (34):14195–14200. https://doi.org/10.1073/pnas.0903542106. 0903542106 [pii]
84. Srisa-Art M, Bonzani IC, Williams A, Stevens MM, deMello AJ, Edel JB (2009) Identification of rare progenitor cells from human periosteal tissue using droplet microfluidics. Analyst 134(11):2239–2245. https://doi.org/10.1039/b910472k
85. Marcoux PR, Dupoy M, Mathey R, Novelli-Rousseau A, Heran V, Morales S, Rivera F, Joly PL, Moy J-P, Mallard F (2011) Micro-confinement of bacteria into w/o emulsion droplets for rapid detection and enumeration. Colloids Surf A Physicochem Eng Asp 377 (1–3):54–62. https://doi.org/10.1016/j.colsurfa.2010.12.013
86. Mazutis L, Gilbert J, Ung WL, Weitz DA, Griffiths AD, Heyman JA (2013) Single-cell analysis and sorting using droplet-based microfluidics. Nat Protoc 8(5):870–891. https://doi.org/10.1038/nprot.2013.046
87. Ramji R, Wang M, Bhagat AAS, Tan Shao Weng D, Thakor NV, Teck Lim C, Chen C-H (2014) Single cell kinase signaling assay using pinched flow coupled droplet microfluidics. Biomicrofluidics 8(3):034104. https://doi.org/10.1063/1.4878635
88. Wang BL, Ghaderi A, Zhou H, Agresti J, Weitz DA, Fink GR, Stephanopoulos G (2014) Microfluidic high-throughput culturing of single cells for selection based on extracellular metabolite production or consumption. Nat Biotechnol 32 (5):473–478. https://doi.org/10.1038/nbt.2857
89. Baret JC, Miller OJ, Taly V, Ryckelynck M, El-Harrak A, Frenz L, Rick C, Samuels ML, Hutchison JB, Agresti JJ, Link DR, Weitz DA, Griffiths AD (2009) Fluorescence-activated droplet sorting (FADS): efficient microfluidic cell sorting based on enzymatic activity. Lab Chip 9(13):1850–1858. https://doi.org/10.1039/b902504a
90. Shim J-u, Olguin LF, Whyte G, Scott D, Babtie A, Abell C, Huck WTS, Hollfelder F (2009) Simultaneous determination of gene expression and enzymatic activity in individual bacterial cells in microdroplet compartments. J Am Chem Soc 131(42):15251–15256. https://doi.org/10.1021/ja904823z
91. Baret JC, Beck Y, Billas-Massobrio I, Moras D, Griffiths AD (2010) Quantitative cell-based reporter gene assays using droplet-based microfluidics. Chem Biol 17 (5):528–536. https://doi.org/10.1016/j.chembiol.2010.04.010
92. Hammar P, Angermayr SA, Sjostrom SL, van der Meer J, Hellingwerf KJ, Hudson EP, Joensson HN (2015) Single-cell screening of photosynthetic growth and lactate production by cyanobacteria. Biotechnol Biofuels 8:193. https://doi.org/10.1186/s13068-015-0380-2
93. Ng EX, Miller MA, Jing T, Chen C-H (2016) Single cell multiplexed assay for proteolytic activity using droplet microfluidics. Biosens Bioelectron 81:408–414. https://doi.org/10.1016/j.bios.2016.03.002
94. Shahi P, Kim SC, Haliburton JR, Gartner ZJ, Abate AR (2017) Abseq: Ultrahigh-throughput single cell protein profiling with droplet microfluidic barcoding. Sci Rep 7:44447. https://doi.org/10.1038/srep44447
95. Love JC, Ronan JL, Grotenbreg GM, van der Veen AG, Ploegh HL (2006) A microengraving method for rapid selection of single cells producing antigen-specific antibodies. Nat Biotechnol 24(6):703–707. https://doi.org/10.1038/nbt1210
96. Bradshaw EM, Kent SC, Tripuraneni V, Orban T, Ploegh HL, Hafler DA, Love JC (2008) Concurrent detection of secreted products from human lymphocytes by microengraving: cytokines and antigen-reactive antibodies. Clin Immunol 129(1):10–18. https://doi.org/10.1016/j.clim.2008.06.009
97. Song Q, Han Q, Bradshaw EM, Kent SC, Raddassi K, Nilsson B, Nepom GT, Hafler DA, Love JC (2010) On-chip activation and subsequent detection of individual antigen-specific T cells. Anal Chem 82(2):473–477. https://doi.org/10.1021/ac9024363
98. Choi J, Love KR, Gong Y, Gierahn TM, Love JC (2011) Immuno-hybridization chain reaction for enhancing detection of individual cytokine-secreting human peripheral mononuclear cells. Anal Chem 83 (17):6890–6895. https://doi.org/10.1021/ac2013916
99. Panagiotou V, Love KR, Jiang B, Nett J, Stadheim T, Love JC (2011) Generation and screening of Pichia pastoris strains with enhanced protein production by use of microengraving. Appl Environ Microbiol 77 (9):3154–3156. https://doi.org/10.1128/aem.00104-11

100. Varadarajan N, Julg B, Yamanaka YJ, Chen H, Ogunniyi AO, McAndrew E, Porter LC, Piechocka-Trocha A, Hill BJ, Douek DC, Pereyra F, Walker BD, Love JC (2011) A high-throughput single-cell analysis of human CD8(+) T cell functions reveals discordance for cytokine secretion and cytolysis. J Clin Invest 121(11):4322–4331. https://doi.org/10.1172/JCI58653
101. Han Q, Bagheri N, Bradshaw EM, Hafler DA, Lauffenburger DA, Love JC (2012) Polyfunctional responses by human T cells result from sequential release of cytokines. Proc Natl Acad Sci U S A 109(5):1607–1612. https://doi.org/10.1073/pnas.1117194109
102. Varadarajan N, Kwon DS, Law KM, Ogunniyi AO, Anahtar MN, Richter JM, Walker BD, Love JC (2012) Rapid, efficient functional characterization and recovery of HIV-specific human CD8+ T cells using microengraving. Proc Natl Acad Sci U S A 109(10):3885–3890. https://doi.org/10.1073/pnas.1111205109
103. Yamanaka YJ, Berger CT, Sips M, Cheney PC, Alter G, Love JC (2012) Single-cell analysis of the dynamics and functional outcomes of interactions between human natural killer cells and target cells. Integr Biol 4(10):1175–1184. https://doi.org/10.1039/c2ib20167d
104. Yamanaka YJ, Szeto GL, Gierahn TM, Forcier TL, Benedict KF, Brefo MS, Lauffenburger DA, Irvine DJ, Love JC (2012) Cellular barcodes for efficiently profiling single-cell secretory responses by microengraving. Anal Chem 84(24):10531–10536. https://doi.org/10.1021/ac302264q
105. Nguyen CQ, Ogunniyi AO, Karabiyik A, Love JC (2013) Single-cell analysis reveals isotype-specific autoreactive B cell repertoires in Sjogren's syndrome. PLoS One 8(3): e58127. https://doi.org/10.1371/journal.pone.0058127
106. Torres AJ, Contento RL, Gordo S, Wucherpfennig KW, Love JC (2013) Functional single-cell analysis of T-cell activation by supported lipid bilayer-tethered ligands on arrays of nanowells. Lab Chip 13(1):90–99. https://doi.org/10.1039/c2lc40869d
107. Torres AJ, Hill AS, Love JC (2014) Nanowell-based immunoassays for measuring single-cell secretion: characterization of transport and surface binding. Anal Chem 86(23):11562–11569. https://doi.org/10.1021/ac4030297
108. Fan R, Vermesh O, Srivastava A, Yen BK, Qin L, Ahmad H, Kwong GA, Liu CC, Gould J, Hood L, Heath JR (2008) Integrated barcode chips for rapid, multiplexed analysis of proteins in microliter quantities of blood. Nat Biotechnol 26(12):1373–1378. https://doi.org/10.1038/nbt.1507
109. Shin YS, Ahmad H, Shi Q, Kim H, Pascal TA, Fan R, Goddard WA 3rd, Heath JR (2010) Chemistries for patterning robust DNA microbarcodes enable multiplex assays of cytoplasm proteins from single cancer cells. Chemphyschem 11(14):3063–3069. https://doi.org/10.1002/cphc.201000528
110. Ahmad H, Sutherland A, Shin YS, Hwang K, Qin L, Krom R-J, Heath JR (2011) A robotics platform for automated batch fabrication of high density, microfluidics-based DNA microarrays, with applications to single cell, multiplex assays of secreted proteins. Rev Sci Instrum 82(9):094301. https://doi.org/10.1063/1.3636077
111. Ma C, Fan R, Ahmad H, Shi Q, Comin-Anduix B, Chodon T, Koya RC, Liu CC, Kwong GA, Radu CG, Ribas A, Heath JR (2011) A clinical microchip for evaluation of single immune cells reveals high functional heterogeneity in phenotypically similar T cells. Nat Med 17(6):738–743. https://doi.org/10.1038/nm.2375
112. Shin YS, Remacle F, Fan R, Hwang K, Wei W, Ahmad H, Levine RD, Heath JR (2011) Protein signaling networks from single cell fluctuations and information theory profiling. Biophys J 100(10):2378–2386. https://doi.org/10.1016/j.bpj.2011.04.025
113. Vermesh U, Vermesh O, Wang J, Kwong GA, Ma C, Hwang K, Heath JR (2011) High-density, multiplexed patterning of cells at single-cell resolution for tissue engineering and other applications. Angew Chem 123(32):7516–7518. https://doi.org/10.1002/ange.201102249
114. Shi Q, Qin L, Wei W, Geng F, Fan R, Shin YS, Guo D, Hood L, Mischel PS, Heath JR (2012) Single-cell proteomic chip for profiling intracellular signaling pathways in single tumor cells. Proc Natl Acad Sci U S A 109(2):419–424. https://doi.org/10.1073/pnas.1110865109
115. Wang J, Tham D, Wei W, Shin YS, Ma C, Ahmad H, Shi Q, Yu J, Levine RD, Heath JR (2012) Quantitating cell-cell interaction functions with applications to glioblastoma multiforme cancer cells. Nano Lett 12(12):6101–6106. https://doi.org/10.1021/nl302748q
116. Kwak M, Mu L, Lu Y, Chen JJ, Wu Y, Brower K, Fan R (2013) Single-cell protein

secretomic signatures as potential correlates to tumor cell lineage evolution and cell-cell interaction. Front Oncol 3. https://doi.org/10.3389/fonc.2013.00010
117. Lu Y, Chen JJ, Mu L, Xue Q, Wu Y, Wu PH, Li J, Vortmeyer AO, Miller-Jensen K, Wirtz D, Fan R (2013) High-throughput secretomic analysis of single cells to assess functional cellular heterogeneity. Anal Chem 85(4):2548–2556. https://doi.org/10.1021/ac400082e
118. Ma C, Cheung AF, Chodon T, Koya RC, Wu Z, Ng C, Avramis E, Cochran AJ, Witte ON, Baltimore D, Chmielowski B, Economou JS, Comin-Anduix B, Ribas A, Heath JR (2013) Multifunctional T-cell analyses to study response and progression in adoptive cell transfer immunotherapy. Cancer Discov 3(4):418–429. https://doi.org/10.1158/2159-8290.CD-12-0383
119. Wei W, Shi Q, Remacle F, Qin L, Shackelford DB, Shin YS, Mischel PS, Levine RD, Heath JR (2013) Hypoxia induces a phase transition within a kinase signaling network in cancer cells. Proc Natl Acad Sci U S A 110(15): E1352–E1360. https://doi.org/10.1073/pnas.1303060110
120. Deng Y, Zhang Y, Sun S, Wang Z, Wang M, Yu B, Czajkowsky DM, Liu B, Li Y, Wei W, Shi Q (2014) An integrated microfluidic chip system for single-cell secretion profiling of rare circulating tumor cells. Sci Rep 4. https://doi.org/10.1038/srep07499
121. Elitas M, Brower K, Lu Y, Chen JJ, Fan R (2014) A microchip platform for interrogating tumor-macrophage paracrine signaling at the single-cell level. Lab Chip 14 (18):3582–3588. https://doi.org/10.1039/C4LC00676C
122. Lu Y, Xue Q, Eisele MR, Sulistijo ES, Brower K, Han L, Amirel AD, Pe'er D, Miller-Jensen K, Fan R (2015) Highly multiplexed profiling of single-cell effector functions reveals deep functional heterogeneity in response to pathogenic ligands. Proc Natl Acad Sci U S A 112(7):E607–E615. https://doi.org/10.1073/pnas.1416756112
123. Xue M, Wei W, Su Y, Kim J, Shin YS, Mai WX, Nathanson DA, Heath JR (2015) Chemical methods for the simultaneous quantitation of metabolites and proteins from single cells. J Am Chem Soc 137(12):4066–4069. https://doi.org/10.1021/jacs.5b00944
124. Zhang Y, Tang Y, Sun S, Wang Z, Wu W, Zhao X, Czajkowsky DM, Li Y, Tian J, Xu L, Wei W, Deng Y, Shi Q (2015) Single-cell codetection of metabolic activity, intracellular functional proteins, and genetic mutations from rare circulating tumor cells. Anal Chem 87(19):9761–9768. https://doi.org/10.1021/acs.analchem.5b01901
125. Ramirez L, Herschkowitz JI, Wang J (2016) Stand-sit microchip for high-throughput, multiplexed analysis of single cancer cells. Sci Rep 6:32505. https://doi.org/10.1038/srep32505
126. Yang L, Wang Z, Deng Y, Li Y, Wei W, Shi Q (2016) Single-cell, multiplexed protein detection of rare tumor cells based on a beads-on-barcode antibody microarray. Anal Chem 88 (22):11077–11083. https://doi.org/10.1021/acs.analchem.6b03086

Check for updates

Chapter 18

Data Analysis in Single-Cell Transcriptome Sequencing

Shan Gao

Abstract

Single-cell transcriptome sequencing, often referred to as single-cell RNA sequencing (scRNA-seq), is used to measure gene expression at the single-cell level and provides a higher resolution of cellular differences than bulk RNA-seq. With more detailed and accurate information, scRNA-seq will greatly promote the understanding of cell functions, disease progression, and treatment response. Although the scRNA-seq experimental protocols have been improved very quickly, many challenges in the scRNA-seq data analysis still need to be overcome. In this chapter, we focus on the introduction and discussion of the research status in the field of scRNA-seq data normalization and cluster analysis, which are the two most important challenges in the scRNA-seq data analysis. Particularly, we present a protocol to discover and validate cancer stem cells (CSCs) using scRNA-seq. Suggestions have also been made to help researchers rationally design their scRNA-seq experiments and data analysis in their future studies.

Key words scRNA-seq, Single-cell transcriptome sequencing, Normalization, Cluster analysis

1 Introduction

Single-cell transcriptome sequencing is used to simultaneously measure the expression levels of genes from a single cell and provides a higher resolution of cellular differences than bulk RNA-seq, which can only produce an expression value for each gene by averaging its expression levels across a large population of cells. Single-cell transcriptome sequencing is also referred to as single-cell RNA sequencing (scRNA-seq) that often uses the next-generation sequencing (NGS) technologies, although it can also use the single molecule sequencing (SMS) technologies [1]. Like bulk RNA-seq, a complete scRNA-seq procedure contains RNA extraction, cDNA amplification, sequencing library preparation, sequencing, and data analysis. In addition, scRNA-seq needs isolation of single cells by manual fluorescence-activated cell sorting (FACS) or by using a microfluidics-based system.

To ensure that scRNA-seq data are fully exploited and interpreted correctly, it is crucial to apply appropriate computational and statistical methods in the data analysis [2]. The scRNA-seq data

Tao Huang (ed.), *Computational Systems Biology: Methods and Protocols*, Methods in Molecular Biology, vol. 1754, https://doi.org/10.1007/978-1-4939-7717-8_18,

analysis includes raw data cleaning with quality control [3], read alignment, generation of read counts, normalization, data filtering with quality control, and downstream analyses. Currently, a major and popular application of downstream analyses is to identify cell types or states using cluster analysis. Although many existing tools for processing bulk RNA-seq data can be used to process scRNA-seq data with or without modification, scRNA-seq data analysis poses several unique computational challenges that necessitate the development of entirely new analytical methods.

In this chapter, we focus on the introduction and discussion of the research status in the field of scRNA-seq data normalization (Subheading 3) and cluster analysis (Subheading 5), which are the two most important challenges in the scRNA-seq data analysis. We also present a schema to generalize four fundamental problems (Subheading 4). Preliminary results from our previous studies of these problems are provided to give directions for researchers in their future studies. Particularly, we present a protocol to discover and validate cancer stem cells (CSCs), which was first implemented by Lin Liu et al. using a colon cancer scRNA-seq dataset (Subheading 2).

2 Experiment Design and Data Quality Control

Six commonly used scRNA-seq protocols are CEL-seq2, Drop-seq, MARS-seq, SCRB-seq, Smart-seq, and Smart-seq2 (Table 1), the performances of which have been evaluated in a comparative study [4]. These performances included sensitivity (i.e., the probability to capture and convert a particular mRNA transcript present in a single cell into a cDNA molecule present in the library), accuracy (i.e., how well the read quantification corresponds to the actual concentration of mRNA), precision (i.e., the technical variation of the quantification), cost, etc. The authors of the comparative study concluded that Smart-seq2 was the most sensitive and accurate protocol with a similar cost efficiency to five other protocols. In this chapter, we demonstrate all the research results from our previous studies using a colon cancer scRNA-seq dataset (SRA: SRP113436) provided by Lin Liu et al. This dataset includes 831 - single-cell samples and 18 bulk samples using the Smart-seq2 scRNA-seq protocol. The 831 single-cell samples are 814 single cells from colon tumor tissues and 17 single cells from distal tissues (>10 cm) as control. The 18 bulk samples are nine samples from colon tumor tissues and nine samples from distal tissues.

Besides six protocols, the scRNA-seq experiment design needs to consider other factors such as sequencing length and depth. The sequence length determines the alignment quality and then affects the accuracy of quantitative analysis. In addition, paired-end (PE) reads have advantages over single-end (SE) reads for genome

Table 1
Overview of scRNA-seq protocols [4]

Method	CEL-seq2/C1	Drop-seq	MARS-seq	SCRB-seq	Smart-seq/C1	Smart-seq2
Single-cell isolation	Automated in the C1 system	Droplets	FACS	FACS	Automated in the C1 system	FACS
ERCC Spike-ins	Yes	No	Yes	Yes	Yes	Yes
UMI	6 bp	8 bp	8 bp	10 bp	No	No
Full-length coverage	No	No	No	No	Yes	Yes
First strand synthesis	oligo-dT	oligo-dT	oligo-dT	oligo-dT	oligo-dT	oligo-dT
Second strand synthesis	RNAseH/ DNA Pol	Template switching	RNAseH/ DNA Pol	Template switching	Template switching	Template switching
Amplification	IVT	PCR	IVT	PCR	PCR	PCR
Imaging of cells possible	Yes	No	No	No	Yes	No
Protocol usable for bulk	Yes	No	Yes	Yes	Yes	Yes
Sequencing	Paired-end	Paired-end	Paired-end	Paired-end	Single-end	Single-end
Library cost/ cell	~9.5€	~0.1€	~1.3€	~2€	~25€	~3/30[a]

[a]In-house produced Tn5/commercial Tn5

alignment, which affects the accuracy of quantitative analysis. The sequencing depth determines the number of detected genes, but it cannot yield additional information over a threshold. One previous study showed that the threshold of sequencing depth could be one million reads [4], but this was not supported by a recent study. In that study, Lin Liu et al. analyzed a colon cancer scRNA-seq dataset (SRA: SRP113436), and the results showed that the one-million-read depth was not enough to acquire all the possible transcripts. The one-million-read depth had been determined as the threshold due to not considering all kinds of noncoding RNAs (ncRNAs). For example, all the previous studies did not consider two novel long ncRNAs (lncRNAs) discovered in human mitochondrial DNA [5]. Longer (e.g., 125 or 150 bp) and PE reads improve the alignment quality and the accuracy of quantitative analysis. However, most of the existing scRNA-seq datasets contain shorter (e.g., 50 bp) or SE reads. In addition, the existing scRNA-seq datasets are

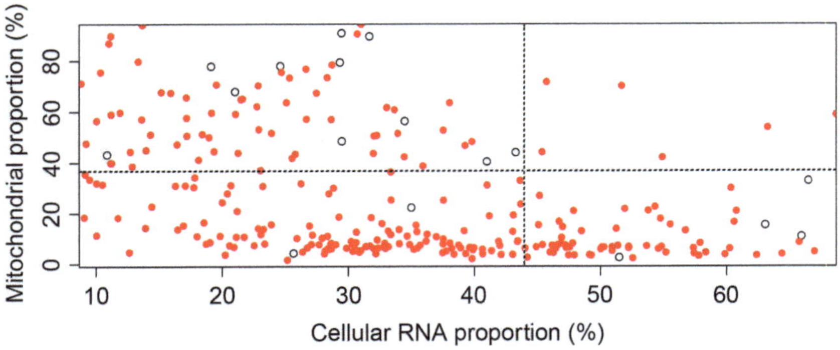

Fig. 1 Total, cellular, nuclear, and mitochondrial RNA. Library size represents total RNA including ERCC RNA and cellular RNA. The latter includes nuclear RNA and mitochondrial RNA. Cellular RNA represents the total count of reads aligned to the nuclear genome (nuclear RNA) and mitochondrial genome (mitochondrial RNA). Cellular RNA proportion is the proportion of cellular RNA to library size. Mitochondrial proportion is the proportion of mitochondrial RNA to cellular RNA. Red solid circles represent single cells from tumor tissues. Hollow circles represent single cells from control tissues. Single cells (in the upper right rectangular) containing high-content cellular RNA and mitochondrial RNA were not likely to undergo apoptosis

from experiments using different experimental parameters (e.g., sequencing length and depth) or quality control criteria. This hinders the integration of these datasets for data mining or knowledge discovery.

To further control data quality, both sample reduction and feature reduction (Subheading 4) need to be performed typically after the normalization process. One commonly used criterion is to remove samples with library sizes less than a threshold. Another criterion is to remove samples containing high-content mitochondrial RNA (usually 30% or above). These samples are usually taken as cells undergoing apoptosis [6], which was not supported by a recent study. In that study, Lin Liu et al. analyzed a colon cancer scRNA-seq dataset (SRA: SRP113436) and discovered that several cells containing high-content mitochondrial RNA were not likely to undergo apoptosis (Fig. 1). These cells were identified to be intact by microexamination, and two of them were CD133-positive (CD133+). The CD133 protein is often used as a marker for CSCs.

3 Normalization Methods

Gene expression raw data from high-throughput technologies (e.g., microarray or RNA-seq) must be normalized to remove technical variation so that meaningful biological comparisons can be made. Currently, both bulk RNA-seq and scRNA-seq only consider to remove the technical variation caused by RNA capture efficiency, cDNA amplification bias, sequencing depth, batch effect, etc. However, it is also necessary to remove uninterested biological variation, which could be confounded with biological variation of interest.

A few researchers have started to consider the removal of some kinds of uninterested biological variation in the scRNA-seq data. For example, the cell-to-cell heterogeneity in gene expression can be caused by stage differences of the cell cycle. A recent study introduced a latent-variable model based on Gaussian processes to account for variation caused by stage differences of the cell cycle [2].

Data normalization is essential as it determines the validity of its downstream analyses. Currently, all the methods are used to normalize a raw gene expression matrix by the multiplication of a factor to each column of it and produce a normalized gene expression matrix (Fig. 2a). This factor is named as normalization factor, or

a

$$\begin{pmatrix} x_{11} & \cdots x_{1j} & \cdots & x_{1n} \\ \vdots & \ddots \ \vdots & & \vdots \\ x_{i1} & \cdots x_{ij} & \cdots & x_{in} \\ \vdots & \vdots & \ddots & \vdots \\ x_{m1} & \cdots x_{mj} & \cdots & x_{mn} \end{pmatrix} \times \begin{pmatrix} f_1 \\ \vdots \\ f_j \\ \vdots \\ f_n \end{pmatrix}$$

b

DESeq: $$1/f_j = exp\left[\underset{i}{\text{median}}\left(\log x_{ij} - \frac{1}{n}\sum_{k=1}^{n}\log x_{ik}\right)\right]$$

Library size: $$s = N_j$$

$$1/f_j = s \, / \, exp\left[\frac{1}{n}\sum_{j=1}^{n}\log(s)\right]$$

RLE: $$s = \underset{i}{\text{median}}\left[x_{ij}/exp\left(\frac{1}{n}\sum_{k=1}^{n}\log x_{ik}\right)\right]/N_j$$

$$10^6 \times (1/f_j) = N_j \times s \, / \, exp\left[\frac{1}{n}\sum_{j=1}^{n}\log(s)\right]$$

Upperquartile: $$s = \underset{i}{Q_{75}}\left(\frac{x_{ij}}{N_j}\right)$$

$$10^6 \times (1/f_j) = N_j \times s \, / \, exp\left[\frac{1}{n}\sum_{j=1}^{n}\log(s)\right]$$

Fig. 2 The commonly used normalization methods. (**a**) A raw gene expression matrix can be transformed into a normalized gene expression matrix by the multiplication of a factor f_j to each column. Each column represents the expression levels of all genes from a cell, and each row represents the expression levels of a gene across all samples. (**b**). N_j represents the library size of the jth sample. Q_{75} represents to obtain the third quartile (75%) of all the gene expression proportions in the *j*th sample. The library size method, RLE, upper quartile, and DESeq are described except TMM. TMM, RLE, and upper quartile have been used to calculate CPM (counts per million) in the bioconductor package edgeR [8] for R environment scaling factor in different contexts. Basically, two classes of methods are available to calculate normalization factors. They are the control-based normalization and the average-bulk normalization. The former class of methods assumes the total expression level summed over a small group of genes is approximately the same across all the samples. The latter class of methods assumes most of genes are not differentially expressed (DE) genes across all the samples. The control-based normalization uses RNA from a group of internal control genes or external spike-in RNA. The commonly used internal control genes are housekeeping genes, and spike-in RNA usually are artificial RNA added to cell lysate. Since internal control genes and spike-in RNA may not be present in some data, the average-bulk normalization is more commonly used for their universality. Five average-bulk normalization methods designed to normalize bulk RNA-seq data are library size, trimmed mean of M values (TMM), relative log expression (RLE), upper quartile, and median of the ratios of observed counts that is also referred to as the DESeq method (Fig. 2b). The DESeq method has been included into the bioconductor package DESeq [7] for R environment. TMM, RLE, and upper quartile have been included into the bioconductor package edgeR [8] for R environment.

Although many methods have been developed and improved, the solution of RNA-seq data normalization is still unsatisfactory. Both the control-based normalization and the average-bulk normalization depend on their assumptions, which cannot be directly validated by experiments. As for internal control genes, the existence of housekeeping genes has been investigated in many previous studies, but none of them sampled human tissues completely. By the integration of these results to remove false-positive genes due to inadequate sampling, Zhang et al. only found 1 common gene across 15 examined housekeeping gene datasets comprised of 187 different tissue and cell types [9]. But the shortcoming of this study is each dataset was normalized based on the highest gene expression level for comparison. It is a logical paradox to use other normalization methods to examine housekeeping genes for their usability in data normalization. The commonly used spike-in RNA is the External RNA Control Consortium (ERCC) RNA set consisting of 92 polyadenylated transcripts with short 3′ polyA tails but without 5′ caps [10]. They are designed to have a wide range of sequence lengths (273–2022 nt) and GC-content percentages (30.79–52.69%). The same quantity of ERCC RNA should be spiked into each sample prior to RNA reverse transcription. Risso et al. evaluated the performance of the ERCC method and concluded that the ERCC method was not reliable enough to be used in standard global scaling or regression-based normalization procedures [11]. Although Risso et al. investigated the ERCC method in two very different datasets, the measures (PCA plot, RLE boxplot, and MA plot) to evaluate the performance were not convictive. The average-bulk normalization assumes that the total amount of RNA in each sample is approximately the same, and most of gene expression changes are less than twofold. However, Loven et al. found that cells with high levels of c-Myc could amplify their gene expression programs, producing two to three times more total RNA and generating cells that were larger than their low-Myc counterparts [12]. In that study, it was recommended that the spike-in normalization should be used as the default standard for all gene expression studies.

The spike-in normalization methods for scRNA-seq data are typically using ERCC RNA or unique molecular identifiers (UMIs) [13], while the average-bulk normalization methods are difficult to be used to process scRNA-seq data due to a high frequency of zeroes. Lun et al. assessed the suitability of three average-bulk normalization methods (library size, TMM, and DESeq) for normalizing scRNA-seq data by simulation [14]. As a result, they introduced a new method using the pooled size factors and claimed that their method outperformed the library size method, TMM, and DESeq. However, this new method had some new assumptions, and the parameters of it need to be arbitrarily set to pool cells of similar library sizes in each group. In addition, they claimed that the pooled size factors were closest to the true factors based on the results using simulated scRNA-seq datasets. Actually, the first reason for difficulty in validating normalization methods is lack of a standard method to estimate the true factors.

Library size is not only a commonly used normalization method but also used to calculate other normalization factors (e.g., TMM, RLE and upper quartile). Library size represents total RNA including spike-in RNA and cellular RNA. The latter includes nuclear RNA and mitochondrial RNA. There are two methods to estimate the library size of one sample. The first one is using the number of all reads, which can be aligned to the spike-in sequences (e.g., ERCC RNA), nuclear genomes, and mitochondrial genomes. The second one is using read count which is usually more than read number due to multiple alignments.

To assess normalization methods using real scRNA-seq data, we compared the ERRC method with six modified average-bulk normalization methods and acquired some new insights into the scRNA-seq data normalization. In that study, 265 samples containing ERCC RNA were selected from a colon cancer scRNA-seq dataset (Subheading 2) to obtain Pearson correlation coefficients (PCCs) between the factors calculated using the ERCC method and those calculated using the library size method, pooled size factors, TMM, RLE, upper quartile, and DESeq. These normalization methods were classified into three groups by hierarchical clustering (Fig. 3). The first group included read number (library size 1), read count (library size 2), and the ERCC method. Although

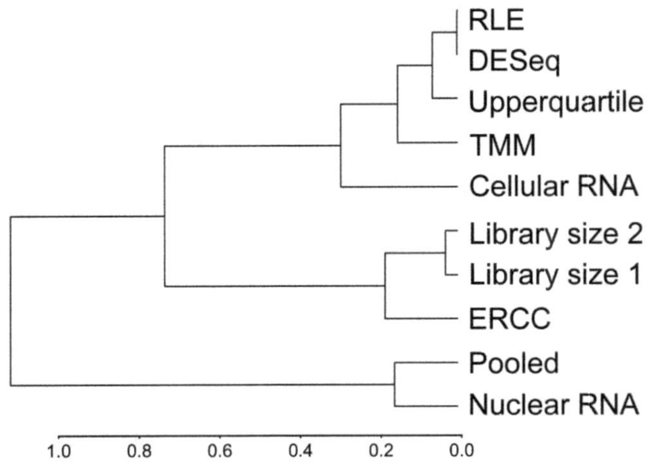

Fig. 3 Correlation of different normalization methods. The hierarchical clustering used correlation distances (1—Pearson correlation coefficients) between 265 samples containing ERCC RNA from Table 2. TMM, RLE, upper quartile, and DESeq were modified to process scRNA-seq data containing a high frequency of zeroes. Pooled represents pooled size factors [14]

the first group of methods produced very similar results, it is more convenient to use read count for data normalization by simply producing column sums of a gene expression matrix. Currently, most of software or R packages (e.g., edgeR and DESeq) produce column sums of a gene expression matrix to obtain library sizes. However, normalization factors calculated using library size (total RNA) have significant differences from those calculated using cellular RNA or nuclear RNA (Table 2). The second group of methods was closer to the first group, compared to the third group. Upper quartile from the second group reached the highest correlation with the ERCC method from the first group. In the third group, nuclear RNA was closest to pooled size factors. In conclusion, the normalization of scRNA-seq or RNA-seq data is still unsettled. Based on our studies, if ERCC data is not available, library size can be used instead of ERCC.

4 Fundamental Problems

Here, we present a schema to generalize four fundamental problems (Fig. 4). Besides data normalization and cluster analysis, sample and feature reduction are two other fundamental problems in the scRNA-seq data analysis. The normalized gene expression matrix is composed of n samples by m features, which can be genes, transcripts or exons (Fig. 2a). The scRNA-seq data from SMS (e.g., PacBio full-length transcriptome [15]) use transcripts as features, while the scRNA-seq data from NGS often use genes as features

Table 2
Comparison of scRNA-seq normalization methods

Method	Pearson correlation coefficient			
ERCC	1.00		1.00#	
Library size 1	0.82		0.83#	
Library size 2	0.89		0.89#	
Cellular RNA	0.29		0.39#	
Nuclear RNA[a]	−0.02		0.36#	
Pooled[a]	−0.09		0.17#	
	Q1 (25%)	Q2 (50%)	Q3 (75%)	
TMM[a]	0.26	0.20	0.29	0.56#
RLE[a]	0.27	0.42	0.53	0.63#
Upper quartile[a]	0.27	0.37	0.50	0.69#
DESeq	0.28	0.43	0.55	0.63#

Pearson correlation coefficients have been calculated between the factors using the ERCC method and those using each bulk normalization method. The PCC calculation used 265 samples containing ERCC RNA from a colon cancer scRNA-seq dataset (Subheading 2). Nuclear RNA represents the total count of reads aligned to the nuclear genome. Cellular RNA represents the total count of reads aligned to the nuclear and mitochondrial genome. Library size 1 and 2 represent total RNA using read number and read count, respectively (Subheading 3). Pooled represents pooled size factors [14] with the parameter sizes = c (15,40,80,130). TMM, RLE, upper quartile, and DESeq were modified to be fit for the scRNA-seq data containing a high frequency of zeroes

[a]These methods use the nuclear RNA as library size for calculation. # From 265 samples, 171 samples containing mitochondrial RNA proportion less than 30% were selected to repeat the calculation

due to their short reads. Using the human genome GRCh38 as reference, 57,992 annotated genes can be used to produce the gene expression matrix, which is a large sparse matrix. So sample reduction and then feature reduction need to be performed to remove as much noise as possible.

Basically, sample reduction is performed based on the library size or the mitochondrial RNA percentage (Subheading 2). Since the gene number and the UMI number correlates well with the library size, they are also used to fill out samples. Using the human genome GRCh38 as reference, the library size can be calculated by counting reads aligned to 57,992 annotated genes (cellular RNA) and 92 ERCC RNA sequences. Cellular RNA can be calculated by counting reads aligned to 57,955 nuclear genes (nuclear RNA) and 37 mitochondrial genes (mitochondrial RNA). In our previous studies, we found that sample reduction greatly affected the results of cluster analysis and the downstream analyses (e.g., differential expression analysis). Instead of library size and cellular RNA, nuclear RNA containing at least 100,000 read counts was suggested as a criterion to filter out samples.

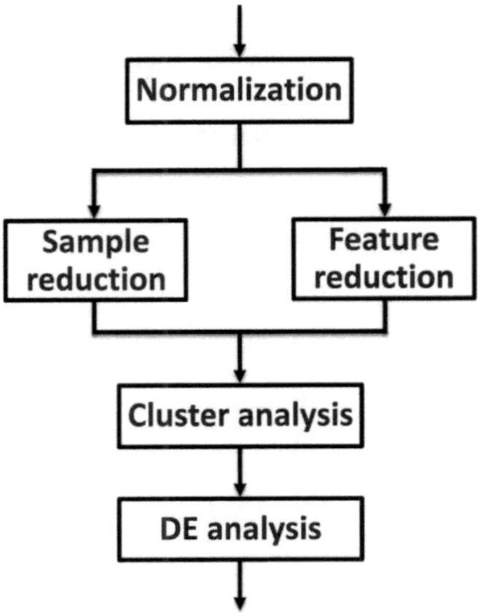

Fig. 4 Fundamental problems in the scRNA-seq data analysis. Normalization, sample reduction, feature reduction, and cluster analysis are fundamental problems in the scRNA-seq data analysis. DE analysis represents differential expression analysis, which is the most typical one of the downstream analyses

Feature reduction is performed often using two criteria. The first one is to remove genes with non-zero read counts in less than three cells. The second one is to use highly variable genes. A commonly used R package Seurat works by calculating the average expression with dispersion to select variable genes [16]. In our previous studies, we introduced another method which calculated Pearson correlation coefficients (PCCs) between normalized expression values and library sizes and used genes with PCCs less than 0.6 for cluster analysis. In addition, we found that feature reduction had little effect on the results of cluster analysis, particularly for data containing samples less than 1000. However, using variable genes improves the cluster analysis.

To overcome the extensive technical noise, principal component (PC) reduction can be used to further remove noise in the scRNA-seq data. Principal component analysis (PCA) is a commonly used dimension reduction method to produce PCs from the gene expression matrix [17]. The R package Seurat clusters cells based on their PCA scores, with each PC essentially representing a "metagene" that combines information across a correlated gene set [16]. Determining how many PCs to include (PC reduction) is therefore an important step. Based on our previous studies, PC reduction made clusters more obviously separated. Currently, PC reduction has been integrated into many software or R packages (e.g., Rt-SNE).

5 Cluster Analysis

Currently, the clustering methods in the scRNA-seq data analysis are actually using dimension reduction methods to visualize high-dimensional data in a low-dimensional space. We assume that the gene expression matrix is composed of n samples by m features ($m \gg n$) after sample and feature reduction (Fig. 2a). The dimension reduction methods transform m-dimensional points $x_1, x_2, \ldots, x_n$ into s-dimensional points $y_1, y_2, \ldots, y_n$ ($m \gg s$). By observation of samples in a two- or three-dimensional space, biologists cluster single cells into different groups. The best-known dimension reduction method is principal component analysis (PCA) [17], and other methods are independent component analysis (ICA) [18], linear discriminant analysis (LDA) [19], multidimensional scaling (MDS) [20], and t-distributed stochastic neighbor embedding (t-SNE) [21]. The most commonly used method t-SNE is a variation of the SNE method. The basic idea of t-SNE is to minimize the Kullback-Leibler divergence (Formula 1) between the joint probability p_{ij} in the high-dimensional space (Formula 2) and the joint probability q_{ij} in the low-dimensional space (Formula 4).

$$KL(P\|Q) = \sum_{i,j} p_{ij} \log \frac{p_{ij}}{q_{ij}} \tag{1}$$

$$p_{ij} = \frac{p_{j|i} + p_{i|j}}{2n} \text{ for } i \neq j \text{ and } p_{ii} = 0 \tag{2}$$

$$p_{j|i} = \frac{\exp\left(-\|x_i\text{-}x_j\|^2/2\sigma_i^2\right)}{\sum_{k \neq i} \exp\left(-\|x_i\text{-}x_k\|^2/2\sigma_i^2\right)} \tag{3}$$

$$q_{ij} = \frac{\left(1 + d_{ij}^2\right)^{-1}}{\sum_{k \neq l} \left(1 + d_{kl}^2\right)^{-1}} \text{ for } i \neq j \text{ and } q_{ii} = 0 \tag{4}$$

$$d_{ij} = \left\|y_i - y_j\right\| \tag{5}$$

$$i, j, k, l \in \{1, \ldots, n\} \tag{6}$$

Here, $x_1, x_2, \ldots, x_n$ represent columns of the gene expression matrix (Fig. 2a), and Formulas 3 and 5 use Euclidean distances. Using the gradient descent method, the solution of Formula 1 can be obtained as the final low-dimensional points $y_1, y_2, \ldots, y_n$. The t-SNE method needs two important user-defined parameters. The first one is the perplexity defined as $Perp(\mathrm{P}_i) = 2^{H(\mathrm{P}_i)}$, where $H(P_i) = -\sum_j p_{j|i} \log_2 p_{j|i}$. The perplexity can be interpreted as a

Table 3
Estimating intrinsic dimensionality of scRNA-seq data

Method	Raw	Raw Gaussian	ERCC	ERCC Gaussian	DESeq	DESeq Gaussian
EigValue	7	6	5	6	7	6
MLE	19.00	19.92	15.27	19.92	22.28	19.92
CorrDim	0.47	2.89	0.56	2.89	0.51	2.89
NearNb	0.05	0	0.03	0	0.01	0
GMST	5.21	28.01	4.71	29.78	6.25	17.36
PackingNumbers	0	0	0	0	0	0

EigValue represents eigenvalue-based estimation. MLE represents maximum likelihood estimation. CorrDim represents correlation dimension. NearNb represents nearest neighbor evaluation. GMST represents geodesic minimum spanning tree. PackingNumbers represents packing numbers. Raw represents gene expression data using read counts of nuclear RNA (Table 2). Raw Gaussian represents raw data processed by the standard Gaussianization. ERCC represents normalized data using the ERCC method. ERCC Gaussian represents ERCC-normalized data processed by the standard Gaussianization. DESeq represents normalized data using DESeq. DESeq Gaussian represents DESeq-normalized data processed by the standard Gaussianization. The standard Gaussianization was performed on each row of the gene expression matrix (Fig. 2a) by the difference of mean and divided by the standard deviation

smooth measure of the effective number of neighbors. The performance of t-SNE is fairly robust to changes in the perplexity and typical values are between 5 and 50. The second parameter is the low dimension s, which is arbitrarily set to two for convenient visualization. However, no research has been conducted to investigate how much the low dimension parameter should be set to.

We proposed that the low dimension parameter can be set as the intrinsic dimensionality of the sparse gene expression matrix. For the first time, we estimated the intrinsic dimensionality of scRNA-seq data using six methods (Table 3). They were eigenvalue-based estimation (EigValue) [17], maximum likelihood estimation (MLE) [22], correlation dimension (CorrDim) [23], nearest neighbor evaluation (NearNb) [24], geodesic minimum spanning tree (GMST) [25], and packing numbers (PackingNumbers) [26]. These methods were implemented using the MATLAB toolbox for dimensionality reduction (available at http://lvdmaaten.github.io/drtoolbox/). Although preliminary results were obtained using only a colon cancer scRNA-seq dataset (Subheading 2), it still suggested that the intrinsic dimensionality could be six estimated by EigValue. Since five other methods are parameter-dependent except EigValue, further estimation needs to be performed using adjusted parameters based on grid search, avoiding the arbitrary settings. The final intrinsic dimensionality must be determined by combined use of six or more methods.

Until this chapter was written, the most popular application of cluster analysis using scRNA-seq data was to identify cells in terms of known or novel types and then perform downstream analyses (e.g., differential expression or regulatory network analysis) based on the clustering results. Cluster analysis is also used to identify cell states, particularly their stages of the cell cycle. The identification of cells by cluster analysis needs to be validated by the known gene signatures or biomarkers that best distinguish cell types or states. These gene signatures were usually discovered using bulk data. Therefore, the cluster analysis results using scRNA-seq data can be used to confirm these gene signatures at the single-cell level. Cluster analysis, as an unsupervised method, can be used to study other biological topics by its combined use with other prior knowledge. For example, the Markov random field (MRF)-based method has been developed to cluster cells using both spatial and expression information [27].

Although scRNA-seq has been successively used to reveal the tissue heterogeneity or discover some new cell types, it has not been used to solve any basic problem or validate any hypothesis well in biological sciences. One well-known and debatable hypothesis is the existence of cancer stem cells (CSCs) that are responsible for tumor initiation and growth, possessing properties such as indefinite self-renewal, slow replication, intrinsic resistance to chemotherapy and radiotherapy, and an ability to give rise to differentiated progeny [28]. Here, we present a protocol to discover and validate CSCs using scRNA-seq. This protocol uses Smart-seq2 150 bp PE sequencing. Sample reduction removes samples with nuclear RNA containing less than 100,000 read counts, and feature reduction (not necessary) removes genes with PCCs less than 0.6 between their normalized expression values and library sizes. The ERCC-normalized data are used to produce clusters by the t-SNE method. Control samples are sequenced to identify the group of CSCs from clusters. Finally, control samples from public scRNA-seq datasets are used to validate the group of CSCs. This protocol was first applied on the colon cancer scRNA-seq dataset (Subheading 2) to identify a special group of cells comprising 4.73% (31/655) of all the single cells from tumor tissues (Fig. 5). This group did not contain any control sample, while other groups contained at least one control sample. In addition, the cells in this group were CD133-positive (CD133+). Therefore, these results suggested that this group of cells could be CSCs. Finally, more control samples from public datasets were used to validate these CSCs.

6 Discussion

Single-cell transcriptome sequencing, often referred to as single-cell RNA sequencing (scRNA-seq), is a powerful tool to investigate cell types, states, and its dynamics. The successful application of scRNA-seq needs a careful experiment design and data analysis. We suggest using Smart-seq2 protocol with PE 150 bp sequencing to

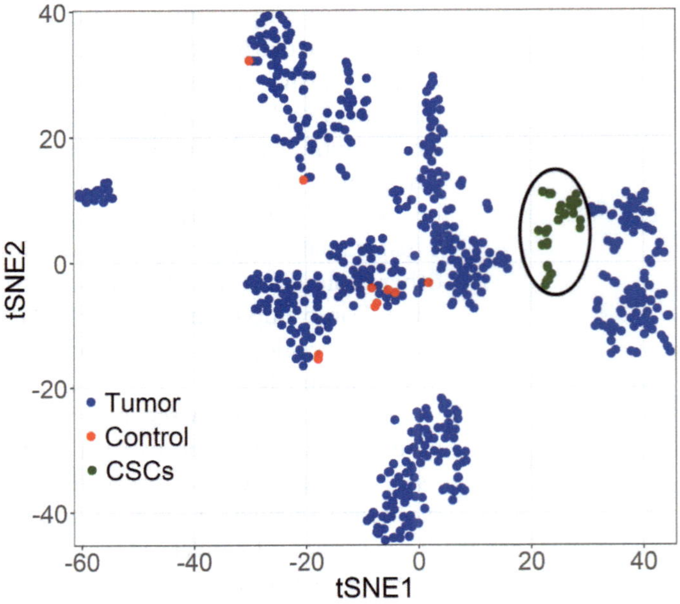

Fig. 5 Using scRNA-seq data to identify cancer stem cells. Samples containing less than 100,000 read counts of nuclear RNA were filtered out without feature selection. The ERCC-normalized gene expression matrix contained 665 samples by 57,955 nuclear genes. A total of ten single cells from distal tissues (in red color) as control and 655 single cells (in blue color) from colon tumor tissues were selected to obtain a group of suspected cancer stem cells (in the green circle). The calculation using the t-SNE method was performed with the package Rt-SNE v0.11 on the R v3.3.2 platform. Input data used Euclidean distances without PCA process. The parameters were set as (is_distance = TRUE, pca = FALSE, perplexity = 12, theta = 0.5, dims = 2, max_iter = 3000) for the t-SNE method

produce high-depth data and using the software Fastq_clean to clean raw sequenced data with quality control. The software STAR [29] needs to be used for paired-end alignment and quantification to produce the raw gene expression matrix. Sample reduction removes samples with nuclear RNA containing less than 100,000 read counts, and feature reduction (not necessary) removes genes with PCCs less than 0.6 between their normalized expression values and library sizes. Data normalization should be conducted based on evaluation [30], and cluster analysis uses t-SNE. Control samples are sequenced to identify the group of CSCs from clusters, and control samples from public scRNA-seq datasets are used to validate the group of CSCs.

Acknowledgments

I appreciate help equally from the people listed below. They are Professor Wenjun Bu; Professor Lin Liu; Ph.D. student Hua Wang; Master's student Yu Sun and Deshui Yu from College of Life Sciences, Nankai University; Professor Jishou Ruan; PhD student Zhenfeng Wu from School of Mathematical Sciences, Nankai University; and Associate Professor Weixiang Liu from Shenzhen University.

References

1. Gao S, Ou J, Xiao K (2014) R language and Bioconductor in bioinformatics applications (Chinese Edition). Tianjin Science and Technology Translation Publishing, Co. Ltd, Tianjin
2. Stegle O, Teichmann SA, Marioni JC (2015) Computational and analytical challenges in single-cell transcriptomics. Nat Rev Genet 16 (3):133–145
3. Zhang M, Sun H, Fei Z, Zhan F, Gong X, Gao S (2014) Fastq_clean: an optimized pipeline to clean the Illumina sequencing data with quality control. 2014 I.E. international conference on bioinformatics and biomedicine, pp 44–48
4. Ziegenhain C, Vieth B, Parekh S, Reinius B, Guillaumet-Adkins A, Smets M, Leonhardt H, Heyn H, Hellmann I, Enard W (2017) Comparative analysis of single-cell RNA sequencing methods. Mol Cell 65(4):631–643
5. Gao S, Tian X, Chang H, Sun Y, Wu Z, Cheng Z, Dong P, Zhao Q, Ruan J, Bu W (2017) Two novel lncRNAs discovered in human mitochondrial DNA using PacBio full-length transcriptome data. Mitochondrion. https://doi.org/10.1016/j.mito.2017.08.002
6. Ilicic T, Kim JK, Kolodziejczyk AA, Bagger FO, Mccarthy DJ, Marioni JC, Teichmann SA (2016) Classification of low quality cells from single-cell RNA-seq data. Genome Biol 17(1):29
7. Anders S, Huber W (2010) Differential expression analysis for sequence count data. Genome Biol 11(10):R106
8. Robinson MD, Mccarthy DJ, Smyth GK (2010) edgeR: a Bioconductor package for differential expression analysis of digital gene expression data. Bioinformatics 26(1):139–140
9. Zhang Y, Li D, Sun B (2015) Do housekeeping genes exist? PLoS One 10(5):e0123691
10. Jiang L, Schlesinger F, Davis CA, Zhang Y, Li R, Salit M, Gingeras TR, Oliver B (2011) Synthetic spike-in standards for RNA-seq experiments. Genome Res 21(9):1543–1551
11. Risso D, Ngai J, Speed TP, Dudoit S (2014) Normalization of RNA-seq data using factor analysis of control genes or samples. Nat Biotechnol 32(9):896–902
12. Lovén J, Orlando DA, Sigova AA, Lin CY, Rahl PB, Burge CB, Levens DL, Lee TI, Young RA (2012) Revisiting global gene expression analysis. Cell 151(3):476–482
13. Islam S, Zeisel A, Joost S, La MG, Zajac P, Kasper M, Lönnerberg P, Linnarsson S (2014) Quantitative single-cell RNA-seq with unique molecular identifiers. Nat Methods 11 (2):163–166
14. Lun AT, Bach K, Marioni JC (2016) Pooling across cells to normalize single-cell RNA sequencing data with many zero counts. Genome Biol 17(1):75
15. Ren Y, Zhang J, Sun Y, Wu Z, Ruan J, He B, Liu G, Gao S, Bu W (2016) Full-length transcriptome sequencing on PacBio platform (in Chinese). Chin Sci Bull 11(61):1250–1254
16. Satija R, Farrell JA, Gennert D, Schier AF, Regev A (2015) Spatial reconstruction of single-cell gene expression data. Nat Biotechnol 33(5):495–502
17. Wold S, Esbensen K, Geladi P (1987) Principal component analysis. Chemometr Intell Lab Syst 2(1–3):37–52
18. Hyvarinen A, Oja E (2000) Independent component analysis: algorithms and applications. Neural Netw 13(4–5):411–430
19. Balakrishnama S, Ganapathiraju A (1998) Linear discriminant analysis – a brief tutorial. Procof Intjoint Confon Neural Networks 3 (94):387–391
20. Carroll JD, Arabie P (1980) Multidimensional scaling. Annu Rev Psychol 31(31):607–649
21. Maaten LVD, Hinton G (2008) Viualizing data using t-SNE. J Mach Learn Res 9 (2605):2579–2605

22. Levina E, Bickel PJ (2004) Maximum likelihood estimation of intrinsic dimension. Adv Neural Inf Proces Syst 17:777–784
23. Camastra F, Vinciarelli A (2002) Estimating the intrinsic dimension of data with a fractal-based method. IEEE Trans Pattern Anal Mach Intell 24(10):1404–1407
24. Pettis KW, Bailey TA, Jain AK, Dubes RC (1979) An intrinsic dimensionality estimator from near-neighbor information. IEEE Trans Pattern Anal Mach Intell PAMI-1(1):25–37
25. Costa JA, Hero AO (2004) Geodesic entropic graphs for dimension and entropy estimation in manifold learning. IEEE Trans Signal Process 52(8):2210–2221
26. Kfgl B (2002) Intrinsic dimension estimation using packing numbers. Adv Neural Inform Process Syst NIPS-02:697–704
27. Pettit JB, Tomer R, Achim K, Richardson S, Azizi L, Marioni J (2014) Identifying cell types from spatially referenced single-cell expression datasets. PLoS Comput Biol 10(9):e1003824
28. O'Flaherty JD, Barr M, Fennell D, Richard D, Reynolds J, O'Leary J, O'Byrne K (2012) The cancer stem-cell hypothesis: its emerging role in lung cancer biology and its relevance for future therapy. J Thorac Oncol 7 (12):1880–1890
29. Dobin A, Davis CA, Schlesinger F, Drenkow J, Zaleski C, Jha S, Batut P, Chaisson M, Gingeras TR (2013) Star: ultrafast universal rna-seq aligner. Bioinformatics 29(1):15–21
30. Wu Z, Liu W, Jin X, Yu D, Wang H, Liu L, Ruan J, Gao S (2018) NormExpression: an R package to normalize gene expression data using evaluated methods. bioRxiv. https://doi.org/10.1101/251140

Chapter 19

Applications of Single-Cell Sequencing for Multiomics

Yungang Xu and Xiaobo Zhou

Abstract

Single-cell sequencing interrogates the sequence or chromatin information from individual cells with advanced next-generation sequencing technologies. It provides a higher resolution of cellular differences and a better understanding of the underlying genetic and epigenetic mechanisms of an individual cell in the context of its survival and adaptation to microenvironment. However, it is more challenging to perform single-cell sequencing and downstream data analysis, owing to the minimal amount of starting materials, sample loss, and contamination. In addition, due to the picogram level of the amount of nucleic acids used, heavy amplification is often needed during sample preparation of single-cell sequencing, resulting in the uneven coverage, noise, and inaccurate quantification of sequencing data. All these unique properties raise challenges in and thus high demands for computational methods that specifically fit single-cell sequencing data. We here comprehensively survey the current strategies and challenges for multiple single-cell sequencing, including single-cell transcriptome, genome, and epigenome, beginning with a brief introduction to multiple sequencing techniques for single cells.

Key words Single-cell sequencing, Single-cell transcriptome, Genome, Epigenome, Multiomics, Allele-specific expression, Single nucleotide variant calling, Clonal structure

1 Multiple Sequencing Protocols for Single Cells

Next-generation sequencing (NGS) technologies allow increasing ability to profile transcriptome [1], genome [2], epigenome [3], and even protein [4, 5] of single cells sampled from heterogeneous cell types and different cellular states. Multiple molecular measurements from single cells enable normal development and disease processes to be studied and dissected at the resolution of individual cell. In this section, we first review the developments, opportunities, and challenges of multiple single-cell sequencing technologies, which have enabled comprehensively understanding and modeling cellular processes regarding development and diseases. Like conventional NGS experiments, the protocols of a single-cell sequencing generally contain the following steps: isolation of single cell, nucleic

Tao Huang (ed.), *Computational Systems Biology: Methods and Protocols*, Methods in Molecular Biology, vol. 1754, https://doi.org/10.1007/978-1-4939-7717-8_19,

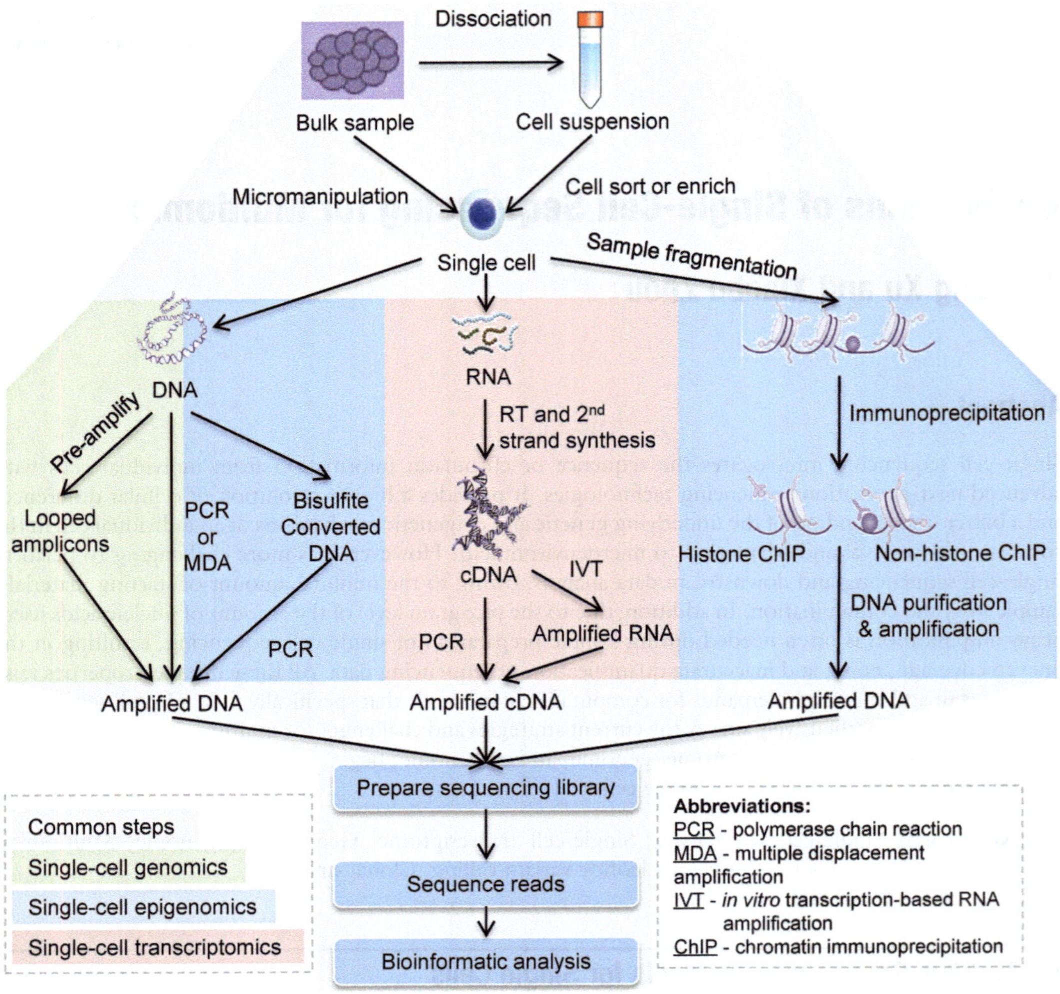

Fig. 1 Schematic overview of multiple single-cell sequencing technologies. Special steps are highlighted by different colors for multiple omics

acid extraction and amplification, sequencing library preparation, and sequencing and bioinformatic data analysis (Fig. 1).

1.1 Single-Cell Isolation

Single-cell analyses rely on various molecular measurements of a single cell. Although many technologies depending on single cells have been developed, ensuring that a sample contains only a single cell remains technically challenging. The first key step is to generate a single-cell suspension, which varies considerably between tissue types, and optimization is required to ensure analysis of a viable and unbiased cell population. When tissue complexity or handling prohibits intact cell isolation, suspensions of single nuclei can be prepared alternatively [6, 7], which have been demonstrated in (epi)genomic and transcriptomic analyses [8]. Various approaches are available for isolating single cells from a suspension, of which we briefly introduce the five most commonly used ones.

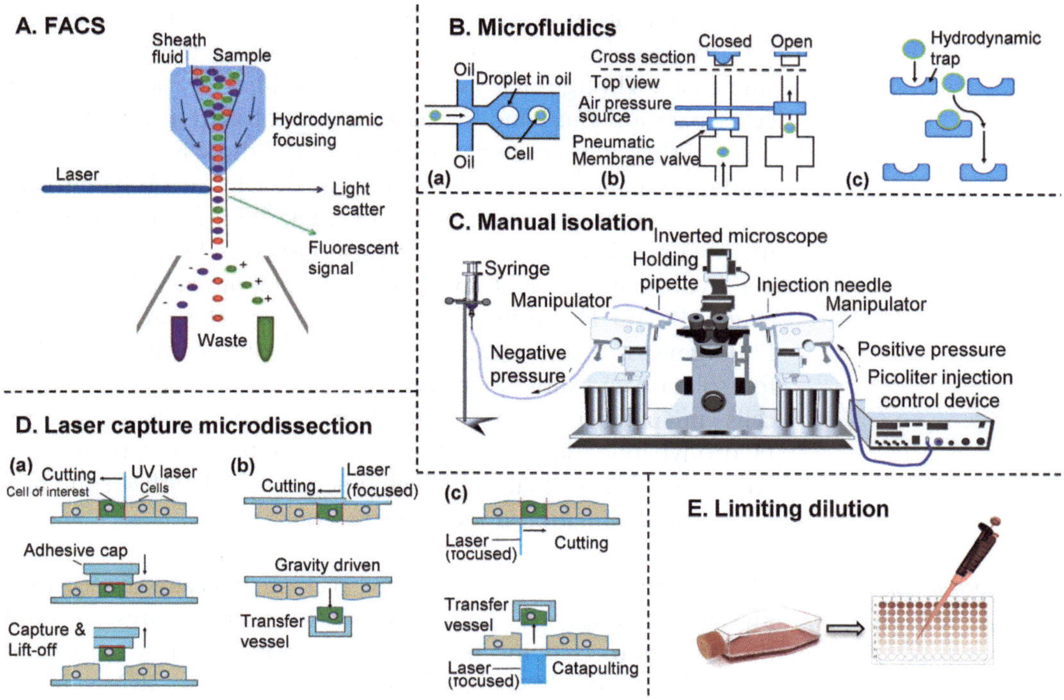

Fig. 2 Schematic overview of single-cell isolation technologies. (**A**) An example of flow cytometry, fluorescence-activated cell sorting (FACS). (**B**) Three microfluidic principles to isolate single cells. (a) An aqueous stream of cells is broken up into individual droplets in oil containing random distribution of cells; (b) pneumatic membrane valves use air pressure to close a microfluidic channel by membrane deflection. This stops the flow and can trap a cell; and (c) hydrodynamic traps are passive elements that only fit single cells and hold them at one position. (**C**) The manual isolation using micromanipulator. (**D**) Schematic view on laser capture microdissection (LCM) methods. (a) Contact-based via adhesive tapes; (b) cutting with a focused laser followed by capture with a vessel. Cutout section extracted by gravity and (c) cutting with a focused laser followed by pressure catapulting with a defocused laser pulse. (**E**) Isolation of individual cells through dilution of the cell suspension using hand pipettes or pipetting robots (adapted from [9] and Google image)

1.1.1 Flow Cytometry

Among the various types of flow cytometers, the fluorescence-activated cell sorting (FACS) systems provide the ability to isolate single cells (Fig. 2A). FACS allows phenotypically distinct cells, and even nuclei, to be sorted into user-defined vessels and lysis buffers, thus enabling diverse single-cell and single-nucleus protocols to be applied at significantly higher throughput [6]. Index sorting [10] additionally allows direct linking of a single cell's phenotype (e.g., surface marker expression, DNA content) with multiomics analysis. However, large numbers of cells are required as input, and because the platform currently offers no opportunity to visualize sorted cells, care must be taken to identify and exclude cell doublets.

1.1.2 Microfluidics

Microfluidic technologies, such as Fluidigm C1 [11], have been widely applied in single-cell omics studies. These technologies isolate single cells in individual capture sites and initiate nucleic acid amplification in nanoliter volumes (Fig. 2B). Once captured, cells can be visualized on the chip, confirming the presence of a single cell. Though many different microfluidic devices for single-cell separation and handling have been published in the literature, most of these devices use at least one of the three following microfluidic principles to isolate single cells: droplet-in-oil-based isolation as, for example, published in [12, 13] (Fig. 2B(a)); pneumatic membrane valving as, for example, published in [14] (Fig. 2B(b)); and hydrodynamic cell traps as, for example, published in [15, 16] (Fig. 2B(c)). The most advanced commercial system C1 from Fluidigm Corp., for example, exploits the third principle and allows for isolation and subsequent genetic analysis, of up to 96 individual cells in parallel (http://cn.fluidigm.com/products/c1-system). Hydrodynamic trapping can even be integrated into handheld pipettes to enable manual single-cell pipetting [17] without the need of micromanipulation under a microscope.

1.1.3 Manual Isolation

Micromanipulators for manual cell picking typically consist of an inverted microscope combined with micropipettes movable through motorized mechanical stages, notably allowing a single cell to be directly visualized during isolation (Fig. 2C). The cell sample is typically provided as suspension in a dish or well plate. Via microscope observation the operator selects a specific cell, moves the micropipette in close proximity, and aspirates the cell by applying suction to the micropipette. The aspirated liquid volume including the selected cell can be transferred to a collection vessel (e.g., a well of a well plate), where it is released by dispensation. This process is commonly performed manually. Micromanipulators enable the controlled separation of selected, living cells from suspension and even allow for isolation of prokaryotic cells [18]. When all of a small number of cells are to be analyzed—for example, daughter cells from a single-cell division—this is often the most suitable option [19]. Nevertheless, it is by necessity low throughput.

1.1.4 Laser Capture Microdissection

Laser capture microdissection (LCM, Fig. 2D) is an advanced technique to isolate individual cell or cell compartments from mostly solid tissue samples [20]. A tissue section is observed through a microscope, and the target cell or compartment is visually identified. The operator marks the section to be cut off on the display by drawing a line around it. Along this trajectory the laser cuts the tissue and the isolated cell (or compartment) is—if required—extracted. Analysis of solid tissue is of great interest when investigating heterogeneous tissue sections regarding their cellular structure as well as physiological and pathological processes [21]. In solid tumor research, linking the molecular information of individual cells to their specific location in the tissue has become an important research field. Particularly, the access to cells in situ is of interest [22]. In combination with immune histological staining, LCM is a powerful tool for solid sample analysis on the single-cell level [23]. In the past years, various applications in single-cell analysis based on LCM-extracted cells have been published: single-cell RT-PCR [24], short tandem repeat (STR) analysis in forensics [25], Western blot, and mass spectrophotometry [26].

1.1.5 Limiting Dilution

Today many laboratories and companies use hand pipettes or pipetting robots to isolate individual cells through dilution of the cell suspension (Fig. 2E). Due to the statistical distribution of the cells in the suspension, the number of cells in a highly diluted sample can be as low as one single cell per aliquot, when the suspension is split into small volumes (aliquots). This process is termed limiting dilution and is well known for decades for the production of monoclonal cell cultures [27]. Besides antibody production (as done by hybridomas), other applications such as cell-based assays, etc. also require cell populations grown from a single cell.

1.2 Sequencing Transcriptomes of Single Cells

RNA sequencing (RNA-seq) enabled transcriptomic profiling at unprecedented sensitivity and breadth, leading to the discovery of new RNA species and deepening our understanding of transcriptome dynamics [28]. In recent years, low-input RNA-seq methods have been adapted to work in single cells [29], which introduced a derived technology called single-cell RNA sequencing (scRNA-seq). scRNA-seq can quantify intrapopulation heterogeneity and enable study of cell states and transitions at very high resolution, potentially revealing cell subtypes or gene expression dynamics that are masked in bulk, population-averaged measurements [30, 31]. Over the past years, numerous scRNA-seq protocols have been developed [29, 32–45], including the widely used Smart-seq2 [37] and CEL-Seq [35]. Currently published scRNA-seq protocols all follow the same general workflow: single cells are isolated (*see* former section); cells are lysed, and the RNA is captured for reverse transcription into cDNA; and the cDNA is preamplified and then used to prepare libraries for sequencing and downstream analysis. Since the technology and protocol of each scRNA-seq approach are out of the scope of this section, readers refer to [46] for a comprehensive review of individual scRNA-seq protocols and their relative strengths and weaknesses. Moreover, two recent studies from the Enard group [47] and the Teichmann group [48] have compared tens of different protocols, which will give the readers more comprehensive insights into the scRNA-seq technologies.

1.3 Sequencing Genomes of Single Cells

Although bulk approaches for studying genetic variation have identified thousands of new unicellular species and determined genetic etiologies for thousands of human disease, most of them have been done at the level of the ecosystem or organism. We know that the cell-to-cell heterogeneity exists not only in transcriptome but also in genome, i.e., the genomes within the cells of an individual multicellular organism are not always the same. Single-cell genomics aims to provide new perspectives to our understanding of genetics by bringing the study of genomes to the cellular level. Sequencing a genome of single cells has four steps, which remain technically challenging: (1) efficient physical isolation of individual cells, (2) amplification of the genome of that single cell to acquire sufficient material for downstream analyses, (3) querying the genome in a cost-effective manner to identify variation that can test the hypotheses of the study, and (4) interpreting the data within the context of biases and error that are introduced during the first three steps. Reference 2 gives a comprehensive review of the state of this field.

1.4 Sequencing Epigenomes of Single Cells

Epigenome is an integrative collection of chemical modifications, associations, and conformations of genomic DNA sequences, including histone modifications and variations, DNA methylations, nucleosome positioning, chromatin conformations, etc. The epigenomics aims to link these with epigenetic memory, cellular identity, and tissue-specific functions. While the average epigenomic features across large cell populations have been largely characterized with the help of current techniques, the tissue complexity and cell-to-cell heterogeneity are driving the development of single-cell epigenomics. We, here, survey emerging methodologies for single-cell epigenomics, which have been comprehensively depicted in Fig. 1 of ref. 49.

1.4.1 Single-Cell DNA Methylation

DNA methylation is an epigenetic mechanism that occurs by the addition of a methyl (CH3) group to DNA, thereby often modifying the function of the genes. The most widely characterized DNA methylation process is the covalent addition of the methyl group at the 5-carbon of the cytosine of the dinucleotide CpG, resulting in 5-methylcytosine (5-mC). For DNA methylation profiling of bulk tissue or cell populations, bisulfite sequencing has enabled high-throughput interrogation of CpG modifications, by generating millions or billions of reads to accurately define the methylation state across the entire genome or within some strategic genomic regions [50]. Single-cell bisulfite sequencing improves our understanding of the methylome intrapopulation distribution by simultaneously defining the methylation states of CpGs within an entire epigenome. Three single-cell methylome assays based on bisulfite sequencing have been reported recently. The first ones include single-cell bisulfite sequencing (scBS-seq) [51] and reduced representation bisulfite sequencing (scRRBS-seq) [52]. The second one, post-bisulfite adaptor tagging (PBAT) [53], was implemented on single cells [51]. The third one implemented another variant of the post-bisulfite library construction procedure, deriving lower coverage while using lower sequencing depth [54]. Importantly, alternative approaches for characterizing DNA methylation are also applicable at the single-cell level. In the single-cell restriction analysis of methylation (SCRAM) approach [55], methylation-sensitive restriction enzymes and single-cell quantitative PCR (qPCR) are combined to facilitate profiling of methylation state across small panels of target sites.

1.4.2 Single-Cell Histone Modifications

Different from DNA methylation that can be measured directly from bisulfite-converted sequences, histone modifications are profiled indirectly using chromatin immunoprecipitation followed by sequencing (ChIP-seq). The measures of histone modifications are defined by the enrichment of reads by distinguishing statically true positive from the genomic background when sequencing binding peaks at adequate depth. It is more problematic if the same approaches are applied to single cells, since the absence of the statistical pooling disables the direct way to distinguish true positives from false positives. A similar effect may restrict the single-cell application of techniques for mapping accessibility patterns using DNase [56], micrococcal nuclease (MNase) [57], or transposases (ATAC-seq) [58]. Nevertheless, some technologies are under development to tolerate high number of false-positive readouts by multiplexing analysis of thousands of single cells [59–62]. Alternatively, pooling data of related loci (e.g., clustering of loci) can be used to derive higher-quality data on generalized single-cell epigenomic features, even if the quality of individual profiles is not optimal [63, 64]. Therefore, in summary, genome-wide single-cell histone modification or TF-binding mapping is still under development, which is far behind the single-cell RNA-seq.

1.4.3 Single-Cell Chromosomal Confirmation Capture

Chromosome conformation capture (3C) and derivative techniques [65–67], such as Hi-C, allow characterization of chromosomal topologies at different size scales: from whole chromosomes to topologically associating domains (TADs) and down to sub-megabase-scale chromosomal loops. Since it is clear as early as from microscopy that chromosome folding is highly nuclei-specific [68, 69], understanding large-scale chromosomal structure inevitably involves considerations of single-cell dynamics. The ability to scale up 3C toward the sequencing of billions of ligation products makes Hi-C on cell populations a powerful experimental tool. Similarly, Hi-C can be scaled down effectively to single nuclei, as it encodes distinct ligation events within individual nuclei if an appropriate nuclei separation and labeling scheme is used. The first development of single-cell Hi-C used manual separation and labeling of amplified ligation products from single nuclei to recover 10,000–30,000 ligation events per cell [70].

1.5 Multiomics of Single Cell

Single-cell genome, epigenome, and transcriptome sequencing methods have been separately developed. However, the measurements and analyses of just one molecular type from individual cells provide only incomplete information because a cellular state is determined by the complex interplay of multiple molecules within its genome, epigenome, transcriptome, and proteome. To more comprehensively and accurately analyze the mechanism by which these molecules regulate each other, these omic methods need to be performed in the same single cell, named multiomics. A few of such multiomics approaches [8, 71–74] have been proposed and enable the generation of mechanistic models relating (epi)genomic variation and transcript/protein expression dynamics, which in turn allow a more detailed exploration of cellular behavior in health and disease.

To date, three types of strategies have been developed that permit different combinations of genome, transcriptome, and methylome sequencing for single cells. The first two approaches allow parallel interrogation of genomes and transcriptomes for single cells, including gDNA-mRNA sequencing (DR-seq) [71] and genome and transcriptome sequencing (G&T-seq) [72, 73]. For DR-seq, genomic DNA (gDNA) and mRNA present in a single cell's lysate are preamplified simultaneously before splitting the reaction in two for parallel gDNA (using a modified multiple annealing and looping-based amplification cycles (MALBAC) approach [75]) and mRNA library preparation (using a modified CEL-seq approach [35]) and subsequent sequencing. For G&T-seq, mRNA is physically separated from gDNA using oligo-dT-coated beads to capture and isolate the polyadenylated mRNA molecules from a fully lysed single cell. The mRNA is then amplified using a modified Smart-seq2 protocol [37, 76], while gDNA can be amplified and sequenced by a variety of methods. The second strategy enables the measures of both DNA methylation and transcript abundance for the same single cells sampled from dynamic and heterogeneous cell populations. The first such method, single-cell methylome and transcriptome sequencing (scMT-seq) [74], is an extension to the G&T-seq protocol in which mRNA is captured, amplified, and sequenced as before. However, the isolated gDNA of the single cell undergoes bisulfite sequencing, rather than whole genome amplification, allowing parallel analysis of genome-wide DNA methylation and transcriptome-wide gene expression from the same single cell. The third strategy extended the second one to simultaneously probe the copy number variation, DNA methylation, and transcript levels of the same single cell, which are implemented in scTrio-seq [8]. The scTrio-seq computationally investigates both the DNA methylation and copy number states from scRRBS sequencing libraries and in parallel measures cytoplasmic transcript levels.

In summary, the integration of genomic, epigenomic, transcriptomic data are emerging as realistic prospect, benefiting from rapidly developing sequencing technology. For example, Pacific Biosciences and nanopore sequencers can interpret more than one analyte in parallel [77, 78]. Thus, we expect that future sequencing and analyzing approaches may provide the crucial advances that expand the refine single-cell multiomics to the insights into generating the comprehensive atlases of cell states and lineages for cellular systems, ranging from tissue microenvironment to whole organisms.

2 Computational Methods and Challenges for Single-Cell Transcriptomics

The development of high-throughput scRNA-seq has already led to profound insights into biology, ranging from the identification of novel cell types or subclones to global patterns of stochastic gene expression for cell-to-cell heterogeneity or reconstructing cell differentiation trajectories. Alongside the technological breakthroughs that have facilitated the large-scale generation of single-cell transcriptomic data, it is crucial to apply appropriate computational and/or statistical methods to ensure that scRNA-seq data are fully exploited and interpreted correctly. Although some tools for analyzing RNA-seq data from bulk samples can be readily applied to scRNA-seq data, many new computational strategies are required to fully exploit this data type and to enable a comprehensive yet detailed study of gene expression at the single-cell level. Generally, as the same as the bulk RNA-seq data, three types of steps are implemented to interrogate scRNA-seq data (Fig. 3): alignment and quality control (QC), normalization and quantification, and applications for biological insights. The first steps (orange) are general for any high-throughput sequencing data. Later steps (blue) require a mix of existing RNA-seq analysis methods and novel methods to address the technical difference of scRNA-seq. The biological interpretation (red) should be analyzed with methods specifically developed for scRNA-seq. In this section, we survey the various computational methods that are applied specifically to the multiple steps of scRNA-seq data analysis aiming at deciphering the transcriptomic dynamics at single-cell level (Table 1).

2.1 Read Alignment and Quality Control

A single cell possesses only a very small amount of RNA, and the sequencing reaction is limited by the amount of starting material. Therefore, scRNA-seq experiments have several unique properties including high technical noise [82], low coverage [116], and 3′ bias [117], requiring the use of methods different from bulk RNA-seq

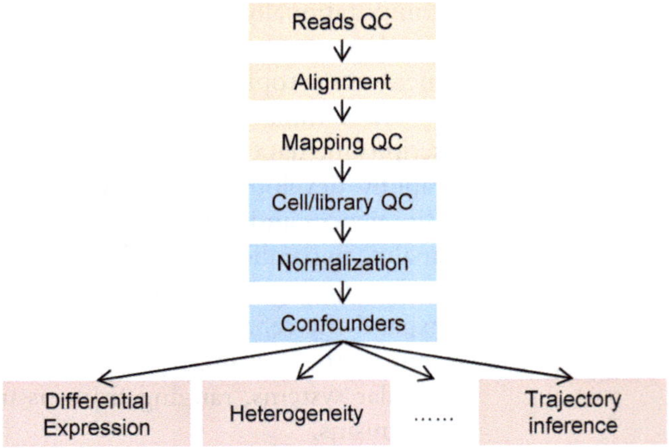

Fig. 3 Flowchart of scRNA-seq data analyses. The first steps (orange) are general for any high-throughput sequencing data. Later steps (blue) require a mix of existing RNA-seq analysis methods and novel methods to address the technical difference of scRNA-seq. The biological interpretation (red) should be analyzed with methods specifically developed for scRNA-seq

experiments for read alignment, transcript quantification, quality control, normalization, and downstream modeling. For some of these steps, pipelines and tools that have been developed for bulk RNA-seq data can be reused. However, some important single-cell-specific aspects and pitfall need to be considered.

2.1.1 Read QC and Alignment

Read QC and alignment is the first computational step in analyzing RNA-seq data sets, no exception for scRNA-seq. In general, most of the methodology developed for bulk RNA-seq, including insights for how to best map the raw sequencing reads, such as TopHat [118], can be reused for scRNA-seq. However, like processing bulk RNA-seq reads, it is important to consider biases such as incomplete knowledge of the target genome or transcriptome annotation [119]. For the specific designation of scRNA-seq protocols, spike-in RNAs, such as ERCC [120] or unique molecular identifiers (UMIs) [121] are always used to help reduce the technical variation and produce more accurate quantification. If the synthetic spike-in RNAs are used, the reference genome should be augmented with the DNA sequence of the spike-in molecules before mapping. Alternatively, if the UMI is used, the barcode attached to each read should be removed for alignment. Specifically, if both of them are used in conjunction, it needs to ensure that the sequences at the ends of the spike-ins are complete. Otherwise, the expression level of spike-ins will be underestimated.

2.1.2 Read Counting

The mapped reads can be summarized to generate read counts using the same approaches that are applied in conventional RNA-seq, such as HTSeq [122]. When UMIs are used, these

Table 1
Computational tools for multistep analyses of scRNA-seq data

Tool	Description	Requirements	Type	Interface	Operating system	Programing languages	References
Noise reduction							
scLVM	Uses a Gaussian Process Latent variable model to estimate the covariance matrix associated with latent factors. Residuals from a linear mixed model with the covariance term represent de-noised expression estimates	Requires genes associated with the latent factor to be identified a priori. Normalization factors are estimated using the median normalization method	Package	Command line	Unix/ Linux	Python	[79]
f-scLVM	Use a factorial single-cell latent variable model (f-scLVM) to dissect and model single-cell transcriptome heterogeneity, thereby allowing to identify biological drivers of cell-to-cell variability and model confounding factors	f-scLVM requires two input files, a gene expression file and an annotation file. The gene expression file is a text file containing the normalized, log-transformed gene expression matrix, with every row corresponding to a cell. The annotation file is a text file with every row containing the name of a gene set, followed by the gene identifiers annotated to that gene set	Package	Command line	Unix/ Linux	Python	[80]
OEFinder	Uses orthogonal polynomial regression to identify genes whose expression is associated with position on the C1 Fluidigm integrated fluidic circuit (IFC)	Gene-specific P values are provided to identify genes affected by the artifact	Package	Command line/ GUI	Unix/ Linux, Mac OS, Windows	R	[81]

(continued)

Table 1 (continued)

Tool	Description	Requirements	Type	Interface	Operating system	Programing languages	References
Normalization							
Single-cell normalization	A quantitative statistical method to distinguish true biological variability from the high levels of technical noise in single-cell experiments. This approach quantifies the statistical significance of observed cell-to-cell variability in expression strength on a gene-by-gene basis	It is a complete workflow to recreate all figures and numbers shown in the paper. It is extensively commented in order to demonstrate how our suggested analysis method is performed in practice and facilitate re-implementation by the user	Package	Command line	Unix/ Linux, Mac OS, Windows	R	[82]
SAMstrt	The resampling-based bulk normalization method in SAMseq is applied to spike-ins	Assumes that an equal number of spike-in control RNA molecules have been added to all samples	Package	Command line	Unix/ Linux, Mac OS, Windows	R	[83]
GRM	Fits polynomial gamma regression model to FPKM data from spike-ins; estimated parameters are used to convert FPKM of endogenous genes to an absolute scale within each cell	Performs within cell normalization and may be used with FPKM, RPKM, or TPM	Package	Command line	Unix/ Linux, Mac OS, Windows	R	[84]

SCnorm	A quantile regression method for accurate and efficient normalization of scRNA-seq data. SCnorm uses quantile regression to estimate the dependence of read counts on sequencing depth for every gene. Genes with similar dependence are then grouped, and a second quantile regression is used to estimate scale factors within each group	SCnorm does not require spike-ins, and performance may be improved if good spike-ins are available	Package	Command line	Unix/ Linux, Mac OS, Windows	R	[85]
NODES	A normalization technique that substantially reduces technical variability and improves the quality of downstream analyses. NODES provides a transformative reduction in computational complexity and execution time, which will be crucial for analyzing the massive single-cell data sets generated by inDrop/Drop-seq and other high-throughput single-cell technologies	NA	Package	Command line	Unix/ Linux, Mac OS, Windows	R	[86]
SCONE	An R package for single-cell RNA-seq data quality control and normalization. This data-driven framework uses summaries of expression data to assess the efficacy of normalization workflows	R (R version 3.3)	Package	Command line	Unix/ Linux, Mac OS, Windows	R	[87]

(continued)

Table 1
(continued)

Tool	Description	Requirements	Type	Interface	Operating system	Programing languages	References
Differential expression detection							
SCDE	Models gene-specific expression as a two-component mixture: a Poisson component describes zero and a negative binomial describes non-zero measurements	Requires un-normalized gene expression counts (performs internal correction) and provides gene-specific posterior probabilities of differential expression (DE) between two biological conditions. Tests for DE are performed on non-zeros	Package	Command line	Unix/ Linux, Mac OS, Windows	R	[88]
powsimR	Simulates and evaluates differential expression from bulk and especially single-cell RNA-seq data making it suitable for a priori and posterior power analyses. Powsim is a flexible tool that can not only estimate sample sizes necessary to achieve a certain power but also informs about the power to detect differential expression (DE) in a data set at hand. It can allow the researcher to distinguish between actual discrepancies and incongruities due to lack of power	NA	Package/ module	Command line	Unix/ Linux, Mac OS, Windows	R	[89]

MAST	A logistic regression model is used to test differential expression rate between groups while a Gaussian generalized linear model (GLM) describes expression conditionally on non-zero expression estimates. Models are corrected for cellular detection rate	Requires normalized gene expression estimates and provides gene-specific P values from summing likelihood ratio or Wald tests from the two components	Framework	Command line	Unix/ Linux, Mac OS, Windows	R	[90]
D3E	A tool for identifying differentially expressed genes, based on single-cell RNA-seq data. D3E consists of two modules: one for identifying differentially expressed (DE) genes and one for fitting the parameters of a Poisson-Beta distribution	D3E takes a read count table as an input, with rows and columns corresponding to transcripts and cells, respectively. The user should split the columns into two or more groups by providing cell labels in the input file. If there are more than two groups of cells, they must be compared one pair at a time	Package	Command line	Unix/ Linux	Python	[91]
scDD	Models expressed counts as a Dirichlet process mixture (DPM) of normals to test for differentially distributed (DD) genes associated with multimodality in the expressed component. Samples from the posterior further characterize the gene-specific distributional difference between two biological conditions to identify genes that are differentially expressed (DE), differ in the	Requires normalized, log-transformed gene expression estimates and provides gene-specific P values (or a false discovery rate (FDR)-controlled list) of DD genes between two biological conditions. Each DD gene is then classified into a specific type of distributional difference	Framework, package	Command line	Unix/ Linux, Mac OS, Windows	R	[92, 93]

(continued)

Table 1
(continued)

Tool	Description	Requirements	Type	Interface	Operating system	Programing languages	References
	proportion of cells within modes (DP), differ in the number of modes (DM), or are both DE and DM (DB)						
TASC	Incorporates the technical parameters, which reflect cell-to-cell batch effects, into a hierarchical mixture model to estimate the biological variance of a gene and detect differentially expressed genes. TASC is a statistical framework, to reliably model the cell-specific dropout rates and amplification bias by use of external RNA spike-ins. It is programmed to be computationally efficient, taking advantage of multi-threaded parallelization	Boost C++ Library, GNU Scientific Library, cmake, gcc	Framework	Command line	Unix/Linux, Mac OS, Windows	C	[94]
SpatialDE	Identifies and characterizes spatially variable genes. SpatialDE is based on Gaussian process regression (GPR) that decomposes the expression variability into a spatial and non-spatial component. It can be used to test for spatial expression variation in cell culture systems, where spatial variation may not be expected a priori	The tool is useful in expression data type with spatial and/or temporal resolution	Package/Module	Command line	Unix/Linux	Python	[95]

Subpopulation detection							
RaceID	Uses k-means applied to a similarity matrix of Pearson's correlation coefficients for all pairs of cells; the number of clusters is chosen using the gap statistic. Outlier cells are those that cannot be explained by a background model that accounts for technical and biological noise. In a second step, rare subpopulations can be identified and outlier cells may be merged to an outlier cluster; new cluster centers are then computed and each cell is assigned to the most highly correlated cluster center	Requires a reduced set of genes. The authors consider genes with a minimum of five transcripts in at least one cell	Package	Command line	Unix/Linux, Mac OS, Windows	R	[96]
BackSPIN	Iteratively splits a two-way sorted (by both genes and cells) expression matrix into two clusters containing independent cells and genes, for a maximum number of splits. The algorithm has a stopping condition to avoid splitting data that are very homogeneous	Requires a reduced set of genes and the maximum number of splits allowed. The authors recommend selecting the top 5000 genes that have the largest residuals after fitting a simple noise model	Package	Command line	Unix/Linux	Python	[97]
ZIFA	Models dropout rate as a function of expression in a factor analysis (linear dimension reduction) framework	Requires normalized, log-transformed estimates of gene expression (zeros are not transformed)	Package	Command line	Unix/Linux	Python	[98]

(continued)

Table 1 (continued)

Tool	Description	Requirements	Type	Interface	Operating system	Programing languages	References
Destiny	Extends diffusion maps (a nonlinear dimension reduction approach) to handle zeros and sampling density heterogeneities inherent in single-cell data	Requires variance-stabilized gene expression estimates; works best with a large number of cells	Package	Command line	Unix/ Linux, Mac OS, Windows	R	[99]
SNN-Cliq	Clusters cells by identifying and merging subgraphs (quasi-cliques) in a shared nearest neighbor (SNN) graph; the number of clusters is chosen automatically	Requires a reduced set of genes. The authors recommend using genes with average RPKM >20 and using a log transformation to reduce the effect of outliers. Relies on a valid choice of graph parameters	Package	Command line	Unix/ Linux	MATLAB, Python	[100]
SCUBA	Uses k-means to cluster data along a binary tree detailing bifurcation events for time course data. Models expression regulation along the tree using bifurcation theory	Requires a reduced set of genes. The authors recommend using the 1000 most variable genes that are expressed in at least 30% of cells	Package	Command line	Unix/ Linux, Mac OS, Windows	MATLAB	[101]
Cell lineage and pseudotime inference							
Monocle	Reduces data using independent component analysis (ICA) and constructs a minimum spanning tree (MST) to order cells in pseudotime	Requires normalized, log-transformed gene expression estimates and a reduced set of genes. The authors recommend	Toolkit	Command line	Unix/ Linux, Mac OS		[30]

		identifying genes that are differentially expressed between time points or, if data at multiple time points are not available, choosing genes above a mean and variance threshold					
Oscope	Uses a paired-sine model and K-medoids clustering to identify groups of oscillatory genes. For each oscillatory group, an extended nearest insertion algorithm is used to construct the cyclic order of cells, defined as the order that specifies each cell's position within one cycle of the oscillation of that group	Identifies groups of oscillatory genes, when present. Requires normalized gene expression and use of only high mean, high-variance genes is recommended	Package	Command line	Unix/ Linux, Mac OS, Windows	R	[102]
TSCAN	A software tool developed to better support in silico pseudotime reconstruction in single-cell RNA-seq analysis. TSCAN uses a cluster-based minimum spanning tree (MST) approach to order cells. TSCAN has a graphical user interface (GUI) to support data visualization and user interaction	Single-cell data should be prepared in a matrix-like data format. Each row corresponds to a gene/feature and each column corresponds to a single cell	Package	Command line/ Web user interface	Unix/ Linux, Mac OS, Windows	R	[103]

(continued)

Table 1
(continued)

Tool	Description	Requirements	Type	Interface	Operating system	Programing languages	References
LEAP	Utilizes the estimated pseudotime of the cells to find gene co-expression that involves time delay. LEAP sorts cells according to the estimated pseudotime and then computes the maximum correlation of all possible time lags. In addition, LEAP can apply a time series inspired lag-based correlation analysis to reveal linearly dependent genetic associations	NA	Package	Command line	Unix/ Linux, Mac OS, Windows	R	[104]
SLICER	A method for inferring cellular trajectories from single-cell RNA-seq data. SLICER uses locally linear embedding to reconstruct cellular trajectories.	NA	Package	Command line	Unix/ Linux, Mac OS, Windows	R	[105]
CellTree	Produces tree structures outlining the hierarchical relationship between single-cell samples, while identifying latent groups of genes that can provide biological insights. CellTree can infer complex underlying hierarchical structures in cell populations from expression data alone and also provide biological backing for the model it creates	A typical use case will require you to run compute.lda on your expression data, to fit an LDA model, followed by compute. backbone.tree to generate a tree structure from the LDA model	Package	Command line	Unix/ Linux, Mac OS, Windows	R	[106]

TASIC	Temporal Assignment of SIngle Cells (TASIC) uses on a probabilistic graphical model to integrate expression and time information making it more robust to noise and stochastic variations. It uses a Hidden Markov Model (HMM) based on a probabilistic Kalman Filter approach to combine time and expression information for determining the branching process associated with time series single-cell studies	Input 1. Gene expression mat file containing normalized gene expression value. The 'expression_matrix' matrix dimension is # of genes * # of cells. 2. Time label mat file containing an array denoting the time assignment of cells, e.g., time_label = [0 0 0 24 24 24], for six cells where first three are from time point 0 and the last three are from time point 24 h	Package	Command line	Unix/ Linux, Mac OS, Windows	MATLAB	[107]
Gpfates	Models transcriptional cell fates as mixtures of the Gaussian Process Latent Variable Model and Overlapping Mixtures of Gaussian Processes (OMGP)	NA	Package	Command line	Unix/ Linux	Python	[108]
SCOUP	A method to analyze single-cell expression data for differentiation. Unlike previous methods, which use dimension reduction approaches and reconstruct differentiation trajectories in reduced space, SCOUP describes gene expression dynamics during differentiation directly, including pseudotime and cell fate	The following two libraries are necessary for pseudotime estimation based on the shortest path on the PCA space: LAPACK, BLAS	Package	Command line	Unix/ Linux	C	[109]

(continued)

Table 1 (continued)

Tool	Description	Requirements	Type	Interface	Operating system	Programing languages	References
switchde	An R package for detecting switch-like differential expression along single-cell RNA-seq trajectories. Switchde assumes genes follow a sigmoidal pattern of gene expression and tests for differential expression using a likelihood ratio test. It also returns maximum likelihood estimates (MLE) for the sigmoid parameters, which allows filtering of genes for up- or downregulation as well as where along the trajectory the regulation occurs	switchde accepts either an SCESet from Scater [110] or a matrix of gene expression measurements. These should ideally be in log(TPM + 1) form, but any logged nonnegative expression measurements will work	Package	Command line	Unix/ Linux, Mac OS, Windows	R	[111]
Ouija	A statistical framework for learning pseudotimes from single-cell RNA-seq data. Ouija using only small panels of marker genes and prior knowledge of gene behavior	Uses Bayesian hierarchical nonlinear factor analysis as implemented in the probabilistic programming language Stan	Package	Command line	Unix/ Linux, Mac OS, Windows	R	[112]

Embeddr	Laplacian Eigenmaps and principal curves for pseudotemporal ordering and clustering of single-cell RNA-seq data. Embeddr creates a reduced dimensional representation of the gene space using a high-variance gene correlation graph and Laplacian Eigenmaps. It then fits a smooth pseudotime trajectory using principal curves	NA	Package	Command line	Unix/ Linux, Mac OS, Windows	R	[113]
Dimensionality reduction							
ZIFA	Models dropout rate as a function of expression in a factor analysis (linear dimension reduction) framework	Requires normalized, log-transformed estimates of gene expression (zeros are not transformed)	Package	Command line	Unix/ Linux	Python	[98]
ZINB-WaVE	ZINB-WaVE is a general and flexible zero-inflated negative binomial model which is able to give a more stable and accurate low dimensional representation of the data than principal component analysis (PCA) and zero-inflated factor analysis (ZIFA), without the need for a preliminary normalization step	NA	Package	Command line	Unix/ Linux, Mac OS, Windows	R	[114]

(continued)

Table 1 (continued)

Tool	Description	Requirements	Type	Interface	Operating system	Programing languages	References
Batch effect removal							
Batch effect removal	Batch effect removal is a method based on a residual neural network, trained to minimize the maximum mean discrepancy (MMD) between the multivariate distributions of two replicates, measured in different batches	NA	Package	Command line	Unix/ Linux	Python, R	[115]

counts can be collapsed by summing the number of unique barcodes associated with all reads mapped to a given gene. When performing this analysis, care must be taken into account for sequencing errors in the UMIs that might result in the appearance of artificial molecules. To overcome this, error correction of the barcodes and/or removal of singleton barcodes may be required [34]. Although scRNA-seq data can, in principle, be used to quantify the expression of individual exons or to resolve isoform abundance, such analyses are currently challenging owing to the large proportion of technical noise and biases compared to the traditional RNA-seq protocols. Till now, only two reports claimed their approaches hold the promise to isoform quantification using scRNA-seq data sets. The first one is SingleSplice [123], which uses a statistical model, hurdle model, to detect gene whose isoform usage shows biological variation significantly exceeding technical noise in a population of single cells. It circumvents the pitfalls of low coverage and 3′ bias which hamper the assembly and quantification of full-length isoforms, by only considering the partial regions that undergo alternative splicing, called alternative splicing modules (ASM). Importantly it requires the spike-ins to model the technical variability. Most recently, an algorithm called Census [124] was developed to convert relative RNA-seq expression level into relative transcripts counts without the need for experimental spike-in controls. It can handle the splicing pattern among the single cells. All these attempts are aiming at the circumvention or reducing the effects of factors specific for single-cell sequencing.

2.1.3 Quality Control

Quality control (QC) is needed for both the raw reads and library size or called cell size (Fig. 3). Similar to the bulk RNA-seq experiments, FastQC or Kraken [125] can be used to for QC of the scRNA-seq raw reads. The resulted data can be visualized using Integrative Genomic Viewer [126, 127]. These steps will help to identify potential sample mix-ups and external contamination or whether there was problem with the sequencing itself as opposed to the single-cell capture and amplification.

After alignment and initial read counts are obtained, quality control needs to be applied to identify poor-quality libraries of individual cells, which is perhaps extremely important for scRNA-seq data, since currently all protocols for single-cell library display a very low capture rate and high amplification bias. This is a more serious problem for primary tissue samples, as the process of extracting a tissue and then isolating individual cells can affect the quality of the RNA collected. Three commonly used metrics are available for this step before going through downstream analyses.

The first metric is the same as was used for bulk RNA-seq data. It is the fraction of reads that map back to the genome of the organism of interest (i.e., the rate of mapped reads), which can be obtained directly from the FastQC output. This metric indicates whether there is a problem with the sequencing library generated from an individual cell. A low fraction might indicate that RNA has degraded, that there is external contamination, or that the cell was inefficiently lysed.

The second metric is available when the spike-in control was used. It is the ratio of the number of read mapped to the endogenous RNA to the number of reads mapped to the extrinsic spike-ins, which can be computed from the FastQC output or directly from the table of counts gained from HTSeq. A high proportion of reads mapped to the spike-ins would be indicative of a low quality of RNA in the cell of interest and might be a reason to exclude these cells for downstream analyses. However, this ratio could vary noticeably from cell-to-cell biological visibilities (e.g., if the cells are of different phases of the cell cycle). Nevertheless, cells for which the ratio of spike-ins is extremely discordant from the remaining population are strong candidates for exclusion.

The last useful approach for identifying problematic cells is to apply principal component analysis (PCA) to the read count matrix or gene expression matrix. The expectation is that good-quality cells cluster together and poor-quality cells are outliers. Note that, in some cases, poor-quality cell may also cluster together to form a second distinct population. For example, it has been observed that poor-quality cells are often enriched in the expression of mitochondrial genes [34], which can cause them to cluster separately. This, therefore, stresses that outlier analyses must be performed carefully to ensure that cells with physiologically relevant differences are not inadvertently discarded. To prevent this, one useful observation is that poor-quality cells typically display extreme values of the two other metrics described above.

2.2 Normalization and Quantification

Besides the QC that is extremely important for scRNA-seq data analyses, normalization is also a computational challenge for scRNA-seq quantification. For bulk RNA-seq data, read counts between different samples are always standardized by the transcript length and sequencing depth, such as FPKM (fragments per kilobase per million fragments mapped) and RPKM (reads per kilobase per million reads mapped) for paired and single ends, respectively. However, standers for normalizing bulk RNA-seq reads make an implicit assumption that the total amount of RNA processed in each library is approximately the same or that the variation is technical noise. This assumption is always useful when relative expression estimates are compared. In scRNA-seq, the normalization procedure can substantially affect the interpretation of the data, and thus special attention should be taken. There are two categories of approaches depending on whether the UMI criterion is used.

2.2.1 Normalization of scRNA-seq Data Without UMIs

Although the bulk-based normalization strategy has to be used without other choices in the cases, mostly at the very beginning of single-cell sequencing emerge, that both UMIs and extrinsic spike-ins are absent, recent strategies are based on spike-ins. Because without external spike-in controls, it is difficult to determine how much RNA is present in a cell, which varies cell to cell. Since the spike-in RNAs are assumed to be constantly added into the libraries across cells, it is possible to accurately estimate relative differences in the total RNA content between cells. Specifically, the ratio of the number of reads mapped to the genome of interest to the number of reads mapped to the spike-ins is easy to be calculated. When compared between cells, this ratio accounts for differences in the amount of RNA within a cell to be inferred. Therefore, if the spike-ins are available, the read counts associated with each gene can be converted into absolute numbers of mRNA molecules based on the level of the spike-ins, which are of known concentrations. However, spike-in control is not a perfect resolution for scRNA-seq read normalization. Most common set of spike-ins, such as ERCC [120], are 500–2000 nucleotides (nts) in length, which is shorter than an average human mRNAs (~2100 nts including untranslated regions [128]). The 3′ bias of scRNA-seq protocols leads the conversion based on the shorter ERCC spike-ins is potentially problematic. Additionally, the spike-ins have comparatively short poly(A) tails and lack 5′ caps, which may result in different degree of degradation and efficiency of reverse transcription of the endogenous mRNAs. Consequently, it is challenging to develop a generally applicable normalization strategy for scRNA-seq data that properly accounts for variability both in sequencing depth and cell size.

In many cases, a sensible and pragmatic approach is to calculate two alternative size factors: one for the spike-ins and one for the endogenous mRNAs [129]. The former accounts solely for sequencing depth whereas the latter for the endogenous mRNAs reflecting the relative differences in cell size. This twofold normalization strategy relies on the assumption that the normalized spike-ins can be used to estimate the degree of the technical variability across the whole dynamic range of expression, which is the basic principle for spike-in control. Even if the technical noise is well accounted for by the spike-in control, the transcript length-based normalization methods, such as FPKM or RPKM, are still problematic. In particular, although improvements have been made recently [37], there is still noticeable 3′ bias to several scRNA-seq protocols, including smart set used by the popular Fluidigm technology. This bias leads to the underestimation of the expression long transcripts and overestimation of the short ones. Therefore, until protocols allow unbiased sampling of reads from across the whole transcript length, using FPKMs to compare the expression of transcripts with different lengths must be paid with more attention. To overcome this shortcoming, the UMIs were designed and used for scRNA-seq protocols.

2.2.2 Normalization of scRNA-seq Data with UMIs

To correct for amplification bias, the UMIs have been described how molecules can be directly counted [34]. For scRNA-seq, UMIs have been used as an internal validation control and also been explored as a direct, quantitative measure of gene expression. When UMIs are used, and assuming that the sample is sequenced at a sufficient depth to ensure that each cDNA molecule is observed at least once, thus the number of UMIs linked to each gene is a direct measure of the number of cDNA molecules associated with that gene. Therefore, it is tempting to use these raw molecular counts—which, unlike the expression estimates from non-UMI protocols, are independent of amplification biases—in all downstream analysis. However, differences in the number of UMIs associated with each gene still can vary between cells for technical rather than biological reasons, such as differences in efficiency of the reverse transcription and other cell-specific technical effects independent of amplification.

One strongly recommended approach that can help to overcome this problem is to combine the spike-in control with the UMIs. As described above, the number of spike-in molecules is theoretically the same across cells, and the variability in the number of UMIs associated with spike-in genes across cells is indicative of differences in amplification efficiency, that is, technical variability. Consequently, a spike-in-based normalization step can be used to convert the number of cDNA molecules (the number of associated UMIs) to the number of mRNA molecules. As the UMI-based scRNA-seq protocols currently sequence only a fragment of each molecule (from either the 5′ or the 3′ end of the transcript), normalization on transcript length is unnecessary. However, these tag-based protocols cannot be used to study isoform variants or allele-specific expression. Most recently, the Census algorithm [124] was introduced to convert conventional measures of relative expression such as transcript per million (TPM) in single cells to relative transcript counts without the need for spike-ins or UMIs. "Census counts" eliminate much of the apparent technical variability in single-cell experiments and are thus easier to model with standard regression techniques than normalized read counts. The authors claimed Census counts could reveal changes in cell fate-dependent gene expression, splicing patterns, and allelic imbalances. Census is implemented in Monocle 2 [30], the second major release of the open-source single-cell toolkit (https://github.com/cole-trapnell-lab/monocle-release).

2.3 Accounting for Technical Noise in scRNA-Seq Experiments

Technical noise is unavoidable, especially for the scRNA-seq experiments, owing to the low amount of starting material and over-amplification, and it must be quantified in order to avoid mistaking it for genuine differences in biological expression levels. Once normalized gene expression levels or molecular counts have been generated, it is important to incorporate technical variability estimates. This applies to any downstream analysis but is particularly important when comparing expression levels between cells or when assessing the variability of individual genes. Because of the typically low capture efficiency of current scRNA-seq protocols, even moderately expressed genes are frequently undetected. Consequently, methods to accurately estimate the extent of this technical variability are crucial in order to differentiate between genuine gene expression changes and experimental artefacts.

The use of spike-ins as control genes is appealing, since the same amount of ERCC (or other) spike-in was added to each cell in our experiment. In principle, all the variability we observe for these genes is due to technical noise, whereas endogenous genes are affected by both technical noise and biological variability. Technical noise can be removed by fitting a model to the spike-ins and "substracting" this from the endogenous genes. There are several methods available based on this premise (e.g., BASiCS [130, 131], scLVM [79], RUVg [132]), each using different noise models and different fitting procedures. Alternatively, one can identify genes which exhibit significant variation beyond technical noise (e.g., distance to median, highly variable genes). However, there are issues with the use of spike-ins for normalization (particularly ERCCs, derived from bacterial sequences), including that their variability can, for various reasons, actually be higher than that of endogenous genes.

Given the issues with using spike-ins, better results can often be obtained by using endogenous genes instead. Where we have a large number of endogenous genes that, on average, do not vary systematically between cells and where we expect technical effects to affect a large number of genes (a very common and reasonable assumption), then such methods (e.g., the RUVs method [132]) can perform well. Although almost all analyzing approaches take the technical noise into account, the readers are recommended to refer to the most common strategies used in refs. 88, 130 for details.

2.4 Getting Biological Insights into Single-Cell RNA Sequencing

In this section, we will give a review of the applications of scRNA-seq to the specific biological questions for which bulk RNA-seq may not give the answers. Thus far, single-cell RNA sequencing has already shown great effectiveness in unraveling complex cell populations, reconstructing developmental trajectories, modeling transcriptional dynamics, and so on. In all the following analyses, we assume that the input data is a matrix of gene expression or transcript counts that have been normalized and cleared the technical visibilities using whatever approaches described above.

2.4.1 Accounting for Heterogeneity: Cell Identity and Cellular State

Solid tissues of human and other eukaryotes comprise of several different types of cells. These different cell types have distinct transcriptomic profiles. Although there are a lot of computational methods aiming at inferring the heterogeneity based on sequencing data from bulk samples [133], single-cell transcriptomics provide valuable ability to characterize a sample in terms of the known and novel cell types it contains, i.e., the heterogeneity [40, 79, 134–138].

scRNA-seq is useful for cell type identification by clustering cells on the basis of their expression profiles. Distinct subsets, potentially corresponding to unknown cell types, can be identified. Especially, the genes that best distinguish different cell types can characterize them. There are two types of clustering methods for cell type identification based on scRNA-seq data, depending on whether there is established information or expectation regarding the relationship between these cells. If there is no prior expectation, unbiased or unsupervised clustering methods, such as hierarchical clustering or PCA-like methods, can be used to group cells according to their individual expression profile. For example, Trapnell et al. use a PCA-based approach to group cells according to their position along the differentiation cascade [30], generating the developmental trajectory. This approach was implemented as a stand-alone tool for public uses, called Monocle (http://cole-trapnell-lab.github.io/monocle-release/). If the prior information is available, PCA-like approach can be combined with knowledge of the expression patterns of a small set of known marker genes, allowing an approximate spatial map of the tissue under study to be obtained [135].

In addition to cell type identification, unsupervised methods such as PCA can also be used to explore cellular state, for example, stage or speed of the cell cycle. Perhaps counterintuitively, slow-cycling cells tend to have clearer transcriptional signatures of G1/S versus G2/M stages, whereas fast-cycling cells tend to be more homogeneous with respect to expression of cell cycle genes. A recent study of single cells obtained from glioblastomas describes a computational strategy for quantifying the speed of the cell cycle in each cell by comparing expression levels of G1/S versus G2/M genes [139].

2.4.2 Differential Expression and Alternative Splicing

Differential expressed gene (DEG) detection is the most useful and first application to distinguish the different bulk samples or distinct cluster of single cells based on their transcriptomic profiles. From computational perspective, approaches based on standard differential expression tools for bulk RNA-seq can be used [140–144], of which Rapaport et al. gave a comprehensive review and evaluation [145]. However, care must be taken that scRNA-seq data are typically noisier than bulk RNA-seq, so the technical variability must be characterized and accounted for before DEG analysis. Recently, alternative approaches [83, 86, 88–92, 94, 111, 124] designed specifically for scRNA-seq data have to be developed

(Table 1), which explicitly take into account all scRNA-seq-specific confounding factors.

Unlike the DEG analysis, the identification of alternative splicing (AS) events or exons is much more challenging, owing to the unique properties of scRNA-seq data as low coverage, 3′ bias, and technical noise. Although the technical noise is optimistically resolvable through computational methods described above, another two peculiarities make AS study problematic. So far, there is only one reported effort aiming at AS detection based on scRNA-seq reads, called SingleSplice [123]. However, this method only enables the detection of partially AS fragments rather than the full-length AS transcripts that are easily obtained from bulk RNA-seq reads. Additionally, SingleSplice is limited to tests for variability in groups of cells, rather than in really individual cell. Alternatively, to circumvent the detection of different types of AS event directly, methods that characterize the transcripts that display differences in exon or isoform usage have been reported [124, 146]. As in studies of differential expression, tools for identifying differentially expressed exons [147, 148] can be used in these cases. Therefore, AS detection based on scRNA-seq data is still problematic largely due to its low coverage and 3′ bias, which gives the great chances for both the technical improvement and computational method development. Our group are now working on the possibility of 5′ coverage imputation for single-cell sequencing, which holds the promise to overcome the 3′ bias computationally for AS or more other analyses which are commonly used for bulk RNA-seq data.

2.4.3 Modeling Transcriptional Dynamics

Growing evidence suggests that genes are not transcribed consistently but rather undergo highly dynamic expression patterns across a population of cells. scRNA-seq can be used to explore transcriptional kinetics of cells but still of challenge without prior knowledge of the underlying cell types. Nevertheless, unlike the population-averaged data from bulk RNA-seq data, scRNA-seq can characterize diversity in transcription between individual cells to the acceptable extent [149, 150]. Identification of the highly dynamic genes requires the application of statistical approaches that account for technical sources of visibilities, such that biological variability in gene expression levels can be quantified accurately and sensitively. Additionally, the high variability of gene expression can also be caused by confounding factors that are not accounted for, such as the cell cycle [139]. One approach is to compute the coefficient of variation for each gene across the population of cells under study and to rank the genes accordingly. Unfortunately, technical variability, which is intrinsic to the experimental protocol and not associated with genuine biological variability, is greater for lowly expressed genes than for highly expressed genes [1]. Consequently, a null estimate of the expected technical variability needs to be computed. This can be done using the extrinsic spike-in molecules as described above—the extent of variability in their expression across cells can be used as an estimate of the null variance. This information allows the expected technical variation to be modeled across the whole dynamic range of expression, which forms the basis of a statistical test to determine the set of genes that show more variability in expression than would be expected by chance [82, 94]. Recently, extrinsic spike-in molecules have been used to further decompose technical variability into two terms that correspond to sampling noise and heterogeneity in sequencing efficiency across cells [151].

2.4.4 Allele-Specific Expression

scRNA-seq could in principle be used to study on the degree to which the two alleles of each gene are regulated in a coordinated manner. Like bulk RNA-seq studies [152], allele-specific expression can be measured and used to determine the extent of allelic bias in gene expression [150]. By exploring the degree of allele-specific expression, stochastic transcription of each allele and the degree of coordination of expression between alleles can be investigated. For example, scRNA-seq has been used to study stochastic allelic expression during early embryogenesis [150]. Specifically, using first-generation intercrosses between two different inbred strains of mice, the extent of stochastic allele-specific expression during early embryogenesis has been quantified transcriptome-wide.

One concern with using scRNA-seq to study random monoallelic expression is that allelic dropout during library preparation might lead to erroneous measurements of monoallelic expression. Previous approaches have addressed this by splitting cell lysates into two and then repeating the experiment to provide a background estimate of allelic dropout [150]. However, this is an area in which more work is required to develop computational methods that can accurately model this feature of scRNA-seq library preparation, such that accurate measures of allele-specific expression can be obtained.

3 Computational Approaches for Single-Cell Genomics

The field of single-cell genomics is developing rapidly and is generating increasingly new insights into complex biological systems. In contrast to RNA-seq, DNA-seq is still costly; thus, to ensure adequate and appropriate data are obtained to address the hypotheses of the study, the balance between the number of cells sequenced and the breadth of the genome of each cell that will be queried needs to be taken into account when designing an experiment. Broadly speaking, for complex eukaryotic genomes such as the human genome, one can choose to query specific loci of interest (typically <1 Mb), sequence all of the protein-coding regions (whole exome, 30–60 Mb), or sequence the whole genome (3 Gb). In this section, we do not focus on the specific techniques and related challenges of sequencing strategies; rather we review the computational methodologies for single-cell genome sequence data analysis, which have specific considerations compared to those for bulk sample genome sequences. Currently, only single nucleotide variants and clonal structures have been reported to be identified from a single cell with targeted or low-pass sequencing.

3.1 Accounting for Noise

Interpreting the data within the context of biases and technical noises is one of the major challenges of analyzing single-cell genomic data. During single-cell isolation, the population of cells being interrogated can be biased through selection of cells based on size, viability, or propensity to enter the cell cycle. Therefore, it is necessary to compare the variant alleles detected in the single cells to the bulk sample to ensure there was no selection bias. This can be done by comparing the percentage of single cells with a variant to the variant allele frequency in the original bulk sample [153]. During whole genome amplification (WGA), numerous errors can be introduced, including loss of coverage, decreased coverage uniformity, allelic imbalance, allelic dropout (ADO), and errors during genome amplification. Most published papers have attempted to quantify rates for some or all of these errors. Two recent studies developed methods to predict the breadth of genome coverage using low-pass sequencing, which could provide a low-cost approach for assessing cell lysate quality in larger eukaryotic genomes [154, 155]. However, comparing single-cell genomic studies is currently difficult, as most studies do not report the total number of cells evaluated, the quality of the data from the discarded cells, or the metrics used for the quality control categorization. A recent study compared errors and assembly performance between species [156]. Hence, more uniform analysis and reporting methods are needed to facilitate data interpretation between single-cell studies and provide accurate performance metrics for each approach.

3.2 Single Nucleotide Variant Calling

Although single-cell genome sequencing introduces numerous errors, tools and strategies are now being developed to overcome the additional technical noise, allowing the identification of true variation. Single nucleotide variant (SNV) calling requires coverage of a variant allele at a rate that exceeds the sum of the amplification and sequencing error rates. More specifically, mutations introduced during the amplification, as well as the allelic imbalances that occur during genome amplification, must be taken into account when calling variants. There are two basic strategies to overcome the false-positive variants introduced as artefacts of the amplification. First, the bulk sample can be used as a reference to reduce the false discovery rate [153]. Second, when using only the single-cell data, two or three cells can be required to have the same variant at the same location, which is unlikely to occur by chance with the several thousand mutations introduced during single-cell WGA in a 3 Gb human genome [75]. However, the actual number of cells required to call a mutation has not yet been rigorously tested based on the size of the genomic region interrogated. To overcome allelic imbalance, we need variant calling algorithms that are designed to take the technical noise into consideration. One strategy is to require that all variant calls be above the level of technical noise in control samples, which should not have variants [153]. Another approach is to decrease the sequencing error rate by using molecular barcoding [157]. Finally, algorithms are beginning to be developed to correct errors in single-cell sequencing data [158]. Nonetheless, more tools that incorporate all single-cell amplification errors are needed to optimally carry out variant calling in single-cell data.

CNV detection relies on algorithms such as hidden Markov models, circular binary segmentation, and rank segmentation, which can normalize noisy coverage data after single-cell WGA to identify regions that are over- or underrepresented compared with a diploid genome [6, 75, 159]. CNV detection algorithms are currently being developed to specifically address the technical artefacts introduced during specific types of single-cell WGA [159, 160]. Chimera formation can create false structural variants, although unless they occur at the beginning of the amplification, they should be much less abundant than the corresponding wild-type sequences. This is important for both identifying structural variation in sequencing data and when constructing contigs for de novo genome assemblies. In addition, assemblies are hampered by loss of coverage and uneven coverage, which results in truncated or artefactual sequences in assembled genomes. Several assemblers have been created to specifically address these challenges [161, 162], and it is likely that further progress will be made in the coming years.

3.3 Characterizing Clonal Structure

Determination of the number of clones and from which a single cell originates based on single-cell genome is currently computationally challenging. This demand is extremely noticeable for tumor evolution [163, 164], which is currently a very active field of research and will be for a while [165]. Typically, general strategies for clustering gene expression and other large data set depend heavily on the distance functions used to provide a quantitative measure of the differences between samples [166] (Fig. 4a). Specific for single-cell sequencing, these functions are required to tolerate to missing data as a result of low coverage and false-negative variant detection. Although the Jaccard distance has proven to be the best for genotype data, the false-negative rate still hinders the statistical determination of the number of clones in a sample. Alternatively, model-based clustering allows the inclusion of false-negative errors

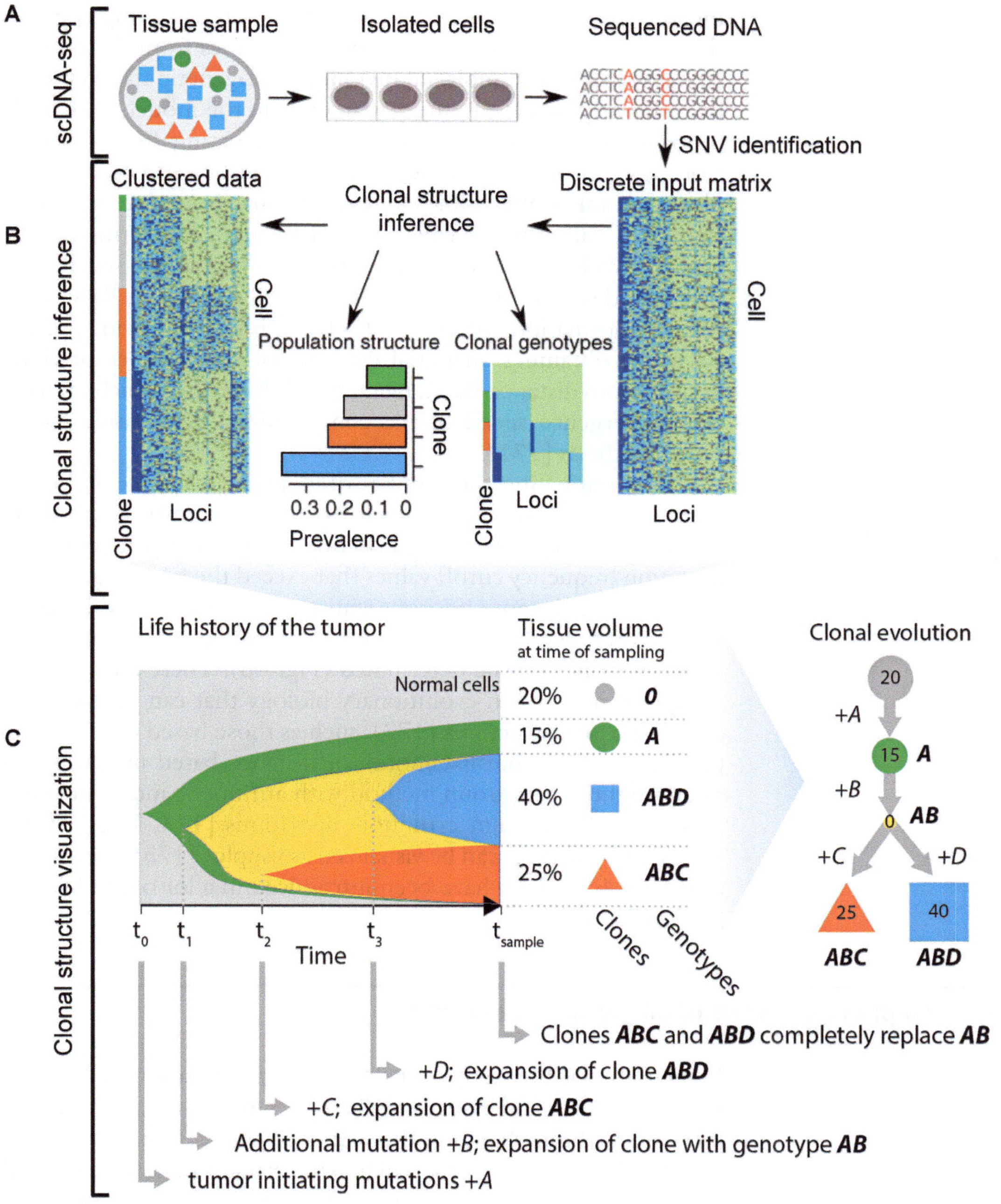

Fig. 4 A typical schematic workflow of clonal structure inference and visualization based on single-cell DNA sequencing. (**a**) The brief processed of single-cell DNA sequencing. (**b**) Clonal structure inferences, including clustering the data, predicting clonal genotypes, and inferring the prevalence of clones. (**c**) Visualize the clonal structure with fishplot or evolution tree (a toy example). **a** and **b** were adapted from ref. 167; **c** was courtesy of Dr. Florian Markowetz

modeled as binomial process [168]. It is a soft clustering approach that provides the probabilities that a cell comes from the different clones rather than simply subdividing phylogenetic tree generated by distance-based clustering. Parameters in the model, such as the probability of a particular single cell originating from a specific clone, as well as the false-negative rate, can be estimated across a distinct number of possible clones using an expectation–maximization (EM) algorithm [169]. The challenge of determining the number of clones is then reduced to selecting the statistical model that best describes the observed single-cell data using Bayesian or Akaike information criteria [170]. There is also a hybrid approach based on obtaining an initial estimate of the number of clones derived from distance-based hierarchical clustering, which increases the convergence speed of the computationally intensive model-based methods [171].

After estimating the number of clones in a sample and determining which clone each cell belongs to, a consensus clonal mutation profile can be established. Bentley et al. have done this using mutation frequency cutoff values that exceed the false-negative rate [172], although more rigorous statistical methods could be developed. After determining the clonal genotype, the relationships between clones can be determined (Fig. 4b). There are a number of algorithms used in evolutionary biology that can be applied to establish clonal structures [173], such as those based on maximum parsimony, maximum likelihood, or distance-based methods such as unweighted pair group method with arithmetic mean, neighbor joining, and minimum evolution algorithms [173, 174]. Finally, the clonal structures can be visualized as fishplot [175] or evolution tree (Fig. 4c), which have been introduced in a lot of publications [167, 176–189].

4 Computational View of Single-Cell Epigenomics

Epigenomics is the study of the chemical modifications of genomic DNA sequences and associated chromatin proteins such as histone proteins, without changing the sequence per se. It aims at dissecting the functions of these epigenetic properties associated with cellular memory, identity, and tissue-specific gene expressions. While the current techniques in the field are characterizing the average epigenomic features of large cell populations of bulk samples, the increasing interest in the epigenetic heterogeneity within complex tissues is driving the development of single-cell epigenomics. In this section, we briefly introduce the computational methods and challenges to analyzing single-cell epigenomic data, including DNA methylation, chromatin accessibility, histone modifications, chromosome conformation, and replication dynamics. Like the scRNA-seq and scDNA-seq, the main computational challenges in single-cell epigenome arise from the technical variability and low capture efficiency.

4.1 Controlling for Technical Noise

Sequencing of single-cell epigenomic libraries is typically saturated, such that a limited set of distinct DNA molecules is sequenced over many reads. The first step in analyzing data from such experiments is therefore the inference of the set of epigenetic features (or molecules) that were successfully extracted from each of the cells. As aggressive amplification is required to transform picograms of source material into sequencing-ready libraries, there are multiple possible sources of noise that must be considered when making such initial inferences. For example, sequencing errors, possible PCR effects, and even mispairing of paired-end reads on the sequencer's flow cells [70] can modify or splice erroneous cell- or well-specific barcodes and thereby distribute reads from one molecule among several cells. Even if such events are rare when evaluating their rate per sequence read, their effect can be considerable when weighted against a limited set of true-positive molecules for each cell [40, 70]. Negative control wells or labels must be used, and the quality of single-cell libraries must be analyzed routinely to ensure individual measurements can be interpreted as coming from a single-cell source. The strategies for accounting for technical noises have been discussed in detail with respect to scRNA-seq where synthetic spike-ins and UMIs are introduced during reverse transcription. Single-cell epigenomic methods would likely benefit from similar strategies.

4.2 Accounting for Missing Data

Properly filtered single-cell epigenomic profiles must be analyzed while assuming they are missing a substantial amount of data. Epigenomic data (or epi-alleles) are present in essentially single copies within each cell, such that any imperfection in the single-cell epigenomic recovery process results in loss of information. With all current technologies, it is therefore not realistic to narrow down the analysis to epigenomic features that are represented in all (or even most) sampled cells. This is because the probability of perfect simultaneous detection of the feature, assuming it is missing with probability P per cell, scales down exponentially as $(1 - P)^n$ with the size of the data set of n cells. Instead, different strategies for imputation of missing values or analysis relying on more coarse-grained features should be used [54, 190, 191]. Imputation can be possible when the epigenomic structure is highly constrained, such that partial observations are sufficient to determine the complete epigenomic state. This may be the case in analyses of large-scale chromosome conformations or replication landscapes [70] but not when observing DNA methylation or local chromatin structure. Alternatively, it is possible to form groups of epigenetic loci (based on prior assumptions or using unsupervised clustering [63]) and extract for each cell the average epigenetic state in each of the groups, thereby overcoming missing data. In such cases, reference epigenomic maps can be used efficiently to form groups of elements, such as promoters, enhancers, or insulators, which can be pooled together to allow comparison of the epigenetic states of individual cells.

References

1. Stegle O, Teichmann SA, Marioni JC (2015) Computational and analytical challenges in single-cell transcriptomics. Nat Rev Genet 16(3):133–145. https://doi.org/10.1038/nrg3833
2. Gawad C, Koh W, Quake SR (2016) Single-cell genome sequencing: current state of the science. Nat Rev Genet 17(3):175–188. https://doi.org/10.1038/nrg.2015.16
3. Schwartzman O, Tanay A (2015) Single-cell epigenomics: techniques and emerging applications. Nat Rev Genet 16(12):716–726. https://doi.org/10.1038/nrg3980
4. Barkla BJ, Vera-Estrella R, Raymond C (2016) Single-cell-type quantitative proteomic and ionomic analysis of epidermal bladder cells from the halophyte model plant Mesembryanthemum crystallinum to identify salt-responsive proteins. BMC Plant Biol 16. https://doi.org/10.1186/S12870-016-0797-1
5. Wu MY, Singh AK (2012) Single-cell protein analysis. Curr Opin Biotechnol 23(1):83–88. https://doi.org/10.1016/j.copbio.2011.11.023
6. Baslan T, Kendall J, Rodgers L, Cox H, Riggs M, Stepansky A, Troge J, Ravi K, Esposito D, Lakshmi B, Wigler M, Navin N, Hicks J (2012) Genome-wide copy number analysis of single cells. Nat Protoc 7 (6):1024–1041. https://doi.org/10.1038/nprot.2012.039
7. Habib N, Li YQ, Heidenreich M, Swiech L, Avraham-Davidi I, Trombetta JJ, Hession C, Zhang F, Regev A (2016) Div-Seq: single-nucleus RNA-Seq reveals dynamics of rare adult newborn neurons. Science 353 (6302):925–928. https://doi.org/10.1126/science.aad7038
8. Hou Y, Guo HH, Cao C, Li XL, Hu BQ, Zhu P, Wu XL, Wen L, Tang FC, Huang YY, Peng JR (2016) Single-cell triple omics sequencing reveals genetic, epigenetic, and transcriptomic heterogeneity in hepatocellular carcinomas. Cell Res 26(3):304–319. https://doi.org/10.1038/cr.2016.23
9. Gross A, Schoendube J, Zimmermann S, Steeb M, Zengerle R, Koltay P (2015) Technologies for single-cell isolation. Int J Mol Sci 16(8):16897–16919. https://doi.org/10.3390/ijms160816897
10. Osborne GW (2011) Recent advances in flow cytometric cell sorting. Methods Cell Biol 102:533–556. https://doi.org/10.1016/B978-0-12-374912-3.00021-3
11. Xin YR, Kim J, Ni M, Wei Y, Okamoto H, Lee J, Adler C, Cavino K, Murphy AJ, Yancopoulos GD, Lin HC, Gromada J (2016) Use of the Fluidigm C1 platform for RNA sequencing of single mouse pancreatic islet cells. Proc Natl Acad Sci U S A 113(12):3293–3298. https://doi.org/10.1073/pnas.1602306113
12. Strey HH, Brouzes E, Kruse T (2013) Droplet microfluidic technologies for high-throughput single-cell gene expression analysis. Biophys J 104(2):676a
13. Brouzes E, Medkova M, Savenelli N, Marran D, Twardowski M, Hutchison JB, Rothberg JM, Link DR, Perrimon N, Samuels ML (2009) Droplet microfluidic technology for single-cell high-throughput screening. Proc Natl Acad Sci U S A 106(34):14195–14200. https://doi.org/10.1073/pnas.0903542106
14. Gomez-Sjoberg R, Leyrat AA, Pirone DM, Chen CS, Quake SR (2007) Versatile, fully automated, microfluidic cell culture system. Anal Chem 79(22):8557–8563. https://doi.org/10.1021/ac071311w
15. Ino K, Okochi M, Konishi N, Nakatochi M, Imai R, Shikida M, Ito A, Honda H (2008) Cell culture arrays using magnetic force-based cell patterning for dynamic single cell analysis. Lab Chip 8(1):134–142. https://doi.org/10.1039/b712330b
16. Di Carlo D, Wu LY, Lee LP (2006) Dynamic single cell culture array. Lab Chip 6 (11):1445–1449. https://doi.org/10.1039/b605937f
17. Zhang K, Han X, Li Y, Li SY, Zu YL, Wang ZQ, Qin LD (2014) Hand-held and

integrated single-cell pipettes. J Am Chem Soc 136(31):10858–10861. https://doi.org/10.1021/ja5053279

18. Frohlich J, Konig H (2000) New techniques for isolation of single prokaryotic cells. FEMS Microbiol Rev 24(5):567–572
19. Voet T, Kumar P, Van Loo P, Cooke SL, Marshall J, Lin ML, Esteki MZ, Van der Aa N, Mateiu L, McBride DJ, Bignell GR, McLaren S, Teague J, Butler A, Raine K, Stebbings LA, Quail MA, D'Hooghe T, Moreau Y, Futreal PA, Stratton MR, Vermeesch JR, Campbell PJ (2013) Single-cell paired-end genome sequencing reveals structural variation per cell cycle. Nucleic Acids Res 41 (12):6119–6138. https://doi.org/10.1093/nar/gkt345
20. Klitgaard K, Jensen TK, Angen O, Boye M (2007) Measurement of bacterial gene expression in vivo by laser capture microdissection and quantitative real-time RT-PCR. J Microbiol Methods 69(2):414–416. https://doi.org/10.1016/j.mimet.2006.12.003
21. Fink L, Kwapiszewska G, Wilhelm J, Bohle RM (2006) Laser-microdissection for cell type- and compartment-specific analyses on genomic and proteomic level. Exp Toxicol Pathol 57:25–29. https://doi.org/10.1016/j.etp.2006.02.010
22. Vannucci FA, Foster DN, Gebhart CJ (2013) Laser microdissection coupled with RNA-seq analysis of porcine enterocytes infected with an obligate intracellular pathogen (Lawsonia intracellularis). BMC Genomics 14. https://doi.org/10.1186/1471-2164-14-421
23. Nakamura N, Ruebel K, Jin L, Qian X, Zhang H, Lloyd RV (2007) Laser capture microdissection for analysis of single cells. Methods Mol Med 132:11–18
24. Huang G, Wang S (2013) Establishment of a new method to detect gene expression by laser capture microdissection-assisted single-cell real time RT-PCR without RNA purification. Mol Biol 47(4):509–514. https://doi.org/10.1134/S0026893313040055
25. Vandewoestyne M, Deforce D (2010) Laser capture microdissection in forensic research: a review. Int J Legal Med 124(6):513–521. https://doi.org/10.1007/s00414-010-0499-4
26. Decarlo K, Emley A, Dadzie OE, Mahalingam M (2011) Laser capture microdissection: methods and applications. Methods Mol Biol 755:1–15. https://doi.org/10.1007/978-1-61779-163-5_1
27. Fuller SA, Takahashi M, Hurrell JG (2001) Cloning of hybridoma cell lines by limiting dilution. Curr Protoc Mol Biol Chapter 11: Unit 11 18. https://doi.org/10.1002/0471142727.mb1108s01
28. Wang Z, Gerstein M, Snyder M (2009) RNA-Seq: a revolutionary tool for transcriptomics. Nat Rev Genet 10(1):57–63. https://doi.org/10.1038/nrg2484
29. Tang FC, Barbacioru C, Wang YZ, Nordman E, Lee C, Xu NL, Wang XH, Bodeau J, Tuch BB, Siddiqui A, Lao KQ, Surani MA (2009) mRNA-Seq whole-transcriptome analysis of a single cell. Nat Methods 6(5):377–U386. https://doi.org/10.1038/Nmeth.1315
30. Trapnell C, Cacchiarelli D, Grimsby J, Pokharel P, Li SQ, Morse M, Lennon NJ, Livak KJ, Mikkelsen TS, Rinn JL (2014) The dynamics and regulators of cell fate decisions are revealed by pseudotemporal ordering of single cells. Nat Biotechnol 32 (4):381–U251. https://doi.org/10.1038/nbt.2859
31. Wills QF, Livak KJ, Tipping AJ, Enver T, Goldson AJ, Sexton DW, Holmes C (2013) Single-cell gene expression analysis reveals genetic associations masked in whole-tissue experiments. Nat Biotechnol 31(8):748. https://doi.org/10.1038/nbt.2642
32. Islam S, Kjallquist U, Moliner A, Zajac P, Fan JB, Lonnerberg P, Linnarsson S (2011) Characterization of the single-cell transcriptional landscape by highly multiplex RNA-seq. Genome Res 21(7):1160–1167. https://doi.org/10.1101/gr.110882.110
33. Islam S, Kjallquist U, Moliner A, Zajac P, Fan JB, Lonnerberg P, Linnarsson S (2012) Highly multiplexed and strand-specific single-cell RNA 5′ end sequencing. Nat Protoc 7(5):813–828. https://doi.org/10.1038/nprot.2012.022
34. Islam S, Zeisel A, Joost S, La Manno G, Zajac P, Kasper M, Lonnerberg P, Linnarsson S (2014) Quantitative single-cell RNA-seq with unique molecular identifiers. Nat Methods 11(2):163–166. https://doi.org/10.1038/nmeth.2772
35. Hashimshony T, Wagner F, Sher N, Yanai I (2012) CEL-seq: single-cell RNA-seq by multiplexed linear amplification. Cell Rep 2 (3):666–673. https://doi.org/10.1016/j.celrep.2012.08.003
36. Ramskold D, Luo SJ, Wang YC, Li R, Deng QL, Faridani OR, Daniels GA, Khrebtukova I, Loring JF, Laurent LC, Schroth GP, Sandberg R (2012) Full-length mRNA-Seq from single-cell levels of RNA and individual circulating tumor cells. Nat Biotechnol 30 (8):777–782. https://doi.org/10.1038/nbt.2282

37. Picelli S, Bjorklund AK, Faridani OR, Sagasser S, Winberg G, Sandberg R (2013) Smart-seq2 for sensitive full-length transcriptome profiling in single cells. Nat Methods 10 (11):1096–1098. https://doi.org/10.1038/Nmeth.2639
38. Sasagawa Y, Nikaido I, Hayashi T, Danno H, Uno KD, Imai T, Ueda HR (2017) Quartz-Seq: a highly reproducible and sensitive single-cell RNA sequencing method, reveals non-genetic gene-expression heterogeneity (vol 14: R31, 2013). Genome Biol 18. https://doi.org/10.1186/S13059-017-1154-X
39. Nakamura T, Yabuta Y, Okamoto I, Aramaki S, Yokobayashi S, Kurimoto K, Sekiguchi K, Nakagawa M, Yamamoto T, Saitou M (2015) SC3-seq: a method for highly parallel and quantitative measurement of single-cell gene expression. Nucleic Acids Res 43(9). https://doi.org/10.1093/nar/gkv134
40. Jaitin DA, Kenigsberg E, Keren-Shaul H, Elefant N, Paul F, Zaretsky I, Mildner A, Cohen N, Jung S, Tanay A, Amit I (2014) Massively parallel single-cell RNA-seq for marker-free decomposition of tissues into cell types. Science 343(6172):776–779. https://doi.org/10.1126/science.1247651
41. Fan HC, Fu GK, Fodor SP (2015) Expression profiling. Combinatorial labeling of single cells for gene expression cytometry. Science 347(6222):1258367. https://doi.org/10.1126/science.1258367
42. Macosko EZ, Basu A, Satija R, Nemesh J, Shekhar K, Goldman M, Tirosh I, Bialas AR, Kamitaki N, Martersteck EM, Trombetta JJ, Weitz DA, Sanes JR, Shalek AK, Regev A, McCarroll SA (2015) Highly parallel genome-wide expression profiling of individual cells using nanoliter droplets. Cell 161 (5):1202–1214. https://doi.org/10.1016/j.cell.2015.05.002
43. Klein AM, Mazutis L, Akartuna I, Tallapragada N, Veres A, Li V, Peshkin L, Weitz DA, Kirschner MW (2015) Droplet barcoding for single-cell transcriptomics applied to embryonic stem cells. Cell 161(5):1187–1201. https://doi.org/10.1016/j.cell.2015.04.044
44. Fan XY, Zhang XN, Wu XL, Guo HS, Hu YQ, Tang FC, Huang YY (2015) Single-cell RNA-seq transcriptome analysis of linear and circular RNAs in mouse preimplantation embryos. Genome Biol 16. https://doi.org/10.1186/S13059-015-0706-1
45. Kang Y, Norris MH, Zarzycki-Siek J, Nierman WC, Donachie SP, Hoang TT (2011) Transcript amplification from single bacterium for transcriptome analysis. Genome Res 21(6):925–935. https://doi.org/10.1101/gr.116103.110
46. Kolodziejczyk AA, Kim JK, Svensson V, Marioni JC, Teichmann SA (2015) The technology and biology of single-cell RNA sequencing. Mol Cell 58(4):610–620. https://doi.org/10.1016/j.molcel.2015.04.005
47. Ziegenhain C, Vieth B, Parekh S, Reinius B, Guillaumet-Adkins A, Smets M, Leonhardt H, Heyn H, Hellmann I, Enard W (2017) Comparative analysis of single-cell RNA sequencing methods. Mol Cell 65 (4):631–643. https://doi.org/10.1016/j.molcel.2017.01.023
48. Svensson V, Natarajan KN, Ly LH, Miragaia RJ, Labalette C, Macaulay IC, Cvejic A, Teichmann SA (2017) Power analysis of single-cell RNA-sequencing experiments. Nat Methods 14(4):381–387. https://doi.org/10.1038/nmeth.4220
49. Clark SJ, Lee HJ, Smallwood SA, Kelsey G, Reik W (2016) Single-cell epigenomics: powerful new methods for understanding gene regulation and cell identity. Genome Biol 17. https://doi.org/10.1186/s13059-016-0944-x
50. Plongthongkum N, Diep DH, Zhang K (2014) Advances in the profiling of DNA modifications: cytosine methylation and beyond. Nat Rev Genet 15(10):647–661. https://doi.org/10.1038/nrg3772
51. Smallwood SA, Lee HJ, Angermueller C, Krueger F, Saadeh H, Peet J, Andrews SR, Stegle O, Reik W, Kelsey G (2014) Single-cell genome-wide bisulfite sequencing for assessing epigenetic heterogeneity. Nat Methods 11(8):817–820. https://doi.org/10.1038/Nmeth.3035
52. Guo HS, Zhu P, Wu XL, Li XL, Wen L, Tang FC (2013) Single-cell methylome landscapes of mouse embryonic stem cells and early embryos analyzed using reduced representation bisulfite sequencing. Genome Res 23 (12):2126–2135. https://doi.org/10.1101/gr.161679.113
53. Miura F, Enomoto Y, Dairiki R, Ito T (2012) Amplification-free whole-genome bisulfite sequencing by post-bisulfite adaptor tagging. Nucleic Acids Res 40(17). https://doi.org/10.1093/nar/gks454
54. Farlik M, Sheffield NC, Nuzzo A, Datlinger P, Schonegger A, Klughammer J, Bock C (2015) Single-cell DNA methylome sequencing and bioinformatic inference of epigenomic cell-state dynamics. Cell Rep 10(8):1386–1397.

https://doi.org/10.1016/j.celrep.2015.02.001

55. Lorthongpanich C, Cheow LF, Balu S, Quake SR, Knowles BB, Burkholder WF, Solter D, Messerschmidt DM (2013) Single-cell DNA-methylation analysis reveals epigenetic chimerism in preimplantation embryos. Science 341 (6150):1110–1112. https://doi.org/10.1126/science.1240617
56. Thurman RE, Rynes E, Humbert R, Vierstra J, Maurano MT, Haugen E, Sheffield NC, Stergachis AB, Wang H, Vernot B, Garg K, John S, Sandstrom R, Bates D, Boatman L, Canfield TK, Diegel M, Dunn D, Ebersol AK, Frum T, Giste E, Johnson AK, Johnson EM, Kutyavin T, Lajoie B, Lee BK, Lee K, London D, Lotakis D, Neph S, Neri F, Nguyen ED, Qu HZ, Reynolds AP, Roach V, Safi A, Sanchez ME, Sanyal A, Shafer A, Simon JM, Song LY, Vong S, Weaver M, Yan YQ, Zhang ZC, Zhang ZZ, Lenhard B, Tewari M, Dorschner MO, Hansen RS, Navas PA, Stamatoyannopoulos G, Iyer VR, Lieb JD, Sunyaev SR, Akey JM, Sabo PJ, Kaul R, Furey TS, Dekker J, Crawford GE, Stamatoyannopoulos JA (2012) The accessible chromatin landscape of the human genome. Nature 489 (7414):75–82. https://doi.org/10.1038/nature11232
57. Schones DE, Cui KR, Cuddapah S, Roh TY, Barski A, Wang ZB, Wei G, Zhao KJ (2008) Dynamic regulation of nucleosome positioning in the human genome. Cell 132 (5):887–898. https://doi.org/10.1016/j.cell.2008.02.022
58. Buenrostro JD, Giresi PG, Zaba LC, Chang HY, Greenleaf WJ (2013) Transposition of native chromatin for fast and sensitive epigenomic profiling of open chromatin, DNA-binding proteins and nucleosome position. Nat Methods 10(12):1213. https://doi.org/10.1038/Nmeth.2688
59. Adli M, Zhu JA, Bernstein BE (2010) Genome-wide chromatin maps derived from limited numbers of hematopoietic progenitors. Nat Methods 7(8):615–U624. https://doi.org/10.1038/Nmeth.1478
60. Lara-Astiaso D, Weiner A, Lorenzo-Vivas E, Zaretsky I, Jaitin DA, David E, Keren-Shaul H, Mildner A, Winter D, Jung S, Friedman N, Amit I (2014) Chromatin state dynamics during blood formation. Science 345 (6199):943–949. https://doi.org/10.1126/science.1256271
61. Brind'Amour J, Liu S, Hudson M, Chen C, Karimi MM, Lorincz MC (2015) An ultra-low-input native ChIP-seq protocol for genome-wide profiling of rare cell populations. Nat Commun 6. https://doi.org/10.1038/Ncomms7033
62. Cao ZN, Chen CY, He B, Tan K, Lu C (2015) A microfluidic device for epigenomic profiling using 100 cells. Nat Methods 12 (10):959–962. https://doi.org/10.1038/Nmeth.3488
63. Cusanovich DA, Daza R, Adey A, Pliner HA, Christiansen L, Gunderson KL, Steemers FJ, Trapnell C, Shendure J (2015) Multiplex single-cell profiling of chromatin accessibility by combinatorial cellular indexing. Science 348 (6237):910–914. https://doi.org/10.1126/science.aab1601
64. Buenostro JD, Wu BJ, Litzenburger UM, Ruff D, Gonzales ML, Snyder MP, Chang HY, Greenleaf WJ (2015) Single-cell chromatin accessibility reveals principles of regulatory variation. Nature 523(7561):486–U264. https://doi.org/10.1038/nature14590
65. Dekker J, Rippe K, Dekker M, Kleckner N (2002) Capturing chromosome conformation. Science 295(5558):1306–1311. https://doi.org/10.1126/science.1067799
66. Lieberman-Aiden E, van Berkum NL, Williams L, Imakaev M, Ragoczy T, Telling A, Amit I, Lajoie BR, Sabo PJ, Dorschner MO, Sandstrom R, Bernstein B, Bender MA, Groudine M, Gnirke A, Stamatoyannopoulos J, Mirny LA, Lander ES, Dekker J (2009) Comprehensive mapping of long-range interactions reveals folding principles of the human genome. Science 326(5950):289–293. https://doi.org/10.1126/science.1181369
67. Simonis M, Klous P, Splinter E, Moshkin Y, Willemsen R, de Wit E, van Steensel B, de Laat W (2006) Nuclear organization of active and inactive chromatin domains uncovered by chromosome conformation capture-on-chip (4C). Nat Genet 38(11):1348–1354. https://doi.org/10.1038/ng1896
68. Lanctot C, Cheutin T, Cremer M, Cavalli G, Cremer T (2007) Dynamic genome architecture in the nuclear space: regulation of gene expression in three dimensions. Nat Rev Genet 8(2):104–115. https://doi.org/10.1038/nrg2041
69. Parada LA, Roix JJ, Misteli T (2003) An uncertainty principle in chromosome positioning. Trends Cell Biol 13(8):393–396. https://doi.org/10.1016/S0962-8924(03)00149-1
70. Nagano T, Lubling Y, Stevens TJ, Schoenfelder S, Yaffe E, Dean W, Laue ED, Tanay A, Fraser P (2013) Single-cell Hi-C reveals cell-to-cell variability in chromosome structure. Nature 502(7469):59. https://doi.org/10.1038/nature12593

71. Dey SS, Kester L, Spanjaard B, Bienko M, van Oudenaarden A (2015) Integrated genome and transcriptome sequencing of the same cell. Nat Biotechnol 33(3):285. https://doi.org/10.1038/nbt.3129
72. Macaulay IC, Teng MJ, Haerty W, Kumar P, Ponting CP, Voet T (2016) Separation and parallel sequencing of the genomes and transcriptomes of single cells using G&T-seq. Nat Protoc 11(11):36–58. https://doi.org/10.1038/nprot.2016.138
73. Macaulay IC, Haerty W, Kumar P, Li YI, Hu TX, Teng MJ, Goolam M, Saurat N, Coupland P, Shirley LM, Smith M, Van der Aa N, Banerjee R, Ellis PD, Quail MA, Swerdlow HP, Zernicka-Goetz M, Livesey FJ, Ponting CP, Voet T (2015) G&T-seq: parallel sequencing of single-cell genomes and transcriptomes. Nat Methods 12(6):519. https://doi.org/10.1038/nmeth.3370
74. Hu YJ, Huang K, An Q, Du GZ, Hu GL, Xue JF, Zhu XM, Wang CY, Xue ZG, Fan GP (2016) Simultaneous profiling of transcriptome and DNA methylome from a single cell. Genome Biol 17. https://doi.org/10.1186/s13059-016-0950-z
75. Zong CH, Lu SJ, Chapman AR, Xie XS (2012) Genome-wide detection of single-nucleotide and copy-number variations of a single human cell. Science 338 (6114):1622–1626. https://doi.org/10.1126/science.1229164
76. Picelli S, Faridani OR, Bjorklund AK, Winberg G, Sagasser S, Sandberg R (2014) Full-length RNA-seq from single cells using Smart-seq2. Nat Protoc 9(1):171–181. https://doi.org/10.1038/nprot.2014.006
77. Laszlo AH, Derrington IM, Brinkerhoff H, Langford KW, Nova IC, Samson JM, Bartlett JJ, Pavlenok M, Gundlach JH (2013) Detection and mapping of 5-methylcytosine and 5-hydroxymethylcytosine with nanopore MspA. Proc Natl Acad Sci U S A 110 (47):18904–18909. https://doi.org/10.1073/pnas.1310240110
78. Laszlo AH, Derrington IM, Manrao EA, Gundlach JH (2013) Detection and mapping of 5-methylcytosine and 5-hydroxymethylcytosine in short strands of ssDNA using nanopore sequencing with MspA. Biophys J 104 (2):211a
79. Buettner F, Natarajan KN, Casale FP, Proserpio V, Scialdone A, Theis FJ, Teichmann SA, Marioni JC, Stegie O (2015) Computational analysis of cell-to-cell heterogeneity in single-cell RNA-sequencing data reveals hidden subpopulations of cells. Nat Biotechnol 33 (2):155–160. https://doi.org/10.1038/nbt.3102
80. Buettner F, Pratanwanich N, Marioni JC, Stegle O (2016) Scalable latent-factor models applied to single-cell RNA-seq data separate biological drivers from confounding effects. bioRxiv 2016:087775
81. Leng N, Choi J, Chu LF, Thomson JA, Kendziorski C, Stewart R (2016) OEFinder: a user interface to identify and visualize ordering effects in single-cell RNA-seq data. Bioinformatics 32(9):1408–1410. https://doi.org/10.1093/bioinformatics/btw004
82. Brennecke P, Anders S, Kim JK, Kolodziejczyk AA, Zhang XW, Proserpio V, Baying B, Benes V, Teichmann SA, Marioni JC, Heisler MG (2013) Accounting for technical noise in single-cell RNA-seq experiments. Nat Methods 10(11):1093–1095. https://doi.org/10.1038/Nmeth.2645
83. Katayama S, Tohonen V, Linnarsson S, Kere J (2013) SAMstrt: statistical test for differential expression in single-cell transcriptome with spike-in normalization. Bioinformatics 29 (22):2943–2945. https://doi.org/10.1093/bioinformatics/btt511
84. Ding B, Zheng LN, Zhu Y, Li N, Jia HY, Ai RZ, Wildberg A, Wang W (2015) Normalization and noise reduction for single cell RNA-seq experiments. Bioinformatics 31 (13):2225–2227. https://doi.org/10.1093/bioinformatics/btv122
85. Bacher R, Chu L-F, Leng N, Gasch AP, Thomson JA, Stewart RM, Newton M, Kendziorski C (2016) SCnorm: a quantile-regression based approach for robust normalization of single-cell RNA-seq data. bioRxiv 2016:090167
86. Sengupta D, Rayan NA, Lim M, Lim B, Prabhakar S (2016) Fast, scalable and accurate differential expression analysis for single cells. bioRxiv 2016:049734
87. Cole M RD, Wagner A, Ngai J, Purdom E, Dudoit S, Yosef N. SCONE: correcting and evaluating the influence of unwanted variation on single-cell RNA-seq data. https://niryosef.wordpress.com/tools/scone/
88. Kharchenko PV, Silberstein L, Scadden DT (2014) Bayesian approach to single-cell differential expression analysis. Nat Methods 11 (7):740–U184. https://doi.org/10.1038/Nmeth.2967
89. Vieth B, Ziegenhain C, Parekh S, Enard W, Hellmann I (2017) powsimR: power analysis for bulk and single cell RNA-seq experiments. bioRxiv 2017:117150
90. Finak G, McDavid A, Yajima M, Deng JY, Gersuk V, Shalek AK, Slichter CK, Miller HW, McElrath MJ, Prlic M, Linsley PS, Gottardo R (2015) MAST: a flexible statistical framework for assessing transcriptional

changes and characterizing heterogeneity in single-cell RNA sequencing data. Genome Biol 16. https://doi.org/10.1186/S13059-015-0844-5

91. Delmans M, Hemberg M (2016) Discrete distributional differential expression ((DE)-E-3) – a tool for gene expression analysis of single-cell RNA-seq data. BMC Bioinformatics 17. https://doi.org/10.1186/S12859-016-0944-6

92. Korthauer KD, Chu LF, Newton MA, Li Y, Thomson J, Stewart R, Kendziorski C (2016) A statistical approach for identifying differential distributions in single-cell RNA-seq experiments. Genome Biol 17(1):222. https://doi.org/10.1186/s13059-016-1077-y

93. Korthauer KD, Chu L-F, Newton MA, Li Y, Thomson J, Stewart R, Kendziorski C (2016) scDD: a statistical approach for identifying differential distributions in single-cell RNA-seq experiments. bioRxiv 2016:035501

94. Jia C, Kelly D, Kim J, Li M, Zhang N (2017) Accounting for technical noise in single-cell RNA sequencing analysis. bioRxiv 2017:116939

95. Svensson V, Teichmann SA, Stegle O (2017) SpatialDE-identification of spatially variable genes. bioRxiv 2017:143321

96. Grun D, Lyubimova A, Kester L, Wiebrands K, Basak O, Sasaki N, Clevers H, van Oudenaarden A (2015) Single-cell messenger RNA sequencing reveals rare intestinal cell types. Nature 525(7568):251. https://doi.org/10.1038/nature14966

97. Zeisel A, Munoz-Manchado AB, Codeluppi S, Lonnerberg P, La Manno G, Jureus A, Marques S, Munguba H, He LQ, Betsholtz C, Rolny C, Castelo-Branco G, Hjerling-Leffler J, Linnarsson S (2015) Cell types in the mouse cortex and hippocampus revealed by single-cell RNA-seq. Science 347 (6226):1138–1142. https://doi.org/10.1126/science.aaa1934

98. Pierson E, Yau C (2015) ZIFA: dimensionality reduction for zero-inflated single-cell gene expression analysis. Genome Biol 16. https://doi.org/10.1186/S13059-015-0805-Z

99. Angerer P, Haghverdi L, Buttner M, Theis FJ, Marr C, Buettner F (2016) destiny: diffusion maps for large-scale single cell data in R. Bioinformatics 32(8):1241–1243. https://doi.org/10.1093/bioinformatics/btv715

100. Xu C, Su ZC (2015) Identification of cell types from single-cell transcriptomes using a novel clustering method. Bioinformatics 31 (12):1974–1980. https://doi.org/10.1093/bioinformatics/btv088

101. Marco E, Karp RL, Guo GJ, Robson P, Hart AH, Trippa L, Yuan GC (2014) Bifurcation analysis of single-cell gene expression data reveals epigenetic landscape. Proc Natl Acad Sci U S A 111(52):E5643–E5650. https://doi.org/10.1073/pnas.1408993111

102. Leng N, Chu LF, Barry C, Li Y, Choi J, Li XM, Jiang P, Stewart RM, Thomson JA, Kendziorski C (2015) Oscope identifies oscillatory genes in unsynchronized single-cell RNA-seq experiments. Nat Methods 12 (10):947–950. https://doi.org/10.1038/Nmeth.3549

103. Ji ZC, Ji HK (2016) TSCAN: pseudo-time reconstruction and evaluation in single-cell RNA-seq analysis. Nucleic Acids Res 44(13). https://doi.org/10.1093/nar/gkw430

104. Specht AT, Li J (2017) LEAP: constructing gene co-expression networks for single-cell RNA-sequencing data using pseudotime ordering. Bioinformatics 33(5):764–766. https://doi.org/10.1093/bioinformatics/btw729

105. Welch JD, Hartemink AJ, Prins JF (2016) SLICER: inferring branched, nonlinear cellular trajectories from single cell RNA-seq data. Genome Biol 17. https://doi.org/10.1186/S13059-016-0975-3

106. duVerle D, Yotsukura S, Nomura S, Aburatani H, Tsuda K (2016) CellTree: an R/bioconductor package to infer the hierarchical structure of cell populations from single-cell RNA-seq data. BMC Bioinformatics 17. https://doi.org/10.1186/S12859-016-1175-6

107. Rashid S, Kotton DN, Bar-Joseph Z (2017) TASIC: determining branching models from time series single cell data. Bioinformatics. https://doi.org/10.1093/bioinformatics/btx173

108. Lönnberg T, Svensson V, James KR, Fernandez-Ruiz D, Sebina I, Montandon R, Soon MS, Fogg LG, Nair AS, Liligeto U (2017) Single-cell RNA-seq and computational analysis using temporal mixture modelling resolves Th1/Tfh fate bifurcation in malaria. Sci Immunol 2(9):eaal2192

109. Matsumoto H, Kiryu H (2016) SCOUP: a probabilistic model based on the Ornstein-Uhlenbeck process to analyze single-cell expression data during differentiation. BMC Bioinformatics 17. https://doi.org/10.1186/S12859-016-1109-3

110. McCarthy DJ, Campbell KR, Lun ATL, Wills QF (2017) Scater: pre-processing, quality control, normalization and visualization of single-cell RNA-seq data in R. Bioinformatics 33(8):1179–1186. https://doi.org/10.1093/bioinformatics/btw777

111. Campbell KR, Yau C (2017) switchde: inference of switch-like differential expression along single-cell trajectories. Bioinformatics 33(8):1241–1242. https://doi.org/10.1093/bioinformatics/btw798
112. Campbell K, Yau C (2016) Ouija: incorporating prior knowledge in single-cell trajectory learning using Bayesian nonlinear factor analysis. bioRxiv 2016:060442
113. Campbell K, Ponting CP, Webber C (2015) Laplacian eigenmaps and principal curves for high resolution pseudotemporal ordering of single-cell RNA-seq profiles. bioRxiv 2015:027219
114. Risso D, Perraudeau F, Gribkova S, Dudoit S, Vert J-P (2017) ZINB-WaVE: a general and flexible method for signal extraction from single-cell RNA-seq data. bioRxiv 2017:125112
115. Shaham U, Stanton KP, Li H, Montgomery R, Kluger Y (2016) Removal of batch effects using distribution-matching residual networks. arXiv 2016:161004181
116. Streets AM, Huang YY (2014) How deep is enough in single-cell RNA-seq? Nat Biotechnol 32(10):1005–1006. https://doi.org/10.1038/nbt.3039
117. Saliba AE, Westermann AJ, Gorski SA, Vogel J (2014) Single-cell RNA-seq: advances and future challenges. Nucleic Acids Res 42 (14):8845–8860. https://doi.org/10.1093/nar/gku555
118. Trapnell C, Pachter L, Salzberg SL (2009) TopHat: discovering splice junctions with RNA-Seq. Bioinformatics 25(9):1105–1111. https://doi.org/10.1093/bioinformatics/btp120
119. Guttman M, Garber M, Levin JZ, Donaghey J, Robinson J, Adiconis X, Fan L, Koziol MJ, Gnirke A, Nusbaum C, Rinn JL, Lander ES, Regev A (2010) Ab initio reconstruction of cell type-specific transcriptomes in mouse reveals the conserved multi-exonic structure of lincRNAs. Nat Biotechnol 28 (5):503–U166. https://doi.org/10.1038/nbt.1633
120. Jiang LC, Schlesinger F, Davis CA, Zhang Y, Li RH, Salit M, Gingeras TR, Oliver B (2011) Synthetic spike-in standards for RNA-seq experiments. Genome Res 21 (9):1543–1551. https://doi.org/10.1101/gr.121095.111
121. Fu GK, Hu J, Wang PH, Fodor SPA (2011) Counting individual DNA molecules by the stochastic attachment of diverse labels. Proc Natl Acad Sci U S A 108(22):9026–9031. https://doi.org/10.1073/pnas.1017621108
122. Anders S, Pyl PT, Huber W (2015) HTSeq-a Python framework to work with high-throughput sequencing data. Bioinformatics 31(2):166–169. https://doi.org/10.1093/bioinformatics/btu638
123. Welch JD, Hu Y, Prins JF (2016) Robust detection of alternative splicing in a population of single cells. Nucleic Acids Res 44(8). https://doi.org/10.1093/nar/gkv1525
124. Qiu X, Hill A, Packer J, Lin D, Ma YA, Trapnell C (2017) Single-cell mRNA quantification and differential analysis with census. Nat Methods 14(3):309–315. https://doi.org/10.1038/nmeth.4150
125. Davis MP, van Dongen S, Abreu-Goodger C, Bartonicek N, Enright AJ (2013) Kraken: a set of tools for quality control and analysis of high-throughput sequence data. Methods 63 (1):41–49. https://doi.org/10.1016/j.ymeth.2013.06.027
126. Thorvaldsdottir H, Robinson JT, Mesirov JP (2013) Integrative genomics viewer (IGV): high-performance genomics data visualization and exploration. Brief Bioinform 14 (2):178–192. https://doi.org/10.1093/bib/bbs017
127. Robinson JT, Thorvaldsdottir H, Winckler W, Guttman M, Lander ES, Getz G, Mesirov JP (2011) Integrative genomics viewer. Nat Biotechnol 29(1):24–26. https://doi.org/10.1038/nbt.1754
128. Krebs JE, Lewin B, Goldstein ES, Kilpatrick ST (2014) Lewin's genes XI. Jones & Bartlett Publishers, Burlington, MA
129. Brennecke P, Anders S, Kim JK, Kolodziejczyk AA, Zhang XW, Proserpio V, Baying B, Benes V, Teichmann SA, Marioni JC, Heisler MG (2014) Accounting for technical noise in single-cell RNA-seq experiments (vol 10: p. 1093, 2013). Nat Methods 11(2):210–210. https://doi.org/10.1038/nmeth0214-210b
130. Vallejos CA, Marioni JC, Richardson S (2015) BASiCS: Bayesian analysis of single-cell sequencing data. PLoS Comput Biol 11 (6). https://doi.org/10.1371/journal.pcbi.1004333
131. Vallejos CA, Richardson S, Marioni JC (2016) Beyond comparisons of means: understanding changes in gene expression at the single-cell level. Genome Biol 17. https://doi.org/10.1186/s13059-016-0930-3
132. Risso D, Ngai J, Speed TP, Dudoit S (2014) Normalization of RNA-seq data using factor analysis of control genes or samples. Nat Biotechnol 32(9):896–902. https://doi.org/10.1038/nbt.2931

133. Yadav VK, De S (2015) An assessment of computational methods for estimating purity and clonality using genomic data derived from heterogeneous tumor tissue samples. Brief Bioinform 16(2):232–241. https://doi.org/10.1093/bib/bbu002
134. Tang FC, Barbacioru C, Bao SQ, Lee C, Nordman E, Wang XH, Lao KQ, Surani MA (2010) Tracing the derivation of embryonic stem cells from the inner cell mass by single-cell RNA-seq analysis. Cell Stem Cell 6 (5):468–478. https://doi.org/10.1016/j.stem.2010.03.015
135. Treutlein B, Brownfield DG, Wu AR, Neff NF, Mantalas GL, Espinoza FH, Desai TJ, Krasnow MA, Quake SR (2014) Reconstructing lineage hierarchies of the distal lung epithelium using single-cell RNA-seq. Nature 509(7500):371. https://doi.org/10.1038/nature13173
136. Durruthy-Durruthy R, Gottlieb A, Hartman BH, Waldhaus J, Laske RD, Altman R, Heller S (2014) Reconstruction of the mouse otocyst and early neuroblast lineage at single-cell resolution. Cell 157(4):964–978. https://doi.org/10.1016/j.cell.2014.03.036
137. Moignard V, Macaulay IC, Swiers G, Buettner F, Schutte J, Calero-Nieto FJ, Kinston S, Joshi A, Hannah R, Theis FJ, Jacobsen SE, de Bruijn MF, Gottgens B (2013) Characterization of transcriptional networks in blood stem and progenitor cells using high-throughput single-cell gene expression analysisNat. Cell Biol 15(4):363. https://doi.org/10.1038/ncb2709
138. Mahata B, Zhang XW, Kolodziejczyk AA, Proserpio V, Haim-Vilmovsky L, Taylor AE, Hebenstreit D, Dingler FA, Moignard V, Gottgens B, Arlt W, McKenzie ANJ, Teichmann SA (2014) Single-cell RNA sequencing reveals T helper cells synthesizing steroids de novo to contribute to immune homeostasis. Cell Rep 7(4):1130–1142. https://doi.org/10.1016/j.celrep.2014.04.011
139. Patel AP, Tirosh I, Trombetta JJ, Shalek AK, Gillespie SM, Wakimoto H, Cahill DP, Nahed BV, Curry WT, Martuza RL, Louis DN, Rozenblatt-Rosen O, Suva ML, Regev A, Bernstein BE (2014) Single-cell RNA-seq highlights intratumoral heterogeneity in primary glioblastoma. Science 344 (6190):1396–1401. https://doi.org/10.1126/science.1254257
140. Anders S, Huber W (2010) Differential expression analysis for sequence count data. Genome Biol 11(10). https://doi.org/10.1186/Gb-2010-11-10-R106
141. Zhou XB, Lindsay H, Robinson MD (2014) Robustly detecting differential expression in RNA sequencing data using observation weights. Nucleic Acids Res 42(11). https://doi.org/10.1093/nar/gku310
142. Robinson MD, McCarthy DJ, Smyth GK (2010) edgeR: a Bioconductor package for differential expression analysis of digital gene expression data. Bioinformatics 26 (1):139–140. https://doi.org/10.1093/bioinformatics/btp616
143. Trapnell C, Roberts A, Goff L, Pertea G, Kim D, Kelley DR, Pimentel H, Salzberg SL, Rinn JL, Pachter L (2014) Differential gene and transcript expression analysis of RNA-seq experiments with TopHat and Cufflinks (vol 7: p. 562, 2012). Nat Protoc 9 (10):2513–2513. https://doi.org/10.1038/nprot1014-2513a
144. Trapnell C, Roberts A, Goff L, Pertea G, Kim D, Kelley DR, Pimentel H, Salzberg SL, Rinn JL, Pachter L (2012) Differential gene and transcript expression analysis of RNA-seq experiments with TopHat and Cufflinks. Nat Protoc 7(3):562–578. https://doi.org/10.1038/nprot.2012.016
145. Rapaport F, Khanin R, Liang YP, Pirun M, Krek A, Zumbo P, Mason CE, Socci ND, Betel D (2013) Comprehensive evaluation of differential gene expression analysis methods for RNA-seq data. Genome Biol 14(9). https://doi.org/10.1186/Gb-2013-14-9-R95
146. Shalek AK, Satija R, Adiconis X, Gertner RS, Gaublomme JT, Raychowdhury R, Schwartz S, Yosef N, Malboeuf C, Lu DN, Trombetta JJ, Gennert D, Gnirke A, Goren A, Hacohen N, Levin JZ, Park H, Regev A (2013) Single-cell transcriptomics reveals bimodality in expression and splicing in immune cells. Nature 498(7453):236–240. https://doi.org/10.1038/nature12172
147. Anders S, Reyes A, Huber W (2012) Detecting differential usage of exons from RNA-seq data. Genome Res 22(10):2008–2017. https://doi.org/10.1101/gr.133744.111
148. Katz Y, Wang ET, Airoldi EM, Burge CB (2010) Analysis and design of RNA sequencing experiments for identifying isoform regulation. Nat Methods 7(12):1009–U1101. https://doi.org/10.1038/Nmeth.1528
149. Kim JK, Marioni JC (2013) Inferring the kinetics of stochastic gene expression from single-cell RNA-sequencing data. Genome Biol 14(1):R7. https://doi.org/10.1186/gb-2013-14-1-r7
150. Deng Q, Ramskold D, Reinius B, Sandberg R (2014) Single-cell RNA-seq reveals dynamic,

random monoallelic gene expression in mammalian cells. Science 343(6167):193–196. https://doi.org/10.1126/science.1245316

151. Grun D, Kester L, van Oudenaarden A (2014) Validation of noise models for single-cell transcriptomics. Nat Methods 11(6):637. https://doi.org/10.1038/Nmeth.2930

152. McManus CJ, Coolon JD, Duff MO, Eipper-Mains J, Graveley BR, Wittkopp PJ (2010) Regulatory divergence in Drosophila revealed by mRNA-seq. Genome Res 20(6):816–825. https://doi.org/10.1101/gr.102491.109

153. Gawad C, Koh W, Quake SR (2014) Dissecting the clonal origins of childhood acute lymphoblastic leukemia by single-cell genomics. Proc Natl Acad Sci U S A 111 (50):17947–17952. https://doi.org/10.1073/pnas.1420822111

154. Zhang CZ, Adalsteinsson VA, Francis J, Cornils H, Jung J, Maire C, Ligon KL, Meyerson M, Love JC (2015) Calibrating genomic and allelic coverage bias in single-cell sequencing. Nat Commun 6:6822. https://doi.org/10.1038/ncomms7822

155. Daley T, Smith AD (2014) Modeling genome coverage in single-cell sequencing. Bioinformatics 30(22):3159–3165. https://doi.org/10.1093/bioinformatics/btu540

156. Clingenpeel S, Clum A, Schwientek P, Rinke C, Woyke T (2015) Reconstructing each cell's genome within complex microbial communities-dream or reality? Front Microbiol 5. https://doi.org/10.3389/Fmicb.2014.00771

157. Wang Y, Waters J, Leung ML, Unruh A, Roh W, Shi X, Chen K, Scheet P, Vattathil S, Liang H, Multani A, Zhang H, Zhao R, Michor F, Meric-Bernstam F, Navin NE (2014) Clonal evolution in breast cancer revealed by single nucleus genome sequencing. Nature 512 (7513):155–160. https://doi.org/10.1038/nature13600

158. Nikolenko SI, Korobeynikov AI, Alekseyev MA (2013) BayesHammer: Bayesian clustering for error correction in single-cell sequencing. BMC Genomics 14(Suppl 1):S7. https://doi.org/10.1186/1471-2164-14-S1-S7

159. Zhang C, Zhang C, Chen S, Yin X, Pan X, Lin G, Tan Y, Tan K, Xu Z, Hu P, Li X, Chen F, Xu X, Li Y, Zhang X, Jiang H, Wang W (2013) A single cell level based method for copy number variation analysis by low coverage massively parallel sequencing. PLoS One 8(1):e54236. https://doi.org/10.1371/journal.pone.0054236

160. Cheng JQ, Vanneste E, Konings P, Voet T, Vermeesch JR, Moreau Y (2011) Single-cell copy number variation detection. Genome Biol 12(8). https://doi.org/10.1186/Gb-2011-12-8-R80

161. Bankevich A, Nurk S, Antipov D, Gurevich AA, Dvorkin M, Kulikov AS, Lesin VM, Nikolenko SI, Pham S, Prjibelski AD, Pyshkin AV, Sirotkin AV, Vyahhi N, Tesler G, Alekseyev MA, Pevzner PA (2012) SPAdes: a new genome assembly algorithm and its applications to single-cell sequencing. J Comput Biol 19(5):455–477. https://doi.org/10.1089/cmb.2012.0021

162. Peng Y, Leung HCM, Yiu SM, Chin FYL (2012) IDBA-UD: a de novo assembler for single-cell and metagenomic sequencing data with highly uneven depth. Bioinformatics 28 (11):1420–1428. https://doi.org/10.1093/bioinformatics/bts174

163. Navin N, Kendall J, Troge J, Andrews P, Rodgers L, McIndoo J, Cook K, Stepansky A, Levy D, Esposito D, Muthuswamy L, Krasnitz A, McCombie WR, Hicks J, Wigler M (2011) Tumour evolution inferred by single-cell sequencing. Nature 472 (7341):90–U119. https://doi.org/10.1038/nature09807

164. Kuipers J, Jahn K, Beerenwinkel N (2017) Advances in understanding tumour evolution through single-cell sequencing. Biochim Biophys Acta 1867(2):127–138. https://doi.org/10.1016/j.bbcan.2017.02.001

165. Nik-Zainal S, Van Loo P, Wedge DC, Alexandrov LB, Greenman CD, Lau KW, Raine K, Jones D, Marshall J, Ramakrishna M, Shlien A, Cooke SL, Hinton J, Menzies A, Stebbings LA, Leroy C, Jia MM, Rance R, Mudie LJ, Gamble SJ, Stephens PJ, McLaren S, Tarpey PS, Papaemmanuil E, Davies HR, Varela I, McBride DJ, Bignell GR, Leung K, Butler AP, Teague JW, Martin S, Jonsson G, Mariani O, Boyault S, Miron P, Fatima A, Langerod A, Aparicio SAJR, Tutt A, Sieuwerts AM, Borg A, Thomas G, Salomon AV, Richardson AL, Borresen-Dale AL, Futreal PA, Stratton MR, Campbell PJ, Consortium ICG (2012) The life history of 21 breast cancers. Cell 149 (5). https://doi.org/10.1016/j.cell.2012.04.023

166. Eisen MB, Spellman PT, Brown PO, Botstein D (1998) Cluster analysis and display of genome-wide expression patterns. Proc Natl Acad Sci U S A 95(25):14863–14868. https://doi.org/10.1073/pnas.95.25.14863

167. Roth A, McPherson A, Laks E, Biele J, Yap D, Wan A, Smith MA, Nielsen CB, McAlpine JN, Aparicio S, Bouchard-Cote A, Shah SP (2016) Clonal genotype and population

structure inference from single-cell tumor sequencing. Nat Methods 13(7):573. https://doi.org/10.1038/Nmeth.3867

168. Fraley C, Raftery AE (2002) Model-based clustering, discriminant analysis, and density estimation. J Am Stat Assoc 97 (458):611–631. https://doi.org/10.1198/016214502760047131

169. Meng XL, Rubin DB (1993) Maximum-likelihood-estimation via the Ecm algorithm–a general framework. Biometrika 80 (2):267–278. https://doi.org/10.2307/2337198

170. Fraley C, Raftery AE (1998) How many clusters? Which clustering method? Answers via model-based cluster analysis. Comput J 41 (8):578–588

171. Fraley C, Raftery AE (1999) MCLUST: software for model-based cluster analysis. J Classif 16(2):297–306

172. Bentley DR, Balasubramanian S, Swerdlow HP, Smith GP, Milton J, Brown CG, Hall KP, Evers DJ, Barnes CL, Bignell HR, Boutell JM, Bryant J, Carter RJ, Cheetham RK, Cox AJ, Ellis DJ, Flatbush MR, Gormley NA, Humphray SJ, Irving LJ, Karbelashvili MS, Kirk SM, Li H, Liu XH, Maisinger KS, Murray LJ, Obradovic B, Ost T, Parkinson ML, Pratt MR, Rasolonjatovo IMJ, Reed MT, Rigatti R, Rodighiero C, Ross MT, Sabot A, Sankar SV, Scally A, Schroth GP, Smith ME, Smith VP, Spiridou A, Torrance PE, Tzonev SS, Vermaas EH, Walter K, Wu XL, Zhang L, Alam MD, Anastasi C, Aniebo IC, Bailey DMD, Bancarz IR, Banerjee S, Barbour SG, Baybayan PA, Benoit VA, Benson KF, Bevis C, Black PJ, Boodhun A, Brennan JS, Bridgham JA, Brown RC, Brown AA, Buermann DH, Bundu AA, Burrows JC, Carter NP, Castillo N, Catenazzi MCE, Chang S, Cooley RN, Crake NR, Dada OO, Diakoumakos KD, Dominguez-Fernandez B, Earnshaw DJ, Egbujor UC, Elmore DW, Etchin SS, Ewan MR, Fedurco M, Fraser LJ, Fajardo KVF, Furey WS, George D, Gietzen KJ, Goddard CP, Golda GS, Granieri PA, Green DE, Gustafson DL, Hansen NF, Harnish K, Haudenschild CD, Heyer NI, Hims MM, Ho JT, Horgan AM, Hoschler K, Hurwitz S, Ivanov DV, Johnson MQ, James T, Jones TAH, Kang GD, Kerelska TH, Kersey AD, Khrebtukova I, Kindwall AP, Kingsbury Z, Kokko-Gonzales PI, Kumar A, Laurent MA, Lawley CT, Lee SE, Lee X, Liao AK, Loch JA, Lok M, Luo SJ, Mammen RM, Martin JW, McCauley PG, McNitt P, Mehta P, Moon KW, Mullens JW, Newington T, Ning ZM, Ng BL, Novo SM, O'Neill MJ, Osborne MA, Osnowski A, Ostadan O, Paraschos LL, Pickering L, Pike AC, Pike AC, Pinkard DC, Pliskin DP, Podhasky J, Quijano VJ, Raczy C, Rae VH, Rawlings SR, Rodriguez AC, Roe PM, Rogers J, Bacigalupo MCR, Romanov N, Romieu A, Roth RK, Rourke NJ, Ruediger ST, Rusman E, Sanches-Kuiper RM, Schenker MR, Seoane JM, Shaw RJ, Shiver MK, Short SW, Sizto NL, Sluis JP, Smith MA, Sohna JES, Spence EJ, Stevens K, Sutton N, Szajkowski L, Tregidgo CL, Turcatti G, vandeVondele S, Verhovsky Y, Virk SM, Wakelin S, Walcott GC, Wang JW, Worsley GJ, Yan JY, Yau L, Zuerlein M, Rogers J, Mullikin JC, Hurles ME, McCooke NJ, West JS, Oaks FL, Lundberg PL, Klenerman D, Durbin R, Smith AJ (2008) Accurate whole human genome sequencing using reversible terminator chemistry. Nature 456(7218):53–59. https://doi.org/10.1038/nature07517

173. Kim KI, Simon R (2014) Using single cell sequencing data to model the evolutionary history of a tumor. BMC Bioinformatics 15. https://doi.org/10.1186/1471-2105-15-27

174. Yang ZH, Rannala B (2012) Molecular phylogenetics: principles and practice. Nat Rev Genet 13(5):303–314. https://doi.org/10.1038/nrg3186

175. Miller CA, McMichael J, Dang HX, Maher CA, Ding L, Ley TJ, Mardis ER, Wilson RK (2016) Visualizing tumor evolution with the fishplot package for R. BMC Genomics 17. https://doi.org/10.1186/S12864-016-3195-Z

176. Ross EM, Markowetz F (2016) OncoNEM: inferring tumor evolution from single-cell sequencing data. Genome Biol 17. https://doi.org/10.1186/s13059-016-0929-9

177. Jahn K, Kuipers J, Beerenwinkel N (2016) Tree inference for single-cell data. Genome Biol 17. https://doi.org/10.1186/s13059-016-0936-x

178. Roth A, Khattra J, Yap D, Wan A, Laks E, Biele J, Ha G, Aparicio S, Bouchard-Cote A, Shah SP (2014) PyClone: statistical inference of clonal population structure in cancer. Nat Methods 11(4):396. https://doi.org/10.1038/Nmeth.2883

179. Jiao W, Vembu S, Deshwar AG, Stein L, Morris Q (2014) Inferring clonal evolution of tumors from single nucleotide somatic mutations. BMC Bioinformatics 15. https://doi.org/10.1186/1471-2105-15-35

180. Ha G, Roth A, Khattra J, Ho J, Yap DM, Prentice LM, Melnyk N, McPherson A, Bashashati A, Laks E, Biele J, Ding JR, Le A, Rosner J, Shumansky K, Marra MA, Gilks CB,

Huntsman DG, McAlpine JN, Aparicio S, Shah SP (2014) TITAN: inference of copy number architectures, in clonal cell populations from tumor whole-genome sequence data. Genome Res 24(11):1881–1893. https://doi.org/10.1101/gr.180281.114

181. Strino F, Parisi F, Micsinai M, Kluger Y (2013) TrAp: a tree approach for fingerprinting subclonal tumor composition. Nucleic Acids Res 41(17). https://doi.org/10.1093/nar/gkt641

182. El-Kebir M, Oesper L, Acheson-Field H, Raphael BJ (2015) Reconstruction of clonal trees and tumor composition from multi-sample sequencing data. Bioinformatics 31 (12):62–70. https://doi.org/10.1093/bioinformatics/btv261

183. Qiao Y, Quinlan AR, Jazaeri AA, Verhaak RGW, Wheeler DA, Marth GT (2014) SubcloneSeeker: a computational framework for reconstructing tumor clone structure for cancer variant interpretation and prioritization. Genome Biol 15(8). https://doi.org/10.1186/S13059-014-0443-X

184. Deshwar AG, Vembu S, Yung CK, Jang GH, Stein L, Morris Q (2015) PhyloWGS: reconstructing subclonal composition and evolution from whole-genome sequencing of tumors. Genome Biol 16. https://doi.org/10.1186/s13059-015-0602-8

185. Malikic S, McPherson AW, Donmez N, Sahinalp CS (2015) Clonality inference in multiple tumor samples using phylogeny. Bioinformatics 31(9):1349–1356. https://doi.org/10.1093/bioinformatics/btv003

186. Davis A, Navin NE (2016) Computing tumor trees from single cells. Genome Biol 17. https://doi.org/10.1186/S13059-016-0987-Z

187. Potter NE, Ermini L, Papaemmanuil E, Cazzaniga G, Vijayaraghavan G, Titley I, Ford A, Campbell P, Kearney L, Greaves M (2013) Single-cell mutational profiling and clonal phylogeny in cancer. Genome Res 23 (12):2115–2125. https://doi.org/10.1101/gr.159913.113

188. Salehi S, Steif A, Roth A, Aparicio S, Bouchard-Cote A, Shah SP (2017) ddClone: joint statistical inference of clonal populations from single cell and bulk tumour sequencing data. Genome Biol 18. https://doi.org/10.1186/s13059-017-1169-3

189. Miller CA, White BS, Dees ND, Griffith M, Welch JS, Griffith OL, Vij R, Tomasson MH, Graubert TA, Walter MJ, Ellis MJ, Schierding W, DiPersio JF, Ley TJ, Mardis ER, Wilson RK, Ding L (2014) SciClone: inferring clonal architecture and tracking the spatial and temporal patterns of tumor evolution. PLoS Comput Biol 10(8). https://doi.org/10.1371/journal.pcbi.1003665

190. Ernst J, Kellis M (2015) Large-scale imputation of epigenomic datasets for systematic annotation of diverse human tissues. Nat Biotechnol 33(4):364–U374. https://doi.org/10.1038/nbt.3157

191. Stevens M, Cheng JB, Li DF, Xie MC, Hong CB, Maire CL, Ligon KL, Hirst M, Marra MA, Costello JF, Wang T (2013) Estimating absolute methylation levels at single-CpG resolution from methylation enrichment and restriction enzyme sequencing methods. Genome Res 23(9):1541–1553. https://doi.org/10.1101/gr.152231.112

Check for updates

Chapter 20

Progress on Diagnosis of Tuberculous Meningitis

Yi-yi Wang and Bing-di Xie

Abstract

Central nervous system (CNS) disease caused by *Mycobacterium tuberculosis* (MTB) is highly devastating. Tuberculous meningitis (TBM) is the most common form of CNS tuberculosis (TB). Rapid, sensitive, and affordable diagnostic tests are not available. Ziehl–Neelsen (ZN) stain has a very low sensitivity in cases of TBM, the sensitivity rates is of about 10–20%.The detection rate can be improved by taking large volume CSF samples (>6 ml) and prolonged slide examination (30 min). Culture of MTB from the CSF is slow and insufficiently sensitive. The sensitivity is different, which varies from 36% to 81.8%. The microscopic observation drug susceptibility (MODS) assay was recommended by the World Health Organization in 2011. The sensitivity is 65%, which is more sensitive and faster than CSF smear. Commercial PCR assays were found to be insensitive at detecting MTB in CSF samples. Many research provided the value of ADA on the TBM diagnosis. Interferon-gamma release assays (IGRAs) are not recommended for diagnosis of active TB disease. Imaging is essential in diagnosis and showing complications of CNS TB. Thwaites criteria and the Lancet consensus scoring system (LCSS) were developed to improve the diagnosis of TBM. Clinicians will continue to make judgment based on clinical examination, inflammatory CSF examinations, imaging studies, and scoring systems.

Key words Tuberculous meningitis, Diagnosis, Progress, Scoring system

1 Introduction

According to WHO global tuberculosis report of 2015, the tuberculosis (TB) remained one of the top ten causes of death worldwide in 2015. There were an estimated 1.4 million TB deaths in 2015. Central nervous system (CNS) disease caused by *Mycobacterium tuberculosis* (Mtb) is highly devastating [1]. CNS tuberculosis accounts for approximately 1–5% of all cases of tuberculosis [2, 3]. The types of CNS TB involve intracranial TB and intraspinal TB. The types of intracranial TB involve TB meningitis (TBM), complications of TBM, sequel of TBM, and parenchymal TB. Complications of TBM include hydrocephalus, tuberculous vasculitis, and cranial nerve involvement. The types of parenchymal TB involve tuberculomas, tuberculous abscess, tuberculous cerebritis, and tuberculous encephalopathy [4]. Tuberculous

Tao Huang (ed.), *Computational Systems Biology: Methods and Protocols*, Methods in Molecular Biology, vol. 1754, https://doi.org/10.1007/978-1-4939-7717-8_20,

meningitis is the most common presentation among CNS tuberculosis, which remains a formidable diagnostic challenge [5]. Mortality and long-term disability remain unacceptably high [6]. Despite antituberculosis chemotherapy, according to literature report, mortality of TBM is high, which varies from 10% to 36.5% [7–10].

Rapid, sensitive, and affordable diagnostic tests are not available. What is the progress on the diagnosis of tuberculous meningitis? The purpose of this review is to discuss recent advances and describe the utility and limitations of current diagnostic methods for TBM (Table 1).

2 Microscopy

2.1 Ziehl–Neelsen (ZN) Stain

Detection of acid-fast bacilli (AFB) in patient samples using Ziehl–Neelsen (ZN) staining is the most practical and universally adopted test for diagnosing TB. The ZN stain, also known as the acid-fast stain, was first described by two German doctors: the bacteriologist Franz Ziehl (1859–1926) and the pathologist Friedrich Neelsen (1854–1898). It is a special bacteriological stain used to identify acid-fast organisms, mainly *Mycobacterium tuberculosis.* The CSF of most patients with TBM contains only 10^0–10^2 organisms/ml, yet approximately 10^4 organisms/ml are required for reliable detection with ZN stains [23]. The limit of detection on microscopy is 100 mycobacteria/ml [24]. Although the sensitivity of ZN stain in different studies varies considerably (0–87%) [3], it has a very low sensitivity in cases of TBM; the sensitivity rate is of about 10–20% [11].

The detection rate of smear microscopy in TBM can be improved by taking large volume CSF samples (>6 ml) and prolonged slide examination (30 min). For example, Thwaites et al. have increased the positive rate to 58% with a prolonged slide examination (median 10 min) [12]. However, these criteria are rarely achieved in practice.

In high-income countries, fluorescence microscopy is the standard diagnostic method in ZN stain, which has improved the sensitivity of microscopy over conventional ZN staining (by approximately 10% in sputum) and significantly decreased the time required to examine each slide [25]. The equipment and bulbs of fluorescent microscopy using fluorochrome dye (auramine-O or auramine-rhodamine) are more expensive [25]. The development of light-emitting diode (LED) fluorescent microscopy (FM) is less expensive than mercury vapor fluorescence microscopes and is now recommended by the World Health Organization [26]. Using mycobacterial culture as a reference standard, the sensitivity of LED-FM is higher than conventional fluorescence microscope in the sputum, other respiratory samples, and extrapulmonary samples. In extrapulmonary samples, the sensitivity of LED-FM is 50%

Table 1
Current laboratory methods for TBM

		Sensitivity	Specificity	Characteristic	Category
Microscopy	Ziehl–Neelsen stain	10–20% [11]	Special to identify MTB	The most practical and universally adopted test	1. Large volume CSF samples (>6 ml) 2. Fluorescence microscopy
	Culture of Mtb	36–81.8% [12–15]	Special to identify MTB	Slow and insufficiently sensitive	1. Solid culture medium 2. Liquid culture medium
	MODS	65% [16]	Special to identify MTB	Sensitive and faster	
NAAT	PCR	56% for commercial assays [17]	98% for commercial assays [17]	Can detect fewer than ten organisms	1. Commercial 2. In-house
	The Xpert MTB/RIF assay	27–86% [18, 19]	95% [20]	1. Entirely automated and faster 2. Detection of rifampicin resistance	
	IGRAs	71% for T-SPOT [21]	57% for T-SPOT [21]	1. A high specificity diagnostic tool in TBM 2. Not recommended for diagnosis of active TB disease	1. QFT-IT 2. T-SPOT.TB
	ADA	29.9–79% [14, 22]	91% [22]	1. ADA values from 1 to 4 U/l helped to exclude TBM 2. Values between 4 and 8 U/l were insufficient to confirm or exclude the diagnosis of TBM 3. Values >8 U/l improved the diagnosis of TBM	

Mtb mycobacteria, *MODS* the microscopic observation drug susceptibility, *NAATs* nucleic acid amplification tests, *IGRAs* interferon-gamma release assays, *QFT-IT* QuantiFERON-TB© Gold In Tube, *ADA* adenosine deaminase

(95% CI 23.0–77.0), which is higher than conventional fluorescence microscope with 35.7% (95% CI 12.8–64.9). Specificity was very similar between conventional fluorescence microscope and LED-FM [26]. However, until now, the evidence is insufficient.

2.2 Culture of Mycobacterium tuberculosis

Culture of *Mycobacterium tuberculosis* from the CSF of TBM patients is slow and insufficiently sensitive. The sensitivity is different, which varies from 36% to 81.8% [12–15]. A further increase of sensitivity was observed when the solid and liquid culture medium were used at the same time.

Traditionally, solid culture media for culture of Mtb are kept for up to 8 weeks before a negative result is reported to the physician. Reduced turnaround times have been achieved using broth-based culture compared with solid media for the isolation of Mtb (13 days vs. 26 days) [27]. Most incubation protocols still require a maximum of 6 weeks [28]. Culture of Mtb is used as a "rule-in" test not a "rule-out" diagnostic test. The research from India suggested that for CSF samples, both liquid and solid culture media should be used for optimal detection and should be incubated for longer period (up to 8–10 weeks) than routine culture [29]. Another research attempted to shorten the incubation time of mycobacterial cultures. The study from Swiss indicated that 58.3% of all mycobacteria were detected within 14 days, 37.5% were detected within 21 days, and 4.2% were detected within 28 days [30]. It seems that a final report can be issued after 4 weeks.

2.3 The Microscopic Observation Drug Susceptibility

The microscopic observation drug susceptibility (MODS) assay was developed by Caviedes in 2000. He found that Mycobacterium tuberculosis form the characteristic cable structure in liquid medium. MODS is a kind of liquid culture [31]. Processed CSF is inoculated into a microtiter plate containing broth media and incubated. Growth is examined by an inverted microscope. In TBM patients, the sensitivity is 65%, which is more sensitive than CSF smear. The detection time is median 6 days, which is faster than commercial liquid/solid culture [16]. MODS also performed well in drug susceptibility testing (DST): isoniazid DST concordance was 95.7% (kappa 0.85); rifampicin DST concordance was 96.8% (kappa 0.91) [32]. MODS was recommended by the World Health Organization in 2011.

3 Nucleic Acid Amplification Tests

The conventional tests including microscopy and culture are often limited in the diagnosis of TBM since TBM is a paucibacillary form of tuberculosis. Nucleic acid amplification tests (NAAT) can detect fewer than ten organisms that can be used to identify *M. tuberculosis* in clinical specimens or cultures [33]. The first NAAT for use on CSF specimens was developed since 1990 [34].

3.1 The Polymerase Chain Reaction

The polymerase chain reaction (PCR) is the most common methodology, but alternatives are heterogeneous including real-time PCR, isothermal, strain displacement, or transcription-mediated amplification and ligase chain reaction [35]. NAATs are categorized as commercial or in-house. Most (>90%) laboratories used commercial kits such as the Amplicor *M. tuberculosis* tests (Roche Molecular Systems, Branchburg, NJ, USA) and the Amplified *M. tuberculosis* Direct Test (MTD; Gen-Probe Inc., San Diego, CA, USA). The literature on NAATs has been extensively reviewed. The significant heterogeneity in sensitivity and specificity of in-house PCRs led to no useful comparative information could be obtained [17, 35]. Commercial assays were found to be insensitive at detecting Mtb in CSF samples (sensitivity 56% and specificity 98%) [17]. The specificity of NAATs was high applied to body fluids, for example, for TB meningitis, but sensitivity was poor, indicating that these tests cannot be used reliably to rule out TBM. NAATs improve diagnostic certainty but do not replace microscopy and culture.

3.2 The Xpert MTB/RIF Assay

The Xpert MTB/RIF assay (Cepheid, Sunnyvale, CA) simultaneously detects the presence of *Mycobacterium tuberculosis* and its susceptibility to the rifampin in less than 2 h [36]. The assay is entirely automated, requiring only two manual steps. In 2010, the WHO endorsed the use of Xpert MTB/RIF for use on sputum specimens. Xpert MTB/RIF has been extensively evaluated for Mycobacterium tuberculosis detection in sputum specimens and performs well on smear-positive samples (sensitivity 98% compared with 68% in smear-negative samples; specificity 98%) [37]. Although the assay is not recommended by the World Health Organization for the diagnosis of TBM, several studies have evaluated the use of Xpert MTB/RIF for the diagnosis of TBM. The significant heterogeneity has also been found in sensitivity of Xpert MTB/RIF from 27% to 86%. An assay from India reported the sensitivity of Xpert MTB/RIF is 86% [18]. Three studies indicated that the sensitivity of Xpert MTB/RIF was about 60%, which is higher than that of smear microscopy [20, 38, 39]. The report from Tortoli in 2012 indicated the sensitivity is only 27% [19]. Further studies still are required as the studies that have been performed have small subject numbers. The other ability of Xpert MTB/RIF is to detect drug resistance. Xpert MTB/RIF is the only WHO-recommended rapid diagnostic test for detection of TB and rifampicin resistance. However the specificity is lower than the conventional DST. All detected rifampicin-resistant isolates should ideally be confirmed with conventional DST to detect false-positive results [40].

4 Interferon-Gamma Activity and Interferon-Gamma Release Assays (IGRAs)

Interferon-gamma release assays (IGRAs) are whole-blood tests that detect immune responses to a panel of *M. tuberculosis* antigens, which include the measurement of interferon-gamma release in whole blood (QuantiFERON-TB© Gold In Tube [(QFT-IT); Cellestis Limited Chadstone, Vic., Australia] and peripheral blood mononuclear cells (T-SPOT.TB; Oxford Immunotec, Abingdon, UK). On infection of M. tuberculosis, macrophages recognize the mycobacteria by toll-like receptor (TLR) followed by phagocytosis and control of mycobacteria. In addition, macrophages also secrete IL-12 to induce IFN-γ production by T cell, which, in turn, increases the phagocytosis and oxidative burst [41]. IGRAs as a high specificity diagnostic tool in TBM received preliminary much attention. However, IGRAs are not recommended for diagnosis of active TB disease. In a meta-analysis, the sensitivity estimates among HIV-infected persons were 76% (95% CI, 45–92%) for T-SPOT and 60% (95% CI, 34–82%) for QFT-GIT [42]. There was no evidence that IGRA was more sensitive than the tuberculin skin test for active tuberculosis diagnosis [42, 43]. The use of IGRAs directly on CSF specimens has been evaluated for the diagnosis of TBM, based on the premise that mononuclear cells localized to infected sites produce more interferon than peripheral blood mononuclear cells PBMC [21]. However, the sensitivity is variable [44, 45]. CSF IGRAs require large volumes of CSF. It is a barrier to perform in practice.

5 Adenosine Deaminase

Adenosine deaminase (ADA) is an enzyme required for the conversion of adenosine to inosine and is found in many tissues, particularly in T lymphocytes from the lymphoid tissue [46]. ADA exists as two isoenzymes: ADA1 and ADA2. It appears that the ADA2 isoenzyme originates mainly from monocytes and macrophages. In tuberculous pleural effusions, most of the ADA activity consists of ADA2 [47]. High ADA levels in tuberculosis appear to be related to the subset of activated T lymphocytes in response to tuberculous antigens. The use of ADA in CSF diagnosis of tuberculosis started from 20 years ago. Many research provided the value of ADA on the TBM diagnosis, but the results are conflicted. According to a meta-analysis from China in 2010, the sensitivity of ADA in the diagnosis of TBM was 0.79 (95% CI 0.75–0.83) and specificity 0.91 (95% CI 0.89–0.93) [22]. A recent study evaluated the performance of ADA tests in 506 patients with microbiologically confirmed TBM. The sensitivity of the ADA was 29.9% [14]. There is a lack of standardization in the ADA cutoff value for diagnosing TBM. Standardized cutoffs of ADA values for the diagnosis of TBM have not been established, and the values used in the various studies ranged from 5.0 to 15 U/l [2]. Accordingly the different cutoff, the sensitivity, and the specificity are different. ADA values from 1 to 4 U/l (sensitivity >93% and specificity <80%) helped to exclude TBM; values between 4 and 8 U/l were insufficient to confirm or exclude the diagnosis of TBM ($P = 0.07$), and values >8 U/l (sensitivity <59% and specificity >96%) improved the diagnosis of TBM ($P < 0.001$). Based on the ROC curve, the ideal cutoff was 5.3 U/l (84% sensitivity and specificity) [48].

6 Cerebral Imaging

Imaging is essential in diagnosis and showing complications of CNS TB and has the advantages of being noninvasive and quick to perform and report, providing the potential for an improved prognosis. MR is superior to CT because it allows earlier detection of the disease, a more exact definition of the spread, and a more detailed representation of complex inflammatory processes [49]. Przybojewski and colleagues identified four features with 100% specificity for TBM: basal enhancement, hydrocephalus, tuberculoma, and infarction in the supratentorial brain parenchyma and brain stem [50]. Four individual criteria had a specificity of 100%, but the sensitivities of these criteria ranged from 15% to 53% only. The above conclusion derived from pediatric case study. In an adult case study, the results showed that of the five major CT features supporting a diagnosis of TBM (hydrocephalus, infarcts, tuberculoma(s), basal meningeal enhancement, and the presence of precontrast basal hyperdensities), hydrocephalus and meningeal enhancement were the most commonly found consensus features in TBM but that the other features were rare [51]. However basal meningeal enhancement is less often detected in adults than in children with TBM. Only 8–34% of cases had this feature and 45% hydrocephalus [1, 52]. So, if CT features are absent, the TBM cannot be ruled out.

Tubercular hydrocephalus is usually communicating. It occurs because thick gelatinous exudate develops around the basal cisterns, the Sylvian fissure, and the brainstem causing obstruction to CSF flow [53]. Noncommunicating or obstructive hydrocephalus can occur because narrowing of the aqueduct and third ventricle by a small tuberculoma causes consequent hydrocephalus [54].

Cerebral infarction occurs in 15–57% of tuberculous meningitis patients, mainly during stage 3 of the illness [55]. Most infarcts involve the thalamus, basal ganglia, and internal capsule regions [56]. Vasculitis and vasospasm are the causes of cerebral infarction in tuberculous meningitis infections [57]. Cerebral infarction is associated with leptomeningeal enhancement in TBM. The exudate at the basal region surrounds the arteries, leading to arterial narrowing and subsequently stroke. The intense inflammation also causes vasculitis and vasospasm in the nearby vessels [58].

Tuberculomas are among the most common intracranial mass lesions and the most common manifestation of parenchymal TB. They usually occur in the absence of TBM but may occur with meningitis. Tuberculomas may be single or multiple and can be seen anywhere in the brain parenchyma. The number of identified lesions per patient may range from 1 to 12 (or more), with the size varying from 1 mm to 8 cm [59]. Tuberculomas show typical granulomatous reaction. Histopathology is characterized by the presence of epithelioid granuloma with Langhans giant cells. In response to the infection, the activated macrophages, cytokine interferon (IFN), and T cell activity produce a type IV reaction. This reaction combined with ischemia results in central caseation necrosis in the tuberculous granuloma [60]. Imaging findings depend on the stage of tuberculoma, whether it is noncaseating or caseating with solid or liquid center [61]. At the early stage of the tuberculomas, caseating has not yet formed. Tuberculoma usually appears hyperintense on T2W and slightly hypointense on T1W images, which show homogenous enhancement on postcontrast T1W images. A solid caseating tuberculoma appears relatively iso- to hypointense on both T1W and T2W images with an iso- to hyperintense rim on T2W images. It shows rim enhancement on postcontrast T1W images. When the solid center of the caseating lesion liquefies, the center appears hyperintense with a hypointense rim on T2W images. The postcontrast T1W images show rim enhancement [62].

Miliary brain tuberculosis is usually associated with TBM. They typically occur in immunocompromised patients. The infection is characterized by a large amount of M. tuberculosis. Miliary tubercles range from 1–5mm in size and have a mean 2mm which are either not visible on conventional SE MRI images or are seen as tiny foci of hyperintensity on T2W acquisitions. The postcontrast T1W images show numerous, round, small, homogeneous, enhancing lesions [63].

7 Scoring System

Given lack of a gold standard, clinicians will have to continue to use their clinical judgment based on clinical examination, inflammatory cerebrospinal fluid (CSF) examinations, imaging studies, and scoring systems, to make the diagnosis and initiate prompt treatment [64]. In 2002, Thwaites GE compared the clinical and laboratory characteristics of tuberous and purulent meningitis and proposed the Thwaites scoring system [65]. In 2005, Thwaites GE modified possible tuberculous meningitis diagnostic criteria [7]. According to Thwaites criteria, definite, probable, or possible TBM is classified based on the clinical findings, CSF criteria, and the evidence of tuberculosis elsewhere. The results in 2005 from Sunbul suggested that the sensitivity of the Thwaites standard is 95.6%, with a specificity of 70.8% [66]. The results from Shanghai Huashan Hospital showed that the sensitivity is 98.2% and specificity 82.9% [67]. Thwaites' score is simple, cost-effectiveness, more effective and rapid diagnostic tests. These are needed in the primary care setting where imaging facilities are lacking.

In 2010, a uniform research case definition—the Lancet consensus scoring system (LCSS) for TBM—was developed to improve standardization of diagnosis [68]. LCSS also classifies cases as definite, probable, or possible. Classification is based on a composite score of clinical findings, CSF criteria, cerebral imaging criteria, and the evidence of tuberculosis elsewhere. Cerebral imaging criteria are recommended in LCSS. The LCSS is more detailed and resource intensive. The study demonstrated that the widely used Thwaites' score compares well with the more detailed and resource intensive Lancet consensus score [64].

8 Conclusions

The best way to improve survival of TBM is by rapid accurate diagnosis and prompt initiation of therapy. There have been encouraging developments in the diagnosis of TBM. However CSF contains low organism numbers, which limit current diagnostic modalities. Because a gold standard is still lacking, clinicians will have to continue make judgment based on clinical examination, inflammatory CSF examinations, imaging studies, and scoring systems. It seems logical that clinicians need to understand the characteristics of the diagnosis, so as to make a comprehensive judgment of the disease.

References

1. Kalita J, Misra UK, Ranjan P (2007) Predictors of long-term neurological sequelae of tuberculous meningitis: a multivariate analysis. Eur J Neurol 14(1):33–37
2. Rock RB et al (2008) Central nervous system tuberculosis: pathogenesis and clinical aspects. Clin Microbiol Rev 21(2):243–261
3. Bhigjee AI et al (2007) Diagnosis of tuberculous meningitis: clinical and laboratory parameters. Int J Infect Dis 11(4):348–354
4. Patkar D et al (2012) Central nervous system tuberculosis: pathophysiology and imaging findings. Neuroimaging Clin N Am 22 (4):677–705
5. Thwaites GE, Tran TH (2005) Tuberculous meningitis: many questions, too few answers. Lancet Neurol 4(3):160–170
6. Bidstrup C et al (2002) Tuberculous meningitis in a country with a low incidence of tuberculosis: still a serious disease and a diagnostic challenge. Scand J Infect Dis 34(11):811–814
7. Thwaites GE et al (2004) Dexamethasone for the treatment of tuberculous meningitis in

adolescents and adults. N Engl J Med 351 (17):1741–1751
8. Sutlas PN et al (2003) Tuberculous meningitis in adults: review of 61 cases. Infection 31 (6):387–391
9. Jha SK et al (2015) Definite (microbiologically confirmed) tuberculous meningitis: predictors and prognostic impact. Infection 43 (6):639–645
10. Li Y et al (2007) An analysts of the early diagnostic criteria for tuberculosis meningitis. J Zhonghuaneikezazhi 46(3):217–219
11. Thwaites G et al (2000) Tuberculous meningitis. J Neurol Neurosurg Psychiatry 68 (3):289–299
12. Thwaites GE, Chau TT, Farrar JJ (2004) Improving the bacteriological diagnosis of tuberculous meningitis. J Clin Microbiol 42 (1):378–379
13. Marx GE, Chan ED (2011) Tuberculous meningitis: diagnosis and treatment overview. Tuberc Res Treat 2011:798764
14. Erdem H et al (2014) The microbiological diagnosis of tuberculous meningitis: results of Haydarpasa-1 study. Clin Microbiol Infect 20 (10):O600–O608
15. Chaidir L et al (2012) Comparison of real time IS6110-PCR, microscopy, and culture for diagnosis of tuberculous meningitis in a cohort of adult patients in Indonesia. PLoS One 7 (12):e52001
16. Caws M et al (2007) Evaluation of the MODS culture technique for the diagnosis of tuberculous meningitis. PLoS One 2(11):e1173
17. Pai M et al (2003) Diagnostic accuracy of nucleic acid amplification tests for tuberculous meningitis: a systematic review and meta-analysis. Lancet Infect Dis 3(10):633–643
18. Vadwai V et al (2011) Xpert MTB/RIF: a new pillar in diagnosis of extrapulmonary tuberculosis? J Clin Microbiol 49(7):2540–2545
19. Tortoli E et al (2012) Clinical validation of Xpert MTB/RIF for the diagnosis of extrapulmonary tuberculosis. Eur Respir J 40 (2):442–447
20. Patel VB et al (2013) Diagnostic accuracy of quantitative PCR (Xpert MTB/RIF) for tuberculous meningitis in a high burden setting: a prospective study. PLoS Med 10(10): e1001536
21. Kim SH et al (2010) Rapid diagnosis of tuberculous meningitis by T cell-based assays on peripheral blood and cerebrospinal fluid mononuclear cells. Clin Infect Dis 50 (10):1349–1358
22. Xu HB et al (2010) Diagnostic value of adenosine deaminase in cerebrospinal fluid for tuberculous meningitis: a meta-analysis. Int J Tuberc Lung Dis 14(11):1382–1387
23. Garcia-Monco JC (1999) Central nervous system tuberculosis. Neurol Clin 17(4):737–759
24. Kennedy DH, Fallon RJ (1979) Tuberculous meningitis. JAMA 241(3):264–268
25. Steingart KR et al (2006) Fluorescence versus conventional sputum smear microscopy for tuberculosis: a systematic review. Lancet Infect Dis 6(9):570–581
26. Minion J et al (2011) Comparison of LED and conventional fluorescence microscopy for detection of acid fast bacilli in a low-incidence setting. PLoS One 6(7):e22495
27. Tortoli E et al (1999) Use of BACTEC MGIT 960 for recovery of mycobacteria from clinical specimens: multicenter study. J Clin Microbiol 37(11):3578–3582
28. Nakanaga K et al (2013) Laboratory procedures for the detection and identification of cutaneous non-tuberculous mycobacterial infections. J Dermatol 40(3):151–159
29. Venkataswamy MM et al (2007) Comparative evaluation of BACTEC 460TB system and Lowenstein-Jensen medium for the isolation of M. tuberculosis from cerebrospinal fluid samples of tuberculous meningitis patients. Indian J Med Microbiol 25(3):236–240
30. Pfyffer GE, Wittwer F (2012) Incubation time of mycobacterial cultures: how long is long enough to issue a final negative report to the clinician? J Clin Microbiol 50(12):4188–4189
31. Caviedes L et al (2000) Rapid, efficient detection and drug susceptibility testing of Mycobacterium tuberculosis in sputum by microscopic observation of broth cultures. The Tuberculosis Working Group in Peru. J Clin Microbiol 38(3):1203–1208
32. Coronel J et al (2010) MODS accreditation process for regional reference laboratories in Peru: validation by GenoType (R) MTBDRplus. Int J Tuberc Lung Dis 14 (11):1475–1480
33. Torok ME (2015) Tuberculous meningitis: advances in diagnosis and treatment. Br Med Bull 113(1):117–131
34. Kaneko K et al (1990) Rapid diagnosis of tuberculous meningitis by polymerase chain reaction (PCR). Neurology 40(10):1617–1618
35. Dinnes J et al (2007) A systematic review of rapid diagnostic tests for the detection of tuberculosis infection. Health Technol Assess 11(3):1–196

36. Helb D et al (2010) Rapid detection of Mycobacterium tuberculosis and rifampin resistance by use of on-demand, near-patient technology. J Clin Microbiol 48(1):229–237
37. Steingart KR et al (2013) Xpert(R) MTB/RIF assay for pulmonary tuberculosis and rifampicin resistance in adults. Cochrane Database Syst Rev 1:CD009593
38. Nhu NT et al (2014) Evaluation of GeneXpert MTB/RIF for diagnosis of tuberculous meningitis. J Clin Microbiol 52(1):226–233
39. Patel VB et al (2014) Comparison of amplicor and GeneXpert MTB/RIF tests for diagnosis of tuberculous meningitis. J Clin Microbiol 52 (10):3777–3780
40. Moore DA, Shah NS (2011) Alternative methods of diagnosing drug resistance—what can they do for me? J Infect Dis 204(Suppl 4): S1110–S1119
41. Khan TA et al (2016) Interferon-gamma improves macrophages function against M. tuberculosis in multidrug-resistant tuberculosis patients. Chemother Res Pract 2016:7295390
42. Diel R et al (2011) Interferon-gamma release assays for the diagnosis of latent Mycobacterium tuberculosis infection: a systematic review and meta-analysis. Eur Respir J 37(1):88–99
43. Metcalfe JZ et al (2011) Interferon-gamma release assays for active pulmonary tuberculosis diagnosis in adults in low- and middle-income countries: systematic review and meta-analysis. J Infect Dis 204(Suppl 4):S1120–S1129
44. Patel VB et al (2010) Cerebrospinal T-cell responses aid in the diagnosis of tuberculous meningitis in a human immunodeficiency virus- and tuberculosis-endemic population. Am J Respir Crit Care Med 182(4):569–577
45. Vidhate MR et al (2011) Diagnostic and prognostic value of Mycobacterium tuberculosis complex specific interferon gamma release assay in patients with tuberculous meningitis. J Infect 62(5):400–403
46. Lee YC et al (2001) Adenosine deaminase levels in nontuberculous lymphocytic pleural effusions. Chest 120(2):356–361
47. Schutte CM et al (2001) Significance of cerebrospinal fluid adenosine deaminase isoenzymes in tuberculous (TB) meningitis. J Clin Lab Anal 15(5):236–238
48. Tuon FF et al (2010) Adenosine deaminase and tuberculous meningitis—a systematic review with meta-analysis. Scand J Infect Dis 42 (3):198–207
49. Schroth G et al (1987) Advantage of magnetic resonance imaging in the diagnosis of cerebral infections. Neuroradiology 29(2):120–126
50. Przybojewski S, Andronikou S, Wilmshurst J (2006) Objective CT criteria to determine the presence of abnormal basal enhancement in children with suspected tuberculous meningitis. Pediatr Radiol 36(7):687–696
51. Botha H et al (2012) Reliability and diagnostic performance of CT imaging criteria in the diagnosis of tuberculous meningitis. PLoS One 7 (6):e38982
52. Hosoglu S et al (2002) Predictors of outcome in patients with tuberculous meningitis. Int J Tuberc Lung Dis 6(1):64–70
53. Chatterjee S (2011) Brain tuberculomas, tubercular meningitis, and post-tubercular hydrocephalus in children. J Pediatr Neurosci 6(Suppl 1):S96–S100
54. Dastur DK, Manghani DK, Udani PM (1995) Pathology and pathogenetic mechanisms in neurotuberculosis. Radiol Clin N Am 33 (4):733–752
55. Misra UK, Kalita J, Maurya PK (2011) Stroke in tuberculous meningitis. J Neurol Sci 303 (1–2):22–30
56. Nair PP et al (2009) MRI pattern of infarcts in basal ganglia region in patients with tuberculous meningitis. Neuroradiology 51 (4):221–225
57. Tai MS et al (2016) Cerebral infarction pattern in tuberculous meningitis. Sci Rep 6:38802
58. Chan KH et al (2005) Cerebral infarcts complicating tuberculous meningitis. Cerebrovasc Dis 19(6):391–395
59. Dastur HM, Desai AD (1965) A comparative study of brain tuberculomas and gliomas based upon 107 case records of each. Brain 88 (2):375–396
60. Kumar SN et al (2013) Granuloma with langhans giant cells: an overview. J Oral Maxillofac Pathol 17(3):420–423
61. Gupta RK et al (1990) Role of magnetic resonance (MR) in the diagnosis and management of intracranial tuberculomas. Clin Radiol 41 (2):120–127
62. Trivedi R, Saksena S, Gupta RK (2009) Magnetic resonance imaging in central nervous system tuberculosis. Indian J Radiol Imaging 19 (4):256–265
63. Gupta RK, Kathuria MK, Pradhan S (1999) Magnetization transfer MR imaging in CNS tuberculosis. AJNR Am J Neuroradiol 20 (5):867–875
64. Kurien R, Sudarsanam TD, Thomas K (2013) Tuberculous meningitis: a comparison of scoring systems for diagnosis. Oman Med J 28 (3):163–166

65. Thwaites GE et al (2002) Diagnosis of adult tuberculous meningitis by use of clinical and laboratory features. Lancet 360 (9342):1287–1292
66. Sunbul M et al (2005) Thwaites' diagnostic scoring and the prediction of tuberculous meningitis. Med Princ Pract 14(3):151–154
67. Zhang YL et al (2014) Validation of thwaites' diagnostic scoring system for the differential diagnosis of tuberculous meningitis and bacterial meningitis. Jpn J Infect Dis 67(6):428–431
68. Marais S et al (2010) Tuberculous meningitis: a uniform case definition for use in clinical research. Lancet Infect Dis 10(11):803–812

Chapter 21

Insights of Acute Lymphoblastic Leukemia with Development of Genomic Investigation

Heng Xu and Yang Shu

Abstract

Treatment outcomes for acute lymphoblastic leukemia (ALL), especially pediatric ALL, have greatly improved due to the risk-adapted therapy. Combination of drug development, clinical practice, as well as basic genetic researches has brought the survival rate of ALL from less than 10% to more than 90% today, not only increasing the treatment efficacy but also limiting adverse drug reactions (ADRs). In this review, we summarized the landscape identification of ALL genetic alterations, which provided the opportunity to increase the survival rate and especially minimize the relapse risk of ALL, and highlighted the importance of the development of new technologies of genomic investigation for translational medicine.

Key words Acute lymphoblastic leukemia, Next-generation sequencing, Microarray, Single nucleotide polymorphism, Mutation, Drug efficacy, Adverse drug reactions, Translational medicine, Genomic landscape, Bioinformatics, Big data

1 Introduction

Leukemia has been described in 1827 for the first time and recognized as a distinct disease in 1845. Progress in description and treatment outcome of leukemia continues to improve followed with the development of new technologies and accumulation of basic biological/medical knowledge. Acute leukemia is more commonly observed in children than adults, especially acute lymphoblastic leukemia (ALL), which is the most common pediatric malignancy and account for ~80% of all acute leukemia or ~30% of all cancers in children [1–3]. Accordingly, around 6000 and over 100,000 new cases of ALL are diagnosed yearly in the USA and worldwide, respectively [4]. ALL is now recognized as a molecular genetic disorder and thus can be divided into more and more subtypes in terms of morphologic, immune-phenotypic, and genomic characteristics [5–8]. Importantly, current therapeutic regimen determinations and the treatment outcomes are closely related to the molecular subtype classifications, and 5-year event-free survival

Tao Huang (ed.), *Computational Systems Biology: Methods and Protocols*, Methods in Molecular Biology, vol. 1754,
https://doi.org/10.1007/978-1-4939-7717-8_21,

rates have risen from less than 10% in the 1960s to around 90% today primarily due to the drug development as well as risk-adapted personalized therapy [9–12]. However, ~10–20% of the pediatric patients and more than 50% of the adults suffer relapse, which is associated with a high rate treatment failure and death [13–16]. Unfortunately, little progress has been made in the treatment of relapsed ALL in either the older (1988–1994) or recent (1995–2002) eras [13]. Therefore, multiple diagnostic or prognostic factors have been taken into consideration for risk classification and used for individualized treatment, including well-known non--genomic (e.g., age and leukocyte count at diagnosis) as well as genomic signatures of ALL (e.g., leukemia cells with structural rearrangements or aneuploidy) [10]. Actually, multiple genomic alterations have also been noticed to be associated with non-genomic factors and have already been considered as strong prognostic factors to guide the clinical regimen determination and predict the treatment outcomes more sensitively [4, 9, 17]. To balance the treatment efficacy and adverse drug reactions risk, patients at high risk for a relapse are treated aggressively, with less toxic treatments for patients with a lower risk. On the other hand, recurrent genomic alterations can also be considered as potential targets for drug development, which already have multiple successful examples including imatinib for patient with BCR-ABL fusion [8, 10]. No doubt current and future genomic profiling investigation identified by the new technologies and in-depth mechanism studies, together with advanced understandings of the non-genomic clinical features, will guild more precious ALL classification, which is helpful to determine the individualized treatment therapy and further improve the survival rate as well as life quality of ALL patients [11]. In addition to the clinical treatment, prevention of ALL has also been drawn attention, especially for those individuals with family history or inherited predisposition. With the development of inherited genetic investigations of ALL susceptibility, many rare and common germline variants have been determined as risk indicators with varied odds ratio [18]. Therefore, preventions or interventions can be taken into consideration for individuals with high risk to develop ALL, including getting away from the risk environment (e.g., formaldehyde-based home decoration, ionizing radiation, and infections), routine physical examinations, early diagnosis, and treatments [4].

On the other hand, genetic analysis is a powerful and obligatory tool to discover and identify the important genomic alterations for leukemogenesis, diagnosis, and treatment outcomes. Lots of genomic technologies have been developed in the past decades, and some old methods have been replaced by more effective techniques, especially for genome-wide screening. Multiple frequently used techniques for genetic investigation include chromosome banding analysis, fluorescence in situ hybridization (FISH), polymerase

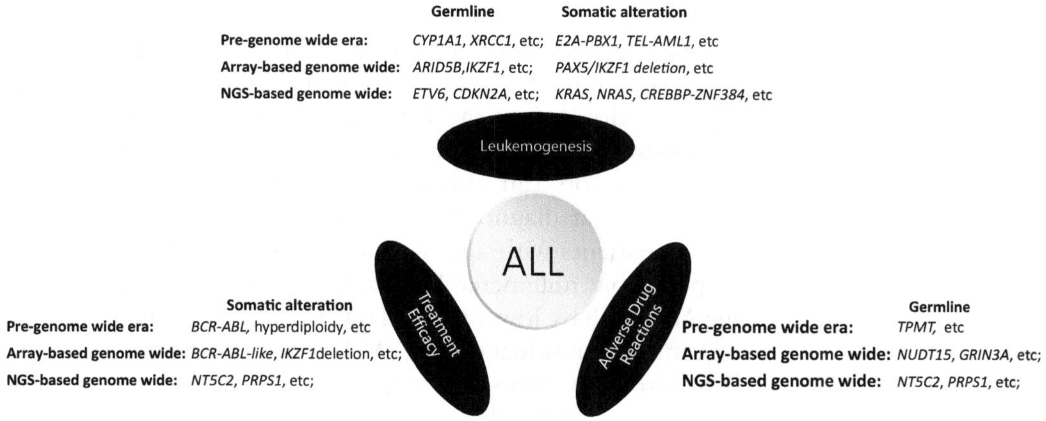

Fig. 1 Genetic alterations of ALL identified in different eras of genetic research

chain reaction (PCR)/real-time PCR, genome/transcriptome microarray technology, and next-generation sequencing [11, 19]. Additionally, methods used in discovery steps for aberrant genomic alterations can be totally different from those used in clinical practice, mainly due to the cost efficiency, being time-consuming, and easy operation. Some non-genomic techniques have been also introduced to detect genomic prognostic factors, such as flow cytometry [19].

Collectively, improvement for ALL treatment is a very good example for translational medicine from basic genomic study to clinical practice; we will demonstrate its history in three parts based on the new genomic technologies that were introduced and illustrate the genetic variants of ALL identified in three eras of genetic research (Fig. 1).

2 Pregenome-Wide Investigation Era

2.1 Nongenomic Diagnostic/Prognostic Factors

It is much earlier for description of leukemia as a distinct disease than introduction of genetic investigation into ALL classification. Multiple interindividual differences have been firstly noticed for several risk factors including age at diagnosis and ethnicity. For instance, the incidence rates for ALL have a peak prevalence between 2 and 5 years old and rank in the order of Hispanics, Caucasians/Asians, and Africans according to the epidemiology investigation [6]. With the use of microscope staining technique, the classification of blood cells and the diagnosis of leukemia can be determined by cytomorphologic and immunophenotypic features, thus introducing diagnostic factors like leukocyte (or white blood cell, WBC) count and immune lineage (separating B-lineage and T-lineage ALL) [20]. Not surprisingly, the most important usage of the diagnostic factors is not only for ALL diagnosis but also for its classifications such as risk group determination. Most of the anti-leukemic drugs are developed from the 1950s through the 1980s. The survival rates for ALL have increased steadily by optimally using these drugs since the 1960s due to the stringent application of prognostic factors for risk-adapted therapy. Therefore, multiple diagnostic factors can also be considered as prognostic factors, including age at diagnosis, WBC count, and ethnicity [10]. For instance, patients with age of 1–9 years old and WBC < 50,000 cells per cubic millimeter have a favorable treatment outcomes, while Africans have lower survival rate than Caucasians [10]. Importantly, minimal residual disease (MRD) and central nervous system (CNS) affection status are two strong non-genomic prognostic factors [10]. MRD indicates the small numbers of leukemic cells that remain in patients in remission after treatment and ranks as one of the major causes of relapse in ALL. Flow cytometry examination and real-time PCR have widely been used for MRD estimation [6, 10].

2.2 Genetic Characteristics and Clinical Relevance of the Molecular Subtypes

In addition to the non-genomic signatures of ALL, genomic characteristics have been widely observed, suggesting recurring gross chromosomal changes are a hallmark of ALL [19]. Indeed, the subsequent studies indicated that primary genetic abnormalities of leukemia cells have important prognostic significance [6, 21, 22]. Before the genome-wide investigation, approximately 75% of childhood ALL cases have been observed to harbor a recurring chromosomal alteration detectable by karyotyping, fluorescence in situ hybridization (FISH), or other low-throughput molecular techniques, such as reverse transcription PCR and Sanger sequencing [23], and most of these alterations (including aneuploidy and translocations) tend to be mutually exclusive in B-ALL. Leukemia cells with aneuploidy were firstly identified through chromosome counting by using microscope, with which the patients can be separated into hyperdiploidy, hypodiploidy, and normal. Afterward, with the observation of abnormal chromosome length as well as the karyotyping, multiple types of translocations were characterized. Interestingly, the most frequent translocation, t(12;21) (TEL-AML1 fusion, also named as ETV6-RUNX1), was not observed at first since the length change is not obvious. After advanced techniques (e.g., FISH) were investigated, such fusion has been revealed in a large proportion of ALL patients.

Frequencies of different types of genomic alterations may vary among age groups. For instance, patients with hyperdiploidy and TEL-AML1 fusion are the most common subtypes, accounting for ~25% and ~22%, respectively, in pediatric patients, but only 10% in total for adult patients [24], patients with t(4;11) translocation (MLL-AF4 fusion) account for approximately 50% in infants, 2% in children, and 5%–6% in adults, while patients with t(9;22) translocation (BCR-ABL1 fusion) increase in frequency with age, from

3% in children to 20% in adults and more than 50% in patients older than 50 years old [6, 21, 25]. However, patients with t(1;19) translocations (E2A-PBX1) have similar frequency among patients with different ages, accounting for ~3–5%. The rest of the patients with no obvious translocations are called as B-others. In T-cell ALL, the presence of t(11;19) translocations (MLL-ENL fusion), overexpression of HOX11 gene [22], and more than half of the cases of T-cell ALL have activating mutations of the NOTCH1 gene [10, 26].

Importantly, the prognosis outcome of patients is strongly associated with their molecular subtypes. For instance, the chromosome numbers in patients with aneuploidy are positively associated with survival outcomes, that is, patients with hyperdiploidy (more than 50 chromosomes per leukemia cell) exhibit a highly favorable prognosis [10, 24], while patients with hypodiploidy (fewer than 45 chromosomes per leukemia cell) confer a poor outcome, especially in subgroups with rare low hypodiploidy (33–39 chromosomes) and getting even worse in near-haploidy (23–29 chromosomes). For translocations, patients with TEL-AML1 fusion have a very high survival rate and low risk of relapse, which is in contrast to those with BCR-ABL1 and MLL-AF4. It is also considered that the age influences the prognostic effect of these genetic lesions. For instance, among patients with BCR-ABL1 fusion, the survival rate decreased in the order of children, adolescents, and adults [21, 25, 27], while for patients with MLL-AF4 fusion, infants fare considerably worse than older children, and adults have an especially poor outcome [21, 28]. Patients with MLL-ENL fusion and HOX11 overexpression confer a good prognosis [28–30].

With long time of investigation, ALL has been classified into several different molecular subtypes (Fig. 2), which is very useful to determine the risk-adapted therapy.

2.3 Inherited Predispositions in ALL

Inherited genetic variants are considered to play important roles on leukemogenesis due to the young age of ALL patients at diagnosis. However, most of the ALL cases are sporadic with very rare familial cases, suggesting common germline variants rather than rare mutations may account for susceptibility of more ALL patients. Therefore, association studies were conducted to identify the common ALL-related SNPs by comparing the frequency of variants in unrelated ALL cases and controls. In the pre-genome-wide investigation era, there were several association studies that examined the possible roles of genes in candidate pathways, and finally multiple SNPs were identified to reach statistical significance in *CYP1A1*, *XRCC1* [31], and *HLA-DP* [32], however, with no validation or inconsistent significance in independent patient cohorts. Therefore, a more powerful and accurate strategy is warranted to identify the top ALL-related common and rare variants [18].

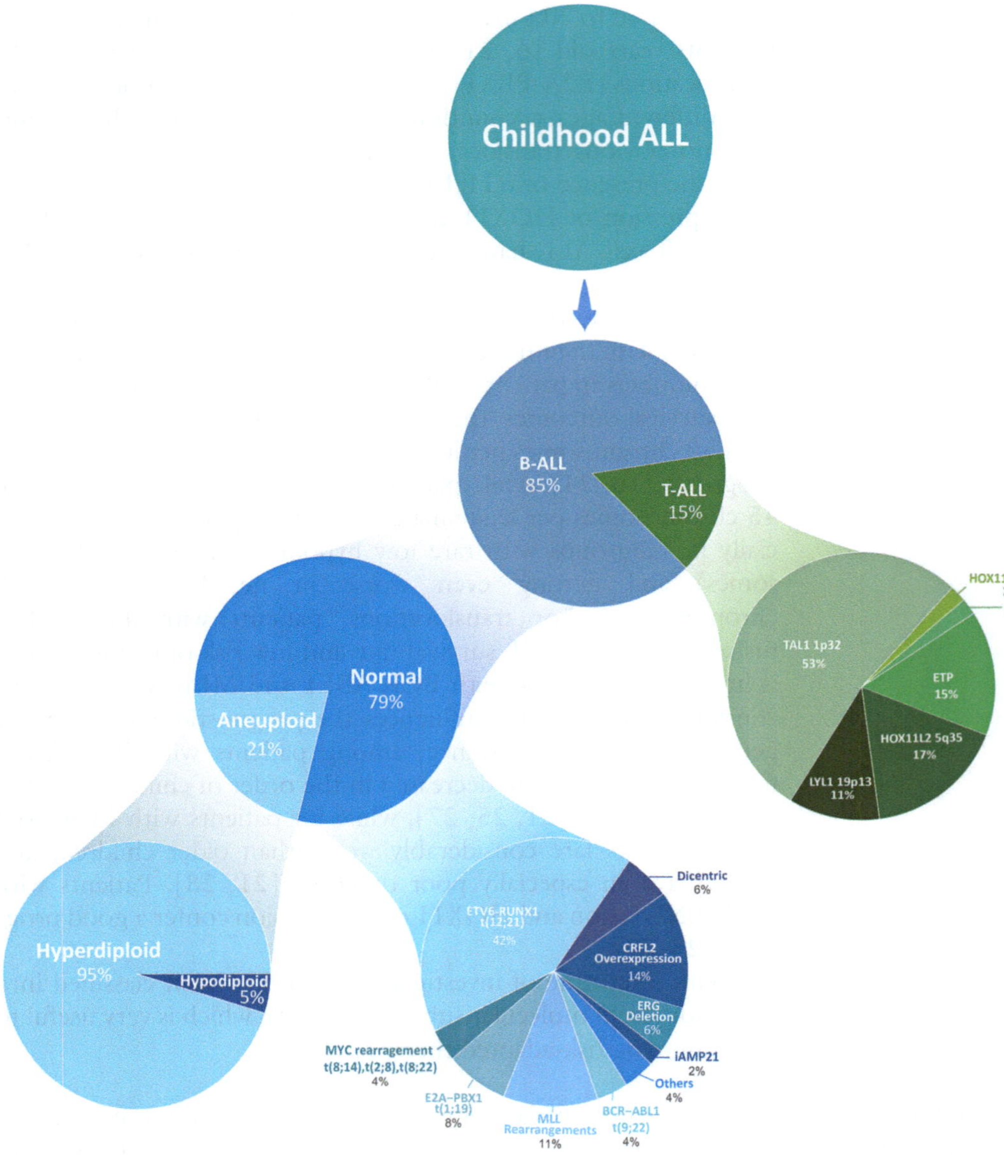

Fig. 2 Molecular subtypes of childhood ALL

On the other hand, inherited genetic variants were also considered to be involved in treatment outcomes, especially for adverse drug reactions. For instance, some patients will suffer life-threatening myelosuppression after mercaptopurine treatment with standard dosage. Thanks to the well understanding of drug metabolic pathway of mercaptopurine [18], TPMT enzyme activity was noticed to be negatively related to the toxic content (i.e., metabolites of mercaptopurine) in plasma in Caucasians, and the germline variants (e.g., TPMT*3C) can largely explain the loss of function in TPMT, which were found through screening of the whole TMPT genes by using Sanger sequencing [33]. This finding has been successfully translated into clinical practice for dosage adjustment of mercaptopurine (e.g., disease ~10-fold in patients with homozygous-inactive *TPMT* SNPs) and remarked as one of the most useful pharmacogenetic markers [34].

Collectively, through the low-throughput techniques in the pre-genome-wide investigation era, most of the frequent abnormal diploid and translations in leukemia cells have been identified and largely improved the classifications of ALL and the following individualized treatment. These techniques (e.g., FISH) are still used in clinical determinations of ALL subtypes because of their high efficiency of cost and time on identifications of limited known alterations.

3 Array-Based Genome-Wide Investigation Era

3.1 Microarray Investigation

The "cDNA microarray" was firstly developed to detect expression of more than 1000 genes simultaneously in the 1990s [35]. This technology has been largely improved and well established by a few companies including Affymetrix and Illumina, which can detect up to 50 thousand genes accurately, and is still popularly used in the study of cancer genomics up to now. "DNA microarray" is mainly used for two purposes: genome-wide investigation of somatic copy number alterations (CNAs) (also for germline copy number variations in some studies) [36] and genotypes of up to two millions germline single nucleotide polymorphisms (SNPs) [37]. Multiple versions of DNA and cDNA microarrays have been used for exploring the unknown cancer-related alterations which can't be identified by the traditional techniques due to the high throughput and resolution of microarrays. This technology consists of the syntheses of nucleic acids at high density on the solid support and allows to investigate thousands of unique nucleic acid fragments simultaneously to genome-widely detect gene expression with RNA/cDNA templates or SNP genotypes and copy number variation with DNA templates [38, 39].

To perform a microarray analysis, nucleic acid molecules will be fragmented and labeled with fluorescent probe; scanning of the microarray will proceed after hybridization in the microarray slide [35, 40]. Expression level, SNP genotypes, and copy number variations can be detected and normalized in a relatively unbiased procedure through bioinformatics analyses afterward [41]. With the development of technology, quantified alternative splicing can be detected by exon-based cDNA microarray [42], while SNP and CNV/CNA detections have been interrogated by using a single DNA microarray (e.g., Affymetrix SNP 6.0 [43]). Different platforms only vary in the nature of the probes used and genomic resolution [44]. Normally, in order to get the informative signals as many as possible, the probes are designed to be nearly evenly distributed across the genome [45] and mostly target the SNPs with relatively higher minor allele frequency. Microarray platforms have evolved rapidly, and current platforms contain up to two million probes that integrate the genome at <1000 resolution for detection of SNPs as well as CNVs/CNAs, which is increased ~200-fold from the first SNP microarrays used for ALL [46]. In another hand, after investigating the genome variants in thousands of individuals by using next-generation sequencing which we will describe in the next section, nearly 90 million SNPs have been identified with around 12 million shared in diverse races/ethnicities. Non-singleton SNPs (i.e., identified in more than one individual) that can alter protein coding (e.g., missense, nonsense) were selected and genotyped by specific probes' integrated microarray, named as exome-array [47], which is widely used for identifying potential functional variants for specific phenotypes through GWAS strategy (e.g., ALL susceptibility).

Moreover, a new type of microarray was developed named as tiling array. This type of microarray probes intensively for sequences to characterize regions that are sequenced but whose location functions are largely unknown. Besides the normal function of cDNA/DNA microarrays, it can be used for transcriptome mapping as well as discovering DNA/protein interaction and fine mapping the break points of copy number variations [48]. However, due to the high cost and, more importantly, the development of next-generation sequencing, tiling array technology has not been well used and almost been abandoned today.

Collectively, transcriptome and genomic microarrays have been widely used for molecular profiling of cancers including leukemia [49, 50]. For instance, ALL-related aberrant gene expressions/mutations/genomic alterations and their related pathways can be revealed genome-widely. Also, cluster-based molecular subtypes or treatment outcome-related alterations can be determined and translated into individualized clinical treatment. Additionally, germline variants can also be detected to find the inherited predispositions to diseases (e.g., cancer) as well as the treatment outcomes [51, 52]. As one of the best models and examples for the translational genetic researches, a lot of novel findings have been revealed for ALL, which has not only largely broadened our insight of understanding for this disease but also greatly contributed to improve the survival rate and life qualities of the patients [53, 54].

3.2 Bioinformatics Analysis

With thousands of signals for genome-wide screening for gene expression, SNP genotypes, and CNVs/CNAs, bioinformatics analyses are largely needed in two steps. The first one is normalization of raw data where technical variations should be removed [55]. Visualization of data would be indispensable to find the appropriate method for normalization [41]. The second one is study-specific analyses which depend on the experiment design. Take microarrays from Affymetrix as examples; SNP genotype calls and CNVs (for Affymetrix SNP 6.0 only) were determined by using BRLMM algorithm for Affymetrix GeneChip Human Mapping 500K Array [56] and Birdseed algorithm for its updated version of Affymetrix Human SNP Array 6.0 [57], while DGE algorithm was used for analyzing expression in Affymetrix GeneChips [58] for the first step; Plink software is used in SNP-based GWAS as for the second step of analyses. For genome-wide association analyses, signals with low quality should be filtered as well. Normally, samples with low SNP call rate (e.g., 95%) are probably due to the low sample quality and should be excluded. In another hand, SNP quality control procedures were performed on the basis of call rate, minor allele frequency, and Hardy–Weinberg equilibrium. After the filter steps, the rest SNPs can be used for association analyses [59]. On each single sites, regression model was applied to conduct GWAS [60].

3.3 Transcriptome Profiling in ALL

Due to the limitation of the genomic determinants or contributions to leukemogenesis as well as the relapse through low-throughput techniques, intensive effort has been taken to use high-resolution genomic profiling to identify novel genetic alterations and may be translated to the clinical practice as new prognostic tools or therapeutic targets ultimately [9]. Several scientific centers have performed genomic profiling of B-progenitor ALL and identified novel genetic alterations associated with high-risk disease [61–68]. In general, most of these studies used microarray-based approaches to investigate gene expression profiling, structural genetic alterations, or individual target genes by profiling leukemic cells obtained at diagnosis or relapse [9].

For expression microarrays, unlike the usage of them in solid cancers, it is impossible to get the cancer samples together with their matched control samples for hematopoietic malignancy. However, transcriptome screenings are conducted to further separate the ALL subtypes or find out the special cases, thus investigating their clinical relevance, revealing the possible related affected pathways, and taking into consideration individualized therapy. Not surprisingly, the well-established subtypes (e.g., hyperdiploidy, TEL-AML1) can be easily separated by expression signatures of a series of genes with high sensitivity and specificity [68]. Of the most importance, a new subtype (i.e., BCR-ABL1-like or Ph-like) can be separated from the previous uncharacterized groups (B-others) according to two independent studies [68, 69]. Patients in this group don't have unique chromosomal changes but can cluster with BCR-ABL1 translation positive cases according to the transcriptome-based hierarchical clustering analyses. Interestingly, BCR-ABL1-like patients also have a significant lower sensitivity for chemotherapy response (e.g., mostly resistant to L-asparaginase and daunorubicin) and higher risk of relapse than other subtypes, which is similar as that for BCR-ABL1 positive patients [68]. However, some of the BCR-ABL1-like patients are sensitive to tyrosine kinase inhibitors (e.g., imatinib) although they don't have the BCR-ABL1 translations. In this case, a newly defined molecular subtype with high risk can be detected by transcriptome profiling and exhibits as a prognostic factor for high risk. BCR-ABL1-like patients are also treated with a specific clinical regimen to increase their survival time. For T-ALL, a subtype has also been observed initially by immunophenotype, which is similar to that of early thymic progenitors (ETP) and has a poor treatment outcome. Besides the aberrant expression of CD1A, CD5, and CD8, ETP T-ALL cases have distinct gene expression signatures and exhibit signals of myeloid and stem cell markers [70, 71].

On the other hand, genome-wide detection of expression for leukemia samples with altered drug sensitivity can also identify the candidate genes or gene pathways for treatment outcomes, especially drug resistance. For instance, significantly higher expression of CASP1 has been noticed in glucocorticoid-resistant leukemia cells through transcriptome profile screening for leukemia cells from more than 4000 ALL patients. Subsequent mechanism studies indicate that overexpression of CASP1 resulted in cleavage of the glucocorticoid receptor, diminished the glucocorticoid-induced transcriptional response, and increased glucocorticoid resistance [72].

3.4 Genomic Profiling for Leukemia Cells

Several microarray-based techniques for the genome-wide analysis of DNA CNA have been well established for cancer genomic studies, especially for ALL [37, 73]. Most ALL samples have relatively fewer alterations than many solid tumors [74] and tend to be focal and limited to one or a few genes, indicating that ALL is not characterized by gross genomic instability [61, 66]. More than 50 genetic regions have been noticed to have recurrent CNAs in ALL patients, containing genes with known or putative roles in lymphoid development or leukemogenesis/tumorigenesis [9, 44, 61], including well-reported tumor suppressors (e.g., *CDKN2A/CDKN2B* and *RB1*), lymphoid development regulators (e.g., *PAX5* and *IKZF1*), and even some rare drug response-related genes (e.g., *NR3C1*) [75]. Interestingly, the frequency of these alterations is also significantly associated with the presence of chromosomal alterations [9, 61].

As the most common CNAs in all types of cancers, large or focal deletion of *CDKN2A/CDKN2B* is also frequently observed in ALL, especially T-ALL [61]. Similar to other solid cancers, loss of three proteins coded by these two genes (i.e., p14 and p16 for *CDKN2A* and p15 for *CDKN2B*), can impact multiple regulatory pathways and induce uncontrolled check of cell cycle, apoptosis escape, and etc. Thus, alteration of this region is considered to occur in the early step of ALL leukemogenesis and mostly cooperate with additional oncogenic events afterward. Meanwhile, alterations in lymphoid development-related genes are unique hallmark for leukemogenesis. For instance, genes regulating B-lymphoid development were mutated in over 40% of B-ALL cases, most involved *PAX5* and *IKZF1* genes [63, 66]. These alterations often impact only a single copy of the gene and are predicted to result in haploinsufficiency or dominant-negative for some focal deletions. *PAX5* is the most common target of genetic alteration in B-ALL and is involved and targeted by focal or broad deletions, translocation, and core domain-located mutation [76]. While *IKZF1* alterations include broad deletions of the gene that result in haploinsufficiency, focal intragenic deletions result in expression of aberrant dominant-negative IKAROS isoforms and sequence mutations [63]. *IKZF1* alterations are present in more than 80% of BCR-ABL1-positive ALL cases in both adults and children [63, 77] but uncommon in ALL subtypes with other genomic alterations, such as TEL-AML1 ALL [62, 78]. Interestingly, *IKZF1* alterations are also enriched in BCR-ABL1-like ALLs, which have been classified as high-risk class. Importantly, with the genome-wide CNA investigation, novel rearrangements have been identified involving the gene *CRLF2*, accounting for one third of BCR-ABL1-like cases and more than 50% of ALL cases with Down syndrome [64, 79]. These rearrangements are either a translocation of *CRLF2* into the immunoglobulin heavy-chain locus or a focal deletion that results in a novel fusion P2RY8-CRLF2 [64, 79, 80], resulting in overexpression of CRLF2 on the surface of leukemic cells. Meanwhile, ~60% of patients with CRLF2 rearrangements have concomitant gain-of-function mutations of *JAK1* or *JAK2* [64, 79, 81]. Genomic profiling has also been successfully investigated to identify new alterations in T-ALL, identifying deletions dysregulating LMO2 [82], amplification of MYB [83, 84], amplification associated with the NUP214-ABL1 rearrangement [85], and deletion and sequence mutation of PTEN [86] and WT1 [87].

3.5 Inherited Predispositions to ALL

SNP microarrays utilize oligonucleotide probes to detect thousands of SNPs simultaneously and are widely used to conduct genome-wide association studies (GWASs) to examine the association of SNP genotype with disease susceptibility or other phenotypes (e.g., interindividual drug response) [88]. Actually, several independent GWASs have been conducted in Caucasians as well as in diverse ethnicities to investigate the inherited predispositions to ALL susceptibility by using Affymetrix or Illumina SNP microarrays, identifying multiple SNPs mainly at six genetic loci: *ARID5B*, *IKZF1*, *CEBPE*, *CDKN2A*, *PIP4K2A-BMI1*, and *GATA3* [59, 72, 89–93]. These associations have been validated in independent cohorts and summarized with meta-analyses [94], however, with varied odds ratio (OR) in terms of different ethnicities, ages, and ALL subtypes. For instance:

1. *ARID5B* and *PIP4K2A* SNPs exhibit consistent significance in all ethnicities, while *CDKN2A* SNPs are only significant in Caucasians, due to the ethnicity-specific allele frequency.
2. Although significant in all ethnicities for *ARID5B* SNPs, both odds ratios and risk allele frequencies of rs10821936 for ALL susceptibility increase in the order of Africans, Caucasians, and Hispanics [59, 95], which is in parallel with ALL incidence among ethnic/racial groups (incidence rate: Africans < Caucasians < Hispanics) [96, 97].
3. Age at diagnosis was also considered to be another impact factor, and higher RAF of top *ARID5B* signal has been observed in younger patients in an ethnicity- and subtype-independent manner [59, 98], which is opposite to SNP in *GATA3* [90, 93, 99].
4. Risk alleles of *ARID5B* SNPs are enriched in hyperdiploidy [95, 100], while *GATA3* SNPs are enriched in BCR-ABL1-like subtype [90].

However, the SNPs described above are all located in intronic or intergenic regions, most of which are considered to be tags for causal variants through lineage disequilibrium, calling for large efforts on post-GWAS [101, 102]. In addition, exome array was also used to identify the ALL-related SNPs that can alter protein coding, only one common missense variant at *CDKN2A* was identified as the top signal in independent studies [92, 103], and patients with heterozygous genotype of rs3731249 tend to loss the expression of wild-type allele in their leukemia cells, either through loss of heterozygosity or posttranscriptional inactivation [92, 104]. Although some rare variants can be detected in the gene-based burden test, no causal variants from exome array can explain the association of the common SNPs in *ARID5B*, etc. New theory indicates that most of the causal variants for GWASs tend to induce the alteration of gene expression through interrupting the transcription factors binding on promoters or enhancers. For ALL, two potential causal variants have been revealed for *ARID5B* and *CDKN2B*, respectively [105, 106].

Array-based GWAS can also be used for identifying the genetic basis of drug response, especially for adverse drug reactions. As described above, variant allele of SNPs in *TPMT* was found to induce loss of enzyme activity, thus resulting in high toxic metabolite content in plasma. However, frequencies of these SNPs are higher in Caucasians and blacks than East Asians, who paradoxically have a higher rate of mercaptopurine-induced myelosuppression [107]. With GWAS studies for mercaptopurine dosage adjustment, two SNPs reached genome-wide significance with one indicating TPMT*3C and a novel missense SNP in *NUDT15* [108]. With functional experiments, NUDT15 was firstly reported to be involved in metabolic pathway of mercaptopurine, and the risk allele of the top GWAS *NUDT15* SNP can largely reduce the enzyme activity of NUDT15 [109]. Interestingly, RAF of such SNP is much higher in East Asians (~10%) than Caucasians (~0.4%) and barely detected in Africans (0%), which perfectly explain the paradox of higher myelosuppression rate *vs.* lower TPMT*3C rate in East Asians than other ethnicities [107]. Actually, array-based GWASs have been successfully used to detect multiple germline markers for ADRs (e.g., cisplatin-induced ototoxicity [110]) and were proved to be powerful tools in pharmacogenomics studies.

3.6 Translational Practice

The array-based genetic investigation has been widely conducted as potential new diagnostic tests in treatment approaches of B-progenitor ALL [9]. For instance, 15%–30% of B-ALL cases have alterations of *IKZF1* and are associated with very poor outcome in ALL, including an up to threefold increased risk of treatment failure in ALL [63, 67], not only because such alterations are enriched in BCR-ABL1 positive ALL [63, 67, 68]. Actually, *IKZF1* alterations is independently associated with poor outcomes in multivariate analysis adjusting for established clinical prognostic factors (e.g., age, WBC, cytogenetic subtypes) [111]. In contrast, multiple studies have failed to observe any association between *PAX5* alterations and outcome suggesting that these changes may only be important in establishing the leukemic clone, rather than influencing responsiveness to therapy [9]. Few associations between individual genetic alterations and outcome in T-ALL have been identified, except for ETP T-ALL, which have poor responsiveness to initial therapy (as evidenced by high levels of minimal residual disease) and very poor outcome. Detection of the ETP immunophenotype is being used to identify these patients and intervene with aggressive therapies, such as bone marrow transplantation.

For clinical applications, as a remarkable molecular prognostic factor, all types of *IKZF1* alterations must be able to detect in clinical assays. However, no single assay is capable of detecting all of these alterations. Alternative approaches include customized arrays, genomic quantitative PCR [62], multiplex ligation-dependent probe amplification [112], and sequence mutation screening methods. Therefore, a simple and one-step assay is largely required to detect *IKZF1* alteration in clinical level. In contrast, *CRLF2* overexpression can be detected by flow cytometry, and CRLF2 rearrangement may be detected by FISH, genomic PCR, or reverse transcriptase PCR (RT PCR) [64]. JAK mutations may be detected by a variety of well-established methods for sequence mutation detection, such as Sanger sequencing. These approaches are being investigated in large prospective clinical trials [44].

3.7 Remained Questions

Although the microarray-based genome-wide investigation is effective and informative, these approaches can't directly identify chromosomal rearrangements, tumor-specific point mutations, small insertions/deletions (INDEL), or structure variations (SVs). Routine diagnostic microarray profiling of DNA CNA in ALL is becoming increasingly widely used, either using oligonucleotide array-based comparative genomic hybridization or SNP microarrays. Data from these arrays must be analyzed and interpreted with great care, as several array platforms, even those with hundreds of thousands of features, lack sufficient resolution to detect very focal alterations. Consequently, there is intense interest in the use of next-generation sequencing technology for comprehensively genomic investigation.

4 Next-Generation Sequencing-Based Investigation Era

4.1 Next-Generation Sequencing Investigation

Since the completion of the Human Genome Project in 2003 [113, 114], genomic research has stepped into the whole genome research era. Of note, the onset of next-generation sequencing (NGS) reduced the cost of whole genome sequencing for one individual from $500 million to around $14 million [115]. With a decade of technology improvement, cost of NGS is still dropping and has only charged $1000 to sequence a human genome in 2016 (Fig. 3); $100 could be even expected in the near future. Although the technologies of sequencing long-read are becoming more mature, the prevalent applications of NGS are still focusing on the short-read sequencing which has advanced cost-effective platforms.

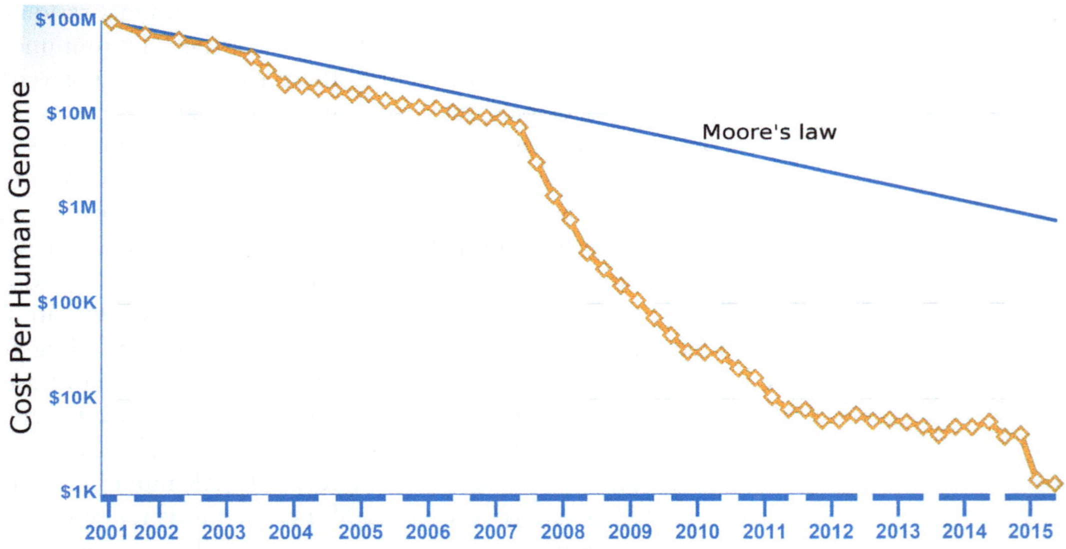

Fig. 3 Cost decrease for sequencing human genome with the development of technologies for genomic investigation

The short-read sequencing platforms can be divided into two major approaches: sequencing by ligation (SBL) and sequencing by synthesis (SBS). SBL depends on the ligation of labeled probes to DNA strands where the probes encode one or two bases; thus the binding of labeled probes and DNA could be recorded for subsequent base reorganization [116]. Among those NGS platforms based on ligation, SOLiD was the most famous one which used two-base-encoded probes for ligation. Although very high accuracy could be achieved by SOLiD, those disadvantages such as short length of reads, underrepresentation of AT-rich region, and long-lasting runtimes undermined its application [117–119]. In contrast, the SBS platforms have better performance. In 2004, Roche developed the first NGS platform: 454 pyrosequencing platform which incorporated single nucleotide into the strand and detects the bioluminescence signal via charge-coupled device each cycle [120]. Unfortunately, 454 platform fell into disuse in 2016 because of its low output and high cost as well as its weakness in sequencing the homopolymer region [121]. In 2010, Thermo Fisher developed Ion Torrent platform where the length of reads could reach almost 400 bp. Unlike the other platforms, Ion Torrent platform detects the H^+ ions which are produced during dNTP incorporation by integrating complementary metal-oxide-semiconductor (CMOS) and ion-sensitive field-effect transistor (ISFET) [122]. Compared to 454 and Ion Torrent platforms which add single nucleotide in each cycle, Illumina has developed series of NGS platforms which terminate the elongation in every cycle [123]. Illumina developed its first SBS platform—Solexa 1G—in 2006 [124]. It was a rememberable event because it made it possible to acquire gigabyte (Gb) data from a single run. Nowadays Illumina has dominated the NGS market for several years, and their products ranged from low-throughput platform (miniSeq) to very-high-throughput platform (HiSeq). Noteworthy, HiSeq X NGS platform could produce 1.8 terabytes in a single run with the cost of only $7 per Gb data [115]. Recently, a new platform called NovaSeq is announcing $100 per human genome in the future.

With NGS platform, we could perform various types of sequencing such as whole genome sequencing (WGS), whole exome sequencing (WES), transcriptome sequencing (RNA-seq) [125], chromatin immunoprecipitation followed by sequencing (ChIP-seq) [126, 127], assay for transposase-accessible chromatin using sequencing (ATAC-seq) [127], DNA methylation sequencing (methyl-seq) [128], etc. Compared with WGS, WES is always performed with high coverage because the specific genomic region for sequencing and this advantage could facilitate the investigation on tumor subclones during relapse [129]. However, the structure variations could be efficiently called in WGS but attenuated identification in WES. RNA-seq was commonly used to quantify the expression of mRNA or noncoding RNA, both of which show important roles in tumor biology; meanwhile, fusion gene can also be identified by using RNA-seq [130].

4.2 Bioinformatics Analyses for NGS

Along with the evolution of NGS technologies, the size of genomic data grows sharply. It is estimated that ~15 petabytes sequencing data are generated every year [131], interpretation of which remains a big challenge. The crux of processing sequencing data mainly lies in the accurate alignment of reads where trade-off between accuracy and efficiency should be considered. At present, plenty of alignment software like BWA [132], Bowtie [133], and SOAP2 [134] which are based on "Burrows-Wheeler transform (BWT)" algorithm are widely used. BWT could compress the reference effectively by building FM-index [135]. Although BWT-based software could bring fast alignment, its accuracy was compromised. In contrast, those hash-based algorithm such as Novoalign and MAQ [136] could provide higher accuracy in the price of longer running time. The accurate alignment would be a good start for processing NGS data, but the subsequent analysis of molecular signatures such as SNVs (single-nucleotide variations), CNVs (copy number variations), and SVs (structural variations) was also crucial for the precise interpretation. In the process of SNVs calling, Bayes' mode was introduced for evaluating quality value (QPhred) of single base. Meanwhile the probabilistic framework was used for genotyping with the information of allele frequencies and linkage disequilibrium was considered in the calculation [137, 138]. GATK [139], SAMtools [140], and SOAP [134] and other genotyping software were broadly applied for calling SNVs. For CNV detection, HMM (hidden Markov model) and sliding window were the two main algorithms for software including CNAseg [141], SegSeq [142], and JointSLM [143]. Recently CONSERTING, a software based on regression tree algorithm, was also developed for calling CNVs and exhibits high sensitivity and accuracy [144]. Besides CNVs, structural variations including insertions, deletions, inversions, and duplications were considered as important molecular signatures. The popular algorithms of detecting SVs could be categorized into four groups: read-pair algorithm like BreakDancer [145]; split-read algorithm like CREST [146], Breakpointer [147], and BreakSeek [148]; read-depth algorithm like GAVPro [149] and GenomeSTRiP [150]; and assembly algorithm like BreaKmer [151], NovelSeq [152], and SOAPindel [59]. Of course, the combinations of two or more algorithms can be used for improving the sensitivity and specificity of SV detection. Noteworthy, the usage of long-read sequencing platform like PacBio could provide remarkable facilitation in SV calling.

4.3 Genetic Characteristics and the Related NGS-Based Genomic Technologies

Extensive efforts have been made to comprehensively understand the genetic basis of leukemogenesis and treatment outcomes through microarray-based profiling of gene expression as well as DNA copy number alterations before 2010. Afterward, next-generation sequencing (NGS) was firstly used to detect ALL in 2009 for sequence alterations and rearrangements to avoid the disadvantages of microarrays [9, 153]. In the past 7–8 years of studies to identify inherited and somatic genetic alterations in ALL including sequencing of gene panels, WES, RNA-seq, and WGS, hundreds of novel alterations and their interactions have been revealed to achieve multiple ALL genomic landscapes [69, 154–160]. Aims to obtain the insights of genomic profiling of ALL are mainly:

1. Identification of novel subtypes of ALL.
2. Characterization of the interactions of genetic alterations in each ALL subtype.
3. Identification of the genetic basis of clonal heterogeneity and the corresponding effects on treatment outcomes.
4. Definition of the role of inherited genetic variants in ALL susceptibility and drug response.
5. Most importantly translation of these findings to improve diagnostic, prognostic, and targeted treatment approaches [11].

Although multiple molecular subtypes have been described in terms of the recurring gross chromosomal changes and chromosomal rearrangements, which are hallmarks of ALL, including multiple subtypes described above, the mechanism involved in their role on leukemogenesis and treatment outcomes is poorly understood. Firstly, some special subtypes with distinct clinical characteristics have been submitted to whole genome profiling with WES, RNA-seq, or even WGS. As an example, ETP T-ALL, a subtype of T-ALL which has poor treatment outcomes, was submitted to be comprehensively sequenced [159]. Genome profiling has shown that ETP-ALL is a genetically heterogeneous disease, not only with distinct gene expression profile but also lacking common chromosomal rearrangement [159]; however, recurrent gene mutations are commonly observed in several pathways, including regulators of hematopoietic development, RAS and cytokine receptor signaling pathways, etc. [159], which are different with other T-ALL (e.g., tumor suppressor pathways including *NOTCH* [26], *FBXW7* [161], and poorly understood *PHF6* in children [153, 162]). Importantly, target therapy strategies (i.e., JAK inhibition and/or chromatin-modifying agents) may be more effective for ETP-ALL due to its involvement in JAK–STAT and PRC2 pathways, and both strategies have been proved to be effective in preclinical models [163].

With the sharply decreased cost of sequencing and the improvement of bioinformatics analyses (both the computer power as well as the development of new algorithms to analyze the big sequencing data), it is possible to conduct sequencing analyses in large patient cohorts and systematically investigate hundreds of target genes or even all genes. For instance, the landscape of somatic mutations in target 633 epigenetic regulator genes across 1000 pediatric cancer genomes has been profiled, identifying the most frequent mutations in *H3F3A*, *PHF6*, *ATRX*, *KDM6A*, *SMARCA4*, *ASXL2*, *CREBBP*, *EZH2*, *MLL2*, *USP7*, *ASXL1*, *NSD2*, *SETD2*, *SMC1A*, and *ZMYM3*. Not surprisingly, different gene mutations tend to enrich in distinct subtypes of ALL [160].

According to the comprehensive investigation of ALL genomes, their somatic mutation load is significantly lower than adult solid tumors probably due to the limited time for mutation accumulation, mostly with less than 10–20 non-silent mutations per patients [164]. For instance, MLL-rearranged leukemia in infancy has one of the lowest somatic mutation frequencies of any described tumor [61, 165], while focal deletions that normally mediate the process of lymphoid antigen receptor gene rearrangement are the most common type of structural genetic alterations observed in ALL [62, 158]. Of the most important findings for classification of molecular subtypes, a comprehensive profiling has been conducted in BCR-ABL1-like ALL (10–15% of pediatric ALL or over 25% of adults ALL) and further divided these patients into several groups in terms of their molecular features, which are screened in both genomic and transcriptome levels. BCR-ABL1-like ALL is also characterized by a diverse range of genetic alterations involving *CRLF2*, *IKZF1*, and *JAK*. With NGS, a lot of novel druggable target fusions were also found, including multiple ABL-class genes. In this case, dasatinib is recommended to be used in patients with fusion involving *ABL1*, *ABL2*, *CSF1R*, and *PDGFRB*, *JAK2* inhibitors can be used in fusion involving *CRLF2*, *JAK2*, *EPOR*, and *TSLP*, while crizotinib and FAK inhibitor may be used in patients with fusion involving *NTRK3* and *PTK2B*, respectively [69, 166]. Actually, such precious regimens have been fully considered to be translated into clinical practice, which highlights the importance of high-throughput sequencing in precision medicine era.

Another important usage of NGS in ALL genomic investigation is to find the relapse-related alteration. Although *CREBBP* mutations have been observed to be related to high risk of treatment failure and relapse through sequencing 300 candidate genes with Sanger methods [167], the single gene can't explain all the relapse cases. Thanks to NGS technology, many lesions novelly acquired in relapse or selected from the initial subclones have been identified, involving *IKZF1*, *TP53* [168], *CREBBP*, *SETD2* [169], and two more drug metabolism-related genes (i.e., NT5C2,

PRPS1) [164, 170–172], which can validate the subsequent drug resistance analyses in cellular level.

NGS-based GWASs become more and more popular as well because all variants (both common and rare) can be detected without array-based imputation, which may induce genotyping error or no call for non-probed variants. Importantly, causal variants can be detected directly without fine-mapping procedure or even missing for post-GWAS effects [102]. Also, burden test can be conducted by combining all potential functional rare variants within one gene and used for gene-based correlation [173]. In research of ALL, rare variants in *ETV6* [174] and *CDKN2A* [92] were identified by target sequencing and illustrated significant association with ALL susceptibility. On the other hand, causal mutations for rare familial ALL cases were also identified by exome sequencing, which is very difficult to achieve due to the limited numbers of patients for family pedigree (e.g., most of the childhood ALL patients died before they have their own children before the 1960s). For instance, germline mutations of *PAX5* were found in all ALL patients from two unrelated kindreds [175], and a nonsense mutation in *ETV6* was found in an ALL patient and her three children, two of which also catch ALL [174]. With the development of the whole genome sequencing, more causal variants were expected to be identified before onset of diseases for prevention, etc.

Collectively, more and more NGS-based studies have been conducted in ALL research, not only broadening our insight to understand the complicated mechanisms of such diseases in terms of leukemogenesis and drug response but also providing the opportunities for us to divide the patients into plenty of different molecular subtypes and design the best individualized regimen. Besides, the NGS technologies are still under rapid development; more and more basic research findings are expected to be translated into the clinical practice.

5 Summary

In this section, we used ALL as a disease model to illustrate the translational researches with the development of technology and strategy used in genetic studies. Plenty of ALL-related alteration has been found in eras of genomic researches (Fig. 4). Thanks to the combination of these basic research effects and the development new drugs/treatment strategies, treatment outcomes of ALL have greatly improved and are still in progress, successful experiences of which are also valuable for treatments of other tumors.

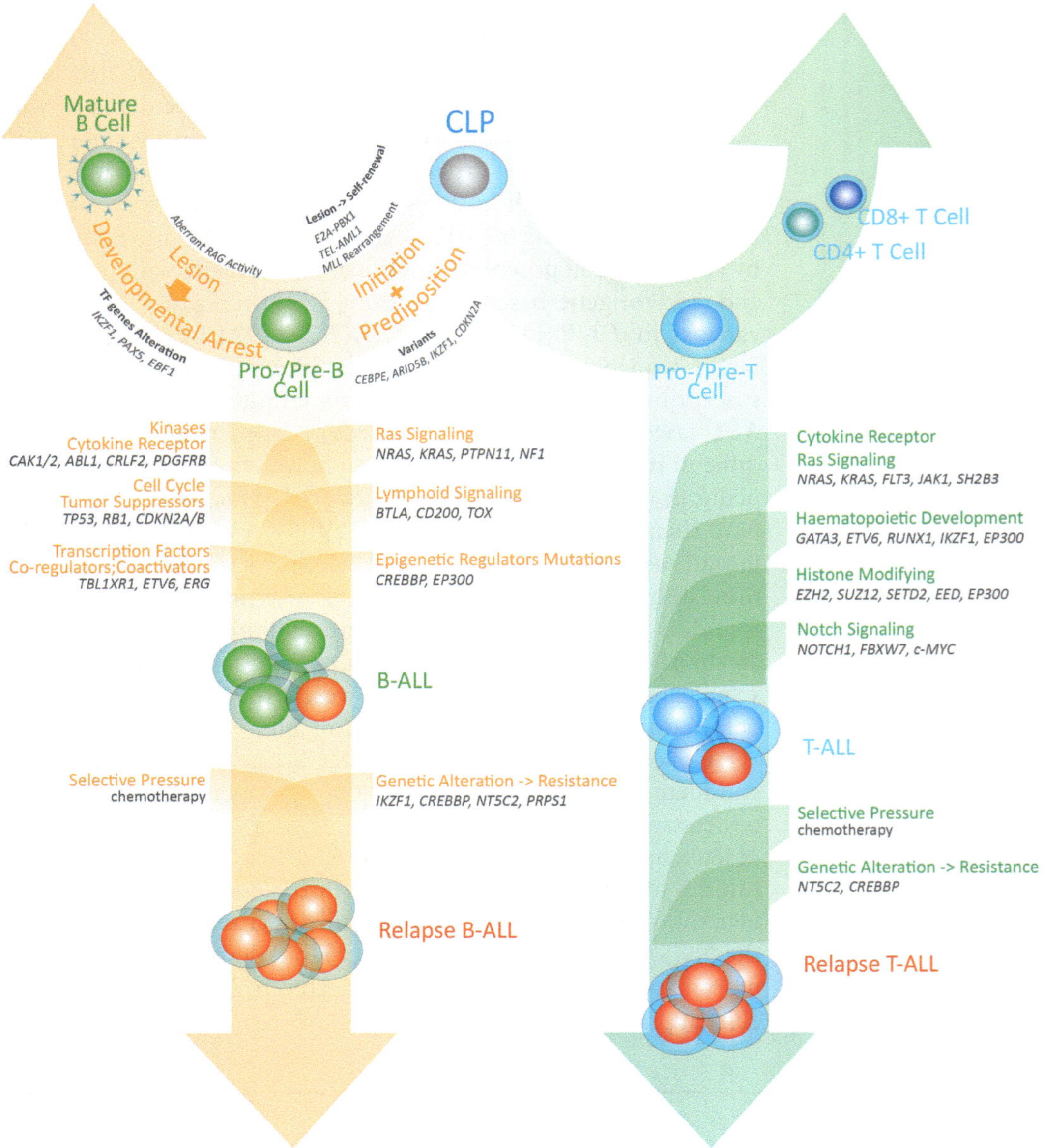

Fig. 4 Alterations in ALL-related pathways of leukemogenesis and relapse stages

References

1. Pui CH (1995) Childhood leukemias. N Engl J Med 332(24):1618–1630
2. Pui CH, Evans WE (1998) Acute lymphoblastic leukemia. N Engl J Med 339(9):605–615
3. Pui CH, Campana D, Evans WE (2001) Childhood acute lymphoblastic leukaemia—current status and future perspectives. Lancet Oncol 2(10):597–607
4. Inaba H, Greaves M, Mullighan CG (2013) Acute lymphoblastic leukaemia. Lancet 381 (9881):1943–1955
5. Pui C-H, Crist WM, Look AT (1990) Biology and clinical significance of cytogenetic abnormalities in childhood acute lymphoblastic leukemia. Blood 76(8):1449–1463

6. Pui CH, Relling MV, Downing JR (2004) Acute lymphoblastic leukemia. N Engl J Med 350(15):1535–1548
7. Mullighan CG (2013) Genomic characterization of childhood acute lymphoblastic leukemia. Semin Hematol 50(4):314–324
8. Mullighan CG (2012) Molecular genetics of B-precursor acute lymphoblastic leukemia. J Clin Invest 122(10):3407–3415
9. Mullighan CG (2011) New strategies in acute lymphoblastic leukemia: translating advances in genomics into clinical practice. Clin Cancer Res 17(3):396–400
10. Pui C-H, Evans WE (2006) Treatment of acute lymphoblastic leukemia. N Engl J Med 354(2):166–178
11. Roberts KG, Mullighan CG (2015) Genomics in acute lymphoblastic leukaemia: insights and treatment implications. Nat Rev Clin Oncol 12(6):344–357
12. Pui C-H, Robison LL, Look AT (2008) Acute lymphoblastic leukaemia. Lancet 371 (9617):1030–1043
13. Bhojwani D, Pui CH (2013) Relapsed childhood acute lymphoblastic leukaemia. Lancet Oncol 14(6):e205–e217
14. Fielding AK et al (2007) Outcome of 609 adults after relapse of acute lymphoblastic leukemia (ALL); an MRC UKALL12/ECOG 2993 study. Blood 109(3):944–950
15. Einsiedel HG et al (2005) Long-term outcome in children with relapsed ALL by risk-stratified salvage therapy: results of trial acute lymphoblastic leukemia-relapse study of the Berlin-Frankfurt-Munster Group 87. J Clin Oncol 23(31):7942–7950
16. Rivera GK et al (2005) Bone marrow recurrence after initial intensive treatment for childhood acute lymphoblastic leukemia. Cancer 103(2):368–376
17. Mullighan CG (2014) The genomic landscape of acute lymphoblastic leukemia in children and young adults. ASH Edu Progr Book 2014(1):174–180
18. Moriyama T, Relling MV, Yang JJ (2015) Inherited genetic variation in childhood acute lymphoblastic leukemia. Blood 125 (26):3988–3995
19. Mullighan CG (2012) The molecular genetic makeup of acute lymphoblastic leukemia. ASH Edu Progr Book 2012(1):389–396
20. Pui C-H, Behm FG, Crist WM (1993) Clinical and biologic relevance of immunologic marker studies in childhood acute lymphoblastic leukemia. Blood 82(2):343–362
21. Mancini M et al (2005) A comprehensive genetic classification of adult acute lymphoblastic leukemia (ALL): analysis of the GIMEMA 0496 protocol. Blood 105 (9):3434–3441
22. Armstrong SA, Look AT (2005) Molecular genetics of acute lymphoblastic leukemia. J Clin Oncol 23(26):6306–6315
23. Mullighan CG (2012) The molecular genetic makeup of acute lymphoblastic leukemia. Hematology Am Soc Hematol Educ Program 2012:389–396
24. Sutcliffe M et al (2005) High concordance from independent studies by the Children's Cancer Group (CCG) and Pediatric Oncology Group (POG) associating favorable prognosis with combined trisomies 4, 10, and 17 in children with NCI Standard-Risk B-precursor Acute Lymphoblastic Leukemia: a Children's Oncology Group (COG) initiative. Leukemia 19(5):734–740
25. Gleißner B et al (2002) Leading prognostic relevance of the BCR-ABL translocation in adult acute B-lineage lymphoblastic leukemia: a prospective study of the German Multicenter Trial Group and confirmed polymerase chain reaction analysis. Blood 99 (5):1536–1543
26. Weng AP et al (2004) Activating mutations of NOTCH1 in human T cell acute lymphoblastic leukemia. Science 306(5694):269–271
27. Aricò M et al (2000) Outcome of treatment in children with Philadelphia chromosome–positive acute lymphoblastic leukemia. N Engl J Med 342(14):998–1006
28. Pui C-H et al (2002) Outcome of treatment in childhood acute lymphoblastic leukaemia with rearrangements of the 11q23 chromosomal region. Lancet 359(9321):1909–1915
29. Ferrando AA et al (2002) Gene expression signatures define novel oncogenic pathways in T cell acute lymphoblastic leukemia. Cancer Cell 1(1):75–87
30. Ferrando AA et al (2004) Prognostic importance of TLX1 (HOX11) oncogene expression in adults with T-cell acute lymphoblastic leukaemia. Lancet 363(9408):535–536
31. Vijayakrishnan J, Houlston R (2010) Candidate gene association studies and risk of childhood acute lymphoblastic leukemia: a systematic review and meta-analysis. Haematologica 95(8):1405–1414
32. Chang JS et al (2010) Genetic polymorphisms in adaptive immunity genes and childhood acute lymphoblastic leukemia. Cancer Epidemiol Biomarkers Prev 19(9):2152–2163

33. Relling MV et al (1999) Mercaptopurine therapy intolerance and heterozygosity at the thiopurine S-methyltransferase gene locus. J Natl Cancer Inst 91(23):2001–2008
34. Relling MV, Evans WE (2015) Pharmacogenomics in the clinic. Nature 526(7573):343
35. Schena M et al (1995) Quantitative monitoring of gene expression patterns with a complementary DNA microarray. Science 270 (5235):467
36. McCarroll SA et al (2008) Integrated detection and population-genetic analysis of SNPs and copy number variation. Nat Genet 40 (10):1166–1174
37. Davies JJ, Wilson IM, Lam WL (2005) Array CGH technologies and their applications to cancer genomes. Chromosom Res 13 (3):237–248
38. Southern E, Mir K, Shchepinov M (1999) Molecular interactions on microarrays. Nat Genet 21:5–9
39. Zhang L, Miles MF, Aldape KD (2003) A model of molecular interactions on short oligonucleotide microarrays. Nat Biotechnol 21 (7):818–821
40. Schena M (1999) DNA microarrays: a practical approach. Oxford University Press, New York, NY
41. Quackenbush J (2002) Microarray data normalization and transformation. Nat Genet 32:496–501
42. Johnson JM et al (2003) Genome-wide survey of human alternative pre-mRNA splicing with exon junction microarrays. Science 302 (5653):2141–2144
43. Matsuzaki H et al (2004) Genotyping over 100,000 SNPs on a pair of oligonucleotide arrays. Nat Methods 1(2):109–111
44. Mullighan CG, Downing JR (2009) Genome-wide profiling of genetic alterations in acute lymphoblastic leukemia: recent insights and future directions. Leukemia 23 (7):1209–1218
45. Yang YH et al (2002) Normalization for cDNA microarray data: a robust composite method addressing single and multiple slide systematic variation. Nucleic Acids Res 30(4): e15–e15
46. Irving JA et al (2005) Loss of heterozygosity in childhood acute lymphoblastic leukemia detected by genome-wide microarray single nucleotide polymorphism analysis. Cancer Res 65(8):3053–3058
47. Huyghe JR et al (2013) Exome array analysis identifies new loci and low-frequency variants influencing insulin processing and secretion. Nat Genet 45(2):197–201
48. Bertone P et al (2004) Global identification of human transcribed sequences with genome tiling arrays. Science 306(5705):2242–2246
49. Rhodes DR, Chinnaiyan AM (2005) Integrative analysis of the cancer transcriptome. Nat Genet 37:S31–S37
50. Bullinger L et al (2004) Use of gene-expression profiling to identify prognostic subclasses in adult acute myeloid leukemia. N Engl J Med 350(16):1605–1616
51. Metzeler KH et al (2011) TET2 mutations improve the new European LeukemiaNet risk classification of acute myeloid leukemia: a Cancer and Leukemia Group B study. J Clin Oncol 29(10):1373–1381
52. Yang JJ et al (2009) Genome-wide interrogation of germline genetic variation associated with treatment response in childhood acute lymphoblastic leukemia. JAMA 301 (4):393–403
53. Bassan R, Hoelzer D (2011) Modern therapy of acute lymphoblastic leukemia. J Clin Oncol 29(5):532–543
54. Pui C-H et al (2004) Improved outcome for children with acute lymphoblastic leukemia: results of total therapy study XIIIB at St Jude Children's Research Hospital. Blood 104(9):2690–2696
55. Simon RM (2003) Design and analysis of DNA microarray investigations. Springer Science & Business Media, New York, NY
56. Rabbee N, Speed TP (2006) A genotype calling algorithm for affymetrix SNP arrays. Bioinformatics 22(1):7–12
57. Korn JM et al (2008) Integrated genotype calling and association analysis of SNPs, common copy number polymorphisms and rare CNVs. Nat Genet 40(10):1253–1260
58. Hubbell E, Liu W-M, Mei R (2002) Robust estimators for expression analysis. Bioinformatics 18(12):1585–1592
59. Xu H et al (2013) Novel susceptibility variants at 10p12.31-12.2 for childhood acute lymphoblastic leukemia in ethnically diverse populations. J Natl Cancer Inst 105 (10):733–742
60. Neale BM et al (2010) Meta-analysis of genome-wide association studies of attention-deficit/hyperactivity disorder. J Am Acad Child Adolesc Psychiatry 49 (9):884–897
61. Mullighan CG et al (2007) Genome-wide analysis of genetic alterations in acute lymphoblastic leukaemia. Nature 446 (7137):758–764
62. Mullighan CG et al (2008) BCR–ABL1 lymphoblastic leukaemia is characterized by the

deletion of Ikaros. Nature 453 (7191):110–114

63. Mullighan CG et al (2009) Deletion of IKZF1 and prognosis in acute lymphoblastic leukemia. N Engl J Med 360(5):470–480
64. Mullighan CG et al (2009) Rearrangement of CRLF2 in B-progenitor–and Down syndrome–associated acute lymphoblastic leukemia. Nat Genet 41(11):1243–1246
65. Kawamata N et al (2008) Molecular allelokaryotyping of pediatric acute lymphoblastic leukemias by high-resolution single nucleotide polymorphism oligonucleotide genomic microarray. Blood 111(2):776–784
66. Kuiper R et al (2007) High-resolution genomic profiling of childhood ALL reveals novel recurrent genetic lesions affecting pathways involved in lymphocyte differentiation and cell cycle progression. Leukemia 21 (6):1258–1266
67. Kuiper R et al (2010) IKZF1 deletions predict relapse in uniformly treated pediatric precursor B-ALL. Leukemia 24(7):1258–1264
68. Den Boer ML et al (2009) A subtype of childhood acute lymphoblastic leukaemia with poor treatment outcome: a genome-wide classification study. Lancet Oncol 10 (2):125–134
69. Roberts KG et al (2012) Genetic alterations activating kinase and cytokine receptor signaling in high-risk acute lymphoblastic leukemia. Cancer Cell 22(2):153–166
70. Wood B et al (2009) Patients with early T-cell precursor (ETP) acute lymphoblastic leukemia (ALL) have high levels of minimal residual disease (MRD) at the end of induction—a Children's Oncology Group (COG) study. Blood 114(22):9–9
71. Coustan-Smith E et al (2009) Early T-cell precursor leukaemia: a subtype of very high-risk acute lymphoblastic leukaemia. Lancet Oncol 10(2):147–156
72. Paugh SW et al (2015) NALP3 inflammasome upregulation and CASP1 cleavage of the glucocorticoid receptor cause glucocorticoid resistance in leukemia cells. Nat Genet 47 (6):607–614
73. Wang Y, Armstrong SA (2007) Genome-wide SNP analysis in cancer: leukemia shows the way. Cancer Cell 11(4):308–309
74. Beroukhim R et al (2010) The landscape of somatic copy-number alteration across human cancers. Nature 463(7283):899–905
75. Irving JA et al (2005) Loss of heterozygosity and somatic mutations of the glucocorticoid receptor gene are rarely found at relapse in pediatric acute lymphoblastic leukemia but may occur in a subpopulation early in the disease course. Cancer Res 65 (21):9712–9718
76. Nebral K et al (2009) Incidence and diversity of PAX5 fusion genes in childhood acute lymphoblastic leukemia. Leukemia 23 (1):134–143
77. Iacobucci I et al (2009) Identification and molecular characterization of recurrent genomic deletions on 7p12 in the IKZF1 gene in a large cohort of BCR-ABL1–positive acute lymphoblastic leukemia patients: on behalf of Gruppo Italiano Malattie Ematologiche dell'Adulto Acute Leukemia Working Party (GIMEMA AL WP). Blood 114 (10):2159–2167
78. Lilljebjörn H et al (2010) The correlation pattern of acquired copy number changes in 164 ETV6/RUNX1-positive childhood acute lymphoblastic leukemias. Hum Mol Genet 19 (16):3150–3158
79. Russell LJ et al (2009) Deregulated expression of cytokine receptor gene, CRLF2, is involved in lymphoid transformation in B-cell precursor acute lymphoblastic leukemia. Blood 114(13):2688–2698
80. Yoda A et al (2010) Functional screening identifies CRLF2 in precursor B-cell acute lymphoblastic leukemia. Proc Natl Acad Sci 107(1):252–257
81. Mullighan CG et al (2009) JAK mutations in high-risk childhood acute lymphoblastic leukemia. Proc Natl Acad Sci 106 (23):9414–9418
82. Van Vlierberghe P et al (2006) The cryptic chromosomal deletion del (11)(p12p13) as a new activation mechanism of LMO2 in pediatric T-cell acute lymphoblastic leukemia. Blood 108(10):3520–3529
83. Lahortiga I et al (2007) Duplication of the MYB oncogene in T cell acute lymphoblastic leukemia. Nat Genet 39(5):593–595
84. Clappier E et al (2007) The C-MYB locus is involved in chromosomal translocation and genomic duplications in human T-cell acute leukemia (T-ALL), the translocation defining a new T-ALL subtype in very young children. Blood 110(4):1251–1261
85. Graux C et al (2004) Fusion of NUP214 to ABL1 on amplified episomes in T-cell acute lymphoblastic leukemia. Nat Genet 36 (10):1084
86. Palomero T et al (2007) Mutational loss of PTEN induces resistance to NOTCH1 inhibition in T-cell leukemia. Nat Med 13 (10):1203–1210

87. Tosello V et al (2009) WT1 mutations in T-all. Blood 114(5):1038–1045
88. Xavier RJ, Rioux JD (2008) Genome-wide association studies: a new window into immune-mediated diseases. Nat Rev Immunol 8(8):631–643
89. Sherborne AL et al (2010) Variation in CDKN2A at 9p21.3 influences childhood acute lymphoblastic leukemia risk. Nat Genet 42(6):492–494
90. Perez-Andreu V et al (2013) Inherited GATA3 variants are associated with Ph-like childhood acute lymphoblastic leukemia and risk of relapse. Nat Genet 45(12):1494–1498
91. Papaemmanuil E et al (2009) Loci on 7p12.2, 10q21.2 and 14q11.2 are associated with risk of childhood acute lymphoblastic leukemia. Nat Genet 41(9):1006–1010
92. Xu H et al (2015) Inherited coding variants at the CDKN2A locus influence susceptibility to acute lymphoblastic leukaemia in children. Nat Commun 6:7553
93. Perez-Andreu V et al (2015) A genome-wide association study of susceptibility to acute lymphoblastic leukemia in adolescents and young adults. Blood 125(4):680–686
94. Liao F et al (2016) Association between PIP4K2A polymorphisms and acute lymphoblastic leukemia susceptibility. Medicine (Baltimore) 95(18):e3542
95. Xu H et al (2012) ARID5B genetic polymorphisms contribute to racial disparities in the incidence and treatment outcome of childhood acute lymphoblastic leukemia. J Clin Oncol 30(7):751–757
96. Linabery AM, Ross JA (2008) Trends in childhood cancer incidence in the U.S. (1992–2004). Cancer 112(2):416–432
97. Dores GM et al (2012) Acute leukemia incidence and patient survival among children and adults in the United States, 2001–2007. Blood 119(1):34–43
98. Burmeister T et al (2014) Germline variants in IKZF1, ARID5B, and CEBPE as risk factors for adult-onset acute lymphoblastic leukemia: an analysis from the GMALL study group. Haematologica 99(2):e23–e25
99. Hou Q et al (2017) Regulatory network of GATA3 in pediatric acute lymphoblastic leukemia. Oncotarget 8(22):36040–36053
100. Trevino LR et al (2009) Germline genomic variants associated with childhood acute lymphoblastic leukemia. Nat Genet 41 (9):1001–1005
101. Huang Q (2015) Genetic study of complex diseases in the post-GWAS era. J Genet Genomics 42(3):87–98
102. Freedman ML et al (2011) Principles for the post-GWAS functional characterization of cancer risk loci. Nat Genet 43(6):513–518
103. Vijayakrishnan J et al (2015) The 9p21.3 risk of childhood acute lymphoblastic leukaemia is explained by a rare high-impact variant in CDKN2A. Sci Rep 5:15065
104. Walsh KM et al (2015) A heritable missense polymorphism in CDKN2A confers strong risk of childhood acute lymphoblastic leukemia and is preferentially selected during clonal evolution. Cancer Res 75(22):4884–4894
105. Studd JB et al (2017) Genetic and regulatory mechanism of susceptibility to high-hyperdiploid acute lymphoblastic leukaemia at 10p21.2. Nat Commun 8:14616
106. Hungate EA et al (2016) A variant at 9p21.3 functionally implicates CDKN2B in paediatric B-cell precursor acute lymphoblastic leukaemia aetiology. Nat Commun 7:10635
107. Yin D et al (2017) Impact of NUDT15 polymorphisms on thiopurines-induced myelotoxicity and thiopurines tolerance dose. Oncotarget 8(8):13575–13585
108. Yang JJ et al (2015) Inherited NUDT15 variant is a genetic determinant of mercaptopurine intolerance in children with acute lymphoblastic leukemia. J Clin Oncol 33 (11):1235–1242
109. Moriyama T et al (2016) NUDT15 polymorphisms alter thiopurine metabolism and hematopoietic toxicity. Nat Genet 48 (4):367–373
110. Xu H et al (2015) Common variants in ACYP2 influence susceptibility to cisplatin-induced hearing loss. Nat Genet 47 (3):263–266
111. Martinelli G et al (2009) IKZF1 (Ikaros) deletions in BCR-ABL1-positive acute lymphoblastic leukemia are associated with short disease-free survival and high rate of cumulative incidence of relapse: a GIMEMA AL WP report. J Clin Oncol 27(31):5202–5207
112. Schwab C et al (2010) Evaluation of multiplex ligation-dependent probe amplification as a method for the detection of copy number abnormalities in B-cell precursor acute lymphoblastic leukemia. Genes Chromosom Cancer 49(12):1104–1113
113. JD MP et al (2001) A physical map of the human genome. Nature 409(6822):934–941
114. Collins FS, Morgan M, Patrinos A (2003) The Human Genome Project: lessons from large-scale biology. Science 300 (5617):286–290
115. Wetterstrand KA (2013) DNA sequencing costs: data from the NHGRI Genome

Sequencing Program (GSP). National Human Genome Research Institute, Bethesda, MD
116. Tomkinson AE et al (2006) DNA ligases: structure, reaction mechanism, and function. Chem Rev 106(2):687–699
117. Harismendy O et al (2009) Evaluation of next generation sequencing platforms for population targeted sequencing studies. Genome Biol 10(3):R32
118. Shendure J, Ji H (2008) Next-generation DNA sequencing. Nat Biotechnol 26 (10):1135–1145
119. Valouev A et al (2008) A high-resolution, nucleosome position map of C. elegans reveals a lack of universal sequence-dictated positioning. Genome Res 18(7):1051–1063
120. Margulies M et al (2005) Genome sequencing in microfabricated high-density picolitre reactors. Nature 437(7057):376–380
121. Loman NJ et al (2012) Performance comparison of benchtop high-throughput sequencing platforms. Nat Biotechnol 30 (5):434–439
122. Rothberg JM et al (2011) An integrated semiconductor device enabling non-optical genome sequencing. Nature 475 (7356):348–352
123. Ju J et al (2006) Four-color DNA sequencing by synthesis using cleavable fluorescent nucleotide reversible terminators. Proc Natl Acad Sci 103(52):19635–19640
124. Solexa I (2006) Illumina: protocol for whole genome sequencing using solexa technology. BioTech Protoc Guide 2006:29
125. Wang Z, Gerstein M, Snyder M (2009) RNA-Seq: a revolutionary tool for transcriptomics. Nat Rev Genet 10(1):57–63
126. Park PJ (2009) ChIP-seq: advantages and challenges of a maturing technology. Nat Rev Genet 10(10):669–680
127. Buenrostro JD et al (2013) Transposition of native chromatin for fast and sensitive epigenomic profiling of open chromatin, DNA-binding proteins and nucleosome position. Nat Methods 10(12):1213–1218
128. Brunner AL et al (2009) Distinct DNA methylation patterns characterize differentiated human embryonic stem cells and developing human fetal liver. Genome Res 19 (6):1044–1056
129. Ding L et al (2012) Clonal evolution in relapsed acute myeloid leukaemia revealed by whole-genome sequencing. Nature 481 (7382):506–510
130. Masetti R et al (2013) CBFA2T3-GLIS2 fusion transcript is a novel common feature in pediatric, cytogenetically normal AML, not restricted to FAB M7 subtype. Blood 121 (17):3469–3472
131. Schatz MC, Langmead B (2013) The DNA data deluge. IEEE Spectr 50(7):28–33
132. Li H, Durbin R (2009) Fast and accurate short read alignment with Burrows–Wheeler transform. Bioinformatics 25 (14):1754–1760
133. Langmead B et al (2009) Ultrafast and memory-efficient alignment of short DNA sequences to the human genome. Genome Biol 10(3):R25
134. Li R et al (2009) SOAP2: an improved ultrafast tool for short read alignment. Bioinformatics 25(15):1966–1967
135. Burrows, M. and D.J. Wheeler, (1994) A block-sorting lossless data compression algorithm. SRC research report
136. Li H, Ruan J, Durbin R (2008) Mapping short DNA sequencing reads and calling variants using mapping quality scores. Genome Res 18(11):1851–1858
137. Ewing B et al (1998) Base-calling of automated sequencer traces usingPhred. I. Accuracy assessment. Genome Res 8 (3):175–185
138. Martin ER et al (2010) SeqEM: an adaptive genotype-calling approach for next-generation sequencing studies. Bioinformatics 26(22):2803–2810
139. McKenna A et al (2010) The genome analysis toolkit: a MapReduce framework for analyzing next-generation DNA sequencing data. Genome Res 20(9):1297–1303
140. Li H et al (2009) The sequence alignment/map format and SAMtools. Bioinformatics 25 (16):2078–2079
141. Ivakhno S et al (2010) CNAseg—a novel framework for identification of copy number changes in cancer from second-generation sequencing data. Bioinformatics 26 (24):3051–3058
142. Chiang DY et al (2009) High-resolution mapping of copy-number alterations with massively parallel sequencing. Nat Methods 6(1):99–103
143. Magi A et al (2011) Detecting common copy number variants in high-throughput sequencing data by using JointSLM algorithm. Nucleic Acids Res 39(10):e65
144. Chen X et al (2015) CONSERTING: integrating copy-number analysis with structural-variation detection. Nat Methods 12 (6):527–530

145. Chen K et al (2009) BreakDancer: an algorithm for high-resolution mapping of genomic structural variation. Nat Methods 6 (9):677–681
146. Wang J et al (2011) CREST maps somatic structural variation in cancer genomes with base-pair resolution. Nat Methods 8 (8):652–654
147. Sun R et al (2012) Breakpointer: using local mapping artifacts to support sequence breakpoint discovery from single-end reads. Bioinformatics 28(7):1024–1025
148. Zhao H, Zhao F (2015) BreakSeek: a breakpoint-based algorithm for full spectral range INDEL detection. Nucleic Acids Res 43(14):6701–6713
149. Sindi SS et al (2012) An integrative probabilistic model for identification of structural variation in sequencing data. Genome Biol 13 (3):R22
150. Handsaker RE et al (2015) Large multiallelic copy number variations in humans. Nat Genet 47(3):296–303
151. Abo RP et al (2015) BreaKmer: detection of structural variation in targeted massively parallel sequencing data using kmers. Nucleic Acids Res 43(3):e19–e19
152. Hajirasouliha I et al (2010) Detection and characterization of novel sequence insertions using paired-end next-generation sequencing. Bioinformatics 26(10):1277–1283
153. Van Vlierberghe P et al (2010) PHF6 mutations in T-cell acute lymphoblastic leukemia. Nat Genet 42(4):338–342
154. De Keersmaecker K (2013) Exome sequencing identifies mutation in CNOT3 and ribosomal genes RPL5 and RPL10 in T-cell acute lymphoblastic leukemia. Nat Genet 45:186–190
155. Holmfeldt L (2013) The genomic landscape of hypodiploid acute lymphoblastic leukemia. Nat Genet 45:242–252
156. Kalender Atak Z (2013) Comprehensive analysis of transcriptome variation uncovers known and novel driver events in T-cell acute lymphoblastic leukemia. PLoS Genet 9:e1003997
157. Li Y (2014) Constitutional and somatic rearrangement of chromosome 21 in acute lymphoblastic leukaemia. Nature 508:98–102
158. Papaemmanuil E (2014) RAG-mediated recombination is the predominant driver of oncogenic rearrangement in ETV6-RUNX1 acute lymphoblastic leukemia. Nat Genet 46:116–125
159. Zhang J (2012) The genetic basis of early T-cell precursor acute lymphoblastic leukaemia. Nature 481:157–163
160. Huether R et al (2014) The landscape of somatic mutations in epigenetic regulators across 1,000 paediatric cancer genomes. Nat Commun 5:3630
161. O'Neil J (2007) FBW7 mutations in leukemic cells mediate NOTCH pathway activation and resistance to gamma-secretase inhibitors. J Exp Med 204:1813–1824
162. Van Vlierberghe P et al (2011) PHF6 mutations in adult acute myeloid leukemia. Leukemia 25(1):130–134
163. Treanor LM (2014) Interleukin-7 receptor mutants initiate early T cell precursor leukemia in murine thymocyte progenitors with multipotent potential. J Exp Med 211:701–713
164. Ma X (2015) Rise and fall of subclones from diagnosis to relapse in pediatric B-progenitor acute lymphoblastic leukemia. Nat Commun 6:6604
165. Andersson AK (2015) The landscape of somatic mutations in infant MLL-rearranged acute lymphoblastic leukaemias. Nat Genet 47(4):330–337
166. Roberts KG (2014) Targetable kinase-activating lesions in Ph-like acute lymphoblastic leukemia. N Engl J Med 371:1005–1015
167. Mullighan CG (2011) CREBBP mutations in relapsed acute lymphoblastic leukaemia. Nature 471:235–239
168. Hof J (2011) Mutations and deletions of the TP53 gene predict nonresponse to treatment and poor outcome in first relapse of childhood acute lymphoblastic leukemia. J Clin Oncol 29:3185–3193
169. Mar BG (2014) Mutations in epigenetic regulators including SETD2 are gained during relapse in paediatric acute lymphoblastic leukaemia. Nat Commun 5:3469
170. Tzoneva G (2013) Activating mutations in the NT5C2 nucleotidase gene drive chemotherapy resistance in relapsed ALL. Nat Med 19:368–371
171. Yang JJ (2012) Genome-wide association study identifies germline polymorphisms associated with relapse of childhood acute lymphoblastic leukemia. Blood 120:4197–4204
172. Li B et al (2015) Negative feedback-defective PRPS1 mutants drive thiopurine resistance in relapsed childhood ALL. Nat Med 21 (6):563–571

173. Ionita-Laza I et al (2013) Sequence kernel association tests for the combined effect of rare and common variants. Am J Hum Genet 92(6):841–853
174. Moriyama T et al (2015) Germline genetic variation in ETV6 and risk of childhood acute lymphoblastic leukaemia: a systematic genetic study. Lancet Oncol 16 (16):1659–1666
175. Shah S et al (2013) A recurrent germline PAX5 mutation confers susceptibility to pre-B cell acute lymphoblastic leukemia. Nat Genet 45(10):1226–1231

Index

A

B

C

D

E

F

G

Tao Huang (ed.), *Computational Systems Biology: Methods and Protocols*, Methods in Molecular Biology, vol. 1754, https://doi.org/10.1007/978-1-4939-7717-8, © Springer Science+Business Media, LLC, part of Springer Nature 2018

R

S

T

U

V

W

X

Z

MIX
Papier aus verantwortungsvollen Quellen
Paper from responsible sources
FSC® C105338

If you have any concerns about our products,
you can contact us on
ProductSafety@springernature.com

In case Publisher is established outside the EU,
the EU authorized representative is:
Springer Nature Customer Service Center GmbH
Europaplatz 3, 69115 Heidelberg, Germany

Printed by Libri Plureos GmbH
in Hamburg, Germany